Case Studies in Geospatial Applications to Groundwater Resources

Case Studies in Geospatial Applications to Groundwater Resources

Edited by

Pravat Kumar Shit

Department of Geography, Raja N. L. Khan Women's College (Autonomous), Midnapore, West Bengal, India

Gouri Sankar Bhunia

Department of Geography, Nalini Prabha Dev Roy College, Bilaspur, Chhattisgarh, India

Partha Pratim Adhikary

ICAR-Indian Institute of Water Management, Bhubaneswar, Odisha, India

Elsevier
Radarweg 29, PO Box 211, 1000 AE Amsterdam, Netherlands
The Boulevard, Langford Lane, Kidlington, Oxford OX5 1GB, United Kingdom
50 Hampshire Street, 5th Floor, Cambridge, MA 02139, United States

ISBN: 978-0-323-99963-2

For Information on all Elsevier publications visit our website at https://www.elsevier.com/books-and-journals

Publisher: Candice G Janco
Acquisitions Editor: Peter Llewellyn
Editorial Project Manager: Maria Elaine D Desamero
Production Project Manager: R. Vijay Bharath
Cover Designer: Vicky Pearson Esser

Typeset by Aptara, New Delhi, India

Contents

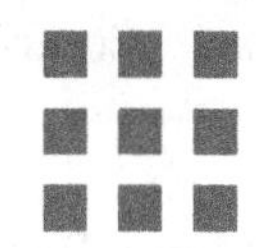

Contributors

Partha Pratim Adhikary
ICAR-Indian Institute of Water Management, Bhubaneswar, Odisha, India

Biswaranjan Behera
ICAR-Indian Institute of Water Management, Bhubaneswar, Odisha, India

Biswajit Bera
Department of Geography, Sidho-Kanho-Birsha University, Purulia, India

Amit Bera
Department of Earth Sciences, Indian Institute of Engineering Science and Technology, Howrah, West Bengal, India

Raj Kumar Bhattacharya
Department of Geography, Vidyasagar University, Midnapore, West Bengal, India

Gouri Sankar Bhunia
Department of Geography, Nalini Prabha Dev Roy College, Bilaspur, Chhattisgarh, India

Baidurya Biswas
Department of Geography and Applied Geography, University of North Bengal, Raja Rammohunpur, Darjeeling, West Bengal, India

Soumen Brahma
Department of Geography, Nalini Prabha Dev Roy College, Bilaspur, Chhattisgarh, India

Ananya Chakraborty
Department of Earth Sciences, Indian Institute of Engineering Science and Technology, Howrah, West Bengal, India

Kunal Chakraborty
Department of Geography and Applied Geography, University of North Bengal, Darjeeling, India

Puja Chowdhury
Department of Earth Sciences, Indian Institute of Engineering Science and Technology, Howrah, West Bengal, India

Kousik Das
Department of Geography, Vidyasagar University, Midnapore, West Bengal, India

Mantu Das
Department of Geography and Applied Geography, University of North Bengal, Raja Rammohunpur, Darjeeling, West Bengal, India

Sandipan Das
Symbiosis Institute of Geoinformatics (SIG), Symbiosis International (Deemed University), Pune, India

Pritiranjan Das
Department of Geography, Vidyasagar University, Midnapore, West Bengal, India

Pulakesh Das
World Resources Institute India, New Delhi, India

Subhrangsu Das
Department of Geography, Utkal University, Odisha, India

Nirmalya Das
Department of Geography, Panskura Banamali College (Autonomous), Panskura, West Bengal, India

Tapan Kumar Das
Department of Geography, Cooch Behar College, Cooch Behar, West Bengal, India

Nilanjana Das Chatterjee
Department of Geography, Vidyasagar University, Midnapore, West Bengal, India

Ch. Jyotiprava Dash
ICAR-Indian Institute of Soil and Water Conservation, Research Centre, Koraput, Odisha, India

Gour Dolui
Department of Geography, Panskura Banamali College (Autonomous), Panskura, West Bengal, India

Adebayo Oluwole Eludoyin
Department of Geography, Obafemi Awolowo University, Ile-Ife, Nigeria

Adewole Abraham Fajiwe
Department of Geography, Obafemi Awolowo University, Ile-Ife, Nigeria

Sanjoy Garai
Institute of Forest Productivity, Lalgutwa, Ranchi, India

Anadi Gayen
Central Ground Water Board, Eastern Region, Kolkata, Department of Water Resources, River Development and Ganga Rejuvenation, Ministry of Jal Shakti, Government of India, India

Arijit Ghosh
Department of Geography, Sidho-Kanho-Birsha University, Purulia, India

Santu Guchhait
Department of Geography, Panskura Banamali College (Autonomous), Panskura, West Bengal, India

Md. Mofizul Hoque
Department of Geography, Aliah University, Kolkata, India

Aznarul Islam
Department of Geography, Aliah University, Kolkata, India

Subrata Jana
Department of Geography, Belda College, Belda, Paschim Medinipur, India

Sriparna Jana
Department of Geography, Bajkul Milani Mahavidyalaya, Bajkul, Purba Medinipur, India

Masjuda Khatun
Institute of Forest Productivity, Lalgutwa, Ranchi, India

Satish S. Kulkarni
Department of Geology, Bharatiya Mahavidyalaya, Amravati, India

Jotirmayee Lenka
ICAR-Indian Institute of Soil and Water Conservation, Research Centre, Koraput, Odisha, India

Sadik Mahammad
Department of Geography, Aliah University, Kolkata, India

Biswajit Maity
Department of Geography, Vidyasagar University, Midnapore, West Bengal, India

Arijit Majumder
Department of Geography, Jadavpur University, Kolkata, India

Suraj kumar Mallick
Department of Geography, Vidyasagar University, Midnapore, West Bengal, India

Tapash Mandal
Department of Geography and Applied Geography, University of North Bengal, Darjeeling, India

S. Mohanty
ICAR-Indian Institute of Water Management, Bhubaneswar, Odisha, India

Swatilekha Parihari
Department of Geography, Vidyasagar University, Midnapore, West Bengal, India

Priyank Pravin Patel
Department of Geography, Presidency University, Kolkata

Pranav Pratik
Department of Geography, Presidency University, Kolkata

Sk Mujibar Rahaman
Institute of Forest Productivity, Lalgutwa, Ranchi, India

Somnath Rudra
Department of Geography, Vidyasagar University, Midnapore, West Bengal, India

Snehasish Saha
Department of Geography and Applied Geography, University of North Bengal, Darjeeling, India

Dipankar Saha
Department of Geography, Coochbehar Panchanan Barma University, Cooch Behar, West Bengal, India

Ujjal Senapati
Department of Geography, Coochbehar Panchanan Barma University, Cooch Behar, West Bengal, India

Pravat Kumar Shit
Department of Geography, Raja N. L. Khan Women's College (Autonomous), Midnapore, West Bengal, India

Debasish Talukdar
Department of Geography, Coochbehar Panchanan Barma University, Cooch Behar, West Bengal, India

Sharad Tiwari
Institute of Forest Productivity, Lalgutwa, Ranchi, India

Sumedh R. Warghat
Department of Geology, Bharatiya Mahavidyalaya, Amravati, India

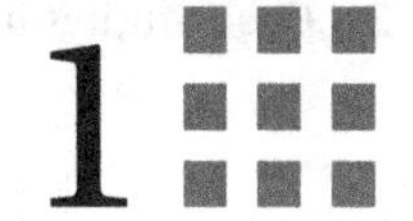

Principle of GIScience and geostatistics in groundwater modeling

Gouri Sankar Bhunia[a] and Pravat Kumar Shit[b]

[a]DEPARTMENT OF GEOGRAPHY, NALINI PRABHA DEV ROY COLLEGE, BILASPUR, CHHATTISGARH, INDIA [b]DEPARTMENT OF GEOGRAPHY, RAJA N. L. KHAN WOMEN'S COLLEGE (AUTONOMOUS), MIDNAPORE, WEST BENGAL, INDIA

1.1 Introduction

The origins of GIScience can be traced back to two keynote speeches by Michael F. Goodchild of the University of California, Santa Barbara at a conference in Europe. "Progress of the GIS Research Agenda" at the 2nd European GIS Conference held in Brussels, Belgium in July 1990 and April 1991. GIScience is an existing technology and research field of geographic information system (GIS), mapping (mapping), geodesy (measurement of the earth itself), surveying (measurement of natural and man-made features on the earth), orphotographs (measurements using photographs), global positioning systems or GPS (accurate and accurate positioning of the ground surface using satellites), digital image processing (processing and analysis of image data), remote sensing (RS) (observation of the Earth from space or underwater), quantitative spatial analysis and modeling (Rouhani and Hall, 1989). Therefore, GIScience covers issues such as spatial data structure, analysis, accuracy, meaning, cognition, and visualization, some traditional dealings with the physical processes of the earth, and the interaction between humans and the earth. Overlapping areas of the discipline (e.g., geography, geology, and geophysics, marine science, ecology, environmental sciences, applied mathematics, spatial statistics, physics), and mutual between humans and computer technology. Fields dealing with action (e.g., computer science, information science, cognitive science, cognitive psychology, artificial intelligence).

It is important to distinguish between GIS and GIScience. While GIS is primarily concerned with hardware and software for capturing, manipulating, and displaying geographic data and information (e.g., GIS as a container for data, maps, and software tools), GIScience is essentially "The science behind GIS" or "the science behind the system." In addition, starting with the basic questions that occur using GIS (such as tracking errors through the system), a systematic survey of geographic information from scientific methods (scale, accuracy, and quantitative analysis of geospatial data). Science performed using GIS (e.g., developing spatial models to predict susceptibility to local landslides, or developing agent-based models) simulating vehicle

Case Studies in Geospatial Applications to Groundwater Resources. DOI: https://doi.org/10.1016/B978-0-323-99963-2.00012-2

movements or interactions within transportation networks A model, table, or spatial statistic that represents the environmental impact that results from a decision to commercialize a property.

The science of geostatistics has grown enormously from its roots in mining around 50 years ago and encompasses a wide range of disciplines. Geoscientists often have interpolation and estimation problems when analyzing sparse data from field observations. or temporary phenomena. Geostatistics originated in the mining and oil industries starting with the work of Danie Krige in the 1950s and was developed by Georges Matheron in the 1960s. Most geological data (e.g., rock properties, pollutant concentrations) often do not meet these assumptions, as they can be heavily biased and/or have a spatial relationship (i.e., data values for locations that are closer together tend to be more similar to data values for locations that are further apart locations). Compared to classic statistics, which examine the statistical distribution of a sample data set, geostatistics take into account both the statistical distribution of the sample data and the spatial correlation between the sample data. Because of this difference, several geoscience-related difficulties can be more efficiently implemented using geostatistical methods. Since then, geostatistics has expanded to many other areas related to the geosciences, such as hydrogeology, hydrology, meteorology, oceanography, geochemistry, geography, soil science, forestry, landscape ecology.

Geostatistics requires a significant amount of computational work, including two important and time-consuming processes: estimating the semivariogram and determining the optimal semivariogram model. However, geostatistics often provides the most accurate estimates because they take into account the spatial structure of the variables and also allow the quantification of the corresponding estimation error. Geostatistics usually includes different types of kriging methods such as simple, universal, probability, indicator, disjunctive, and kriging. Kriging quantifies the spatial correlation of data, called variography, and presents predictions of where there is no measurement data. The intention of geostatistics is to expect the viable spatial distribution of a property. Such prediction regularly takes the shape of a map or a sequence of maps. Two simple sorts of prediction exist estimation and simulation (Lee at al., 2010). On the other hand, the simulation creates many similar maps (sometimes called "images") of the property distribution using the same spatial correlation model that is required for kriging. Based on the above discussion, the present chapter described about the role of GIScience and geostatistics in groundwater mapping and modeling.

1.2 GIS and groundwater

With the dawn of geographic information systems, especially after the 1990s, it has greatly improved the display, interpretation, and presentation of groundwater quality assessments at large spatial scales (Lo and Yeung, 2003). GIS is capable of collecting, storing, analyzing, manipulating, retrieving, and displaying large volumes of spatial data for rapid organization, quantification, and interpretation for decision-making in areas such as science and technology, engineering, and environment. It has proven to be a powerful tool for analyzing and mapping hydrogeological/hydrogeological data on spatial and temporal scales to provide useful information on spatial variability, helping ultimate benefit in decision-making (Machiwal and Jha, 2014). GIS applications are useful for studies assessing groundwater quality, especially mapping spatial variation in water quality, modeling groundwater flow and pollution, and designing groundwater quality monitoring networks (Jha et al., 2007). In addition, GIS-based water quality mapping is essential for pollution hazard modeling, assessment, and conservation planning, and detection of environmental changes (Chen et al., 2004).

In fact, water quality can be defined for different uses (drinking water, agricultural irrigation, livestock, industry, etc.), at different times and spaces, and by different parameters (chemical, physical, microbiological, radioactive). Some parameters are more problematic than others in terms of health issues. When calculating WQI using GIS, you can implement a number of applications that lead to the proper and sustainable management of water resources. The next subsection highlights the application of WQI in the field of hydrogeology. Here, the groundwater quality index (Machiwal et al., 2011), the pollution index (Backman et al., 1998), and the metal pollution index (Giri et al., 2010), and the aquifer water quality index (Melloul and Collin, 1998) were developed to define the water quality of groundwater. With the further development of computing systems, WQI is now integrated into GIS and provides quantitative maps of groundwater quality in different geographic regions and sizes (Sadat Noori et al., 2014).

In an effort to provide a general overview of groundwater pollution in an area, Backman et al. (1998) tested the applicability of groundwater pollution index (C_d) mapping in Finland and in Slovakia. Over the past decade, a number of studies have integrated the GWQI concept into GIS to support different groundwater quality assessment strategies, as well as proper management and monitoring of aquifers and groundwater resources. Babiker et al. (2007) proposed a GIS-based GWQI with the aim of summarizing available water quality data in easy-to-understand maps. They used GIS to implement the proposed GWQI and to test the sensitivity of the model. In their GIS-based GWQI space-time study, Machiwal et al. (2011) developed, following the GWQI map, an optimal index factor (OIF) to generate a potential GWQI (PGWQI) map of western India. Vulnerability maps can be calculated using GIS, which enables spatial data acquisition, while providing average values for data processing such as georeferenced, integration, aggregation, or spatial analysis (Burrough and McDonnell, 1998). Many approaches have been developed to assess aquifer vulnerability and can be divided into three categories: (1) overlay and indexing techniques, (2) a method using a process-based simulation model, and (3) a statistical method (Tesoriero et al., 1998). Many methods of GMM can distinguish the degree of fragility at the regional level where various lithology exists and are mainly used for groundwater protection of porous aquifers. DRASTIC (Aller et al., 1987), GOD (Foster, 1987), AVI (Van Stempvoort et al., 1993) and SINTACS (Civita, 1994). A complete overview of existing methods can be found in Vrba and Zaporozec (1994) and Gogu and Dassargues (2000).

1.3 Remote sensing and groundwater

The availability of groundwater in any terrain is largely determined by the prevalence and orientation of the primary and secondary porosity. The exploration of groundwater includes the delineation and mapping of various lithological, structural, and geomorphological units. Satellite-based RS data facilitate the creation of lithological, structural, and geomorphological maps, especially at the regional level. RS typically produces data in the form of grids or regions, which can be transmuted into distribution models through numerous processing approaches, such as machine learning algorithms. By applying the features of RS data to groundwater resources, the point hydrological model of groundwater can be extended globally.

Visual explanation of RS images is attained in a competent and effective manner using keys or basic interpretation elements (Sabins, 1987). Investigations of the spectral reflectivity of rock-forming minerals deliver the physical basis for the remote purpose of terrestrial materials. Data are an imperative part of investigations associated with tectonics, engineering, geomorphology, and the investigation of natural resources for instance groundwater, oil, and minerals. The mapping of lineaments from different RS images is a frequently used step in groundwater exploration in hard rock areas, taking the form of lineaments in aerial images or RS data. The surface appearance of geological structures, for example, fissures (faults, joints, dikes, and veins), shear zones, and foliations are often exposed or characterized as lineaments in aerial photographs or RS data.

1.4 Geostatistics and groundwater

Geostatistics has played a growing role in the characterization and modeling of the oil tank and modeling, mainly promoted by recognition that heterogeneity in petrophysical properties (i.e., permeability and porosity) dominates the water flow of the water flow, the transport of soldered and multifocal migration in the substrate. Rouhani and Hall (1989) applied space-time kriging to geohydrology by using intrinsic random functions (polynomial space-time covariance) for spatiotemporal geostatistical analyzes of piezometric data. More recently, spatio-temporal kriging has been used to estimate the water level of the Querétaro-Obrajuelo aquifer (Mexico) using a product sum model with spherical components in a large spatio-temporal dataset (Júnez Ferreira and Herrera, 2013) and the seasonal fluctuations in water depths. In Dutch, nature reserves using an exponential space-time variogram metric model (Hoogland et al., 2010). In addition, the space-time ordinary kriging was used to design precipitation networks and to analyze precipitation variations in space and time (Biondi, 2013; Raja et al., 2016) and was tested in a comparative study to estimate runoff time series at uncalibrated locations (Skøien and Blöschl, 2007).

Sparsely monitored watersheds are not frequently monitored via area and time, and consequently, statistics availability is a thing restricting in simple terms spatial or temporal analysis (Fig. 1.1). This problem and the related demanding situations round uncertainty of boundary conditions, the way that it is difficult to set up a dynamic numerical model. GIS-based geostatistical techniques help to create surfaces that incorporate the statistical properties of the data being measured. Many methods are associated with geostatistics, but they all belong to the Kriging family. Simple, universal, probability, indicators, and disjunctive kriging are usually some of the available geostatistical methods (ESRI, 2016). Chaudhry et al. (2019) used integrated exploratory

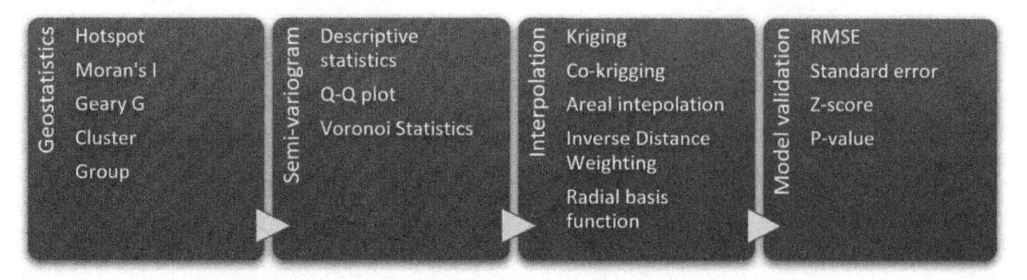

FIGURE 1.1 Geostatistical methods and techniques used in groundwater estimation.

factor analysis and conventional kriging (OK) approaches to identify sources of groundwater pollution in the Lupunagar district of Punjab. A five-factor model has been proposed that explains more than 89.11% of the total variation in groundwater quality. Three semivariogram models, exponent, gauss, and sphere, fit the dataset well and are cross-validated with predictive statistics. The ASCE Task Committee (1990) has applied (1) mapping, (2) simulation of hydrological variables, (3) estimation using flow equations, and (4) sampling of the application of geostatistical modeling techniques in groundwater hydrology. Reviewed in five major sections of design, and (5) geostatistics modeling application in groundwater systems management.

1.5 Geocomputational modeling and groundwater

Groundwater management models are powerful for aquifer management using optimization and simulation methods such as linear programming and quadratic programming that combine groundwater regulated flows and transport equations to solve groundwater management problems. For many years, groundwater hydrogeologists have tried to appraise groundwater resources using numerical imitation models. The application of numerical simulation models by researchers in the field of groundwater hydrology has facilitated to improve the understanding of aquifer functions in the region regarding specific aspects of the groundwater system and to test the hypothesis. The groundwater management model can be divided into two main groups. These are a physical classification model and a data-driven classification model. Physical classification models are reliant upon the use of physical constraints of the groundwater bed to govern changes in water level; however, these models are difficult to implement, expensive, and must be shared to obtain numerical information. The groups of data-driven models are differentiated according to the objective function, whereby the decision is based only on the hydraulic functions of the groundwater and the other, whose management decision is based on the evaluation of the policy, as well as an assignment of the economy of the groundwater. The groundwater model, based on data in its primitive structure, has four basic components: it is not linear in terms of its decision variables; requires the solution of nonlinear partial differential equations to describe groundwater transport and flow; it is stochastic as its primary uncertain source is related to the aquifer simulation mode, and; it is a mixed-integer programming decision because it contains both discrete and continuous objective functions (Yeh, 2015; Wada, 2016).

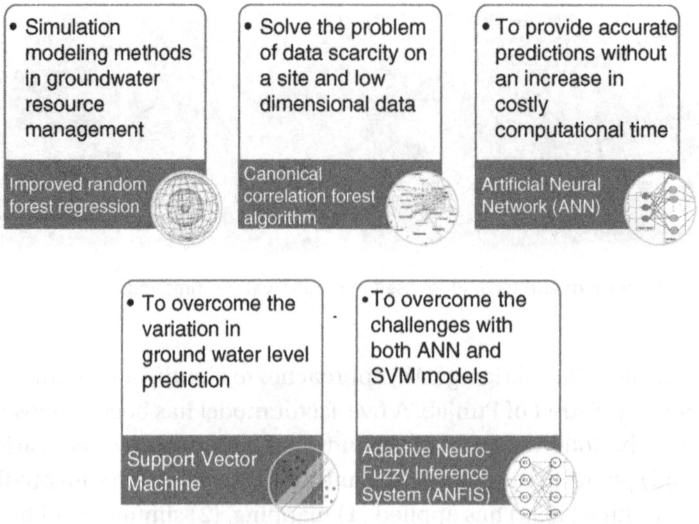

FIGURE 1.2 Data-driven groundwater resource management modeling methods.

The latest data-based classification models, such as artificial neural network technology, genetic programming, the Adaptive NeuroFuzzy Inference System and adaptive neuro-fuzzy inference system and the support vector machine as well as time series methods such as the autoregressive integrated moving average, the multi-objective function approach and the autoregressive moving average are alternatives tested on physical models and treated as standard nonlinear estimators that can overcome the difficulties associated with physical models and are less expensive (Diersch, 2005; Aderemi et al., 2021). In addition, there are numerical groundwater models that have been developed from a conceptual model. However, these models often ignore the complexity and focus only on the basic rationale of groundwater systems (Hosseini and Mahjouri, 2016). With the advances in data mining for modeling, optimization, and simulation techniques for groundwater resource management, the use of finite differences and finite elements has increased exponentially (Lee and Cheng, 1974; Tyson and Weber, 1963). Consequently, both the finite element modeling technique and finite-difference model technique were widely used for the groundwater flow model, the hydro-economic model, calibration (C), sensitivity analysis, as well as validation/verification (V). Fig. 1.2 shows a summary of the latest data-driven modeling methods for groundwater resource management.

1.6 Geospatial intelligence and groundwater modeling

The effective management of the groundwater resources, as well as the modeling, depends on the availability of high-quality data on the observation well information. Information about the aquifer properties may include changes in the water table, storage, flow rate, replenishment, and runoff, among others. Furthermore, information on groundwater resources is lacking due to a dearth of proper integration between the equipment deployed, irrelevant, and inconsistent data due to the lack of large-scale stationary flow obstacles, a process of nonautomated groundwater analysis, and absence of interoperability in previous systems (Su et al., 2020; Laraichi et al., 2016). There are several systems for monitoring the water table. These systems differ in terms of technology, monitoring, and management tasks, scalability, the solution they provide, and the impact on costs. In addition, there is a risk that most of the groundwater level monitoring networks will be regularly abandoned due to a decline in global groundwater monitoring.

In the past few years, the Internet has changed the way people live. This concept of IoT has been adopted in many areas of human activity, including intelligent water level and groundwater management. Hence, the techniques of the IoT are used to collect, transmit, and analyze necessary data about water table data. The main advantage of IoT implementation is that it can be combined with various technologies such as wireless sensors, cloud computing, ubiquitous computing, RFIP, and software to manage groundwater level data in one environment. IoT involves the combination of intelligent technologies, such as sensors for collecting data in a network area with a combination of IDE software on the cloud server (Vijayakumar and Ramya, 2015).

RS is an example of a classic way of obtaining urgently needed hydrological data for groundwater level measurements via the Internet. Although SR can be used to obtain certain parameters of groundwater resources (Zhou et al., 2013), these parameters are usually not useful for modeling groundwater management. As a result, another model is required to manipulate the captured data into usable or verifiable data as input into spatially distributed models. The essential and most relevant data for modeling groundwater resource management is information on recharge and runoff (Xiao et al., 2017). IoT and machine learning techniques can be used to solve these challenges (Faunt et al., 2010). However, the difference in each measurement well depends on the technology used and the frequency of the measurement data. The application of IoT to monitor the daily fluctuations in the water table and the safety quality in the mining environment was carried out by Reddy et al. (2016) using sensor technologies. In addition, Neyens et al. (2018), the quality and quantity of the groundwater from a desktop using the IoT-enabled environmental data management interface (EMI) technology.

1.7 WebGIS and groundwater resource

Starting with the development of web technologies in 1993, various database administrators have begun to develop web-based geographic information systems (WebGIS) to store real-time, aggregated, high-speed data streams. Hence, the WebGIS technique works best in terms of user quality of Service, usable by several users, cost reduction, global reach, and cross-compatibility. The GIS software is known as ArcView and the Groundwater Model (MODFLOW) was combined for the numerical modeling of groundwater resources by Chennai and Mammou (Chenini and Mammou, 2010). The combination of managed aquifer recharge and the global groundwater information service of the International Groundwater Resources Assessment Center (IGRAC's GGIS) has been successfully implemented using advanced historical data from approximately 1200 site surveys in approximately 62 countries (Stefan and Ansems, 2018).

1.8 Conclusion and future direction

In the past, most of the existing aquifer resource management models have been combined with optimization and simulation techniques using appropriate mathematical programming to offer solutions to challenges within the aquifer. The data are important but part of the general constraints are the uncertainties in the input parameters for modeling the system. Applying the IoT-based technique of groundwater resources is a very useful tool in data collection, monitoring, manipulation, and management of groundwater resources. This technology in combination with GIS has great potential in the field of water management. Computing resources further away from the data center via the Internet or cloud computing. With huge amounts of IoT data being transferred to the cloud in bulk, an efficient and scalable IoT platform is required to extract valuable information in real-time for the management of groundwater resources. This will enable the resource management model at the groundwater level to achieve computational efficiency and scalability. In addition, current IoT-enabled automated data processing systems for transferring the data generated by IoT sensors to the centralized cloud are not scalable and efficient, so an alternative model for the management model of the groundwater table resources activated for IoT must be developed. An open research direction should be explored. New technologies and geospatial approaches add significantly to the communicative approaches of geoscientists and allow us to highlight human impacts on the geosphere at multiple levels, including the global level.

References

Aderemi, B.A., Olwal, T.O., Ndambuki, J.M., Rwanga, S.S., 2021. Groundwater level resources management modelling: a review. Preprints, 2021070227 doi:10.20944/preprints202107.0227.v1.

Aller, L., Bennett, T., Lehr, J.H., Petty, R.J., Hackett, G., 1987. DRASTIC: a standardized system for evaluating ground water pollution potential using hydrogeologic settings. United States Environ. Prot. Agency, Ada, Oklahoma 74820, 455p EPA 600/2-85/018.

ASCE Task Committee, 1990. Review of geostatistics in geohydrology. II: Applications. J. Hydraulic Eng. 116 (5), 612. doi:10.1061/(ASCE)0733-9429, ASCE116:5(633).

Babiker, I.S., Mohamed, M.M.A., Hiyama, T., 2007. Assessing groundwater quality using GIS. Water Resources Manage. 21, 699–715.

Backman, B., Bodiš, D., Lahermo, P., Rapant, S., Tarvainen, T., 1998. Application of a groundwater contamination index in Finland and Slovakia. Environ. Geol. 36 (1-2), 55–64.

Biondi, F., 2013. Space-time kriging extension of precipitation variability at 12 km spacing from tree-ring chronologies and its implications for drought analysis. Hydrol. Earth Syst. Sci. Discuss. 10, 4301–4335.

Burrough, P., McDonell, R., 1998. Principles of Geographical Information Systems. Oxford University Press, Oxford.

Chaudhry, A.K., Kumar, K., Alam, MA., 2019. Groundwater contamination characterization using multivariate statistical analysis and geostatistical method. Water Supply 19 (8), 2309–2322. https://doi.org/10.2166/ws.2019.111.

Diersch, H., 2005. FEFLOW Finite Element Subsurface Flow and Transport Simulation System Reference Manual. WASY Institute for Water Resources Planning and Systems Research, Berlin, p. 292.

ESRI, 2016. ArcGIS Desktop Software. ArcGIS Desktop 10.4. ESRI, Redlands, CA.

Foster, S.S.D., 1987. Fundamental concepts in aquifer vulnerability, pollution risk and protection strategy. In: Duijvenbooden, W, Waegeningh, HG (Eds.). In: Vulnerability of Soil and Groundwater to Pollutants, 38. TNO Committee on Hydrological Research, The Hague, Proc Info, Netherlands, pp. 69–86.

Giri, S., Singh, G., Gupta, S.K., Jha, V.N., Tripathi, R.M., 2010. An evaluation of metal contamination in surface and groundwater around a proposed uranium mining site, Jharkhand, India. Mine Water Environ. 29 (3), 225–234.

Gogu, R., Carabin, G., Hallet, V., Peters, V., Dassargues, A., 2001. GIS-based hydrogeological databases and groundwater modelling. Hydrogeol. J. 9 (6), 555–569. doi:10.1007/s10040-001-0167-3.

Hoogland, T., Heuvelink, G.B.M., Knotters, M., 2010. Mapping water-table depths over time to assess desiccation of groundwater-dependent ecosystems in the Netherlands. Wetlands 30 (1), 137–147.

Júnez-Ferreira, H.E., Herrera, G.S., 2013. A geostatistical methodology for the optimal design of space–time hydraulic head monitoring networks and its application to the Valle de Querétaro aquifer. Environ. Monit. Assess. 185 (4), 3527–3549.

Laraichi, S., Hammani, A., Bouignane, A., 2016. Data integration as the key to building a decision support system for groundwater management: case of Saiss Aquifers, Morocco. Groundwater Sustain. Develop. 2, 7–15.

Lee, S.J., Wentz, E.A., Gober, P., 2010. Space-time forecasting using soft geostatistics: a case study in forecasting municipal water demand for Phoenix, Arizona. Stoch. Environ. Res. Risk Assess. 24 (2), 283–295.

Lee, C.H., Cheng, R.T.S., 1974. On seawater encroachment in coastal aquifers. Water Resources Res. 10 (5), 1039–1043.

Lo, C.P., Yeung, A.K.W., 2003. Concepts and Techniques of Geographic Information Systems. Prentice-Hall of India Pvt. Ltd., New Delhi, p. 492.

Machiwal, D., Jha, M.K., 2014. Role of geographical information system for water quality evaluation. In: Nielson, D. (Ed.), Geographic Information Systems (GIS): Techniques, Applications and Technologies. Nova Science Publishers, USA, pp. 217–278.

Machiwal, D., Jha, M.K., Mal, B.C., 2011. GIS-based assessment and characterization of groundwater quality in a hard-rock hilly terrain of Western India. Environ. Monitoring Assess. 174, 645–663.

Melloul, AJ., Collin, M., 1998. A proposed index for aquifer water quality assessment: the case of Israel's Sharon region. J. Environ. Manage. 54, 131–142.

Neyens, D., Baïsset, M., Lovighi, H., 2018. Monitoring the groundwater quality/quantity from your desktop–application to salt water intrusion monitoring EMI: environmental data management interface. In: E3S Web of Conferences, 54. EDP Sciences, Poland, p. 00021.

Raja, N.B., Aydin, O., Türkoğlu, N., Çiçek, I., 2016. Space-time kriging of precipitation variability in Turkey for the period 1976–2010. Theor. Appl. Climatol. 129 (1–2), 1–12.

Reddy, N.S., Saketh, M.S., Dhar, S., 2016. Review Of Sensor Technology For Mine Safety Monitoring Systems: A Holistic Approach. In: 2016 IEEE First International Conference on Control, Measurement and Instrumentation (CMI). IEEE, pp. 429–434.

Rouhani, S., Hall, T.J., 1989. Space–time kriging of groundwater data. In: Armstrong, M. (Ed.), Geostatistics. Kluwer Academic Publishers, Dordrecht, pp. 639–651.

Sadat-Noori, S.M., Ebrahimi, K., Liaghat, A.M., 2014. Groundwater quality assessment using the Water Quality Index and GIS in Saveh-Nobaran aquifer, Iran. Environ. Earth Sci. 71, 3827–3843.

Skøien, J.O., Blöschl, G., 2007. Spatiotemporal topological kriging of runoff time series. Water Resour. Res. 43 (9), 1–21 W09419.

Su, Y.-S., Ni, C.-F., Li, W.-C., Lee, I.-H., Lin, C.-P., 2020. Applying deep learning algorithms to enhance simulations of large-scale groundwater flow in IoTs. Appl. Soft Comput. 92, 106298.

Tesoriero, A.J., Inkpen, E.L., Voss, F.D., 1998. Assessing ground-water vulnerability using logistic regression. In: Proceedings for the Source Water Assessment and Protection 98 Conference, Dallas, TX, pp. 157–165.

Van Stempvoort, D., Ewert, L., Wassenaar, L., 1993. Aquifer vulnerability index (AVI): a GIS compatible method for groundwater vulnerability mapping. Can. Water Res. J. 18, 25–37.

Vijayakumar, N., Ramya, 2015. The real time monitoring of water quality in IoT environment. In: International Conference on Innovations in Information, Embedded and Communication Systems (ICIIECS). IEEE, pp. 1–5.

Yeh, W.W., 2015. Optimization methods for groundwater modeling and management. Hydrogeology 23 (6), 1051–1065.

Wada, Y., 2016. Modeling groundwater depletion at regional and global scales: present state and future prospects. Surveys Geophys. 37 (2), 419–451.

Xiao, M., et al., 2017. How much groundwater did California's Central Valley lose during the 2012–2016 drought? Geophys. Res. Letters 44 (10), 4872–4879.

Zhou, Y., Dong, D., Liu, J., Li, W., 2013. Upgrading a regional groundwater level monitoring network for Beijing Plain, China. Geosci. Frontiers 4 (1), 127–138.

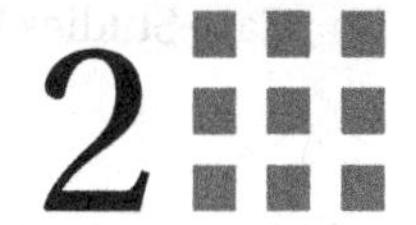

Indicator kriging and its usefulness in assessing spatial suitability of groundwater for drinking

Partha Pratim Adhikary[a], Ch. Jyotiprava Dash[b], Biswaranjan Behera[a], S. Mohanty[a] and Pravat Kumar Shit[c]

[a]ICAR-INDIAN INSTITUTE OF WATER MANAGEMENT, BHUBANESWAR, ODISHA, INDIA [b]ICAR-INDIAN INSTITUTE OF SOIL AND WATER CONSERVATION, RESEARCH CENTRE, KORAPUT, ODISHA, INDIA [c]DEPARTMENT OF GEOGRAPHY, RAJA N. L. KHAN WOMEN'S COLLEGE (AUTONOMOUS), MIDNAPORE, WEST BENGAL, INDIA

2.1 Introduction

The two common features of spatial datasets that arise due to the presence of extreme values are (1) hotspots (data more than detection limit) and (2) black spots (data lower than detection limit), affects the spatial pattern and predictions. There are numerous approaches available for handling highly skewed datasets (Saito and Goovaerts, 2000). One such approach is data transformation (e.g., percent, normal, Box–Cox, ratio or lognormal), where analysis has been done with the transformed data, and the results are back-transformed. However, the presence of many black spot data used to create hindrance in solving a particular problem. The limitations in the use of the data transformation approaches include the production of several similar transformed values, making of a sequence of observations having same value, doubtfulness in the normality of the transformed histogram, acceptance of the normal-score transformation (Deutsch and Journel, 1998), and the possibility of source biasness in the back-transformation of the estimated moments (Saito and Goovaerts, 2000). Therefore, there is a need to explore other approaches to handle skewed datasets.

The use of nonparametric statistics and estimators like indicator kriging (IK) is another approach to handle the extreme values (Journel, 1983; Goovaerts, 2001). In IK, the main concept is the grouping of the variations in data using a set of thresholds and conversion of the data into an indicator of nonexceedance/exceedance of each threshold. The set of indicators is then undergone kriging and a conditional cumulative distribution function is formed using the estimated values. The mean or median of the probability distribution can be used as an estimate

Case Studies in Geospatial Applications to Groundwater Resources. DOI: https://doi.org/10.1016/B978-0-323-99963-2.00011-0

of the concentration of the material in question (Barabas et al., 2001; Cattle et al., 2002; Goovaerts et al., 2005; Adhikary et al., 2011).

2.2 Basic theory of indicator kriging

The approach in IK is that each observation is converted into a binary digit, either 0 or 1, based on its relationship to a threshold value. The procedure is that, firstly, the indicator function will generate indicator codes, which may be below or above a desired threshold value z_{th}. It can be expressed as below:

$$I(x; z_{th}) = \begin{cases} 1, \text{ if } z(x) \geq z_{th} \\ 0, \text{ otherwise} \end{cases} \tag{2.1}$$

After that, the semivariogram$\gamma_I(h)$ will quantify the spatial correlation of the indicator codes, $I(x_i, z_k)$. This can be expressed as follows:

$$\gamma_I(h) = \frac{1}{2N(h)} \sum_{i=1}^{N(h)} [I(x_i; z_{th}) - I(x_i + h; z_{th})]^2 \tag{2.2}$$

At a location x_o, the estimator of IK $I^{\wedge}(x_o; z_{th})$, can be calculated by the following equation:

$$[^{\wedge}(x_o; z_{th}) = \sum_{i=1}^{n} \lambda_i I(x_i; z_{th}) \tag{2.3}$$

Whereas the IK system has given $\Sigma\lambda_i = 1$ is

$$\sum_{j=1}^{n} \lambda_j \gamma_I (x_j - x_i) = \gamma_I (x_o - x_i) - \mu \tag{2.4}$$

Where γ_I is the semivariance of the indicator codes at the respective lag distance, λ_j is the weighted coefficient, and μ is the Lagrange multiplier.

Indicator code is a nonlinear transformation of the data value, into either 1 or 0. Similar indicator value is given to a group of data based on their direction of deviation from the threshold/cut-off value. For example, same indicator value is given to all data which are greater than the threshold value irrespective of the magnitude. Similarly, same indicator value is given to all data which are smaller than the threshold value irrespective of the magnitude. Therefore, the effects of extreme values can be effectively handled through indicator transformation. The indicator-transformed values will provide a value between 0 and 1 when interpolated by ordinary kriging. Thus, IK combining with ordinary kriging will provide an estimation of values that are higher or lower than a particular threshold value.

The product of the IK is the conditional cumulative distribution function. These cumulative distribution grades are useful to get the average value of the dataset. The probability of getting the values higher or lower than a threshold can be very easily obtained by IK. We generally use IK by modeling and calculating indicator variograms at the thresholds of our requirement which generally is called as multiple IK. The variograms of the median values of the whole dataset can be inferred and can be used for all the threshold values. In this case, the threshold values will not influence the kriging weights, so that the interpolation will become very fast. This is the main benefit of Multiple IK.

2.3 Criticisms of indicator kriging

The most common criticism that indicator approach faces is that some of the information is discarded in this process. Although theoretically, the calculation of indicator values at different cut-offs, like indicator cokriging can compensate for this information loss, practically the improvement is very little over IK because substantial information is carried by the cumulative indicator from one threshold to the other (Pardo-Igúzquiza and Dowd, 2005). Using of a large number of threshold points can discretize the distribution of the samples neatly and can increase the resolution of the discrete conditional cumulative distribution function. For instance, Lark and Fergusson (2004) have mapped the nutrient deficiency of soil in the Nebraska region of the USA by using 15 indicator thresholds. Similarly, 22 cut-off values were used by Goovaerts et al. (2005) for modeling the spatial distribution of arsenic concentrations in groundwater of Southeast Michigan probabilistically. For characterization of the spatial distribution of lead contamination in urban soils, 100 threshold values were used by Cattle et al. (2002). The identification of as many cut-off values as the number of observations in a sample dataset is a challenging situation. To overcome this challenge, the observations closest to the interpolated locations were used as thresholds. This manipulation of thresholds based on the local observation will increase the resolution of the discrete conditional cumulative distribution function. Here low thresholds are selected in the low-valued parts of the study area and high thresholds are selected in the high-valued parts of the study area (Saito and Goovaerts, 2000; Lloyd and Atkinson, 2001; Cattle et al., 2002).

There are some assumptions and drawbacks in Median IK. The IK should be nondecreasing and ranged between 0 and 1 as it generates a cumulative distribution at each point. Many times due to nonfulfillment of these requirements, violation of order relations used to happen. To get a finer resolution of the conditional cumulative distribution function, modeling of multiple indicator semivariograms is required, where the probabilities remain in between 0 and 1. If we cannot meet these requirements, a posteriori correction of the set of estimated probabilities will be required (Deutsch and Journel, 1998).

There are several methods available to overcome the order relations issue. Direct correction of the indicator values is one of the commonly used methods (Deutsch and Journel, 1998). Nested indicator approach proposed by Dimitrakopoulos and Dagbert (1993) is another method to deal with order relation problem. In this approach, the datasets are halved successively to achieve the thresholds and termed as nested indicator variables. Although the nested IK approach reduces order relations problem, it suffers from other problems due to data deficiency, especially at high thresholds.

Handling with a very large number of thresholds is a difficult task. Therefore, in most of the cases less than 9–15 thresholds are recommended (Deutsch and Lewis, 1992) so that the deviations remain within the limit. There are other indicator approaches to reduce the order relation problem and to maintain the high resolution of the conditional cumulative distribution function. For instance, the process which requires the solution of only one IK system at each location may reduce the order related problems as compared to the traditional IK approach (Pardo-Igúzquiza and Dowd, 2005). Goovaerts (1997) emphasized that there should not be any sudden change in indicator semivariograms between thresholds and there should at least be one datum from each class within each search neighborhood (z_{k-1}, z_k) while selecting the thresholds (z_k).

In this regard, for all global and locally adaptive thresholds, the same semivariogram model should be used (Saito and Goovaerts, 2000; Lloyd and Atkinson, 2001). A program was developed by Cattle et al. (2002) to locally compute and model indicator semivariograms by using global thresholds across the entire dataset. The indicator transform as well as the actual data were included under the IK algorithm. This approach is first conceived by Sullivan (1984) and later explored by Isaaks (1984) and termed as probability kriging. In essence, probability kriging is the cokriging of indicators between the indicator-transformed data and the uniform ($0 \rightarrow 1$) transform of the sample data. The cokriging and univariate indicator variograms involve in the calculation and modeling of the cross-variograms of the data types.

The underlying processes which influence the data distribution were analyzed by Rivoirard (1993) through the correlations among the cross-variograms of indicators at adjacent thresholds. With these, nonlinear algorithms like the mosaic and diffusion models can be developed, among others. But, practically it can be considered as straightforward implementation of multiple IK, with non-nested indicator transforms of data at multiple thresholds to define local distributions of natural resource concentration.

2.4 Merits of indicator kriging

The primary reason for using IK in most earth science applications is its nonparametric nature and ability to handle mixed data populations. Since IK partitions the overall distribution of samples according to a number of thresholds, it is not necessary to fit or assume any particular analytical distribution model for the data. Multiple IK requires a variogram model to be inferred for each cut-off, and it can handle different anisotropies for each cut-off. The violations of order relations become large when there is high change of anisotropy observed between adjacent thresholds, but in the case of gradual change of anisotropy, the situation can be easily handled.

IK is preferred for highly skewed datasets. It can practically deal with the upper tail of the distribution that is not entirely dependent on an arbitrary uppercut value. It allows the researchers to use features of the actual data to define the upper tail treatment. Although unimpeded estimation is not recommended wherever there is the chance to define constricting domains, multiple IK is an approach that minimizes smoothing under certain conditions when it is a necessity. Under indicator transformation, we can go for a consensus coding for all the data types and their amalgamation into a single process. Uncertainty at unsampled locations can also be estimated through the inference of a distribution of values.

2.5 Practical corrections to use indicator kriging

2.5.1 Treatment of upper and lower tails

The proportion of resources above each of the indicator cut-offs or thresholds analyzed can be estimated using IK. Each indicator class interval must be assigned a threshold in order to convert this data into an estimate of mean quantity or quantity over a cut-off. When determining the class interval threshold, a number of sensitivities must be taken into account. The distribution of cut-offs within multiple classes will be approximately linear if indicator thresholds have been appropriately adjusted with adequate attention for the input distribution. In these classes, the average value of the supplied data will usually serve for threshold assignment.

The distribution of thresholds in the distribution's upper and lower classes will not be linear in most cases. In the event of a positively skewed distribution, the threshold applied to the uppermost class has the highest estimated sensitivities. The threshold distribution in this class is influenced by both distribution skew and outliers, necessitating a more complicated approach of mean threshold selection to minimize parameter overestimation or underestimation. For expressing the parameter distribution above the topmost threshold, Deutsch and Journel (1998) presented a model based on a hyperbolic distribution. The sample parameter distribution can be used to estimate both variables.

2.5.2 The data dilemma

When using multiple IK to estimate resources, a predetermined number of thresholds must be chosen that sufficiently describe the input data distribution form. The chosen threshold number and the availability of time for the required study are always in competition. To distinguish between components of mixed population distributions, additional indicators may be used. Many of the indicator thresholds will be clustered at the lower end of the positively skewed distribution, which is a drawback of these deciles. Higher values will be represented by fewer indicator thresholds. The way out to this conundrum is to acquire more data. For better assessment of high-value and lower-value, enough closely spaced data from representative locations must be collected.

2.5.3 Initially use the median indicator kriging

In median IK, the median indicator variogram is used to determine whether the entire indicators are continuous or not. The median IK is very simple and can widely be used at the early stages of any project when the data availability is scarce in nature and in this stage, the continuity of indicator thresholds may not be defined easily. Among the entire indicator, the median indicator variogram is the most consistent. It shows very high range of continuity and can yield value with high degree of accuracy with a very limited volume of data.

The primary assumption linked with the median indicator approach is shown by applying variograms from a single indicator to all thresholds. This means that the direction and range of continuity do not change as the thresholds change. Continuity virtually always changes with indicator threshold, and invariably falls with rising indicator threshold, according to comprehensive indicator variography studies. As a result, the quantiles of the upper-grade classes will be overestimated with Median IK, resulting in a higher-than-normal anticipated grade. In practice, when the data allows for full estimate of a set of indicator variograms, median IK is not recommended.

2.5.4 Change of support

The development of a distribution of thresholds for nonpoint data is known as change of support (Dowd, 1992, Matheron, 1982). The indicator transform is a nonlinear transforms like uniform transform, logarithmic transform, and normal scores transform. Ordinary kriging, inverse distance, and radial basis functions are linear estimation methods. As a result, one cannot linearly average indicators and also cannot produce a geographical distribution by simply averaging the point distributions consequential at a lower scale. In the linear interpolation technique, the consequence is to discretize the space with a series of points, estimate the value at each point, then derive the spatial pattern using the arithmetic mean. If the median or mean value of IK at each position is known, the space can be divided into points, the conditional cumulative distribution function estimated by multiple IK at each location, the desired statistic deduced, and the result then can be averaged.

The usual method for changing support in IK has been to apply a variance correction factor to the point statistics on a global basis, similar to how one might do with point krigged data. There are a variety of approaches that can be used to accomplish this. The affine correction, a simple factorization of variance from point to (theoretical) block variance, is perhaps the most extensively utilized.

The use of indirect lognormal correction is another alternative (Isaaks and Srivastava, 1989). Moving beyond estimate to sequential indicator simulation gives a truly local change of support, conditional solely on values in the neighborhood, and is the most elegant and theoretically valid solution to generating spatial values.

2.6 Applications in water science

IK can be utilized in water science applications to assess the water quality characteristics that affect sprinkler irrigation performance. IK can also be used for standard irrigation water quality mapping and management. Using IK, heavy metals in groundwater may be quantified with great precision (Adhikary et al., 2011). Because agricultural-related groundwater nitrate pollution is a significant environmental issue, its management requires special attention. As a result, particular measures targeted at reducing the danger of nitrate pollution must be implemented in order to reduce the impact on the environment and the potential damage to human health. Managers and decision-makers would be benefitted from the spatial probability distribution of nitrate concentrations exceeding a threshold value defined to maintain the quality for a desired use. On the other hand, categorical IK can be used to assess the risk of nitrate pollution in groundwater (Chica-Olmo et al., 2014). IK is a method for evaluating projected spatial distributions of risk

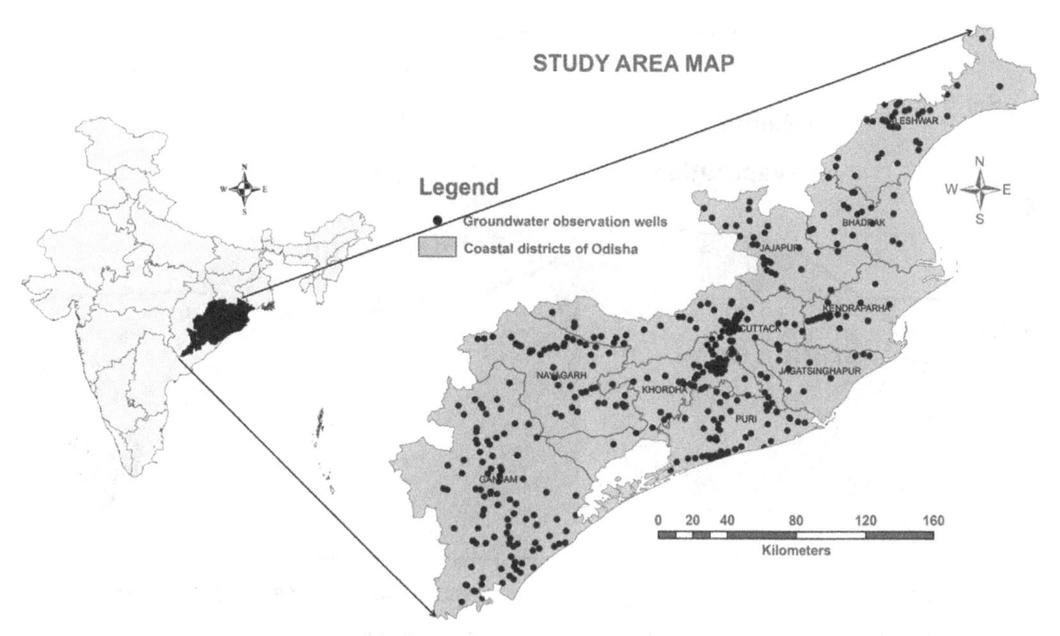

FIGURE 2.1 Map of the study area showing the location of the groundwater observation wells.

of production of excess runoff using routinely used water table data, notably for the recently developed models to predict water quality that define runoff as saturation excess. The beauty of IK lies in the fact that the spatial variables can be transformed into binary indicators (i.e., 1 if the variable is above a given threshold value and 0 if it is below the threshold), and the resulting indicator semivariogram can be modeled, thus producing the probability of exceedance of the measured variable from the threshold value. With the depth to the water table as the variable and the threshold set at the soil surface, IK provides a quantifiable likelihood of saturation or, in line with saturation excess runoff theory, runoff generation risk (Lyon et al., 2006).

2.6.1 Case study

To demonstrate the application of IK in water resource management, groundwater quality in terms of total dissolved solids (TDS) was spatially interpolated using IK. The study area is the costal districts of Odisha, India (Fig. 2.1) comprising 10 districts such as Balasore, Bhadrak, Kendrapada, Cuttack, Jajpur, Jagatsinghpur, Puri, Khordha, Nayagarh, and Ganjam. This coastal plain covers a length of nearly 480 km and the width varies between 10 and 50 km. This plain consists of many deltas of rivers Subarnarekha, Budhabalanga, Brahmani, Baitarani, Mahanadi, and Rushikulya. The total population of this region is 19.3 million with the density of 570 person/km^2 (Census, 2011).

The dominant climate of the study area is hot and humid. The mean annual rainfall is 1456.9 mm and more than 70% of rainfall occurs during monsoon (June–September) season. August is the wettest and December is the driest month. The mean temperature ranges from 23.06

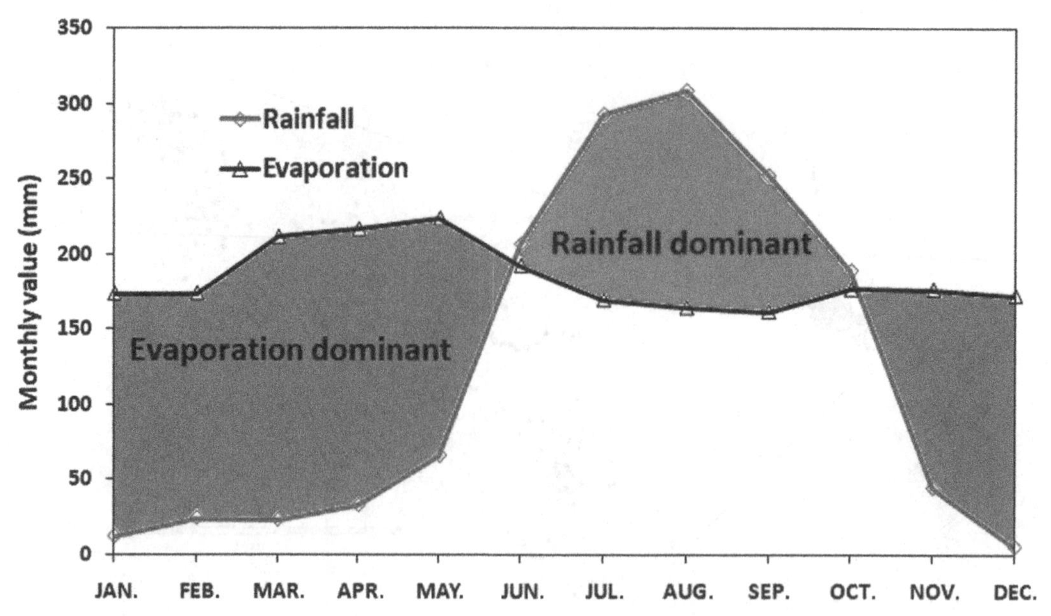

FIGURE 2.2 Long-term variation of rainfall and evaporation in the study area.

to 33.15°C. In May, potential evaporation is as high as 7.22 mm/day, whereas during August it comes down to 5.30 mm/day (Fig. 2.2). Soil texture varies from loamy sand to clay and depends on geomorphology and the types of alluvial deposits. Soil structure is granular or platy. The presence of excess salinity in the soil can be attributed to the shallow saline groundwater table and inundation of seawater in the low-lying areas (Srinivasan et al., 2018).

2.6.1.1 Data description and semivariogram parameter

To get a preliminary idea about the level of TDS in the study area, descriptive statistics have been put forth. Table 2.1 indicates that there is a wide variation in terms of TDS in different coastal districts of Odisha. The TDS is as low as 35 mg/L in Jajpur district and as high as 2765.7 mg/L in Bhadrak district. Although Bhadrak district shows highest TDS, but the average value is highest in Kendrapara district followed by Ganjam, Bhadrak, and Puri districts. The average TDS of whole coastal Odisha is 416.4 mg/L with CV value of 78.1%. The CV value indicates high spatial variation of TDS within the coastal part of Odisha.

Kolmogorov and Smirnov test indicated that groundwater TDS is not normally distributed but distributed log-normally. Log transformed semivariogram parameters were generated and the values of nugget, sill, and range (Fig. 2.3) of the best-fit models were recorded. The spatial dependence of variables can be inferred from nugget-to-sill ratio. According to Liu et al., less than 0.25 nugget-to-sill ratios indicates strong spatial dependence, between 0.25 and 0.75 indicates moderate spatial dependence and more than 0.75 indicates weak spatial dependence. In this study, the groundwater salinity in terms of TDS is moderately spatially correlated. The exponential model was fitted well for the groundwater salinity. Dash et al. (2010); and

Table 2.1 Descriptive statistics of TDS (mg/L) in groundwater in the study area.

District	Mean	Maximum	Minimum	SD	CV (%)	SE
Balasore	349.3	773.0	50.0	199.9	57.2	34.3
Bhadrak	525.6	2765.7	165.0	566.3	107.7	120.7
Cuttack	329.6	727.0	42.0	176.4	53.5	21.1
Jajpur	266.7	760.0	35.0	207.5	77.8	40.7
Jagatsinghpur	396.8	1586.0	162.0	402.7	101.5	116.2
Kendrapada	590.9	1325.0	246.0	275.9	46.7	63.3
Khordha	216.8	1858.0	43.0	215.7	99.5	25.2
Nayagarh	422.3	1296.0	81.7	275.1	65.1	39.7
Puri	514.5	1808.0	90.0	335.8	65.3	38.8
Ganjam	552.1	2561.0	104.0	366.3	66.3	36.4
Coastal Odisha	416.4	2765.7	35.0	325.3	78.1	14.8

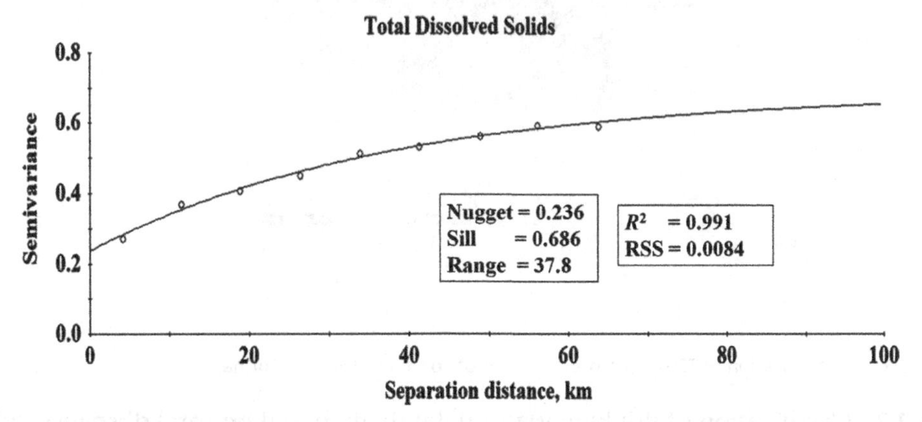

FIGURE 2.3 Best fit semivariogram and model parameters for groundwater TDS in coastal belt of Odisha6.1.2. Groundwater TDS and ordinary kriging.

Adhikary et al. (2011) studied spatial variation of groundwater salinity using geostatistics in different parts of world and reported different best-fitted semivariogram models. Therefore, it can be inferred that spatial patterns of groundwater TDS vary from place to place depending on the climatic, soil, and geologic conditions of the area under consideration.

The spatial variation of groundwater TDS is presented in Fig. 2.4. The TDS values ranged from 35 to 2765 mg/L. The entire study area is categorized into four classes depending on the WHO classification (Table 2.2). Groundwater with low salinity (TDS < 300 mg/L) and very high salinity (TDS > 1000 mg/L) level was confined to only a few pockets in the study area, whereas the groundwater salinity in the study area was mostly medium to high in nature. The area-wise distribution of groundwater salinity level within the study area is presented in Table 2.3. The area under low TDS groundwater is 3623 km^2 which is slightly more than 10% of the study area. More than 40% of the study area is having groundwater TDS ranged between poor and unacceptable. Therefore, the part of the study area is fit for drinking purpose. Less than 1% of

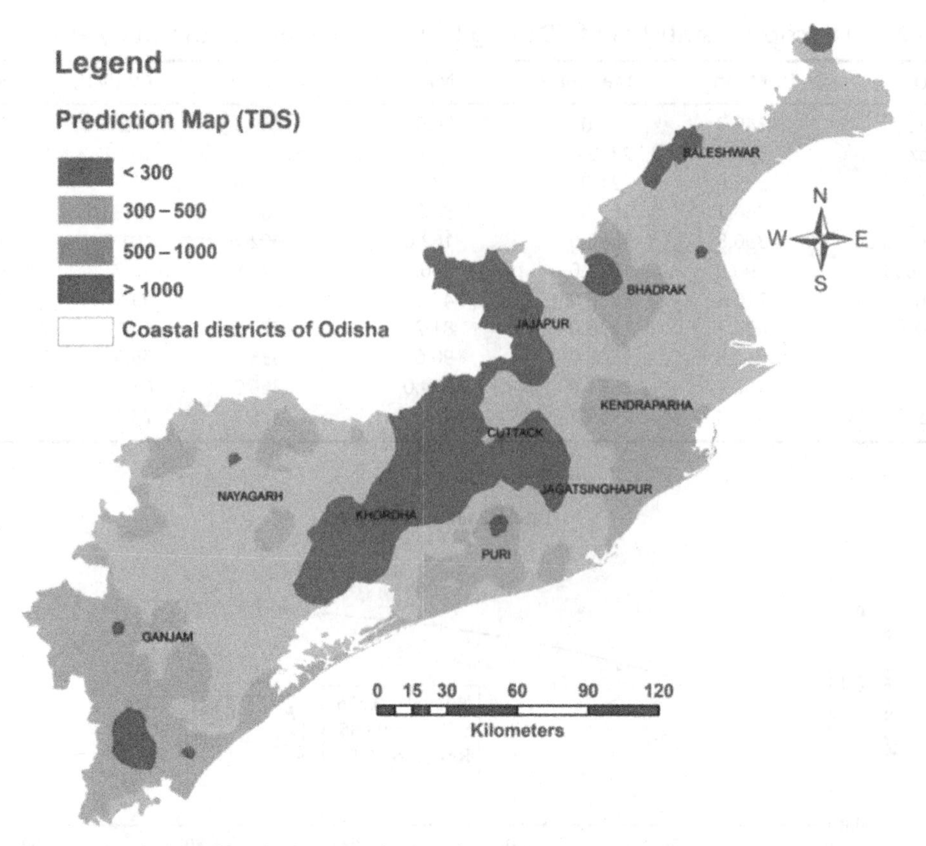

FIGURE 2.4 Spatial variation of TDS in the groundwater of coastal districts of Odisha, India.

Table 2.2 Classification of drinking water palatability based on total dissolve solids.

Classification code	Total dissolve solids (mg/L)	Palatability rating	Uses
T1	<300	Excellent	Very good taste for drinking
T2	300–500	Good	Good taste for drinking
T3	500–1000	Poor	Poor taste for drinking
T4	>1000	Unacceptable	Unacceptable for drinking

the area contains unacceptable level of TDS. The performance accuracy of ordinary kriging in predicting groundwater TDS was computed. Both error terms ME and RMSE were close to 0, which indicated that ordinary kriging's predictability for the coastal belt of Odisha was good and acceptable. The estimated MRE value for TDS is 0.1755, which indicates that the predicted data has a very low relative error in comparison to the observed data.

Among the geostatistical interpolation techniques, kriging is the most widely used technique (Adhikary et al., 2010; Dash et al., 2010). It incorporates the spatial autocorrelation among the measured values of a spatial random variable to estimate a value for an unsampled location.

Table 2.3 Area wise distribution of TDS in the groundwater of coastal districts of Odisha.

Classification code	TDS (mg/L)	Groundwater quality	Area (km^2)	Area (%)	Uses
T1	<300	Excellent	3623	10.11	Can be used for drinking for human and animal without any ill affect
T2	300–500	Good	17826	49.73	Can be used for drinking for human and animal with some degree of cautious
T3	500–1000	Poor	14293	39.87	Restricted use for drinking for human and animal
T4	>1000	Unacceptable	103	0.29	Not suitable for drinking purpose

It includes different approaches, such as ordinary kriging, simple kriging, universal kriging, discjunctive kriging, regression kriging, and IK. In recent years, several researchers applied ordinary kriging for spatial analysis of groundwater depth and salinity for its accurate prediction and simplicity. Ordinary kriging is a remarkably popular geospatial interpolation technique used to interpolate groundwater quality (Adhikary and Biswas, 2011), groundwater salinity (Adhikary et al., 2015), groundwater fluoride (Adhikary et al., 2014), and metal content in groundwater (Adhikary et al., 2011).

2.6.1.2 Indicator kriging: groundwater TDS probability

IK was used to generate the probability map of groundwater TDS exceeding a threshold level. The probability map generated using the IK method where the TDS threshold was 500 mg/L is shown in Fig. 2.5. The probability map showed that high TDS concentration was found in the southern and south-western parts of the study area covering the most part of the district Ganjam. A higher probability of TDS found in groundwater was also observed in the districts of Baleswar, Kendrapada, Puri, and Nayagarh. Groundwater TDS level of more than 1000 mg/L has been considered as unsuitable for drinking purpose. Therefore, we have also considered 1000 mg/L as the threshold level of pollution and prepared the probability of exceedance map using IK (Fig. 2.6). Fig. 2.6 showed that Bhadrak, Puri, and Ganjam districts are more prone to risk of groundwater salinity. It was also found from Table 2.4 that about 39% and 78% area is very much safe from groundwater TDS when we consider the threshold as 500 and 1000 mg/L, respectively. More than 14% area (5158 km^2) may not be suitable for drinking purposes, where the probability of exceedance from the threshold value of TDS was the highest (>0.50) when we considered 500 mg/L as the threshold value. When we considered the threshold value as 1000 mg/L, this figure changed to less than 1 % area, that is, 118 km^2 area. This may be attributed to the facts of seawater intrusion and marine transgression. The use of IK in groundwater application is not widely used but several workers like Adhikary et al. (2011) have used this to interpolate metal pollution in groundwater. The study area is associated with high agricultural activity, and therefore, groundwater nitrate pollution can be considered as an associated happening. Thus,

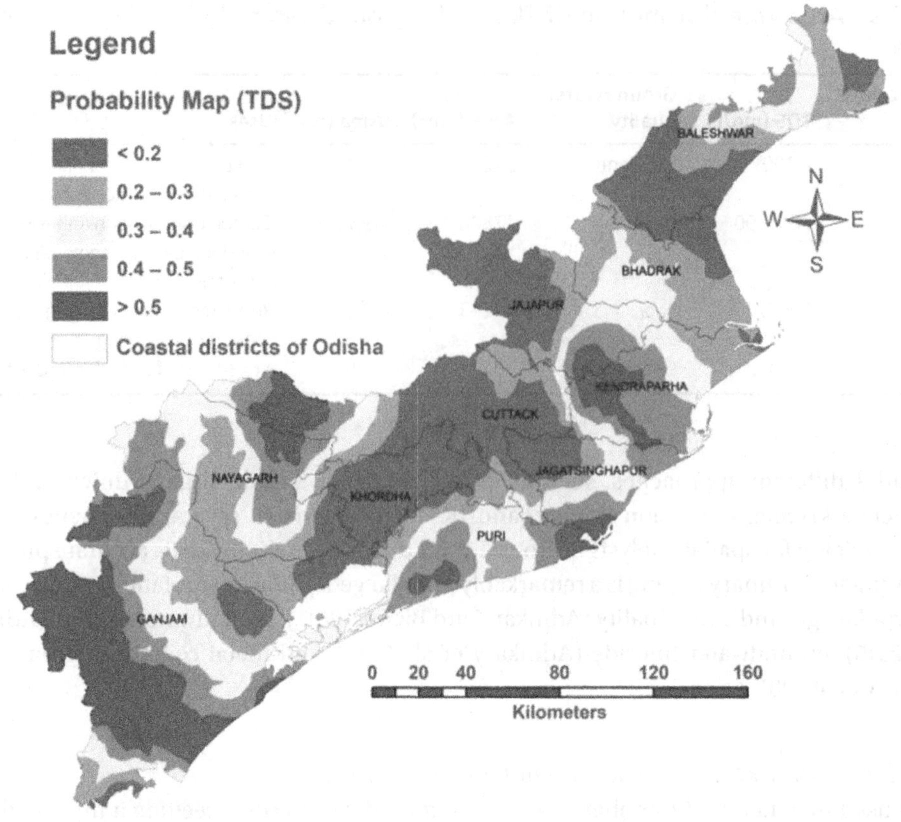

FIGURE 2.5 Prabability map TDS to exceed 500 mg/L concentration in the groundwater of coastal districts of Odisha, India.

Table 2.4 Delineated area under different probability ranges of threshold concentration limits of groundwater TDS in the study area obtained using indicator kriging.

	TDS threshold 500 mg/L		TDS threshold 1000 mg/L	
Probability range	Area (km^2)	Area (%)	Area (km^2)	Area (%)
< 0.20	13711	38.25	28027	78.19
0.20–0.30	8697	24.26	2158	6.02
0.30–0.40	2383	6.65	3206	8.94
0.40–0.50	5896	16.45	2336	6.52
> 0.50	5158	14.39	118	0.33

the probability distribution map of nitrate exceeding a threshold value will be helpful to policy makers. Categorical IK can be used for assessing the risk of groundwater nitrate pollution (Chica-Olmo et al., 2014).

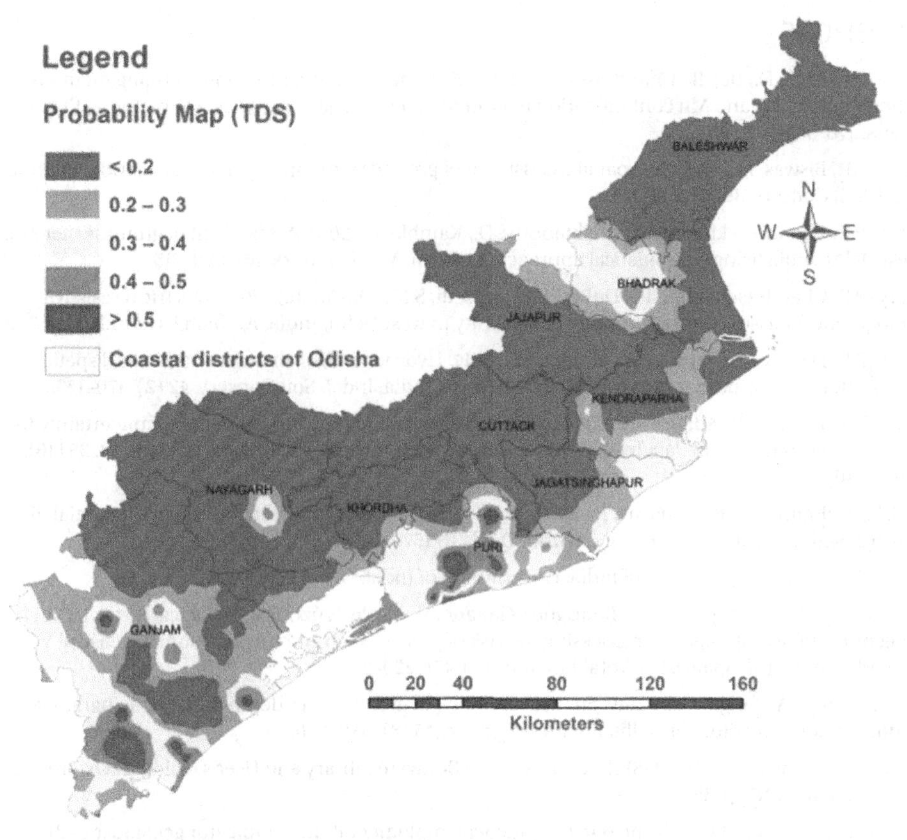

FIGURE 2.6 Prabability map TDS to exceed 1000 mg/L concentration in the groundwater of coastal districts of Odisha, India.

2.7 Conclusions

In natural resource management sciences like soil science, geological science, hydrological science, and environmental science, the use of IK is gradually increasing because of its practical solutions for common problems. Being nonparametric in nature has added an extra advantage in the acceptance and popularity of IK. It has the ability to generate "recoverable" resources, which has made this approach highly popular. Complexity in obtaining appropriate distributions, the annoyance of order relations, and the difficulty and hard work required in surmising variogram models at multiple thresholds are the few roadblocks that need to be addressed. But the researchers and statisticians have been working hard to come up with solutions to almost all of the above-mentioned problems. The case study which used IK to map groundwater suitability for drinking is a classic example of the use of IK in natural resource management.

References

Adhikary, P.P., Ch.J., D., Bej, R., Chandrasekharan, H., 2011. Indicator and probability kriging methods for delineating Cu, Fe, and Mn contamination in groundwater of Najafgarh Block, Delhi, India. Environ. Monit. Assess. 176, 663–676.

Adhikary, P.P., Biswas, H., 2011. Geospatial assessment of ground water quality in Datia district of Bundelkhand. Ind. J. Soil Conserv. 39 (2), 108–116.

Adhikary, P.P., Chandrasekharan, H., Chakraborty, D., Kamble, K., 2010. Assessment of groundwater pollution in West Delhi, India using geostatistical approach. Environ. Monit. Assess. 167, 599–615.

Adhikary, P.P., Chandrasekharan, H., Dubey, S.K., Trivedi, S.M., Dash, Ch.J., 2015. Electrical resistivity tomography for assessment of groundwater salinity in west Delhi, India. Arabian J. Geosci. 8 (5), 2687–2698.

Adhikary, P.P., Dash, Ch.J., Sarangi, A., Singh, D.K., 2014. Hydrochemical characterization and spatial distribution of fluoride in groundwater of Delhi state, India. Ind. J. Soil Conserv. 42 (2), 170–173.

Barabás, N., Goovaerts, P., Adriaens, P., 2001. Geostatistical assessment and validation of uncertainty for three-dimensional dioxin data from sediments in an estuarine river. Environ. Sci. Technol. 35 (16), 3294–3301.

Cattle, J.A., McBratney, A.B., Minasny, B., 2002. Kriging method evaluation for assessing the spatial distribution of urban soil lead contamination. J. Environ. Qual. 31, 1576–1588.

Census, 2011. Census Organization of India. Government of India, New Delhi.

Chica-Olmo, M., Luque-Espinar, J.A., Rodriguez-Galiano, V., Pardo-Igúzquiza, E., Chica-Rivas, L., 2014. Categorical indicator kriging for assessing the risk of groundwater nitrate pollution: the case of Vega de Granada aquifer (SE Spain). Sci. Total Environ. 470-471, 229–239.

Dash, J.P., Sarangi, A., Singh, D.K., 2010. Spatial variability of groundwater depth and quality parameters in the National Capital Territory of Delhi. Environ. Manage. 45 (3), 640–650.

Deutsch, C.V., Journel, AG., 1998. GSLIB: Geostatistical Software Library and User's Guide. 2. Oxford University Press, New York, NY, p. 369.

Deutsch, C.V., Lewis, R., 1992. Advances in the practical implementation of indicator geostatistics. In: Proceedings of the 23rd International APCOM Symposium; Tucson, AZ, Society of Mining Engineers, pp. 169–179.

Dimitrakopoulos, R., Dagbert, M., 1993. Sequential modelling of relative indicator variables: dealing with multiple lithology types. In: Soares, A. (Ed.), Geostatistics Troia '92, 5. Kluwer Academic Publishers, Dordrecht, pp. 413–424. https://doi.org/10.1007/978-94-011-1739-5_33.

Dowd, PA., 1992. A review of recent developments in geostatistics. Computers and Geosciences 17 (10), 1481–1500.

Goovaerts, P., AvRuskin, G., Meliker, J., Slotnick, M., Jacquez, G.M., Nriagu, J., 2005. Geostatistical modeling of the spatial variability of arsenic in groundwater of Southeast Michigan. Water Resour. Res. 41 (7) W07013 10.1029.

Goovaerts, P., 1997. Geostatistics for Natural Resources Evaluation. Oxford University Press, New York, NY, p. 483.

Goovaerts, P., 2001. Geostatistical modelling of uncertainty in soil science. Geoderma 103, 3–26.

Isaaks, E.H., Srivastava, R.M., 1989. An Introduction to Applied Geostatistics. Oxford University Press, New York, NY.

Isaaks, EH., 1984. Risk qualified mappings for hazardous waste sites: a case study in distribution-free geostatistics Master's thesis. Stanford University, Stanford, CA.

Journel, AG., 1983. Nonparametric estimation of spatial distributions. Math. Geol. 15 (3), 445–468.

Lark, R.M., Ferguson, R.B., 2004. Mapping risk of soil nutrient deficiency or excess by disjunctive and indicator kriging. Geoderma 118 (1), 39–53.

Lloyd CDand Atkinson, PM., 2001. Assessing uncertainty in estimates with ordinary and indicator kriging. Comput. Geosci. 27 (8), 929–937.

Lyon, S.W., Lembo Jr, A.J., Walter, M.T., Steenhuis, T.S., 2006. Defining probability of saturation with indicator kriging on hard and soft data. Adv. Water Res. 29, 181–193.

Matheron, G., 1982. La destructuration des hautes teneurs et le krigeage des indicatrices. Centre de Geostatistiqueet de Morphologie Mathematique, Note N-761, 33 p.

Pardo-Igúzquiza, E., Dowd, PA., 2005. Multiple indicator cokriging with application to optimal sampling for environmental monitoring. Comput. Geosci. 31 (1), 1–13.

Rivoirard, J., 1993. Relations between the indicators related to a regionalised variable. In: Soares, A. (Ed.), Geostatistics Troia '92, 5. Kluwer Academic Publishers, Dordrecht, pp. 273–284. https://doi.org/10.1007/978-94-011-1739-5_23.

Saito, H., Goovaerts, P., 2000. Geostatistical interpolation of positively skewed and censored data in a dioxin contaminated site. Environ. Sci. Technol. 34 (19), 4228–4235.

Srinivasan, R., Singh, S.K., Nayak, D.C., Dharumarajan, S., 2018. Assessment of soil and water salinity and alkalinity in coastal Odisha—a case study. J. Soil Salin. Water Qual. 10 (1), 14–23.

Sullivan, J., 1984. Conditional recovery estimation through probability kriging: theory and practice. In: Verly, G (Ed.), Geostatistics for Natural Resources Characterisation. Riedel, Dordrecht, pp. 365–384.

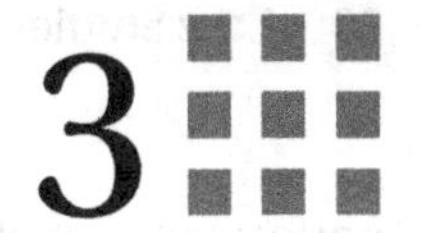

GI Science application for groundwater resources management and decision support

Gouri Sankar Bhunia[a], Pravat Kumar Shit[b] and Soumen Brahma[a]

[a]DEPARTMENT OF GEOGRAPHY, NALINI PRABHA DEV ROY COLLEGE, BILASPUR, CHHATTISGARH, INDIA [b]DEPARTMENT OF GEOGRAPHY, RAJA N. L. KHAN WOMEN'S COLLEGE (AUTONOMOUS), MIDNAPORE, WEST BENGAL, INDIA

3.1 Introduction

Sustainable groundwater development poses significant challenges in terms of maintaining the Earth system's resilience, involving science, technology, and socioeconomic factors (Okello et al., 2015). Groundwater sustainability on a regional and international scale necessitates the evaluation and comprehension of multiple interacting environmental processes operating at numerous spatio-temporal scales (Cumming et al., 2006). Furthermore, groundwater interferences on the earth system necessitate a multidisciplinary approach needed to analyze and modeling geosphere–anthroposphere interrelations for analysis and management. Numerous components of the sustainability challenges, including the formulation and execution of groundwater development policies, rely heavily on earth scientists. GIScience (geographic information science) is a scholarly scientific discipline concerned with the underlying issues surrounding the use of a wide range of digital technologies to deal with geographic information, which includes data about places, events, and natural events on and near the Earth's surface that is sequestered in maps or images. GIScience and technological innovations play a critical role in groundwater management in this context (Trevisani and Omodeo, 2021). Furthermore, understanding the complex dynamics of the groundwater system and its interconnections with the anthroposphere requires the acquisition and computational analysis of geoenvironmental data. On the other hand, geoenvironmental data collection, interpretation, and modeling, along with their application, have taken on a political implication never before seen, as they have become critical for intelligent decision (Rapacciuolo and Blois, 2019).

The retrieval, processing, distribution, and application of Earth remote sensing data are becoming more difficult by the day, necessitating a more thorough examination on a regular basis (Jolliffe and Cadima, 2016). Traditional data processing and distribution methods have poor

Case Studies in Geospatial Applications to Groundwater Resources. DOI: https://doi.org/10.1016/B978-0-323-99963-2.00014-6

performance, making them unsuitable for a wide range of end-users and real-time forecasting applications that require high-performance spatial datasets. Numerous users require custom-made geospatial data with an immersive graphical user interface in a GIS/Web-GIS readable format. On the other hand, better management of future Earth-observing satellite systems will result from more widely disseminated Earth-observing satellite data in various formats via diverse protocols. Furthermore, ground truthing is required for proper analysis of geospatial data. With reference data, the two most important phases, calibration (developing training areas) and validation (accuracy assessment), can be completed. As a result, while satellite systems have unrivaled functionalities for gathering valuable information, the relevance of several remotely sensed data and the accomplishment of most environmental studies may still rely heavily on the availability of high-quality data from nonremote sensing inputs.

The appropriation of frequency components with physicochemical characteristics in satellite remote sensing is challenged because the ideal cause and effect in the system can never be measured in isolation (Gholizadeh et al., 2016). Numerous challenges exist and must be overcome in the remote sensing of waterbodies from satellite-borne instruments for accurate mapping of waterbodies and their characteristics, including bidirectional effect, atmospheric effect, plant structure effect, background effect of soil or vegetation, nonlinear scattering effects, impacts of spatial variability, clustering effect, and nonlinear mixing. The electromagnetic and microwave features of the surface, the physics of the sensor in conjunction with the integrated air–sea interface, knowledge of geophysical processes, and proper testing with widely available technologies can all be used to accomplish precise measurement of the factors and expected output. The retrieval algorithm converts electromagnetic signals from satellite sensors into climate variable measurements. Sensor placement is inherently nondeterministic, and there is an extent of randomness attributed with sensor placement, which leads to uncertainty in climate trends and variable retrieval. Unresolved displacements in the sensor's sensitivity have been identified as a leading cause of the apparent spectrum of alteration. Sensors gradually lose radiometric stability and sensitivity during execution, so proper calibration is essential. Due to a lack of precise on-board or on-orbit calibrations, a few satellite sensors cannot be re-evaluated after launch. Incorporating observations from distinct satellites can be used to create long-term records, but this increases the uncertainty. There would be a significant increase in uncertainties if the output of combining datasets from various systems during growth and calibration was lacking. Interinstrument correction can be used to determine relative bias and can be used to mitigate such concerns.

New technologies and simulation options are available to hydrologists and earth scientists, but they must be re-evaluated in order to be used accurately critically—to achieve global objectives. Our concerns are highly relevant to water science and geoenvironmental research, especially when it focuses on topics like groundwater resource protection/management, various geoengineering-related issues, and decision-making. To begin with, these challenges directly affect the vital zone, a surficial earth layer attributed to the high intensity of relationships between the geosphere (defined broadly to include the hydrosphere), the biosphere, and the anthroposphere. Second, emerging technologies (e.g., geospatial data) for geoenvironmental data collection, analysis, and modeling have a significant impact on this setting. Third, humanities-related information sources can provide valuable insight into the complex relationships between humans and the environment. Based on the above consideration, the present paper aims to study the various aspects that are related to the role of GIScience and advanced technologies for geoenvironmental analysis in the context of sustainable development of groundwater.

3.2 Hydrosphere–geosphere–anthroposphere interlinked dynamics

Different variables, including population global dynamics, will probably increase the relevance of local and global sustainability challenges in the coming years (Galvani et al., 2016). Not only will the world population presumably continue to go up, with a projected population of more than 9 billion by 2050, but it will also be characterized by a significant imbalance, both geographically and socioeconomically, among the world's regions. These basic demographic criteria point to an increase in human–hydrological interactions in the future (Fang and Jawitz, 2019). Land-use changes, natural calamities, pollution, ecological alteration, climate change, groundwater resource depletion, geoengineering dilemmas, and so on are all well-known indications of such relationships, both in terms of human and natural consequences.

Dissolution, hydrolysis, and precipitation interactions; adsorption and ion exchange; oxidation and reduction; and gas exchange between groundwater and the atmosphere are all practices that contribute to the composition of groundwater (Wu et al., 2020). The forms of rock minerals that groundwaters interaction during their journey through the subsurface are reflected in their constituents. Calcium, sodium, and magnesium are the most mobile components in groundwater, meaning they are the components that are most conveniently liberated by the weathering of rock minerals. Aluminum and iron are primarily static and locked up in solid phases, while silicon and potassium have transitional mobilities.

3.3 Spatial and machine learning model for groundwater mapping

The presence of groundwater varies greatly from one location to the next. The presence of groundwater in a specific area is not coincidental (Taghizadeh-Mehrjardi et al., 2021); it is the result of complex interactions of factors including natural geography, climate, hydrology, geology, ecology, topography, soil type and subsurface layers, fracture density, ground slope, and land (Abd Manap et al., 2013). For groundwater-potential mapping, statistical models such as the frequency ratio (Ozdemir et al., 2011), logistic regression (Park et al., 2017), Shannon entropy (Zabihi et al., 2016), weight of evidence (Madani and Niyazi et al., 2015), and the evidential belief function were used. The machine learning techniques used in groundwater potentiality mapping so far have been categorized as random forest (RF) (Al-Abadi et al., 2016), support vector machine (SVM) (Naghibi et al., 2017), boosted regression trees (Naghibi and Pourghasemi, 2015), classification and regression trees (Choubin et al., 2019), linear discriminant evaluation

(Naghibi et al., 2017), multivariate adaptive regression spline (Golkarian et al., 2018), and naïve Bayes (Miraki et al., 2018), respectively.

However, groundwater specialists could not concede on a model for assessing groundwater potency until recently (Mallick et al., 2021). As a result, ensemble techniques have recently gained popularity in the field of geohazard susceptibility and potentiality mapping (Mallick et al., 2021). Various hybrid ensemble machine learning models have recently been suggested for attaining greater consistency in groundwater potential mapping by combining a base model with optimization algorithms or ensemble procedures. To strengthen forecast accuracy, ensemble modeling integrates two or more machine learning techniques (Islam et al., 2021). Ensemble modeling can help a single model overcome its flaws (Talukdar et al., 2021). Miraki et al. (2019) established an ensemble model (RS-RF) for assessing groundwater potential in the Qorveh–Dehgolan plain, Kurdistan province, Iran, using an amalgamation of RF, and random subspace ensemble technique, and concluded that the RS-RF model is an enticing tool for mapping groundwater potential. Al-Fugara et al. (2020) used a hybrid model that integrated SVM and Genetic Algorithm (GA) to map groundwater recharge in Jordan's Jerash and Ajloun regions.

3.4 Big data analytics and groundwater mapping

Big data analytics is an innovative new buzzword that describes the implementation of appropriate and conventional analytical techniques to massive amounts of heterogeneous data in order to gain valuable insights that can be used to drive enhancement, development, and information retrieval (Adamala, 2017; Kitchin and McArdle, 2016). Conventional analytics such as data mining, statistical methods, SQL queries (structured query language queries), and data visualization are examples of big data analytical approaches that work well with structured data. With heterogeneous unstructured data, advanced analytical strategies like natural language processing, text analytics, video analytics, audio analytics, artificial intelligence, and machine learning work well (Gandomi and Haider, 2015).

Standard information systems can easily maintain and analyze field monitoring hydrological data. Remote sensing data are true big data, with datasets that are strongly dimensional, heterogeneous, and growing in size (Liu, 2015). Hundreds of earth observation satellites have been initiated since then, some of which were designed specifically to gather data on Earth's hydrological systems, including Landsat or the gravity recovery and climate experiment (GRACE). Some of these remote-sensing missions have been operational since the early 1970s, like Landsat. New remote-sensing missions have been launched and advanced over time, resulting in an ever-growing large dataset. Table 3.1 shows some of the remote-sensing products that are pertinent to groundwater. Furthermore, hydrological studies based on remote sensing data have typically been conducted at regional or global scales. This is due to the fact that much of the remote-sensing data are at a spatial scale that makes local or site-specific assessment impossible.

In the field of groundwater science, social media offers an unusual new source of big data (Fig. 3.1). Beneficial features can be retrieved from these data sources using advanced analytical techniques including natural language processing or video analytics. Twitter posts containing

Table 3.1 Application of satellite sensor in groundwater mapping and modeling.

Satellite/sensor	Groundwater component	Pixel size	Temporal resolution
GRACE	Terrestrial water storage	110–330 km	Monthly
GRACE-FO	Terrestrial water storage	110–330 km	Monthly
SMAP	Soil moisture	3–36 km	1–7 days
SMOS	Soil moisture	35–50 km	1–3 days
GPM	Precipitation	5–15 km	30 min-monthly
TRMM	Precipitation	5–550 km	3 hours—monthly
Terra/MODIS	Evapotranspiration, LST, NDVI	0.5 km	8 days—annual
Sentinel 3 and 3B	LST, NDVI, GVI	Various	Various
Landsat -TM/ETM, OLI	LST, NDVI, evapotranspiration, soil moisture, geological mapping	30 m—VSIS, NIR, SWIR 60 m, 120 m, 100 m—TIR	16 days

GRACE, gravity recovery and climate experiment; GRACE-FO, gravity recovery and climate experiment-follow on; GPM, global precipitation measurement; SMAP, soil moisture active and passive; SMOS, soil moisture and ocean salinity; TRMM, tropical rainfall measuring mission.

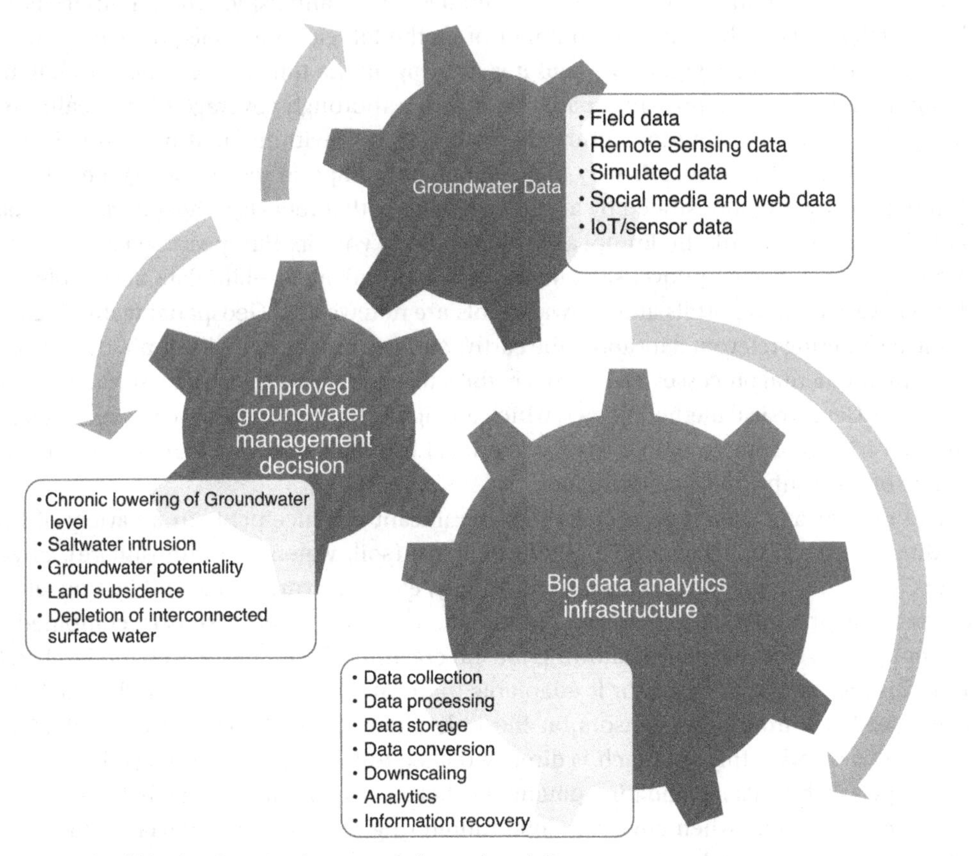

FIGURE 3.1 Role of big data in groundwater mapping and modeling.

geo-located information on water depth and extent were combined with digital elevation models and a flood-fill algorithm to infer near real-time flood extent maps (Lampos and Cristianini, 2012). The real-time spatial and temporal distributions of data from these sources enable a level of data insight previously unattainable (Sun and Scanlon, 2019). This is where the Internet of Things (IoT) comes into play. The use of IoT systems in groundwater science can obtain more information on local groundwater conditions much more quickly than traditional or checklist data collection, allowing for better groundwater resource management (Cecchinel et al., 2014). To enhance sustainable groundwater management, real-time IoT groundwater monitoring, and data management systems have been crewed in different regions, like California and India (Malche and Maheshwary, 2017).

3.5 Geospatial intelligence and information communication technology

Technological, hardware, software, and methodological advancements have a massive effect on geoenvironmental information extraction, management, and exploitation, both in terms of field and laboratory techniques. The number of methodologies and tools that can be used to define the environment is enormous, and it is growing all the time (Yang et al., 2017). Remote sensing imagery, which represents spatial data with a thorough coverage of the realm under investigation, has the ability to capture the evolution of environmental processes in action over large areas and with surprisingly high spatial and temporal resolution (Qiu et al., 2016). This functionality is demonstrated by a series of imagery that reports atmospherics or oceanic circulation. In this regard, the European Union's efforts with its European Space Agency and Copernicus in establishing novel satellite sensors and making satellite data accessible to the public via various web portals and software tools are noteworthy. Geospatial technologies are helpful in gathering relevant data about the earth's surface, but they provide limited information on geoenvironmental processes and factors in the subsoil or below the water's surface (Tsatsaris et al., 2021). Geophysical methodologies, which are tightly linked to geospatial technologies and combined with geocomputational tools, are critical in this context for enhancing our understanding of earth subsurface mechanisms.

Field devices and data loggers have seen significant advancements in a variety of fields, including geochemical sensors for pollution control (soil, water, and atmosphere), physical sensors (pressure, temperature, strain, conductivity, etc.) for hydrological and hydrogeological measurement, proximal sensing, tracers, and so on. Afterward, modern sensors combined with web applications become smart, allowing for the creation of "geosensor webs" (Nittel, 2009); within this framework, each sensor is adaptable to the authorized signal as well as taking into account feedback from other sensors on the web, allowing for the creation of self-adaptive monitoring networks. This approach is directly related to the IOT, a technology that opens up new prospects while also insinuating potential cybersecurity threats (Meneghello et al., 2019), which can be essential when environmental monitoring is focused on crucial economic and strategic assets like natural resources. Furthermore, advances in Information Communication

Technologies, both software (e.g., web) and hardware (e.g., smartphones, microcontrollers, etc.) have made it easier to collect environmental data through participatory approaches and immediately in the field using digital technologies (e.g., digital geological mapping and references). Even in the field of GIS, there is a constant push toward WebGIS services and online GIS, from both proprietary and open-source solutions. Many environmental groups, research institutions, associations, and other organizations use cloud services to gather reliable environmental data, usually adhering to various data and metadata standards (e.g., inspire). Furthermore, big data relating to human activity and consumption is critical for investigating potential geosphere-anthroposphere interactions.

3.6 Expert knowledge and GIScience

In the context of sustainability initiatives, GIScience plays a critical role. These are essential for the quantitative method of the earth system's main processes and their interconnections. For two main tasks: data exploration and prediction, geoenvironmental data are typically exploited using supervised or unsupervised instructional strategies (Wiemken and Kelley, 2020). The main goal of data exploration is to find some "interesting" underlying structure competent of shedding light on phenomenon under investigation (e.g., detection of forcing factors) and supervising the diagnostic techniques that will be used in the future (Pereira et al., 2018). The main goal of predictive model is to determine the value (continuous) or state (discrete) of an environmental asset (or an ensemble of environmental characteristics) in "locations" of the spatiotemporal domain of interest where assessments are lacking or inadequate (Daya et al., 2018). In explorative analysis, where the goal is to find interesting structures in data, the responsibility of expert knowledge becomes even more complicated. This type of exploratory analysis is connected in some ways to the ancestral human trait of seeking hidden meaning in the environment.

Statistical predictive methods including geostatistics, Bayesian modeling, and machine learning can be used when data are prevalent in comparison to expert knowledge. Expert knowledge impacts the assessment in a semiquantitative way in these perspectives, for example, during the exploratory data analysis phases and the collection of vital user-defined settings (e.g., selection of the domain analysis, selecting a specific anisotropy parameter, etc.). In the wake of the recent trends toward the formalization of causal thought, for example, implication of causation from data in Earth system sciences, including by means of machine learning, has gotten a lot of attention (Kanevsky and Maignan, 2004). These trends are corroborated by claims that automated observations are a step forward from previous vetoes against deriving causation from correlation—which helped lead to a blanket ban on causal explanation from facts and figures (Hastie et al., 2009). The problem is enticing in the human sciences. The fathers of modern economics, particularly after David Ricardo, recognized that labor is the source of national wealth, and that its structuring and organization in specific societal formations is influenced by a variety of social and political factors. Patterns can always be architecturally adjusted by "black swans" that irreparably change the paradigms to be modeled, making assertions based on historical cointegrating suspicious.

3.7 Data imbalances and new professionalism

Technological advancements and heightened awareness of geoenvironmental issues are propelling the selection of geoenvironmental data to new heights. If we concentrate on surface geoenvironmental mechanisms, we can see a jagged and cumulative increase in data coverage and spatiotemporal density in the 1970s, which corresponds to the start of NASA's Landsat program (Williams and Carter, 1976). The degeneration in information density, coverage, and reliability over time, termed to as a "data imbalance" for simplicity's sake, is evidently not new, and it is a repetitive curse in several fields of study, including history, archaeology, and, of course, geology. In the context of earth surface procedures, the "bloom" in geoenvironmental data is fairly extreme (Coffer et al., 2020); as a result, when interacting with subsurface data, another extreme data discrepancy exists, categorized by a marked decrease in information moving downward. When analyzing geoenvironmental dynamics over long time periods (or in 3D), such as studying temporal trends of particular environmental parameters, the data imbalance becomes important (e.g., atmospheric temperature, sea level, subsidence, etc.). The data imbalance must be carefully considered by geoscientists in their evaluation (Runge et al., 2019). The existing way to acquire a comprehensive and thorough spatial mapping of environmental variables of interest, even if confined to the earth's surface, is a favorable condition because it allows us to decently guesstimate the influence of severe undersampling and/or weakening in data quality.

Quantitative geoenvironmental data from historical and archaeological records are a difficult task that necessitates a truly balanced approach in which earth scientists (e.g., geologists, soil scientists, ecologists, geographers, etc.) collaborate with historians, archaeologists, philologists, historians of architecture, and philosophers of science. As a result, more interdisciplinary scientific networks and collaborative partnerships are required. Most pertinently, this hybrid skillful figure should serve as a bridge between various academic societies and specializations, facilitating between differing perspectives and applications of concepts that may appear to be equivalent on the surface. Geocomputing issues, such as data imbalance and the integration of historical information, as well as the modeling of societal data, are extremely important in politics and decision-making (Trevisani and Omodeo, 2021). In fact, while pursuing a "green economy," including geoenvironmental models in financial computations reconceptualizes and operationalizes natural and cultural phenomena from the perspective of resources, facilities, and assets.

3.8 Conclusion

As data and GIScience become more widely available, our ability to grasp and model the environment improves. The progression of programming languages and programming environments is a key component in the suitability and improvement of numerical methods in the context of software. Nonetheless, the wide range of data analysis and modeling methodologies available, as well as technologies that encourage participatory approaches, presents an opportunity to develop collective intelligence strategies. These, as well as increased transparency and a diversity of perspectives, are required to illuminate the many aspects of the earth system and its relations with the geosphere. Data exploration, on the other hand, plays an important role in understanding and modeling earth system dynamics, feasibly from a number of viewpoints and with expert knowledge. Today's geological and environmental dilemmas necessitate a closer examination of human history, particularly economics, technology, and science. Furthermore, scientific investigations (the broad range of disciplines that portray on science at philosophical, historical, and sociological levels) foster critical thinking, which is extremely significant in finding a balance between the political and techno-scientific elements that are co-implemented in development strategies.

References

Abd Manap, M., Sulaiman, W.N.A., Ramli, M.F., Pradhan, B., Surip, N., 2013. A knowledge-driven GIS modeling technique for groundwater potential mapping at the upper Langat basin, malaysia. Arab. J. Geosci. 6, 1621–1637.

Adamala, S., 2017. An overview of big data applications in water resources engineering. Mach. Learn. Res. 2, 10–18.

Al-Abadi, A.M., Shahid, S., 2016. Spatial mapping of artesian zone at Iraqi southern desert using a GIS-based random forest machine learning model. Model. Earth Syst. Environ. 2, 1–17.

Al-Fugara, A.K., Ahmadlou, M., Al-Shabeeb, A.R., AlAyyash, S., Al-Amoush, H., Al-Adamat, R., 2020. Spatial mapping of groundwater springs potentiality using grid search-based and genetic algorithm-based support vector regression. Geocarto. Int. 1-20. doi:10.1080/10106049.2020.1716396.

Cecchinel, C., Jimenez, M., Mosser, S., Riveill, M., 2014. An architecture to support the collection of big data in the Internet of Things. In: Proceedings of the International Workshop on Ubiquitous Mobile Cloud. Anchorage, AK.

Choubin, B., Rahmati, O., Soleimani, F., Alilou, H., Moradi, E., Alamdari, N., 2019. Regional groundwater potential analysis using classification and regression trees. In: Spatial Modeling in GIS and R for Earth and Environmental Sciences. Elsevier, Amsterdam, pp. 485–498.

Coffer, M.M., Schaeffer, B.A., Darling, J.A., Urquhart, E.A., Salls, W.B., 2020. Quantifying national and regional cyanobacterial occurrence in US lakes using satellite remote sensing. Ecol. Indic. 111, 105976. https://doi.org/10.1016/j.ecolind.2019.105976.

Cumming, G.S., Cumming, D.H.M., Redman, C.L., 2006. Scale mismatches in social-ecological systems: causes, consequences, and solutions. Ecol. Society 11 (1), 14.

Daya, S.B.S., Cheng, Q., Agterberg, F., 2018. Handbook of Mathematical Geosciences (Fifty Years of IAMG). Springer International Publishing, New York, NY, pp. 28–914.

Fang, Y., Jawitz, J.W., 2019. The evolution of human population distance to water in the USA from 1790 to 2010. Nat. Commun. 10, 430. https://doi.org/10.1038/s41467-019-08366-z.

Galvani, A.P., Bauch, C.T., Anand, M., Singer, B.H., Levin, SA., 2016. Interactions in population and ecosystem health. In: Proceedings of the National Academy of Sciences, 113, pp. 14502–14506. doi:10.1073/pnas.1618138113.

Gandomi, A., Haider, M., 2015. Beyond the hype: Big data concepts, methods, and analytics. Int. J. Inf. Manag. 35, 137–144.

Gholizadeh, M.H., Melesse, A.M., Reddi, L., 2016. A Comprehensive review on water quality parameters estimation using remote sensing techniques. Sensors 16 (8), 1298. https://doi.org/10.3390/s16081298.

Hastie, T., Tibshirani, R., Friedman, J., 2009. The Elements of Statistical Learning: Data Mining, Inference, and Prediction, 2nd ed. Springer, New York, NY.

Islam, A.R.M.T., Talukdar, S., Mahato, S., Kundu, S., Eibek, K.U., Pham, Q.B., Kuriqi, A., Linh, N.T.T., 2021. Flood susceptibility modelling using advanced ensemble machine learning models. Geoscience Front. 12 (3), 101075.

Jolliffe, I.T., Cadima, J., 2016. Principal component analysis: a review and recent developments. Phil. Trans. R. Soc. A. 374 (2065), 20150202. 3742015020220150202. doi:10.1098/rsta.2015.0202.

Kanevsky, M., Maignan, M., 2004. Analysis and Modelling of Spatial Environmental. Data EPFL Press, Basel.

Kitchin, R., McArdle, G., 2016. What makes Big Data, Big Data? Exploring the ontological characteristics of 26 datasets. Big Data Soc., Vol. 3, 1st ed. SAGE. https://doi.org/10.1177/2053951716631130.

Lampos, V., Cristianini, N., 2012. Nowcasting events from the social web with statistical learning. ACM Trans. Intell. Syst. Technol. 3, 1–22.

Liu, P., 2015. A survey of remote-sensing big data. Front. Environ. Sci. doi:10.3389/fenvs.2015.00045.

Madani, A., Niyazi, B., 2015. Groundwater potential mapping using remote sensing techniques and weights of evidence GIS model: A case study from Wadi Yalamlam basin, Makkah Province, Western Saudi Arabia. Environ. Earth Sci. 74, 5129–5142.

Malche, T., Maheshwary, P., 2017. Internet of Things (IoT) Based Water Level Monitoring System for Smart Village. In: Proceedings of the International Conference on Communication and Networks. Silicon Valley, California.

Mallick, J., Talukdar, S., Alsubih, M., Almesfer, M.K., Shahfahad Hoang, T.H., Rahman, A., 2021a. Integration of statistical models and ensemble machine learning algorithms (MLAs) for developing the novel hybrid groundwater potentiality models: a case study of semi-arid watershed in Saudi Arabia. Geocarto. Int. 1–35.

Mallick, J., Talukdar, S., Alsubih, M., Ahmed, M., Islam, A.R.M.T., Shahfahad, Thanh, N.V, 2021. Proposing receiver operating characteristic-based sensitivity analysis with introducing swarm optimized ensemble learning algorithms for groundwater potentiality modelling in Asir region, Saudi Arabia. Geocarto. Int. 37 (15), 4361–4389. doi:10.1080/10106049.2021.1878291.

Meneghello, F., Calore, M., Zucchetto, D., Polese, M., Zanella, A., 2019. IoT: Internet of Threats? A survey of practical security vulnerabilities in real IoT devices. IEEE Internet Things J 6, 8182–8201.

Miraki, S., Zanganeh, S.H., Chapi, K., Singh, V.P., Shirzadi, A., Shahabi, H., Pham, B.T., 2019. Mapping groundwater potential using a novel hybrid intelligence approach. Water Resour. Manag. 33, 281–302. doi:10.1007/s11269-018-2102-6.

Naghibi, S.A., Ahmadi, K., Daneshi, A., 2017. Application of support vector machine, random forest, and genetic algorithm optimized random forest models in groundwater potential mapping. Water Res. Manag. 31, 2761–2775.

Naghibi, S.A., Pourghasemi, H.R., 2015. A comparative assessment between three machine learning models and their performance comparison by bivariate and multivariate statistical methods in groundwater potential mapping. Water Res. Manag. 29, 5217–5236.

Naghibi, S.A., Pourghasemi, H.R., Abbaspour, K., 2017. A comparison between ten advanced and soft computing models for groundwater qanat potential assessment in Iran using R and GIS. Theor. Appl. Clim. 131, 967–984.

Nittel, S., 2009. A survey of geosensor networks: advances in dynamic environmental monitoring. Sensors 9 (7), 5664–5678. https://doi.org/10.3390/s90705664.

Okello, C., Tomasello, B., Greggio, N., Wambiji, N., Antonellini, M., 2015. Impact of population growth and climate change on the freshwater resources of Lamu Island, Kenya. Water. 7 (3), 1264–1290. https://doi.org/10.3390/w7031264.

Ozdemir, A., 2011. GIS-based groundwater spring potential mapping in the Sultan Mountains (Konya, Turkey) using frequency ratio, weights of evidence and logistic regression methods and their comparison. J. Hydrol. 411, 290–308.

Park, S., Hamm, S.-Y., Jeon, H.-T., Kim, J., 2017. Evaluation of logistic regression and multivariate adaptive regression spline models for groundwater potential mapping using R and GIS. Sustainability 9, 1157.

Pereira, P., Brevik, E., Trevisani, S., 2018. Mapping the environment. Sci. Total. Environ. 610–611, 17–23.

Qiu, J., Wu, Q., Ding, G., et al., 2016. A survey of machine learning for big data processing. EURASIP J. Adv. Signal Process. 2016, 67. https://doi.org/10.1186/s13634-016-0355-x.

Rapacciuolo, G., Blois, J.L., 2019. Understanding ecological change across large spatial, temporal and taxonomic scales: integrating data and methods in light of theory. Ecography 42, 1247–1266. https://doi.org/10.1111/ecog.04616.

Runge, J., Bathiany, S., Bollt, E., Camps-Valls, G., Coumou, D., Deyle, E., Glymour, C., Kretschmer, M., Mahecha, M.D., Muñoz-Marí, J., et al., 2019. Inferring causation from time series in Earth system sciences. Nat. Commun. 10, 1–13.

Sun, Y., Scanlon, BR., 2019. How can Big Data and machine learning benefit environment and water management: a survey of methods, applications, and future directions. Environ. Res. Lett. 14, 073001.

Taghizadeh-Mehrjardi, R., Schmidt, K., Toomanian, N., Heung, B., Behrens, T., Mosavi, A., Band, S.S., Amirian-Chakan, A., Fathabadi, A., Scholten, T., 2021. Improving the spatial prediction of soil salinity in arid regions using wavelet transformation and support vector regression models. Geoderma 383, 114793.

Talukdar, S., Eibek, K.U., Akhter, S., Ziaul, S., Islam, A.R.M.T., Mallick, J., 2021. Modeling fragmentation probability of land-use and land-cover using the bagging, random forest and random subspace in the Teesta River Basin, Bangladesh. Ecol. Indic. 126, 107612.

Trevisani, S., Omodeo, P.D., 2021. Earth scientists and sustainable development: geocomputing, new technologies, and the humanities. Land 10, 294. https://doi.org/10.3390/land10030294.

Tsatsaris, A., Kalogeropoulos, K., Stathopoulos, N., Louka, P., Tsanakas, K., Tsesmelis, D.E., Krassanakis, V., Petropoulos, G.P., Pappas, V., Chalkias, C., 2021. Geoinformation technologies in support of environmental hazards monitoring under climate change: an extensive review. ISPRS Int. J. Geo-Information 10 (2), 94. https://doi.org/10.3390/ijgi10020094.

Williams Jr., R.S., Carter, W.D, 1976. ERTS-1, a New Window on Our Planet. Professional Paper 929, U.S. Geological Survey: Washington, DC, p. 362.

Wiemken, T.L., Kelley, R.R., 2020. Machine learning in epidemiology and health outcomes research. Annual Rev. Public Health 2020 (41), 21–36 1.

Wu, WY., Lo, MH., Wada, Y., et al., 2020. Divergent effects of climate change on future groundwater availability in key mid-latitude aquifers. Nat. Commun. 11, 3710. https://doi.org/10.1038/s41467-020-17581-y.

Yang, C., Huang, Q., Li, Z., Liu, K., Hu, F., 2017. Big Data and cloud computing: innovation opportunities and challenges. Int. J. Digital Earth 10 (1), 13–53. doi:10.1080/17538947.2016.1239771.

Zabihi, M., Pourghasemi, H.R., Pourtaghi, Z.S., Behzadfar, M., 2016. GIS-based multivariate adaptive regression spline and random forest models for groundwater potential mapping in Iran. Environ. Earth Sci. 75, 1–19.

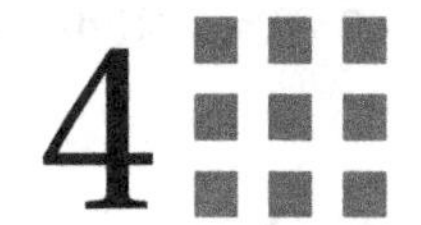

Role of groundwater potentiality and soil nutrient status on agricultural productivity: A case study in Paschim Medinipur District, West Bengal

Swatilekha Parihari, Nilanjana Das Chatterjee, Kousik Das and Raj Kumar Bhattacharya

DEPARTMENT OF GEOGRAPHY, VIDYASAGAR UNIVERSITY, MIDNAPORE, WEST BENGAL, INDIA

4.1 Introduction

In developing nations like India, agriculture is the prime source of income for maximum people (Acharya, 2006; Dalin et al., 2017). Agriculture is India's economic backbone. It makes a significant contribution to India's GDP (20%) and is critical to the country's overall economic development through expanding income options (Sharma et al., 2013). Agriculture maintains a link with the global food system, as well as managed country food production and food security (Parihari et al., 2021). However, most of agricultural activities in India entirely depend on irrigation potentiality that means surface water and groundwater both are crucial requirements for agricultural production (Zaveri et al., 2016). After the Green revolution (since 1960), groundwater-based irrigation has been rapidly grown up in particular north-western India (Shah, 2010; Dalin et al., 2017). On the other hand, nutrient status of soil plays a significant role to determine the soil microbial capacity (Allen and Schlesinger, 2004; Alster et al., 2013), potential environmental pollution (Finzi et al., 2011; Yang et al., 2016), and crop growth (Gao et al., 2017) in one hand while crop absorption capacity, microbial deposition, and migration of fertilizer are controlled by the availability of soil nutrients (Pandey et al., 2010; Ren et al., 2019). In spite of 16 essential elements in soil, three macronutrients of nitrogen (N), phosphorus (P), potassium (K) usually known as NPK that play significant role in the agricultural productivity and development of plant growth;

Case Studies in Geospatial Applications to Groundwater Resources. DOI: https://doi.org/10.1016/B978-0-323-99963-2.00008-0

thus NPK are needed in large volume for that purpose (Basu, 2011). Moreover, nutrition status has greatly influences on agricultural productivity as well as overall aspect of ecosystem.

Several geospatial techniques including multicriteria decision-making (MCDM) are used to determine the groundwater and surface water potentiality (Bhattacharya et al., 2020a). Recently, analytical hierarchical process (AHP) (Patra et al., 2017; Thomas and Duraisamy, 2018), fuzzy-AHP (Sahoo et al., 2017), and frequency ratio (Das, 2019) are widely applied to determine the groundwater potentially at large scale. On the other hand, soil nutrition status in soil is estimated with the considering of texture, availability of N, P, K, pH, respectively. Then many physical (optical) and chemical (electrochemistry) methods are used to actual measure the amount of N, P, and K. To support the country's food production system, land, water, and energy are required. Agriculture is a human-induced process through which natural ecosystem is transformed into the production of food, fiber, and fuel. Topography, climate, soil quality, and technology all play a role in production. The quality of soil varied from location to place due to the varied physiographic circumstances. Agriculture requires greater inputs in locations where land is less productive to overcome the limits of agriculture caused by low soil quality, irrigation facilities, fertilizers, and other technology (Parihari et al., 2021). Thus, two critical gaps have been identified, that is, actual balance state of water utilization and its potentiality in response to agricultural activities; resilience of soil properties incorporating with land productivity. Therefore, we have highlighted two main objectives in this present chapter as follows: (1) analyze water distributional pattern of the study area, (2) estimate soil properties such as soil texture, nitrogen, potassium, phosphorous and soil pH, etc., and (3) identify the crop combination region and agricultural productivity of the area. Moreover, the present study also helps to accelerate the sustainable agricultural practices over the world.

4.2 Study area

The tribal-dominated districts of Paschim Medinipur and Jhargram situated at the south-western part of West Bengal, which extended longitudinal 86°12′40″E to 86°33′50″E and latitudinal 22°57′10″N to 21°36′35″N and spread over 9750 km^2, respectively (Fig. 4.1). Geologically this study area comprises with lateritic hilly upland (north and western parts) to alluvium Gangetic delta (south and eastern parts). Geomorphic set up in Paschim Medinipur and Jhargram districts divided into three major parts, that is, (1) Chotonagpur flanks with hill, mounds and rolling land in the western most part, (2) Rahr plain with lateritic upland in the middle part, and (3) alluvial plain in the east with recent deposits (Bhattacharya et al., 2021). Names of the dominant rivers in the study area are Silabati, Kangsabati, and Subarnarekha which flow from west to east direction as follow of geological and geomorphic structure. Average rainfall in the two districts ranges 1200–1400 mm and temperature varies from 42°C to 16°C (Maximum) and 10.3°C to 27.6°C (minimum). Maximum rainfall and temperature recorded in the month from July to September; thus climatic condition falls under mostly tropical and monsoon type (Parihari et al., 2021). Agriculture is the prime land resource that also act as the livelihood support system.

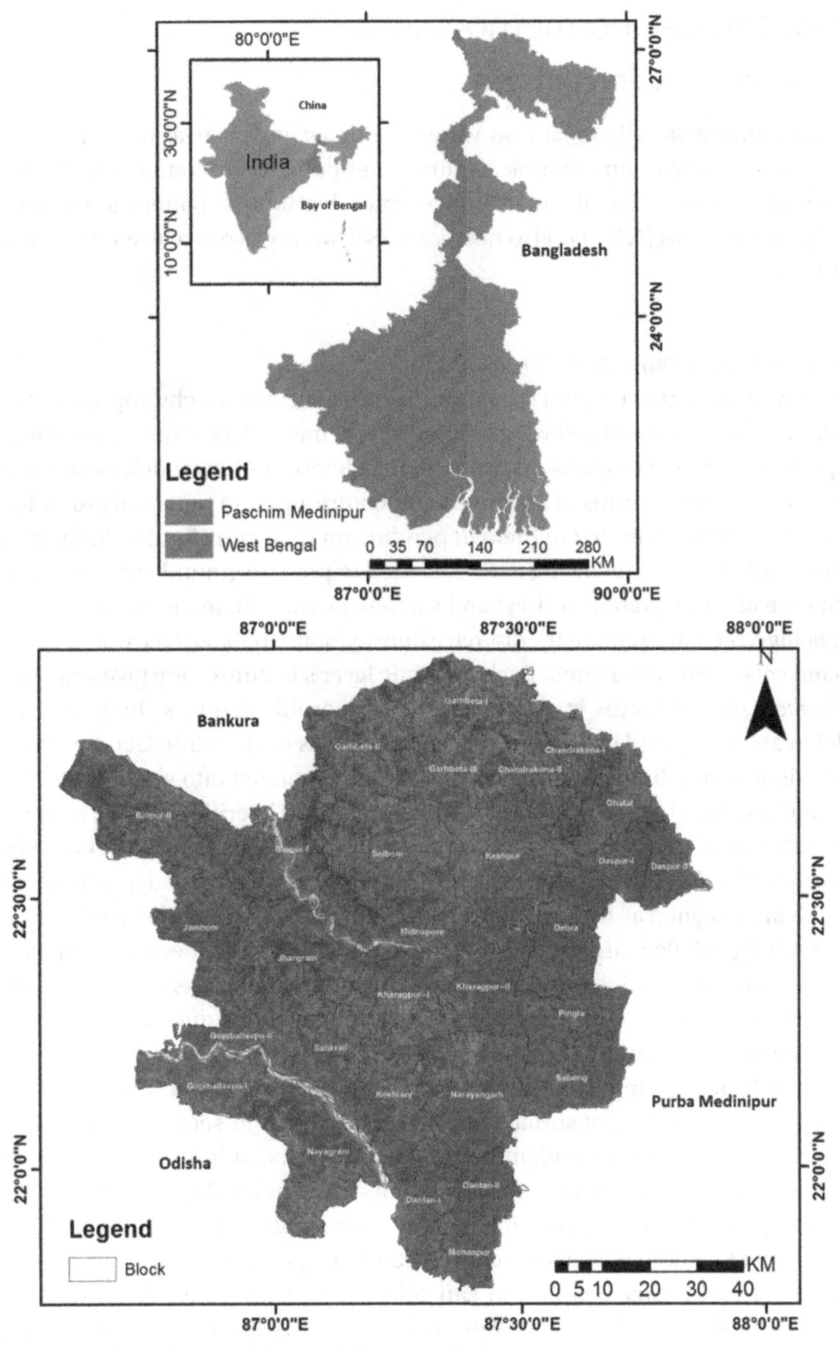

FIGURE 4.1 Study area in Paschim Medinipur and Jhargram districts.

4.3 Data source and methodology

4.3.1 Water distribution pattern

Because agriculture is heavily reliant on water resources, it is critical to understand the current state of water distribution patterns. Groundwater potentiality zonation is required for this purpose. For water availability, the real surface water distribution pattern in the research area nearest neighbor analysis (NNA) is also necessary. Secondary data are used to do this as follow in Table 4.1.

4.3.1.1 Measuring groundwater potentiality

Nine theme layers are retrieved from the image. The analytical hierarchy approach (Saaty, 2008) was used to determine normalized weights for each theme and its features, resulting in a pairwise comparison. To draw comparisons, we must first determine how much more important one stratum is than another in terms of attributes and contribution to potential groundwater. On a scale of 1–6, the weight attributed to them depending on their contribution to the groundwater possibilities (Table 4.2). The following dataset is used to prepare groundwater potential map

With the use of a topographical sheet and satellite picture, thematic layers such as geomorphology, geology, recharge (cm/year), drainage density, soil, slope, surface water, elevation, land use, and land cover map are created. Each thematic layer's features were given varied weights.

Geomorphology: Landforms structure under geomorphic setup is the essential thematic layer to delineate the groundwater potential zones (Biswas et al., 2020). Geomorphic feature in Paschim Medinipur and Jhargram districts has been categorized into seven units, that is, deep buried pediments, deep to moderately buried pediment with lateritic capping, moderate buried pediment with lateritic capping, pediments, denudational terraces and rocky outcrop, flood plain deposits, and valley fill deposits; respectively (Fig. 4.2A). Valley fill deposits and floodplain deposits both are assigned as highest weights for high potential groundwater due to maximum recharge rate and peak flow of river during monsoon period, while deep to moderately buried pediment with lateritic capping, moderate buried pediment with lateritic capping, pediments, denudation terraces, and rocky outcrop gives lowest weights value due to presence of hard rock terrain, maximum surface runoff (Bhattacharya et al., 2020a).

Geology: Geological settings have large influence on occurrence and distribution of groundwater through the controlling of surface runoff, infiltration, and seepage (Maity and Mondal, 2019). Six geological structures are identified in two districts, that is, basement crystalline complex, platform margin conglomerates, fluvial deltaic sediment overland by primary laterite, fluvial deltaic sediment overland by secondary laterite, young alluvium, and old alluvium, respectively (Fig. 4.2B). Maximum weights for groundwater prospective are given to young alluvium, old alluvium caused by huge oxidized in situ caliche groups to allow the deeper aquifer and moderate water level (Central Ground Water Board (CGWB), 2008; 2014). Contrastingly, minimum weights are given to basement crystalline complex, platform margin conglomerates due to higher deformation and foliation leading to more infiltration and low water level (Chowdhury et al., 2010).

Table 4.1 Data used for groundwater potentiality.

Data type	Details	Used for	Year	Source
Advanced space borne thermal emission and reflection radiometer (Aster) data	Entity ID:ASTGDEMV2_0N22E087, ASTGDEMV2_0N22E086, ASTGDEMV2_0N21E087 Resolution of 1-arc seconds	DEM, slope, river density	October 17, 2011	http://earthexplorer.usgs.gov/
Block map of Paschim Medinipur	Scale 1:50000	Area digitization	2008	NRDMS GIS Centre Development Section, Office of the District magistrate, Paschim *Medinipur*
Land use and Land cover map	Scale 1:50000	Area digitization	2009	Land and land reform department, Government of West Bengal
Geomorphology, geology, soil data	Geological quadrangle map, second edition, Scale: 1:50000	Geomorphology, geology, soil map	2002	Geological Survey of India
Monsoon and post monsoon water level data	Well water level station in pre and post monsoon	Water level isoline	2005–2013	Agricultural Department of West Bengal

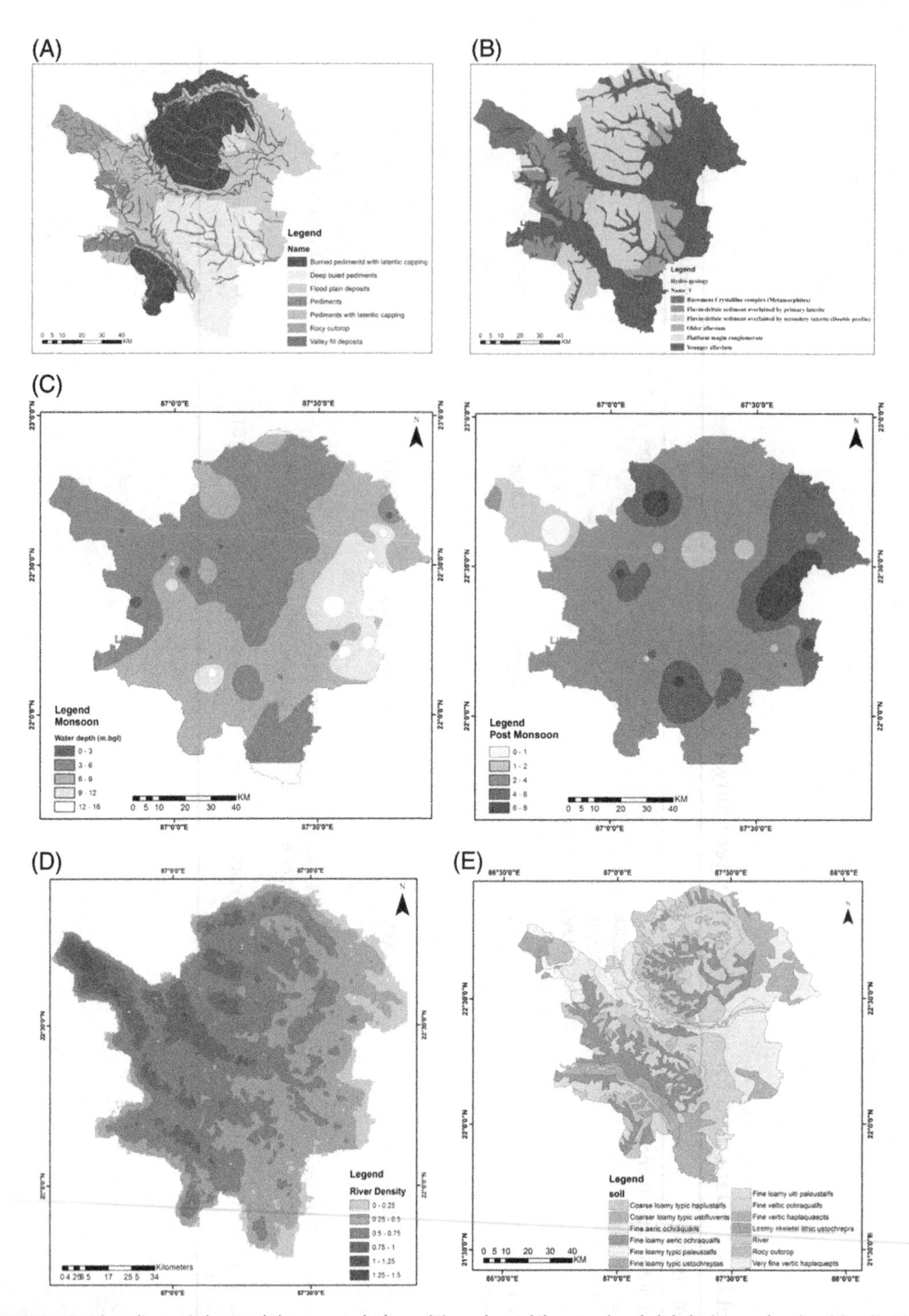

FIGURE 4.2 Nine thematic layers: (A) geomorphology, (B) geology, (C) water level, (D) drainage density, (E) soil, (F) stream water potentiality (G) elevation, and (H) land use land cover map.

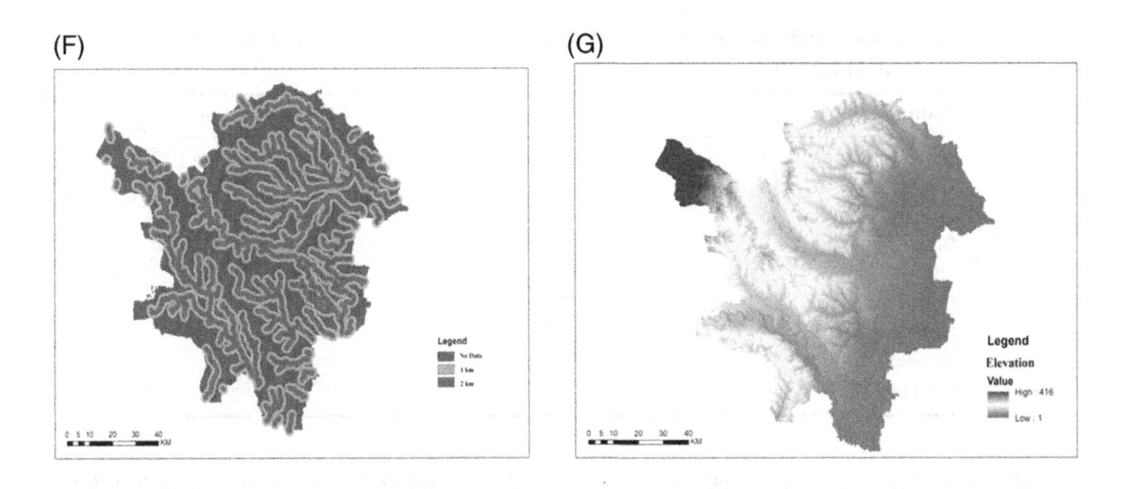

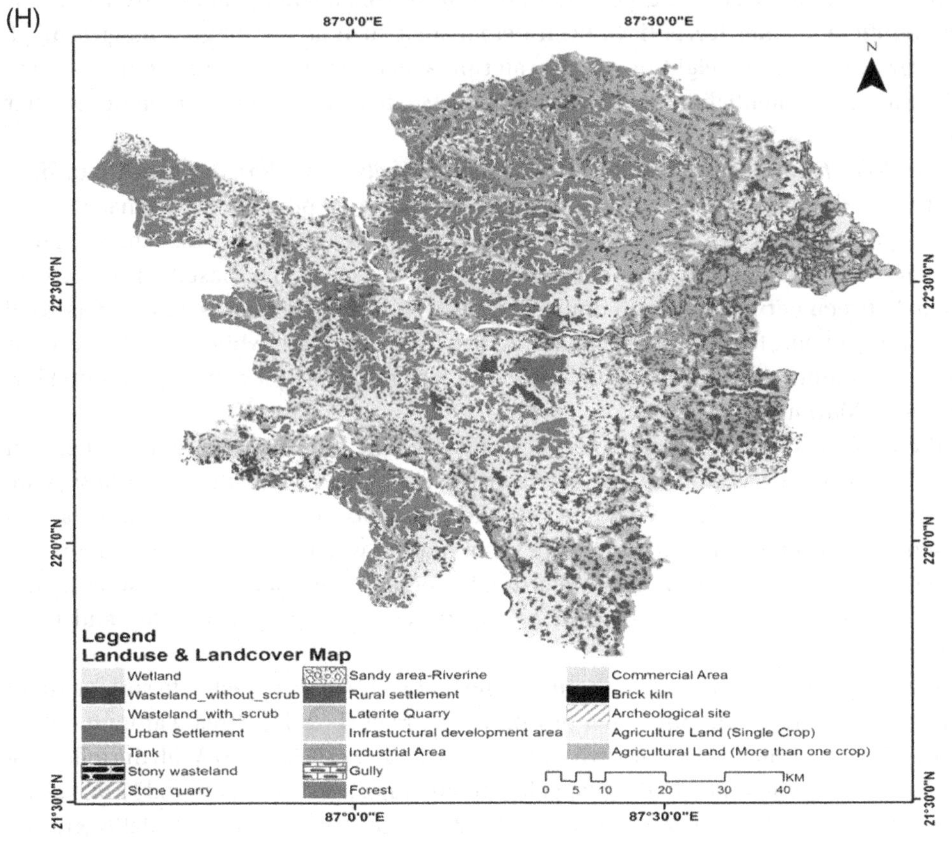

FIGURE 4.2, cont'd.

Table 4.2 Weights assignment of nine themes for groundwater potential zoning.

Themes	Weights score
Geomorphology	5
Geology	4
Recharge(cm/year)	4.5
Drainage density (km)	4
Soil	4
Slope (%)	3.5
Surface water	1
Elevation	4
Land use and land cover	6

Water level: Water level is considered as essential component for groundwater potentiality during premonsoon, monsoon, and postmonsoon season (Bhattacharya et al., 2020a). In this study, three different water levels have been demarcated such as >9 mbgl, 9 mbgl–6 mbgl, <3 mbgl; however, highest weight of groundwater prospects are assigned to >9 mbgl causes for maximum water potentiality and lowest weight assigned to 3 mbgl for minimum water potentiality (Fig. 4.2C).

Drainage Density: Surface runoff and permeable capacity of underlying rock-strata both are reflected by drainage density in assessing of groundwater potentiality (Avinash et al., 2014). Drainage density in two districts classified into four classes, that is, <0.5 km/km^2, 0.5–0.755 km/km^2, 0.75–1.05 km/km^2, >15 km/km^2; respectively (Fig. 4.2D). Based on the inverse relationship between permeability and drainage density, low-density class is assigned as good groundwater potentiality for more permeable and high infiltration rate, while high-density class is assigned as low groundwater potentiality causes for les permeable rock strata promoting low infiltration rate (Murmu et al., 2019).

Soil: The types of soil and its characteristics and covering area can greatly influences on rainfall infiltration as well as groundwater recharge (Biswas et al., 2020). Fifteen soil classes are observed in this study area where very fine vertic haplaquepts, fine vertic haplaquaepts, and coarse loamy typic ustifluvents soil classes are given highest weights score for infiltration rate exceeds the runoff volume, while loamy skeletal lithic ustochreprs, fine loamy typic ustochreptas, fine loamy typic ustifluvents are given lowest weights score for less permeability and more surface runoff (Fig. 4.2E).

Stream water potentiality: Potentiality around the major stream is reflected to surface water potentiality for measuring of groundwater prosperity (Patra et al., 2017). Stream around potentiality in two districts is categorized into two types of 1 km and 2 km (Fig. 4.2F). With the following of infiltration intensity and water holding capacity, stream water potentiality around 1 km is given maximum weight score for very good groundwater recharge storage but, potentiality around 2 km given as average score for moderate groundwater storage.

Elevation: Elevation is another significant factor for groundwater potentiality through the controlling of stream gradient, percolation flow, and hydraulic conductivity at a large basin scale (Das and Pal, 2020). In this study area, elevation ranges from >300 m to <30 m above mean sea level (amsl) with contained of seven classes, that is, <30, 30–60, 60–120, 120–180, 180–240, 240–300, and >300 m (Fig. 4.2G). Elevation classes of 180–240, 240–300, and >300 m are given the lowest weight score for low infiltration capacity and maximum gradient ratio whereas <30, 30–60, and 60–120 m are assigned as the highest weight score for large water logging and least gradient ratio.

Land use and Land cover (LULC): Various information likes soil moisture, infiltration rate, runoff volume, surface water dependency, and groundwater requirements are directly or indirectly extracted from LULC map in every region (Yeh et al., 2016). This is prepared using of thematic data as collected from land and land reform department, Government of West Bengal. LULC of Paschim Medinipur and Jhargram districts comprises into six classes, that is, more than one crop agricultural land, single-crop agricultural land, wasteland or scrub, forest, rural settlement, urban area (Fig. 4.2H). Agricultural land, wasteland, and forest have been assigned high to moderate weights score for low surface runoff and more infiltration rate (Ghosh et al., 2020) whereas rural settlement and urban area gives low weights score for the low rate of infiltration (Saranya and Saravanan, 2020)

4.3.1.2 Groundwater potentiality index

The groundwater potentiality index (GWPI) is a nondimensional number that aids in the identification of groundwater potentiality zones in the research region. The GWPI is calculated by superimposing all of the thematic and normalized weights of various features.

$$\begin{aligned} \text{GWPI} = {} & [(\text{GMw})(\text{GMwi}) + (\text{GGw})(\text{GGwi}) + (\text{WPw})(\text{WPwi}) + (\text{DDw})(\text{DDwi}) + (\text{Sw})(\text{Swi}) \\ & + (\text{SWw})(\text{SWwi}) + (\text{Ew})(\text{Ewi}) + (\text{LU/LCw})(\text{LU/LCwi})] \end{aligned}$$

where GM is geomorphology, GG is geology, WP is water potentiality, DD is drainage density, S is soil, and SW, surface water, E is elevation, LU and LC is land use and land cover and subscripts “w” and “wi” denotes to the normalized weight of a theme and normalized weight of individual features of each layer.

4.3.1.3 Surface water distribution

NNA is a standard approach for determining the pattern of settlement distribution, but in this study, the distribution pattern of water bodies inside the study area must be examined. NNA is a technique for determining whether a distribution is clustered, random, or regular. This two-dimensional measurement of phenomena over the earth’s surface was proposed by Clark and Evan. Landstat 8 “OLI-TIRS” image with green and near-infrared bands dated 22.22.2014 was utilized. Erdas Imagine Software was used to extract the water body using normalized difference water index (McFeeters 2013).

$$\text{NDWI} = \frac{\text{Green} - \text{NIR}}{\text{Green} + \text{NIR}}$$

Following the vector delineation of the water bodies, it is then converted to a polygon with midpoints and the index of randomness is calculated (Rn). The ratio of mean observed distribution (rO) in a random setting is represented by the index of randomness (Rn) (rE).

$$\text{Rn} = \frac{\text{rO}}{\text{rE}}$$

$$\text{rO} = \sum\nolimits_{i}^{n^{r}} i/\text{nrE} = \frac{1}{\sqrt[2]{\frac{n}{A}}} = \pi r^2 n$$

where ri is the nearest neighbor distance of different points and n is the total frequency observed in a block. A is denote the block ground area. Rn value range (<1 = clustered pattern, 1= Random, 1-2 = regular pattern).

4.3.2 Estimation of soil nutrient index

A map of nitrogen, phosphorus, and potassium for the study area was created using the National Bureau of Soil Survey and land use planning data to estimate the district's soil nutrient index. The number of samples is taken in a consistent manner, with 9248 sites on the map being chosen. Parker's (1951) method of calculating the nutrition index (NI) is used to determine soil fertility. The following equation is used to solve the problem:

$$\mathrm{NI} = \frac{(\mathrm{Nl} \times 1) + (\mathrm{Nm} \times 2) + (\mathrm{Nh} \times 3)}{\mathrm{Nt}}$$

where Nh is the number of samples falling in the high category of nutrient status, Nm is the number of samples falling in medium category, Nl is the number of samples falling in low category, and Nt is the total number of samples analyzed for a nutrient.

4.3.2.1 Soil test categories for nutrients of the study area

Separate indices are calculated for their different component on the basis of Parker's formula mentioned above. The soil category of the study area is in the optimum level.

Estimation of SNI using Parker's formula:

$$\text{Phosphorus } (P) = \textbf{Nutrient Index} = \frac{(\mathrm{Nl} \times 1) + (\mathrm{Nm} \times 2) + (\mathrm{Nh} \times 3)}{\mathrm{Nt}}$$

$$\text{Phosphorus } (P) = \text{Nutrient Index} = ((\mathrm{Nl} \times 1) + (\mathrm{Nm} \times 2) + (\mathrm{Nh} \times 3))/\mathrm{Nt}$$

Nh = Number of samples falling in the high category of nutrient status.
Nt = Total number of samples analyzed for a nutrient in any given area.
Nl = Number of samples falling in the low category of nutrient status.
Nm = Number of samples falling in the medium category.

$$\textbf{Nutrient Index} = \frac{(5510 \times 1) + (2419 \times 2) + (1290 \times 3)}{9248} = 1.71$$

$$\textbf{Nitrate (N)} = \textbf{Nutrient Index} = \frac{(\textbf{Nl} \times 1) + (\textbf{Nm} \times 2) + (\textbf{Nh} \times 3)}{\textbf{Nt}}$$

$$\textbf{Nutrient Index} = \frac{(5115 \times 1) + (2916 \times 2) + (1104 \times 3)}{9248} = 1.54$$

$$\textbf{Potassium (K)} = \textbf{Nutrient Index} = \frac{(\textbf{Nl} \times 1) + (\textbf{Nm} \times 2) + (\textbf{Nh} \times 3)}{\textbf{Nt}}$$

$$\textbf{Nutrient Index} = \frac{(4079 \times 1) + (3686 \times 2) + (1455 \times 3)}{9248} = 1.5374$$

4.3.3 Crop combination and agricultural productivity

The study is based on primary and secondary information. The secondary data were collected from District Planning Committee Report 2015–2016, Paschim Medinipur, District Census Handbook, and National Bureau of Soil Survey and Land Use Planning. Primary data were collected from household level survey through the prestructured questionnaire. Weaver's crop combination method has been applied for crop-combination region.

The Formula of Weaver's crop combination is as follows:

$$\delta = \sqrt{\frac{\sum d^2}{n}}$$

where *d* is the difference between the percentage of actual cropped area and hypothetical percentage of the cropped area.

Agricultural productivity: Total agricultural production/total agricultural area.

4.4 Result and discussions

4.4.1 Groundwater potentiality zoning

Groundwater is one of the most important natural resources and supports human health and economic development. It acts as an important water source. In India, 90% of the rural population depends on groundwater for their household and drinking needs (Reddy et al., 1996, Zaveri et al., 2016). Various software functions are derived from satellite images to know the actual groundwater situation (Table 4.1). Geology, topography, soil, water potential, drainage density, etc., serve as indicators of the presence of groundwater.

The groundwater potential map shows three different classes of good, moderate, and poor groundwater potential in the study area (Fig. 4.3). Amount of poor potential zone (45%) is highly dominated followed of medium (30%) and good potential zone (25%) throughout the districts. Hydro-geological conditions in the eastern part of the district are more favorable for groundwater storage and indicate zones with high potential for groundwater storage. The central part of the district is classified as a moderate groundwater potential that controls the area, and the tribal-dominated western part has low groundwater potential due to high slopes and poor geological and geomorphological conditions. This area is an extension of the Chotonagpur Plateau where physiographic conditions are not very good for aquifer storage. Moreover, steep slope and rocky terrain are responsible for low water retention.

4.4.2 Groundwater potentiality index

Weights were assigned for nine themes after analyzing their hydrological importance that effects for groundwater occurrence in the study area.

The process of obtaining of quantitative weights of different themes is presented in Table 4.2. A pair-wise comparison matrix and normalized index of different thematic layer are given in

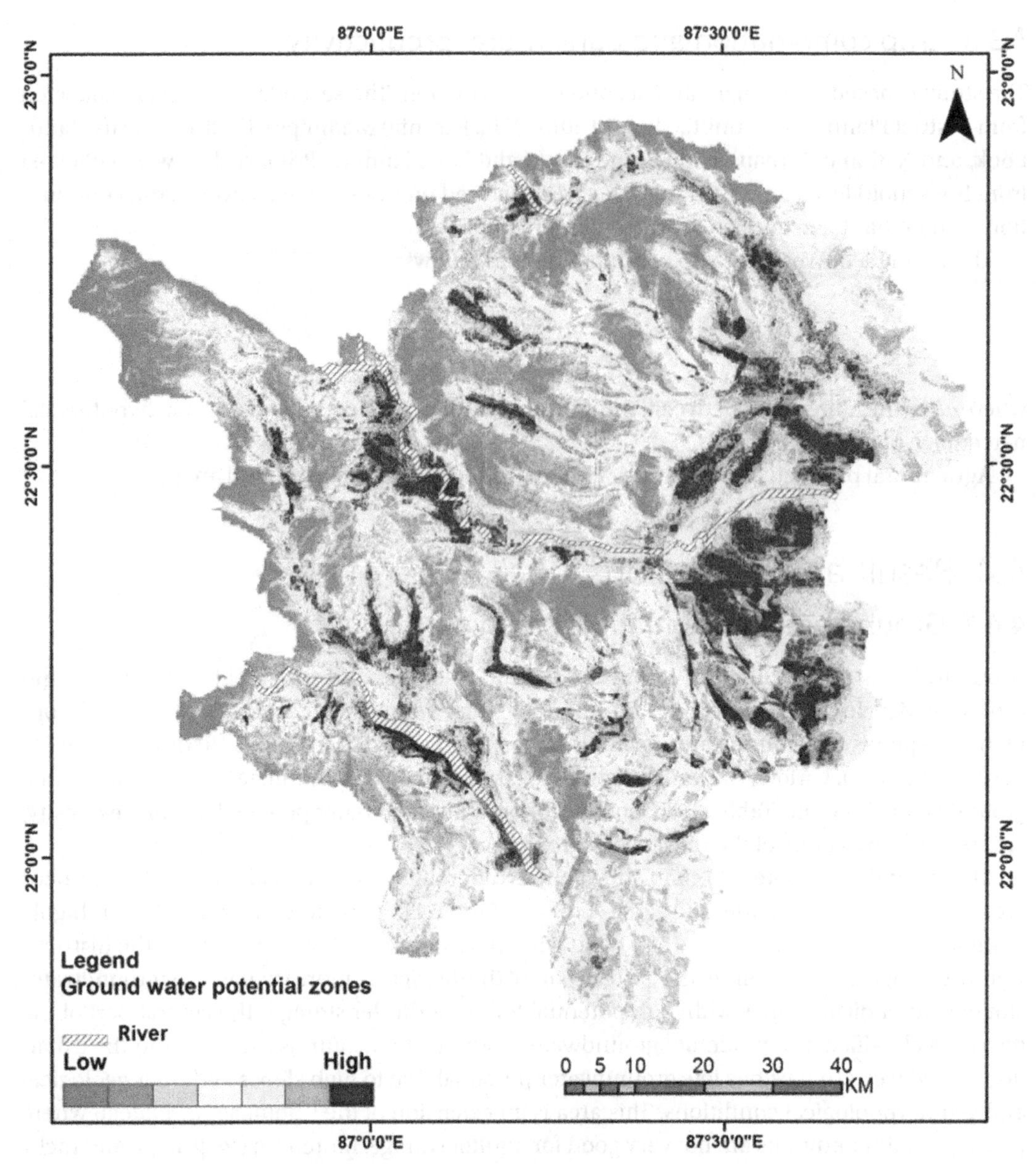

FIGURE 4.3 Groundwater potential zone.

Tables 4.3 and 4.4 . After assigning the normal weight of thematic layers, all thematic layers were integrated with each other for demarcating groundwater potential zone in the district.

where GM is geomorphology, GG is geology, WP is water potentiality, and DD is drainage density, *S* is soil, SW is surface water, *E* is elevation, LU and LC is land use and land cover.

Table 4.3 Quantitative weights score of nine different theme.

Theme	GM	GG	WP	DD	S	Slope	SW	E	LU LC	Mean	TM	Normalized weight
GM	5	5/4	5/4.5	5/4	5/4	5/3.5	5/3	5/4	5/4	11.040	83.9	0.132
GG	4	4/4	4/4.5	4/4	4/4	4/3.5	4/3	4/4	4/4	8.832	83.9	0.105
WP	4.5	4.5/4	4.5/4.5	4.5/4	4.5/4	4.5/3.5	4.5/3	4.5/4	4.5/4	9.936	83.9	0.118
DD	4	4/4	4/4.5	4/4	4/4	4/3.5	4/3	4/4	4/4	8.832	83.9	0.105
S	4	4/4	4/4.5	4/4	4/4	4/3.5	4/3	4/4	4/4	8.832	83.9	0.105
Slope	3.5	3.5/4	3.5/4.5	3.5/4	3.5/4	3.5/3.	3.5/3	3.5/4	3.5/4	7.728	83.9	0.092
SW	3	3/4	3/4.5	3/4	3/4	3/3.5	3/3	3/4	3/4	6.624	83.9	0.079
E	4	4/4	4/4.5	4/4	4/4	4/3.5	4/3	4/4	4/4	8.832	83.9	0.105
LU LC	6	6/5	6/4.5	6/4	6/4	6/3.5	6/3	6/4	6/6	13.248	83.9	0.158
Column total										83.902		1

4.4.3 Surface water distribution and availability

Normalized differential water index is effective surface water detection method (Chowdary et al., 2008; McFeeters, 2013). It shows the actual distribution pattern of surface water and established a relationship with existing settlement patterns. This will also verify the water availability of the study area. The main sources of water in the study area are rivers, streams, and numerous ponds. Remote sensing has been integrated with the vector data layer in the GIS to identify which settlement areas have surface water and which settlement areas do not have water. Since the farmland is within the boundaries of the site, a detailed concept of surface water conditions is also of paramount importance to agriculture. A random index has been calculated. Here are three distribution patterns such as clustered, random, and uniform pattern (Fig. 4.4). The uniform pattern of surface water sources is evenly distributed throughout the study area (Table 4.5). The NN index is arranged in a random pattern with no identifiable pattern. A single area is ideal for the concentration of surface water points and is also the area where people gather, defined as a clustered distribution pattern.

4.4.4 Characterizing soil of the study area

4.4.4.1 Soil texture

Soil texture is more important because it determines the growth of plants. It directly affects various soil characteristics such as water retention, permeability, and workability. These are three important properties of soil that affect plant growth. The most important way soil conditions affect plant growth is through water where nutrients are supplied. Soils with a high proportion of sand are called sandy or loose soils, which have low water retention and therefore low nutrients for plants and are called barren soils. Highly clay soils are characteristic of heavily structured soils with little or inadequate water penetration but good water storage capacity and are called loamy soils. Soil containing 35% clay, 30% silt, and 35% sand is called clay loam, and 40% sand, 40% silt

Table 4.4 Pair-wise comparison matrix and normalized weights of different thematic layers.

Theme	Class	Groundwater prospects	Weight assigned	Normalized weight
Geomorphology	Valley fill deposits	Very good	6.5	0.250
	Flood plain deposits	Very good	7	0.270
	Deep buried pediments	Poor	1.5	0.060
	Deep to moderately buried pediment with lateritic capping	Moderate	2.5	0.096
	Moderately buried pediment with lateritic capping	Moderate	3	0.115
	Pediment	Poor	1	0.039
	Denudational terraces and rocky outcrop	Moderate	2.5	0.096
Geology	Young alluvium	Moderate	2	0.093
	Older alluvium	Very good	6	0.279
	Fluvio deltaic sediment over land by secondary laterite	Good	4.5	0.209
	Fluvio deltaic sediment over land by primary laterite	Good	5	0.233
	Platform margin conglomerates	Poor	1	0.047
	Basement crystalline complex	Moderate	3	0.140
Water level in monsoon season (mbgl)	>9	Very good	7	0.467
	6–9	Good	5	0.333
	<3	Moderate	3	0.200
Drainage density	<0.5	Good	5	0.455
	0.5–0.75	Moderate	3	0.273
	0.75–1.0	Moderate	2	0.182
	>1	Poor	1	0.091
Soil	Very fine vertic haplaquepts		7	0.135
	Fine vertic haplaquaepts		6	0.115
	Coarse loamy typicustifluvents		5	0.096
	Fine loamy typicustifluvents		5	0.096
	Fine vertic ochraqualfs		5	0.096
	Coarse loamy typic haplustalfs		4	0.077
	Fine aericochraqualfs		4	0.077
	Fine loamy aericochraqualfs		4	0.077
	Fine loamy ultipaleustalfs	Moderate	3	0.058
	Rocky outcrops		2	0.038
	Fine loamy typicpaleustalfs		2	0.038
	Loamy lithic ustochrepts		2	0.038
	Fine loamy typicustifluvents		1	0.019
	Fine loamy typicustochreptas		1	0.019
	Loamy skeletal lithic ustochreprs		1	0.019

(continued on next page)

Table 4.4 Pair-wise comparison matrix and normalized weights of different thematic layers—cont'd

Theme	Class	Groundwater prospects	Weight assigned	Normalized weight
Slope	Level to nearly level	Very good	6	0.339
	Very gently sloping (1%–3%)	Good	5	0.282
	Gently sloping (3%–5%)	Moderate	3.5	0.198
	Moderately sloping (5%–10%)	Poor	2	0.113
	Moderate steeply sloping (10%–30%)	Poor	1.2	0.068
Potentiality around the major stream	1 Km	Very good	6	0.598
	2 km	Moderate	4	0.400
Elevation	<30	Very good	7	0.250
	30–60	Very good	6	0.214
	60–120	Good	5	0.179
	120–180	Good	4	0.143
	180–240	Moderate	3	0.107
	240–300	Poor	2	0.071
	>300	Poor	1	0.036
Land use & Land cover	Agricultural land (more than one crop)	Very good	7	0.225
	Agriculture land (single crop)	Very good	6	0.193
	Wasteland/scrub	Very good	6	0.193
	Forest	Good	5	0.161
	Rural settlement	Good	4	0.129
	Urban area	Moderate	3	0.097

and 20% soil is called sandy loam. Considering the soil in the Paschim Medinipur area, a total of 64,709 hectares of land is beneath the sandy soil and 130,313 hectares of land is clayey soil. In total, 198073, 73791, and 106689 hectares of land are located under sandy loam, loam, loam, respectively (District Planning Commission, 2011). Clay soil is highly concentrated (51.80%) in the Karagpur district. Dantan II, (80.22%) Dantan I (70.13%), Pingla (80.19%), Savan (77.61%) are under clay soil. In Daspur II, Ghatal, and KGP II, more than 60% loamy clay soil are found. This area is dominated by sandy loam. Over 60% of the sandy loam soil is concentrated in Galbeta III, Nayagram, Sankryle, Jumboni, Gopibarabpur I and II. Only Midnapore (22.19%) and Jargram subdivision (18.54%) are rich in sandy soil.

4.4.4.2 Nitrogen (N) availability

Nitrogen is the most important nutrient for plants. Nitrogen is very important because it is an important component of chlorophyll, a component that plants use the energy of sunlight to produce sugar from water and CO_2. It is also the main component of amino acids, free of protein wilting and dying plants. Most of the nitrogen in the soil remains in organic form. When organic matter is decomposed, nitrogen produces ammonium, which is converted to ammonia

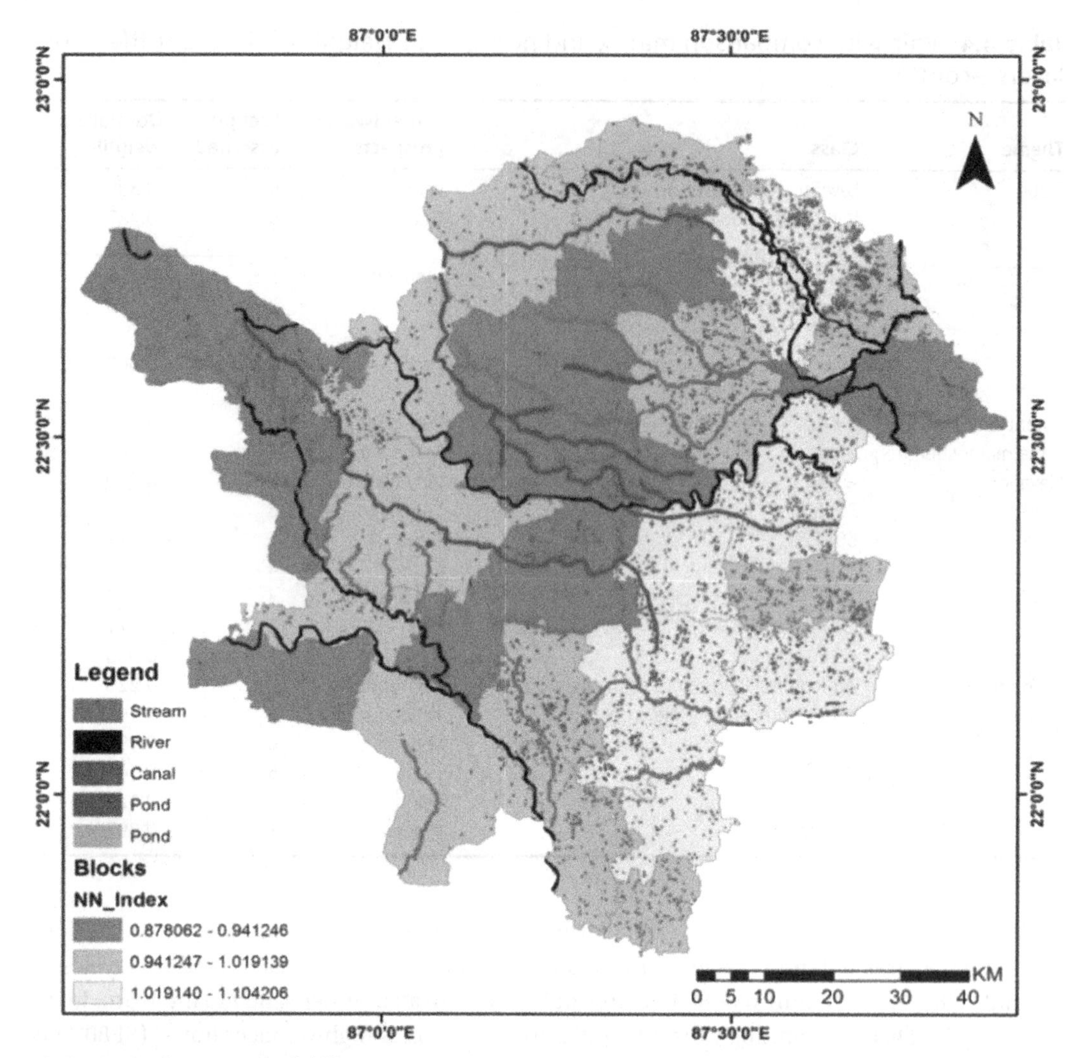

FIGURE 4.4 Surface water availability.

gas and nitrates under alkaline conditions (Chang et al., 2012). The nitrogen concentration is high at Dantan I, Dantan II, Kharagpur I, Kharagpur II, and north western part of Midnapore block that are Gopiballavpur I, Gopiballavpur II, in the eastern part of Midnapore district, blocks, namely, Ghatal, Sabang, Daspur, where the different pattern of nitrogen concentration is found in which some part is high, some medium and some portion of the blocks is under low nitrogen concentration (Fig. 4.5).

4.4.4.3 Phosphorous (P) availability

Phosphorus is another essential component of plant growth, root development, harvest maturity, and seed production, and is an important component of plant DNA and RNA (Ball, 2007). Except for the northeastern part of the district, the entire district is covered with low phosphorus.

Table 4.5 Block wise distribution of index of randomness (Rn) value.

Rn value	Pattern	Blocks	Rn value	Pattern	Blocks
<0.941	Cluster	Garhbeta-II, Gopiballavpur-I, Kharagpur-I, Salboni, Medinipur, Jamboni, Daspur-II Daspur-I Sankrial Binpur-II	0.941–1.02	Cluster to random	Dantan-I Garhbeta-III Pingla Narayangarh
0.941–1.02	Cluster to random	Nayagram Jhargram Garhbeta-I Keshiary Mohanpur Gopiballavpur-II Binpur-I Keshpur Ghatal	>1.02	Random to regular	Debra Chandrakona-II Sabang Dantan-II Kharagpur-II Chandrakona-I

High concentrations of phosphorus blocks are Chndrokona I, II, Ghatal, Dasput I, Daspur II, Mohonpur, Garhbeta II and Sabang. The low concentration blocks in the Pashim Medinipur area are Binpur II, Jamboni and Gopiballavpur I, II, Nayagram, Sankrai, Keshiyari, Dantan I, Dantan II, Kharagpur I, Narayangarh. Medium concentrations are mainly found in Binpur II, Jargram, part of Narayangal, Garbeta III block, etc. (Fig. 4.6).

4.4.4.4 Available potassium (K)

Potassium is responsible for activating more than 80 enzymes throughout the plant. It is important that plants fight cold and very hot, drafts, and pests (Ball et al., 2007). Potassium deficiency causes the edges of the leaves to become brownish and dry (Dhayalan, 2016). In Fig. 4.7, high concentration potassium that is above 350 kg/hectare are found at Daspur I, north eastern part of Binpur I, eastern, north eastern and western part of Jhargram, Sankrail and Kharagpur 1 block. Most part of the district, that is, Binpur I, Binpur II, Jamboni, Gopiballavpur I, II, Sankrail, Nayagram, Jhargram, and the small area of Mohanpur, scattered patches of Debra, Pingla, Sabang, Daspur I, Keshpur, Garhbeta II, Salboni, and Kharagpur II have medium potassium concentration (average 250 kg/Hectare). Low concentration that is below 200 kg/hectare potassium are found at Narayangarh, Keshiyari, western part of Jamboni, southern portion of Nayagram and some small patches of the entire eastern part of the district except Daspur I.

4.4.4.5 PH value

The pH value of soil is largely controlled by the soil water ratio, the electrolyte content and the carbon dioxide level (Husain, 2004). Soil varies in pH from about 4–10 for acidic to alkaline,

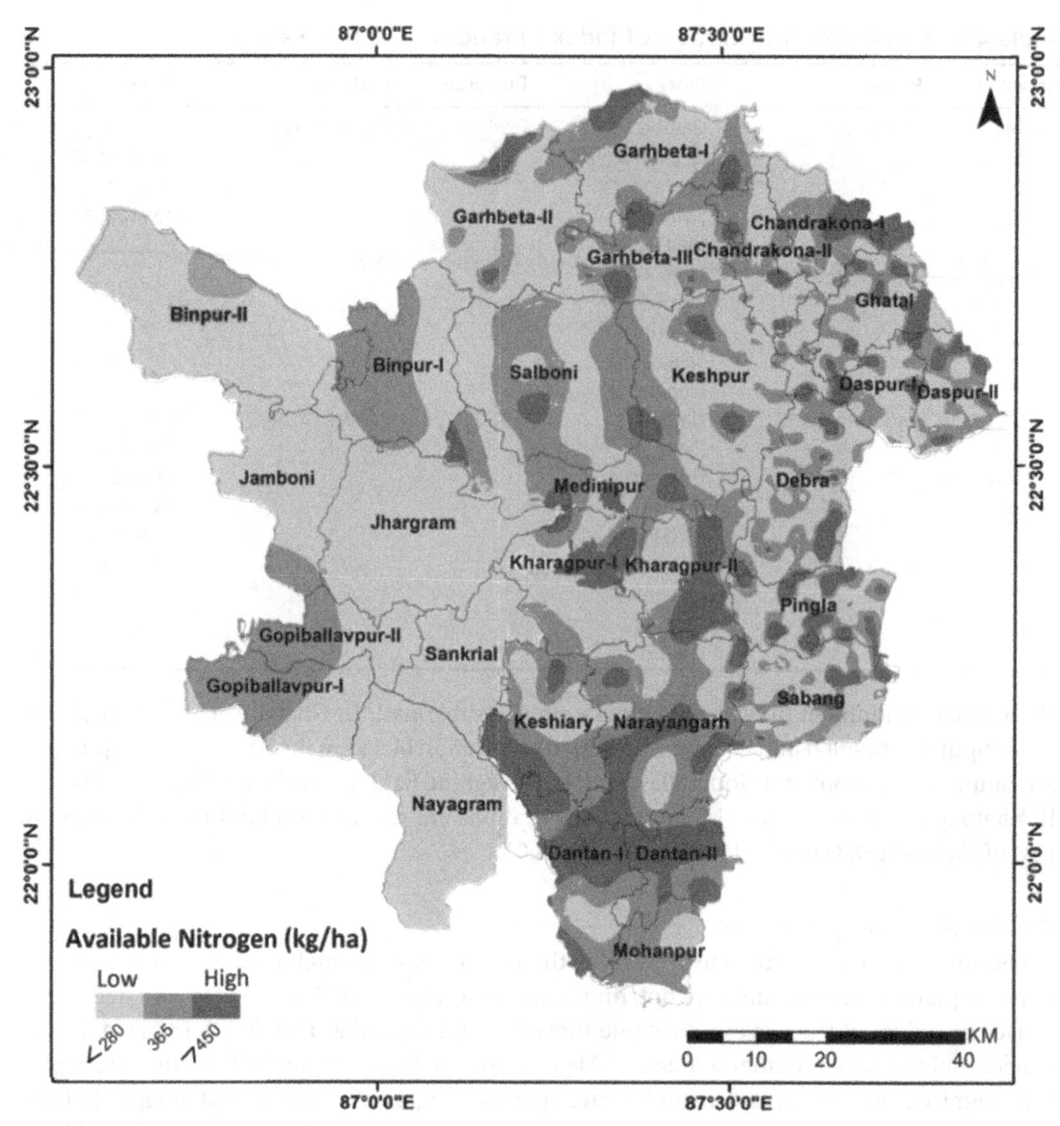

FIGURE 4.5 Available nitrogen of Paschim Medinipur district.

respectively, soil. The pH range for most agricultural soil is 5–8.5. The pH 7 is the neutral value. The following map (Fig. 4.8) shows the pH value of different blocks. The pH value above 8.5 is concentrated over the areas of Binpur II, Jamboni, Jhargram, Gopiballavpur I, II, Nayagram, Pingla, Sabang, and Daspur I block. pH 7.5 is concentrated in Binpur II, Jamboni, Goplballavpur II, Nayagram, Debra, Pingla, Sabang, Daspur I, II, Ghatal, GhatbeteI, II, Kharagpur 1, north-eastern part of Nayagram, Keshiyari block. The pH of the entire central part of the district's contents is 5.5.

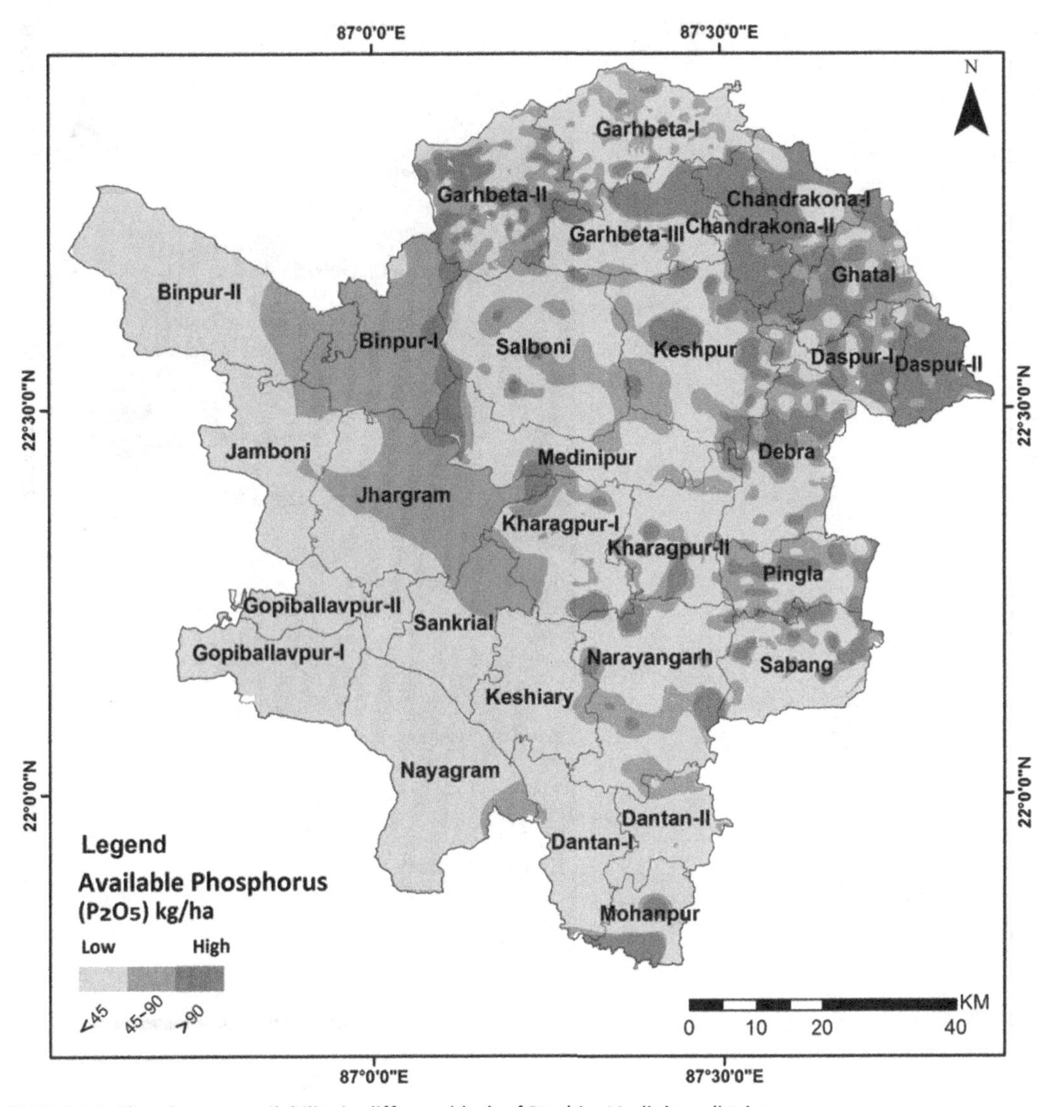

FIGURE 4.6 Phosphorous availability in different block of Paschim Medinipur district.

4.4.4.6 Soil nutrient status as a base for agriculture

Soil is of paramount importance to all life on earth. Soil formation is a dynamic process related to climate, terrain, organisms, parent material importantly time that also influences the fertility of the soil (Jenny, 1941; Kormondy, 1996). Life depends not only on the energy supply but also on the availability of 20 and even 10 chemical elements required for the life processes of most living organisms (Kormondy, 1996). In general, soil nutrients N, P, and K have become essential resources in recent decades (Dhayalan et al., 2016). Nutrient index values were established according to Parker's formula above. The soil in the Paschim Medinipur area can be divided

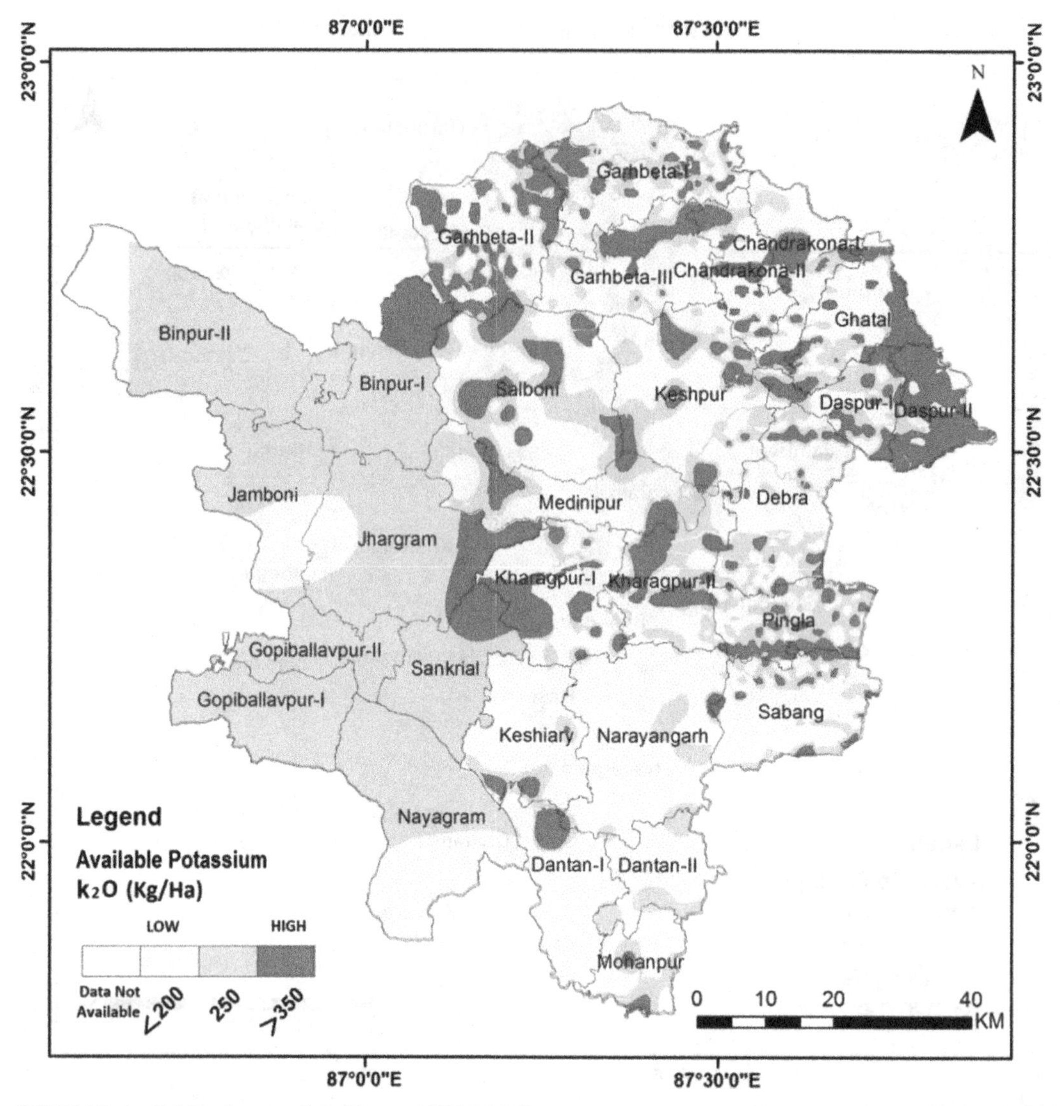

FIGURE 4.7 Availability in potassium (Source: NBSS & LUP).

into three categories, that is, deficiency (low), optimal (high), over-plant requirements (very high) for Phosphorus, potassium, nitrogen (Table 4.6). The analyzed phosphorus samples are 45–90 kg/hectare, potassium is 200–300 kg/hectare, and nitrate is 280–450 kg/hectare (Table 4.7).

4.4.4.7 Agricultural productivity at block level

Agricultural productivity patterns in the Paschim Medinipur area vary from place to place (Fig. 4.9). Agricultural productivity is an input-output ratio. This is a function of the interaction between physical and cultural variables in a particular area. By dividing the zones of very high,

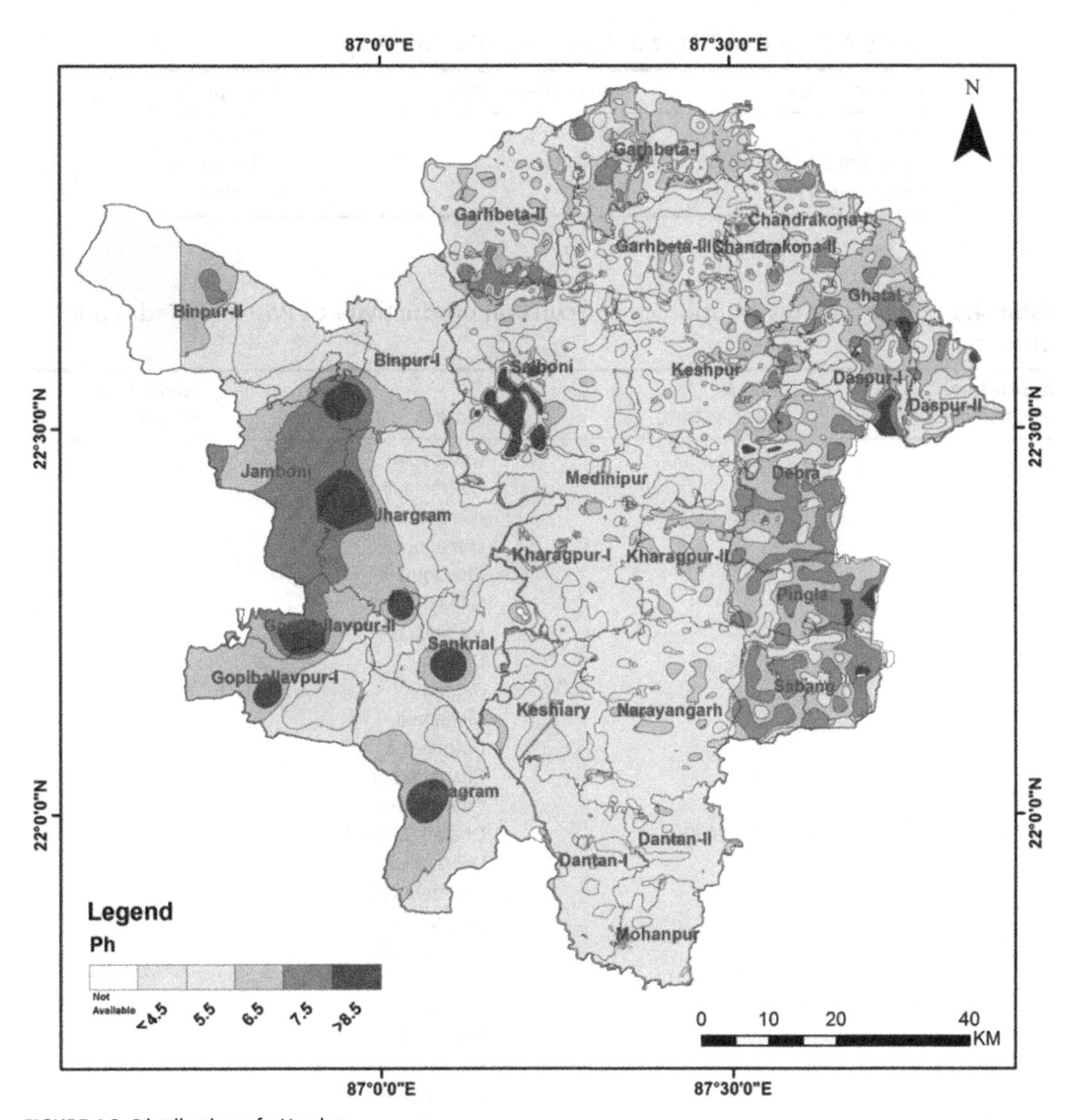

FIGURE 4.8 Distribution of pH value.

Table 4.6 Calculation for soil nutrition index.

Soil test category	Phosphorus (P) in kg/ha	Potassium (K) in kg/ha	Nitrate (N) in kg/ha
Deficient (low)	<45	<200	<280
Optimum (high)	45–90	200–350	280–450
Exceeds crop needs (Very high)	>90	>350	>450

Table 4.7 Soil nutrient index in Paschim Medinipur.

Nutrient index	Soil nutrient index	Description
Nitrate (N)	1.54	Optimum
Potassium (K)	1.71	Optimum
Phosphorus (P)	1.5374	Optimum

Table 4.8 Block wise distribution of agricultural productivity of Paschim Medinipur district.

Agricultural productivity	Value (kg/hectare)	Blocks	Number of blocks
Very high	Above 5500	Garhbeta-I, Garhbeta-III	2
High	4500–5500	Jhargram, kharagpur-I, Kharagpur-II, Salbani, Medinipur, Dantan-I, Dantan-II, Mohanpur, Daspur-I, Chandrakona-I, Chandrakana-II	11
Medium	3500–4500	Ghatal, Daspur-II, Debra, Keshpur, Garhbeta-II, Binpur-I, Narayangarh, Keshiary, Sankrail, Nayagram, Gopiballavpur-II, Gopiballavpur-I, Jamboni,Binpur-II	13
Low	Below 3500	Pingla, Sabang	2

high, medium, and low agricultural productivity, we are developing an agricultural plan to minimize or eliminate regional disparities. In Table 4.8, very high agricultural productivity is observed in Garhbeta 1 and Garhbeta III, with moderate productivity concentrated in Jhargram, Kharagpur I, Kharagpur II, Salbani, Medinipur, Dantan I, Dantan II, Mohanpur, Daspur I, Chandrakana I, Chandrakana II. In addition, low productivity is concentrated in the Pingla block and the Savant block.

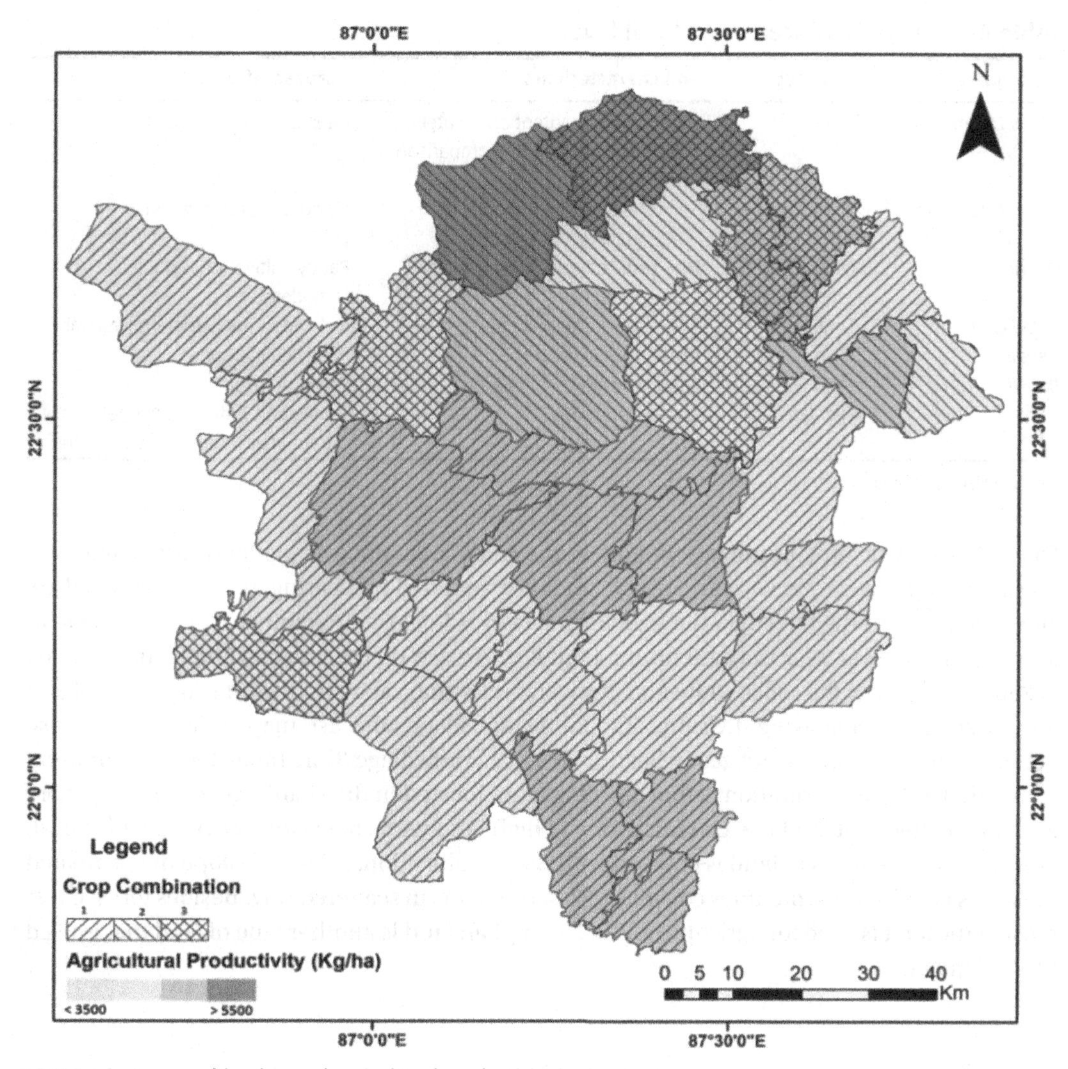

FIGURE 4.9 Crop combination and agricultural productivity in the district.

For the study of agriculture as an economic base for the tribal people, Bhadule village of Monidaha Gram Panchayet under Midnapore Sadar block and Dhobakachha and Mohulbonin village of Simulpal Gram Panchayet under Belpahari block of Paschim Medinipur District and Sonarimara village of Kendugari Gram Panchayet of Gopiballavpur 1 block has been selected where more than 80% people belong to Santhal community. The soil of this region is generally associated with alternative dry and rainy weather conditions with one part thick natural vegetation cover and another part open bare land with sparse vegetation.

The western part of the district is covered with lateritic soil. Due to excessive leaching, sheet and rill erosion minerals and other soluble nutrients washed away and made the soil unfertile.

Table 4.9 Land classification at local level.

Topo name	Land type	Soil characteristics	Purpose of use
Dhani Doem	Plain	High organic content and water retention capacity in comparison to others.	Intense paddy cultivation
Dhani Chaharam	Plain	Soil fertility is good handing adequate water supply	Good paddy cultivation
Dhani Soem	Moderately undulated	Fine textured, moderately fertile	Paddy cultivation is done moderately
Kala Doem	Undulated	Low water retention capacity	Cultivated Rabi crop and vegetable
Kanali	Plain	Land is fertile	Paddy cultivation
Bahali	Plain	Land is fertile	Paddy cultivation
Dahi	Highland	Land is infertile	Cultivation potentiality is zero, only Babui cultivation has been done

Source: Village level study.

Due to the lack of agricultural potentiality in the study area, the existing lands are being used for single cropping. Moreover, agricultural lands are extended by clearing the forest area and are close proximity to the forest. So the agricultural land is in close proximity to the forest. Due to less vegetative cover near the agricultural land it is open to direct erosion (Bhattacharya et al., 2020b). The color of the soil is white and low water retention capacity. The total agricultural and are classified into following divisions. The following village land use map shows the land use pattern of the concerned villages (village 1, village 2, and village 3) in Table 4.9. One common feature in land use distribution is that all villages are located in the vicinity of forest area. There are three patterns of land use classification (1) high land with steep slope covered with forest, grassland, and patches of land is covered with Babui cultivation, (2) gentle slope that is treated mostly as fallow land sometimes covered with settlements in scattered way. Besides forest cover most of the land is used for agriculture. The flat or plain land is another type of land that is used for settlement.

4.5 Conclusion

This chapter focused on relationship amongst the groundwater potentiality, surface water availability, soil nutrient status, and agricultural productivity in tribal livelihood dominant districts of Paschim Medinipur and Jhargram. Groundwater potentiality distribution in this study area was prepared using of AHP-based weights assigning from nine themes of geomorphology, geology, recharge, drainage density, soil, slope, surface water, elevation, and LULC under GIS platform. Soil NI was applied to determine the agricultural productivity considering of NPK and pH using of geospatial datasets. Then, crop combination was measured to find out the crop cultivation region in the entire study area applying of Weaver method. The results demonstrated that good groundwater prospects, availability of surface water, maximum water holding capacity, and rich nutrition status increased the agricultural productivity, and crop combination region in mostly eastern blocks of the study area but, poor groundwater resource, scarcity of surface water, minimum water holding capacity, and insignificant nutrition property declined the crop productivity and minimized the crop combination region in most of the western blocks. Therefore, present study recommended the sustainable agriculture practices in high potential alluvium tract for checking of high groundwater exploitation, and low water consumes crops more cultivated in lateritic tract for protecting from agricultural drought.

Conflict of interest

Authors have no conflict of interest of this chapter.

References

Acharya, S.S., 2006. Sustainable agriculture and rural livelihoods. Agric. Econ. Res. Rev. 19, 205–217.

Allen, A.S., Schlesinger, W.H., 2004. Nutrient limitations to soil microbial biomass and activity in loblolly pine forests. Soil Biol. Biochem. 36 (4), 581–589.

Alster, C.J., German, D.P., Lu, Y., Allison, S.D., 2013. Microbial enzymatic responses to drought and to nitrogen addition in a southern California grassland. Soil Biol. Biochem. 64, 68–79.

Avinash, K., Deepika, B., Jayappa, K.S., 2014. Basin geomorphology and drainage morphometry parameters used as indicators for groundwater prospect: insight from geographical information system (GIS) technique. J. Earth Sci. 25 (6), 1018–1032. https://doi.org/10.1007/s12583-014-0505-8.

Ball, B., Watson, C., Baddeley, J., 2007. Soil physical fertility, soil structure and rooting conditions after ploughing organically managed grass/clover swards. Soil Use Manage. 23 (1), 20–27.

Basu, P.K. (2011). Methods manual: soil testing in India. Department of Agriculture & Cooperation, Ministry of Agriculture Government of India New Delhi. Krishi Bhawan, New Delhi, 110001.

Bhattacharya, R.K., Chatterjee, N.D., Acharya, P., Das, K., 2021. Morphometric analysis to characterize the soil erosion susceptibility in the western part of lower Gangetic River basin, India. Arab. J. Geosci. 14 (6), 1–22.

Bhattacharya, R.K., Chatterjee, N.D., Das, K., 2020a. An integrated GIS approach to analyze the impact of land use change and land cover alteration on ground water potential level: a study in Kangsabati basin, India. Groundw. Sustain. Dev. 11, 100399. https://doi.org/10.1016/j.gsd.2020.100399.

Bhattacharya, R.K., Chatterjee, N.D., Das, K., 2020b. Estimation of erosion susceptibility and sediment yield in ephemeral channel using RUSLE and SDR model: tropical plateau Fringe Region, India gully erosion studies from India and surrounding regions, Advances in Science, Technology & Innovation, P.K. Shit et al. (Eds.). Springer. https://doi.org/10.1007/978-3-030-23243-6_10.

Biswas, S., Mukhopadhyay, B.P., Bera, A., 2020. Delineating groundwater potential zones of agriculture dominated landscapes using GIS based AHP techniques: a case study from Uttar Dinajpur district, West Bengal. Environ. Earth Sci. 79 (12), 1–25.

Central Ground Water Board (CGWB), (2008). Ground water information booklet Paschim Medinipur District, West Bengal, Ministry of Water Resources, Government of India. cgwb.gov.in/District_Profile/WestBangal/Hughli.pdf. (Accessed April 7, 2017).

Central Ground Water Board (CGWB), 2014. Dynamic Ground Water Resources of India (As on 31st March 2011) Ministry of Water Resources. River Development & Ganga Rejuvenation. Government of India. Retrieved from: http://www.cgwb.gov.in/Documents/Dynamic-GW-Resources-2011.pdf, 11th April, 2017. Central Groundwater Board, 2014. http://cgwb.gov.in/GW-Year-Book-State.html.

Chowdary, V.M., Chandran, R.V., Neeti, N., Bothale, R.V., Srivastava, Y.K., Ingle, P., Ramakrishnan, D., Dutta, D., Jeyaram, A., Sharma, J.R., Singh, R., 2008. Assessment of surface and Sub-surface waterlogged areas in irrigation command areas of Bihar state using remote sensing and GIS. Agric. Water Manage. 95 (7), 754–766.

Chowdhury, A., Jha, M.K., Chowdary, V.M., 2010. Delineation of groundwater recharge zones and identification of artificial recharge sites in West Medinipur district, West Bengal, using RS, GIS and MCDM techniques. Environ. Earth Sci. 59 (6), 1209. https://doi.org/10.1007/s12665-009-0110-9.

Dalin, C., Wada, Y., Kastner, T., Puma, M.J., 2017. Groundwater depletion embedded in international food trade. Nature 543 (7647), 700–704.

Das, B., Pal, S.C., 2019. Combination of GIS and fuzzy-AHP for delineating groundwater recharge potential zones in the critical Goghat-II block of West Bengal, India. Nord. Hydrol. 2, 21–30. https://doi.org/10.1016/j.hydres.2019.10.001.

Das, B., Pal, S.C., 2020. Assessment of groundwater vulnerability to over-exploitation using MCDA, AHP, fuzzy logic and novel ensemble models: a case study of Goghat-I and II blocks of West Bengal, India. Environ. Earth Sci. 79 (5), 1–16. https://doi.org/10.1007/s12665-020-8843-6.

Dhayalan, V., Selvamm, M., Ramaraj, M., 2016. Mapping and analysis of soil fertility using Remote Sensing and GIS; A Case study of Tharangambadi Taluk, Nagappatinum District. Int. J. Eng. Res. General Sci. 4 (3), 218–230.

Finzi, A.C., Austin, A.T., Cleland, E.E., Frey, S.D., Houlton, B.Z., Wallenstein, M.D., 2011. Responses and feedbacks of coupled biogeochemical cycles to climate change: examples from terrestrial ecosystems. Front. Ecol. Environ. 9 (1), 61–67.

Gao, D., Zhou, X., Duan, Y., Fu, X., Wu, F., 2017. Wheat cover crop promoted cucumber seedling growth through regulating soil nutrient resources or soil microbial communities. Plant Soil 418 (1), 459–475.

Ghosh, D., Mandal, M., Karmakar, M., Banerjee, M., Mandal, D., 2020. Application of geospatial technology for delineating groundwater potential zones in the Gandheswari watershed, West Bengal. Sustain. Water Resour. Manage. 6 (1), 1–14.

Husain, M., 2004. Systematic Agricultural Geography. Rawat Publication, Jaipur and New Delhi.

Jenny, H., 1941. Factors of soil formation, Newyork, Macgrahill.inc. In: Kormondy, E.J. (Ed.), Concepts of Ecology, Fourth Edition, 1996. Prentice-Hall of India Private Limited, New Delhi.

Kormondy, E.J., 1996. Concepts of Ecology, Fourth Edition. Prentice Hall of India Private Limited, New Delhi.

Maity, D.K., Mandal, S., 2019. Identification of groundwater potential zones of the Kumari river basin, India: an RS & GIS based semi-quantitative approach. Environ. Dev. Sustain. 21 (2), 1013–1034. https://doi.org/10.1007/s10668-017-0072-0.

McFeeters, S.K., 2013. Using the normalized difference water index (NDWI) within a geographic information system to detect swimming pools for mosquito abatement: a practical approach. Remote Sens. 5 (7), 3544–3561.

Murmu, P., Kumar, M., Lal, D., Sonker, I., Singh, S.K., 2019. Delineation of groundwater potential zones using geospatial techniques and analytical hierarchy process in Dumka district, Jharkhand, India. Groundw. Sustain. Dev. 9, 100239. https://doi.org/10.1016/j.gsd.2019.100239.

Pandey, R.N., Rani, R., Yeo, E.J., Spencer, M., Hu, S., Lang, R.A., Hegde, R.S., 2010. The Eyes Absent phosphatase-transactivator proteins promote proliferation, transformation, migration, and invasion of tumor cells. Oncogene 29 (25), 3715–3722.

Parihari, S., Das, K., Chatterjee, N.D, 2021. Land suitability assessment for effective agricultural practices in Paschim Medinipur and Jhargram districts, West Bengal, India. In: Modern Cartography Series, 10. Academic Press, Elsevier, pp. 285–311. https://www.sciencedirect.com/science/article/abs/pii/B9780128238950000348.

Patra, S., Mishra, P., Mahapatra, S.C., 2017. Delineation of groundwater potential zone for sustainable development: a case study from Ganga Alluvial Plain covering Hooghly district of India using remote sensing, geographic information system and analytic hierarchy process. J. Clean. Prod. 172, 2485–2502. https://doi.org/10.1016/j.jclepro.2017.11.161.

Reddy, P.R., Vinod, K., Seshadri, K., 1996. Use of IRS-1C data in groundwater studies. Curr. Sci. 70 (7), 600–605 cited in Hutti, B. and Nijagunappa. R. (2011), Identification of groundwater potential zone using geo informatics in Ghataprabha basin, North Karnataka, India.

Ren, T., Bu, R., Liao, S., Zhang, M., Li, X., Cong, R., Lu, J., 2019. Differences in soil nitrogen transformation and the related seed yield of winter oilseed rape (Brassica napus L.) under paddy-upland and continuous upland rotations. Soil Tillage Res. 192, 206–214.

Saaty, T.L., 2008. Decision making with the analytic hierarchy process. Int. J. Serv. Sci. 1(1), 83–98.

Sahoo, S., Dhar, A., Kar, A., Ram, P., 2017. Grey analytic hierarchy process applied to effectiveness evaluation for groundwater potential zone delineation. Geocarto. Int. 32 (11), 1188–1205. https://doi.org/10.1080/10106049.2016.1195888.

Shah, T., 2010. Taming the anarchy: Groundwater governance in South Asia. Routledge. https://www.routledge.com/Taming-the-Anarchy-Groundwater-Governance-in-South-Asia/Shah/p/book/9781138339187.

Sharma, P., Borua, S., Talukdar, R.K., 2013. Agriculture based livelihood options of inhabitants of tea gardens in Jorhat District of Assam. J. Acad. Indust. Res. 1 (8), 497–500.

Thomas, R., Duraisamy, V., 2018. Hydrogeological delineation of groundwater vulnerability to droughts in semi-arid areas of western Ahmednagar district. Egypt J. Rem. Sens. Space Sci. 21 (2), 121–137. https://doi.org/10.1016/j.ejrs.2016.11.008.

Yang, L., Huang, B., Mao, M., Yao, L., Niedermann, S., Hu, W., Chen, Y., 2016. Sustainability assessment of greenhouse vegetable farming practices from environmental, economic, and socio-institutional perspectives in China. Environ. Sci. Pollut. Res. 23 (17), 17287–17297.

Yeh, H.F., Cheng, Y.S., Lin, H.I., Lee, C.H., 2016. Mapping groundwater recharge potential zone using a GIS approach in Hualian River, Taiwan. Sustain. Environ. Res. 26 (1), 33–43.

Zaveri, E., Grogan, D.S., Fisher-Vanden, K., Frolking, S., Lammers, R.B., Wrenn, D.H., Prusevich, A., Nicholas, R.E., 2016. Invisible water, visible impact: groundwater use and Indian agriculture under climate change. Environ. Res. Lett. 11 (8), p.084005.

Groundwater potential zones identification using integrated remote sensing and GIS-AHP approach in semiarid region of Maharashtra, India

Sumedh R. Warghat[a], Satish V. Kulkarni[a] and Sandipan Das[b]

[a]DEPARTMENT OF GEOLOGY, BHARATIYA MAHAVIDYALAYA, AMRAVATI, INDIA

[b]SYMBIOSIS INSTITUTE OF GEOINFORMATICS (SIG), SYMBIOSIS INTERNATIONAL (DEEMED UNIVERSITY), PUNE, INDIA

5.1 Introduction

Groundwater plays a vital role in environmental stability, human well-being, global food security, ecological balance, and economic development (Snyder, 2019; Famiglietti, 2014; Gleeson et al., 2012; IPCC, 2001). Globally, groundwater contributes approximately 30% of the total resources used for human utilization (Siebert et al., 2010). Nowadays, groundwater resources face severe stress due to rapidly growing population, overexploitation, contamination, poor recharge, lack of infiltration, climate change, industrialization, irrigation, and LULC changes (Jesiya and Gopinath, 2019; Adeyeye et al., 2018; Taweesin et al., 2018; Chang et al., 2017). The groundwater consumption rate in India has exponentially increased from 20 km^3/year in 1950 to 230 km^3/year in 2009 (Fienen and Arshad, 2016). In India, groundwater contributes approximately 45% for household, 62% for irrigation, and 85% for agricultural purposes, respectively (Patra et al., 2018; CGWB, 2013). CWC and CGWB (2016) report observed that groundwater resources depletion across the country due to natural and human development activities.

Water security is expected to be one of the biggest challenges for India due to the rising predicted population of 1.6 billion by 2050 (Venkatesan et al., 2019; Asoka et al., 2017). Therefore, identification and assessment of groundwater potential zone is crucial for sustainable water resource management and planning (Chezgi et al., 2016). The occurrence of groundwater could be assessed by several hydrological, atmospherical, and geoenvironmental factors (Ghosh et al., 2016; Chowdhury et al., 2010). The traditional groundwater exploration involves field surveys, drilling, along with various geophysical, geological, and hydrogeological techniques.

Case Studies in Geospatial Applications to Groundwater Resources. DOI: https://doi.org/10.1016/B978-0-323-99963-2.00013-4

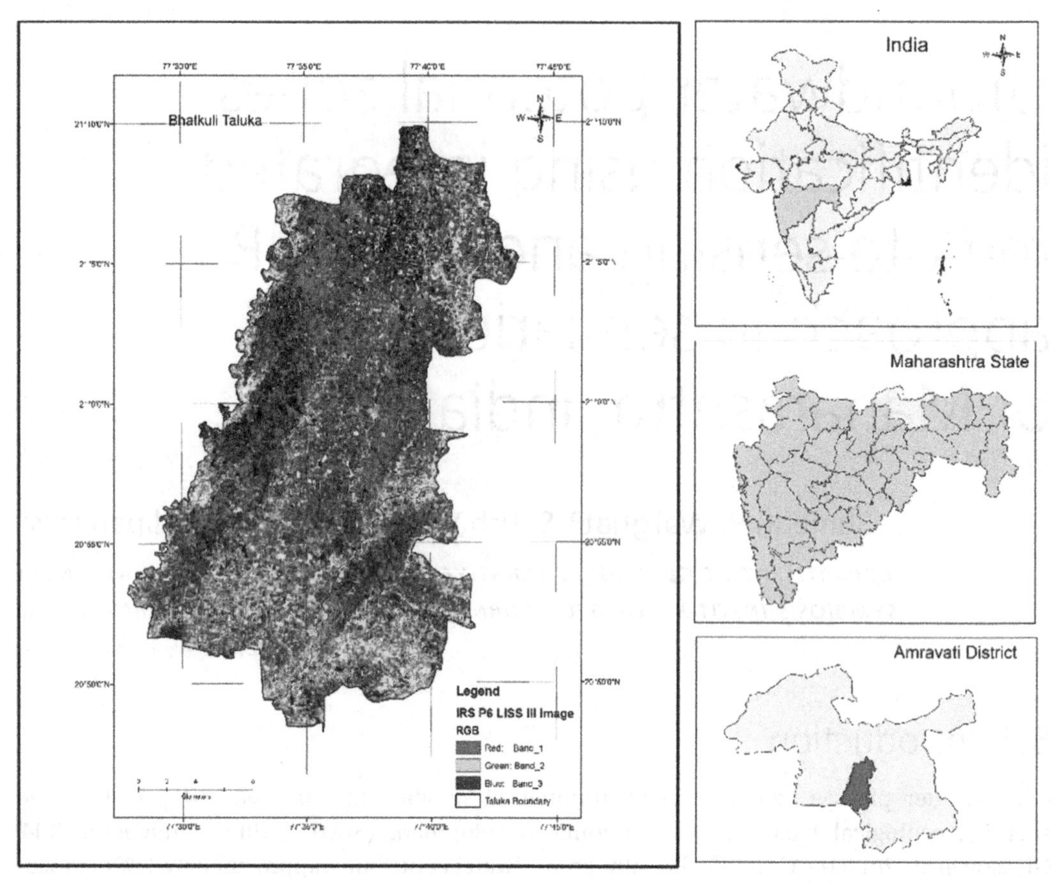

FIGURE 5.1 Location map of study area.

These methods are expensive, time-consuming, labor intensive, and require expert knowledge (Nhamo et al., 2020; Kirubakaran et al., 2016). The recent development in remote sensing and geospatial technology is effective in the identification, monitoring, and preservation of groundwater resources (Ahmad et al., 2020; Silwal and Pathak, 2018; Dadgar et al., 2017; Jasrotia et al., 2013). In recent years, many researchers have extensively used GIS-based analytic hierarchy process (AHP) technique with remote sensing data to delineate the GWPZs (Moges et al., 2019; Kaliraj et al., 2014; Chowdhury et al., 2010). AHP is a strong decision-making method used for incorporating multiple parameters for finding decisions related to earth-environmental challenges based on assigned expert's weights and ratings (Pinto et al., 2017; Kumar and Pandey, 2016).

5.2 Study area

The study area, part of Purna sub-basin of the Tapti main basin is geographically located in 20° 48′ 25″ N to 20° 9′ 54″ N latitude and 77° 27′51″ E to 77° 43′ 27″ E longitude in Bhatkuli taluka, Amravati district, Maharashtra, India (Fig. 5.1). The landscape encompasses an area of approximately

581.62 km^2 and contained within Survey of India (SOI) toposheet numbers 55 H/5, 55 H/9, 55 G/12 on 1:50,000 topographic scale. The climate of this basin is characterized by subtropical to semiarid climates. Geologically, the basin is covered by a quaternary alluvial complex (516.31 km^2) and Deccan trap formation (65.31 km^2). The basins have 137 villages and are depended on the groundwater for drinking and agricultural irrigation activities.

5.3 Methodology

In the present study, various geoenvironmental parameters used to derive the groundwater potential zones map in Bhatkuli Taluka, Amravati District, Maharashtra include geology, geomorphology, drainage density, slope, rainfall, lineament density, groundwater fluctuation, and groundwater quality. The geology map has been prepared from the district resource map of Geological SOI and cross-verified with field visits.

The geomorphological map has been mapped from visual interpretations of IRS-P6 LISS-III satellite imagery and SOI topographic maps at 1:50,000 scale. The slope map is generated based on the Shuttle Radar Topography Mission DEM with 30 m resolution data. Drainage density map was obtained from drainage network generated from topographic map of SOI, and updated with satellite imageries. The soil map has been obtained from Indian Council of Agricultural Research—NBSS & LUP. The lineaments prepared from hillshade maps are used to generate lineament density maps in GIS environment. The IRS-P6 LISS-III satellite data have been used to obtain accurate land use/land cover mapping information of the study area.

The average rainfall data from 2000 to 2013 are obtained from the office of Joint Director of Agriculture, Amravati, Government of Maharashtra, India. The spatial distribution of rainfall map is developed by the Inverse Distance Weighting interpolation method in the ArcGIS environment. The 66 well inventory data obtained from Central Groundwater Board were used to prepare the groundwater fluctuation map. The seasonal variation for summer 2012 to winter 2013 observation wells data were utilized to prepare the spatial variation of groundwater fluctuation depth map. Groundwater quality has been determined from analyzing total dissolve solid (TDS) value for 36 water samples collected in the study area.

GIS-AHP model is a widely used multicriteria analysis technique for groundwater management (Razandi et al., 2015). The AHP assesses several datasets in a pairwise comparison matrix, in order to measure rank or weight for each criterion with respect to the scale based on the priority (Saaty, 1980). In this analysis, various geoenvironmental parameters were assigned normalized weights and ratings to delineate the groundwater potential index (GWPI) zone using the following equation (Eq. 5.1). The methodology adopted in this study is illustrated in Fig. 5.2. The multiple thematic layers assigned with weights and ratings were subjected to weighted overlay analysis in ArcGIS environment to delineate the GWPI of the study area (Table 5.1).

$$\text{Groundwater potential index(GWPI)} = \Sigma X1 \times Y2 + X2 \times Y2 + \ldots..Xn \times Yn \tag{5.1}$$

where X is ranking of thematic variables and Y is normalized weight of various geoenvironmental factors.

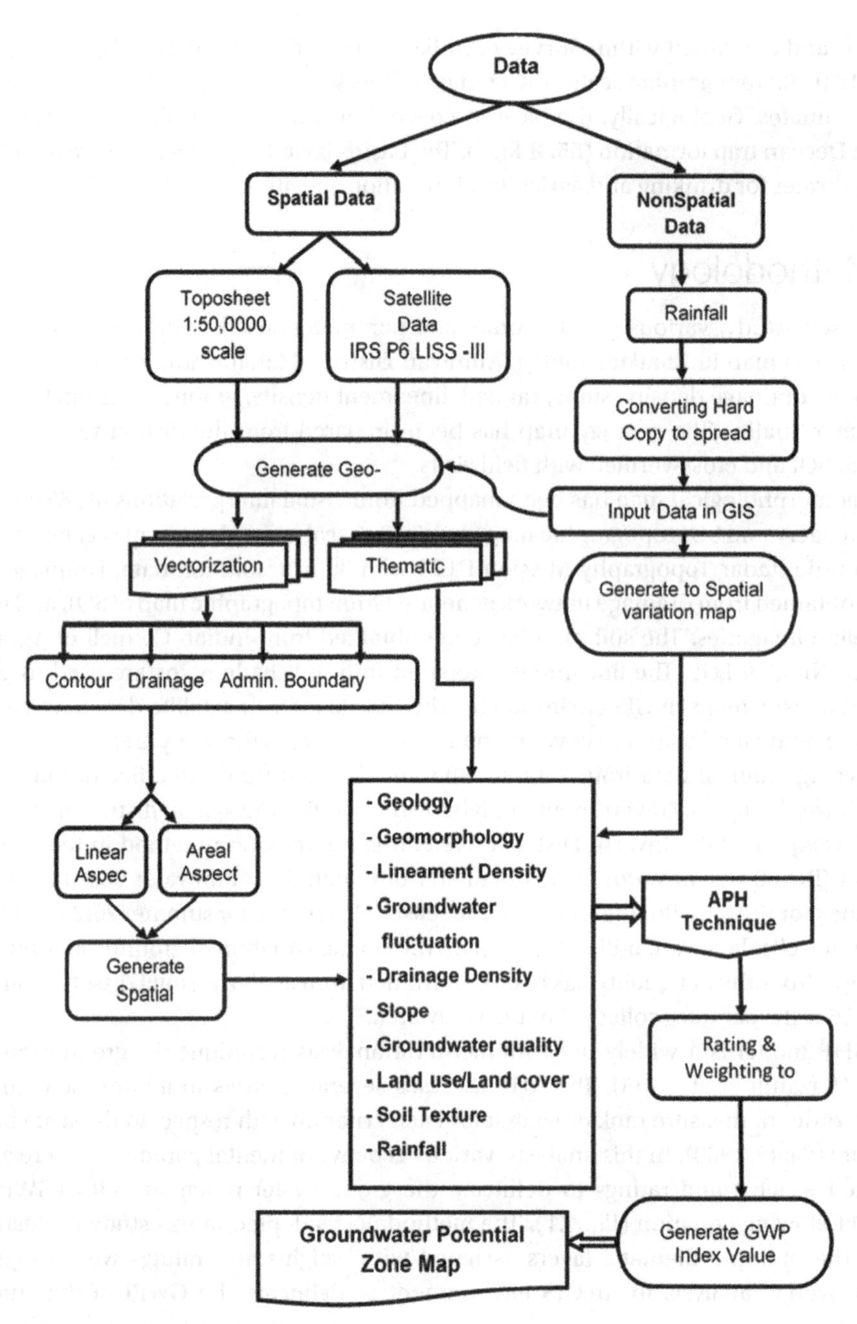

FIGURE 5.2 Methodology flow chart of the study.

Table 5.1 Weight and rank assigned as per the influence of parameters.

Sr. no.	Criterion	Criterion classes	Rating	Weightage
1.	Geology	Younger alluvial plain	5	2
		Older alluvial plain	3	
		Deccan trap	1	
2.	Geomorphology	Younger alluvial plain	5	3
		Older alluvial plain	3	
		Residual hill	2	
		Upper plateau	1	
3.	Groundwater fluctuation (in Meter)	Poor (below >1.50)	5	3
		Low (1.50–2.50)	3	
		Moderate (2.50–3.50)	2	
		High (3.50–8.10)	1	
4.	Groundwater quality	Fresh	5	1
		Slight saline	3	
		Moderate saline	2	
		Very saline	1	
5.	Slope	1%–3%	5	3
		3%–6%	4	
		9%–13%	3	
		13%–17%	2	
		17%–26%	1	
6.	Drainage density	Very low (0.0–0.50)	5	2
		Low (0.50–0.75)	4.5	
		Medium (0.75–1)	4	
		Moderate high (1.0–1.25)	3	
		High (1.25–1.50)	2	
		Very High (1.50–1.75)	1	
7.	Soil texture	Clay loam	3	3
		Clayey	2	
		Gravelly clay loam	5	
		Gravelly sandy loam	5	
		Silty loam	4	
8.	Land use/land cover	Agriculture	5	2
		Built-up (habitation)	1	
		Forest	3	
		Wastelands	2	
		Water bodies	5	
		Agriculture	5	
9.	Rainfall (in mm.)	151–152	1	2
		153–154	2	
		155–155	3	
		156–157	4	
		158–161	5	
10.	Lineament density	0.0–0.2	1	2
		0.21–0.5	3	
		0.51–0.8	4	
		0.81–1.6	5	

5.4 Results & discussion

The various geoenvironmental factors geomorphology, geology, slope, land use/land cover, rainfall, lineament density, drainage density, groundwater drinking quality, groundwater fluctuation, and soil texture, used for identifying the groundwater potential zone for the Bhatkuli taluka, Amravati, are described in the following sections.

5.4.1 Geology

Geology is considered to be a vital factor, as it primarily controls the occurrence and movement of groundwater (Aggarwal et al., 2019; Ghorbani Nejad et al., 2016). The basin area geologically comprises of alluvial deposit ranges in age from lower Pleistocene to recent age and Deccan trap in age from upper cretaceous to lower Eocene age (Fig. 5.3). The majority of the basin is underlined by alluvial plains (younger and older) (88.77%). Deccan trap (weathered basalt) covered 11.23%, prominently seen in the southeastern part of the study area. The groundwater from the hard rock area and younger alluvium is portable and also suitable for irrigational purposes. The older alluvium aquifer has rendered water saline and unsuitable for both human consumption and irrigation.

5.4.2 Geomorphology

Geomorphology has a significant impact on the occurrence water percolation and recharging of water into the earth's subsurface (Donselaar et al., 2017). Geomorphologically, the satellite image has been classified into four geomorphic units based on visual interpretation of the satellite imagery. The study area is characterized by different types of landforms including older alluvial plain, younger alluvial plain, residual hill, and upper plateau (Fig. 5.4). The majority of the basin is dominated by the older alluvial plain (~75.9%). The study area is covered by younger alluvial and older alluvial sediments and the region has very good groundwater recharge due to good porosity and infiltration rate.

5.4.3 Drainage density

The drainage density system provides significant indication to the natural infiltration of the soil as it depends on the nature of underlying lithology (Roy et al. 2020). The drainage density is estimated as the total stream length of all orders in the unit area (Deepa et al. 2016). The low drainage density areas have more infiltration and less surface runoff, thus ideal for the development of groundwater (Sarker et al., 2020; Mumtaz et al., 2019; Nasir et al., 2018). The drainage density in the present study ranges from 0.24 to 1.55 km/km^2, and has been categories into six classes: very low (0–0.50 km/km^2), low (0.50–0.75 km/km^2), medium (0.75–1 km/km^2), high (1.25–1.5 km/km^2), and very high (1.50–1.75 km/km^2) (Fig. 5.5).

5.4.4 Slope map

The topography is an essential parameter in the replenishment of groundwater, as the gradient of the slope controls the surface water runoff and rainfall infiltration (Rahman et al., 2012). The

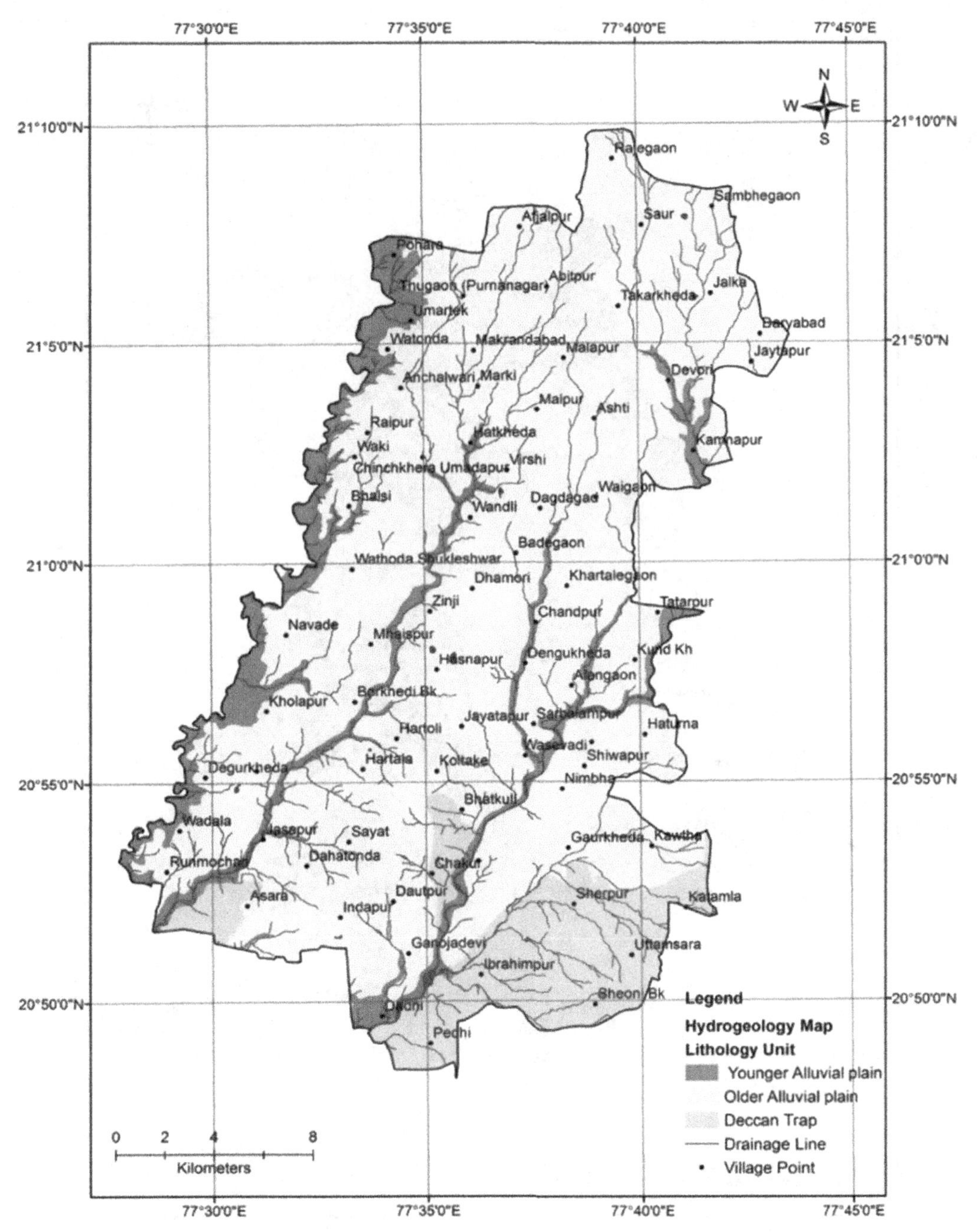

FIGURE 5.3 Hydrogeology map of study area.

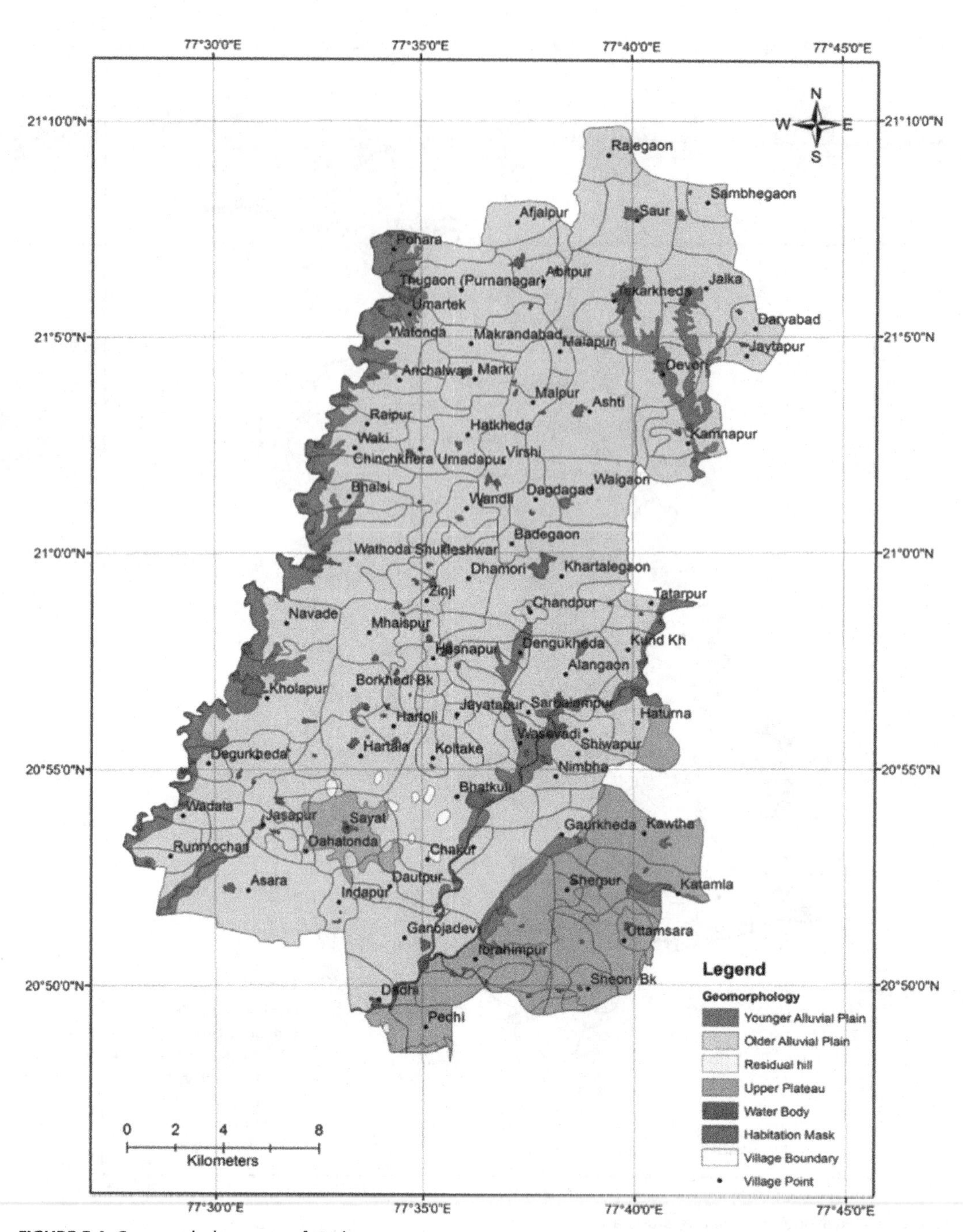

FIGURE 5.4 Geomorphology map of study area.

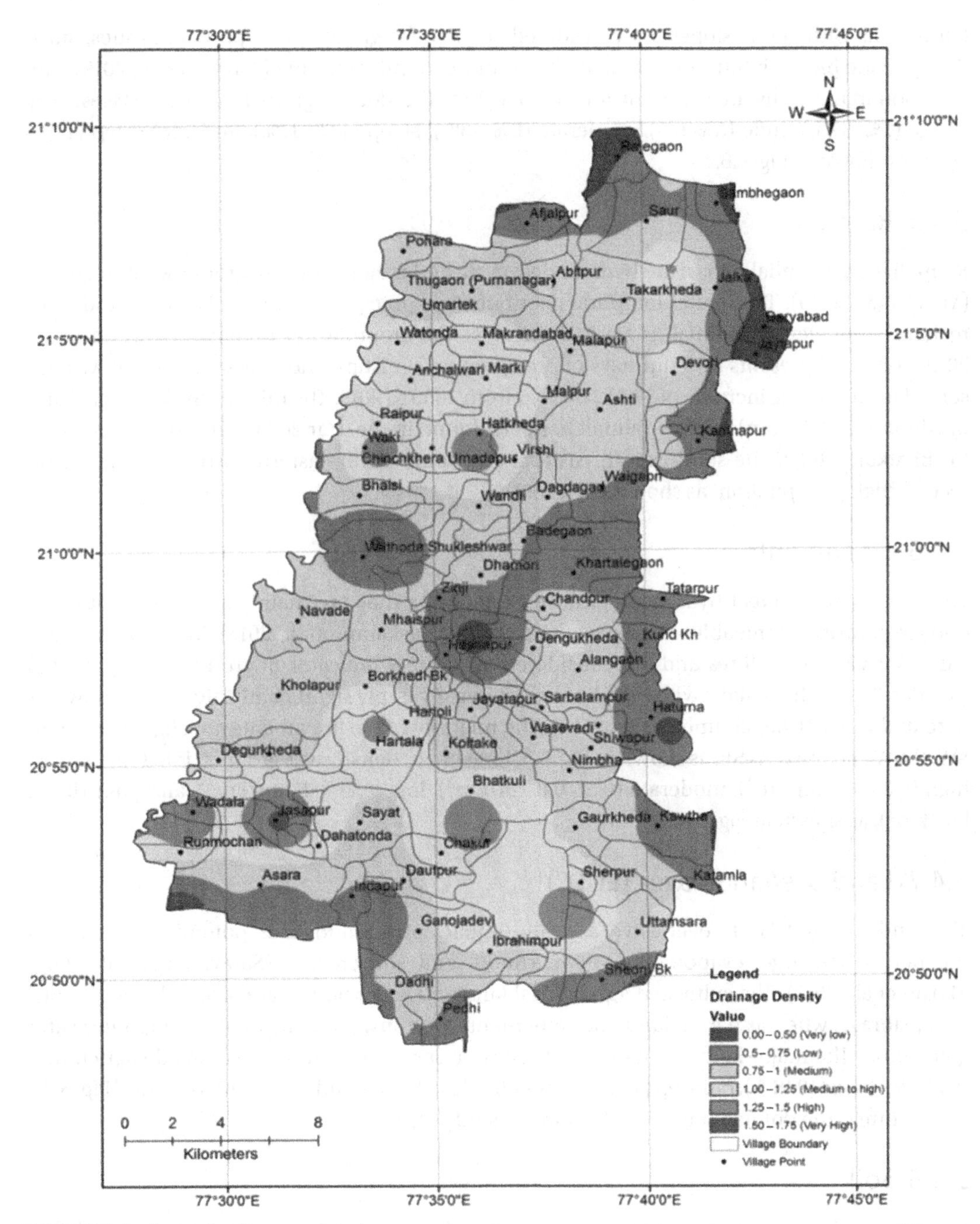

FIGURE 5.5 Drainage density map of study area.

flat regions with gentle slopes are considered suitable for groundwater potential zones, since they provide higher infiltration capacity due to more inundation time (Nayyer et al., 2019). The elevation in the study area varies from 266 to 352 m. The slope is graded into five classes: very gentle (1%–3%), gentle (3%–6%), moderate (6%–13%), steep (13%–17%), and very steep (17%–26%), as shown in Fig. 5.6.

5.4.5 Rainfall

Rainfall is an essential part of the hydrological cycle and major source of groundwater recharge (Yu and Lin, 2015). The spatial-temporal distribution of precipitation controls the groundwater recharge. The elevated rainfall is likely to increase the groundwater potentiality (Minh et al., 2019). Table 5.2 presents the summary of precipitation statistics. The monsoon rainfall was observed to an average increase of 99.01 cm/year from 2000 to 2013. The minimum (93.78 mm) and maximum (254.62 mm) average rainfall has been found in the year 2004 and 2007, respectively, for Bhatkuli taluka. The south, south-eastern, and some north-eastern portions of the region receive high precipitation, as shown in Fig. 5.7.

5.4.6 Lineaments

Lineaments are of great hydrogeological importance as they are mainly associated with secondary porosity, permeability, and weathered zone (Baskaran et al., 2013). These are the linear or curvilinear features and are considered as a direct indicator of groundwater potential (Sander, 2007). The regions with high lineament density imply significantly higher groundwater potential zones (Chepchumba et al., 2019). The majority of the lineaments are aligned in NE to SW, E-SW and the NW–SE, as shown in Fig. 5.6. Lineament density was grouped into four classes: high (0.81–1.6 km/km^2), moderate (0.51–0.8 km/km^2), low (0.21–0.5 km/km^2), and poor (0–0.2 km/km^2), as shown in Fig. 5.8.

5.4.7 Land use/land cover (LULC)

The land use and land cover have a strong impact on groundwater potential as it affects the rate of recharge, evapotranspiration, runoff, and soil erosion (Saravanan et al., 2019; Martin et al., 2017). The cultivated agricultural land and water bodies are favorable for groundwater storage, whereas barren land and settlements are considered unsuitable for groundwater generation. The major land use/land cover types present in the study area include agriculture (94.77%), forest (0.81%), built-up (1.58%), water bodies (0.91%), and wastelands (1.93%) (Fig. 5.9). Agriculture is the dominant type of LULC in the study area.

5.4.8 Soil

Soil is an essential component in determining groundwater occurrence as it controls the infiltration, percolation, and permeability of surface water into the aquifer system (Ahmad et al., 2020; Mogaji et al. 2014). The soil texture of the study area has been classified into five classes namely; clay loam (0.06%), clayey (89.98%), gravelly clay loam (1.77%), gravelly sandy loam (0.20%), and Silty loam (5.50%) (Fig. 5.10). Sandy loam soil is extremely suitable for groundwater storage due

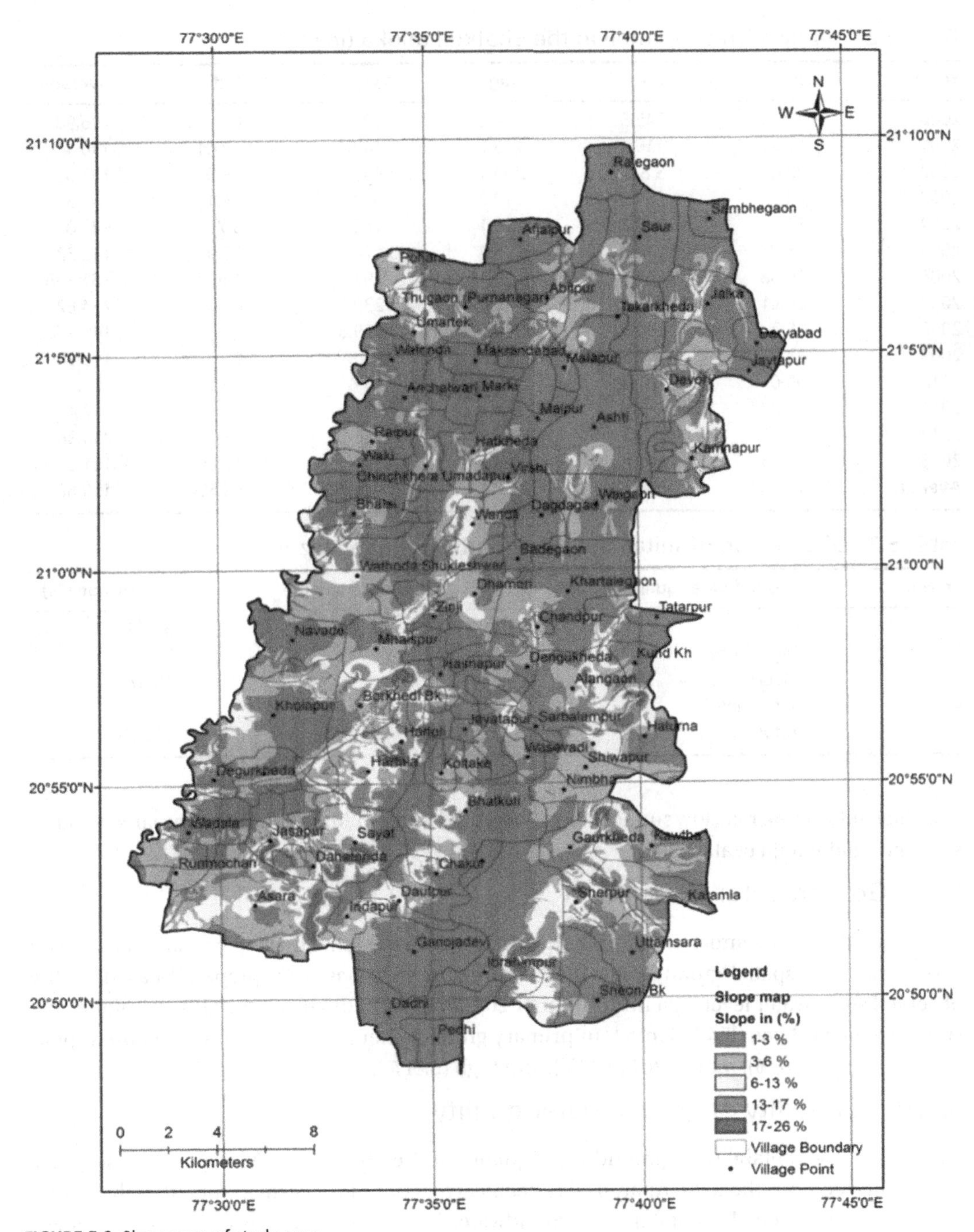

FIGURE 5.6 Slope map of study area.

Table 5.2 Yearly average rainfall in the Bhatkuli taluka (in mm).

Years	Jun.	Jul.	Aug.	Sep.	Oct.	Average
2000	120.1	245.5	139.2	30.1	0	106.98
2001	178.6	116.4	145.6	14.5	197.1	130.44
2002	221.2	8.8	209.1	44.6	53.9	107.52
2003	63.7	257.9	113.2	68.6	14.1	103.5
2004	61.7	138.2	119.1	146.2	3.7	93.78
2005	15.9	232.9	225.4	174.5	62.9	142.32
2006	19.38	287.71	180.1	193.4	126.2	161.358
2007	295.1	361.2	386.3	230.5	0	254.62
2008	85.7	135.3	169.5	359.3	0	149.96
2009	33.4	279.3	193.8	68.8	0	115.06
2010	99.6	391.2	307.3	95.4	60.1	190.72
2011	130.3	218.4	244.9	193.6	0	157.44
2012	174.8	398.7	298.6	256	43.7	234.36
2013	216.5	209.8	168.3	94.3	118.6	161.5
Average	**122.57**	**234.38**	**207.17**	**140.70**	**48.59**	**150.68**

Table 5.3 Distribution of suitability of groundwater quality zone.

Sr. No.	Groundwater quality classes	Number of sample	Percentage (%)
1	Fresh	14	38.89
2	Slight saline	12	33.33
3	Moderate saline	6	16.67
4	Very saline	4	11.11
	Total	**36**	**100.00**

to its high infiltration rate, low surface runoff, and high water holding capacity compared to other soil types (Siddi Raju et al., 2019).

5.4.9 Groundwater fluctuation

The detailed well inventory studies for the 66 observation wells in the given area have been carried out. The spatial groundwater fluctuation zone map has been prepared based on the observed static water level data from summer 2012 to winter 2013 using the GIS technique. The study region has been divided into four primary groundwater fluctuation classes, namely, poor (45.75%), low (37.68%), moderate (10.96%), and high (5.61%), as shown in Fig. 5.11.

5.4.10 Suitability of groundwater quality

To evaluate the suitability of groundwater quality of the study area, 36 water samples have been collected from the field. Analysis has been carried out in the term of total dissolved solid (TDS) value to decide the suitability of groundwater for drinking purposes as per the standard proposed by the Bureau of Indian Standards (BIS, 2003). The suitability of groundwater quality is categorized into four classes such as fresh, slight saline, moderate saline, and very saline zone, as shown in Table 5.3 and its corresponding map in Fig. 5.12.

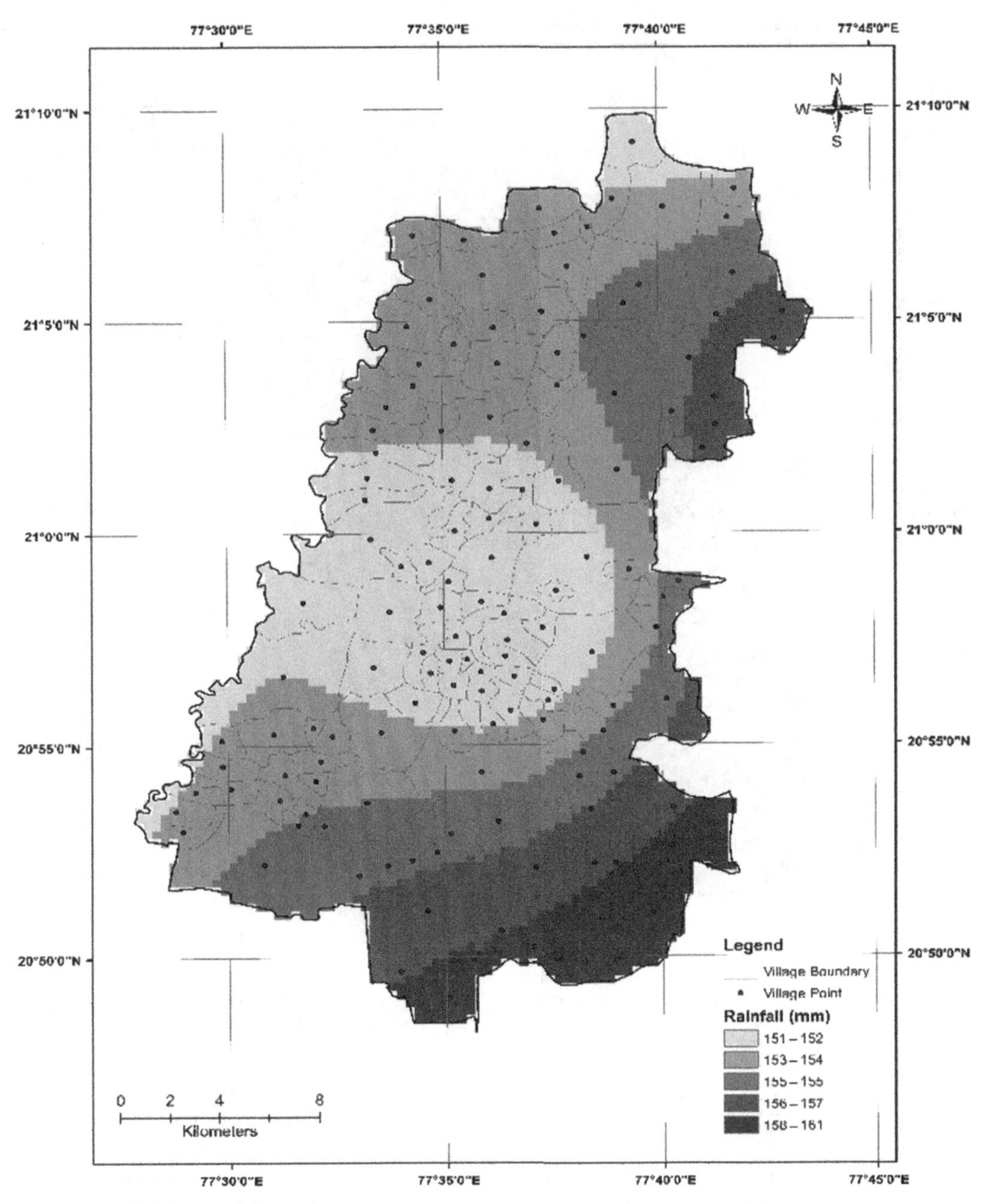

FIGURE 5.7 Rainfall map of the study area.

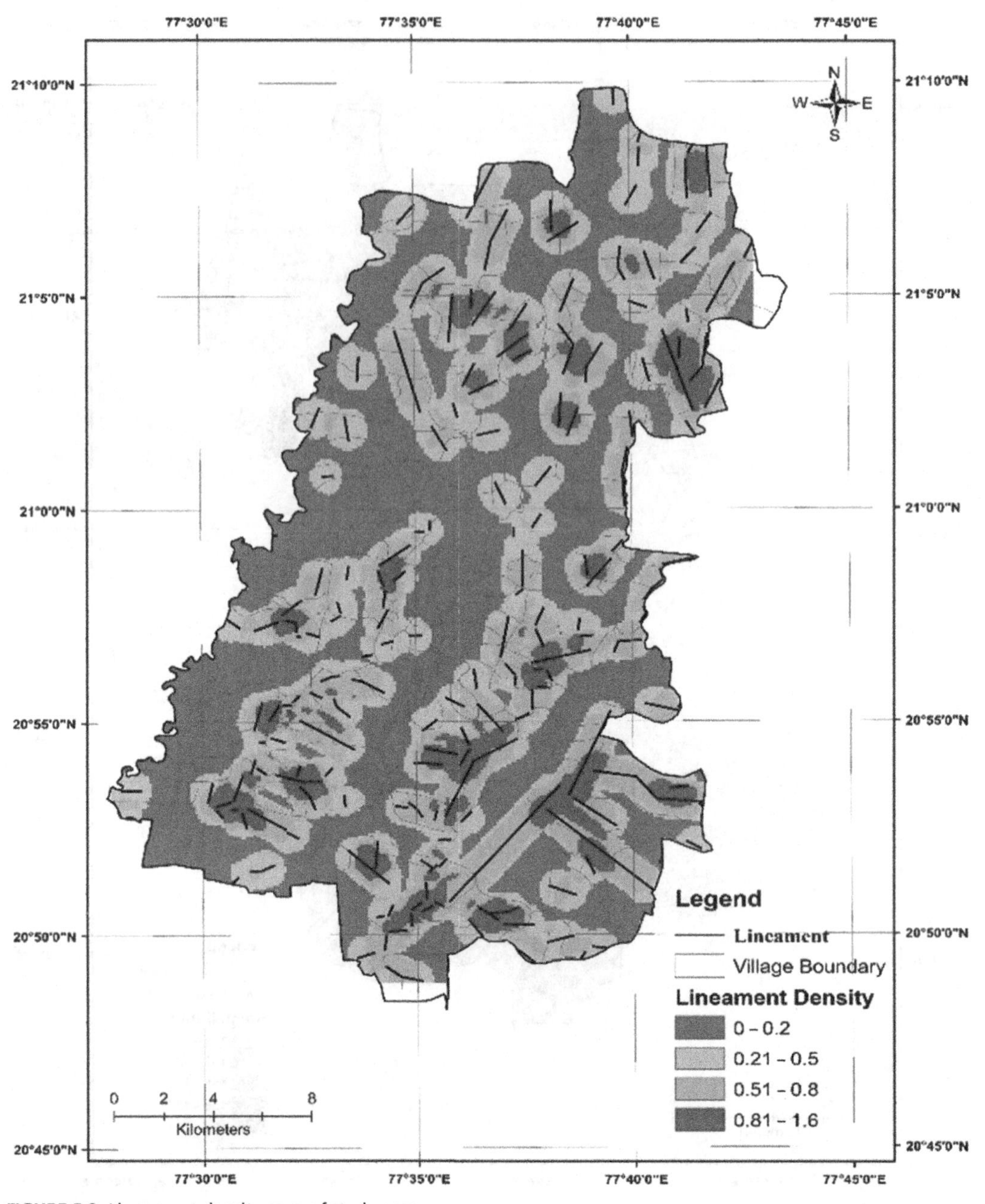

FIGURE 5.8 Lineament density map of study area.

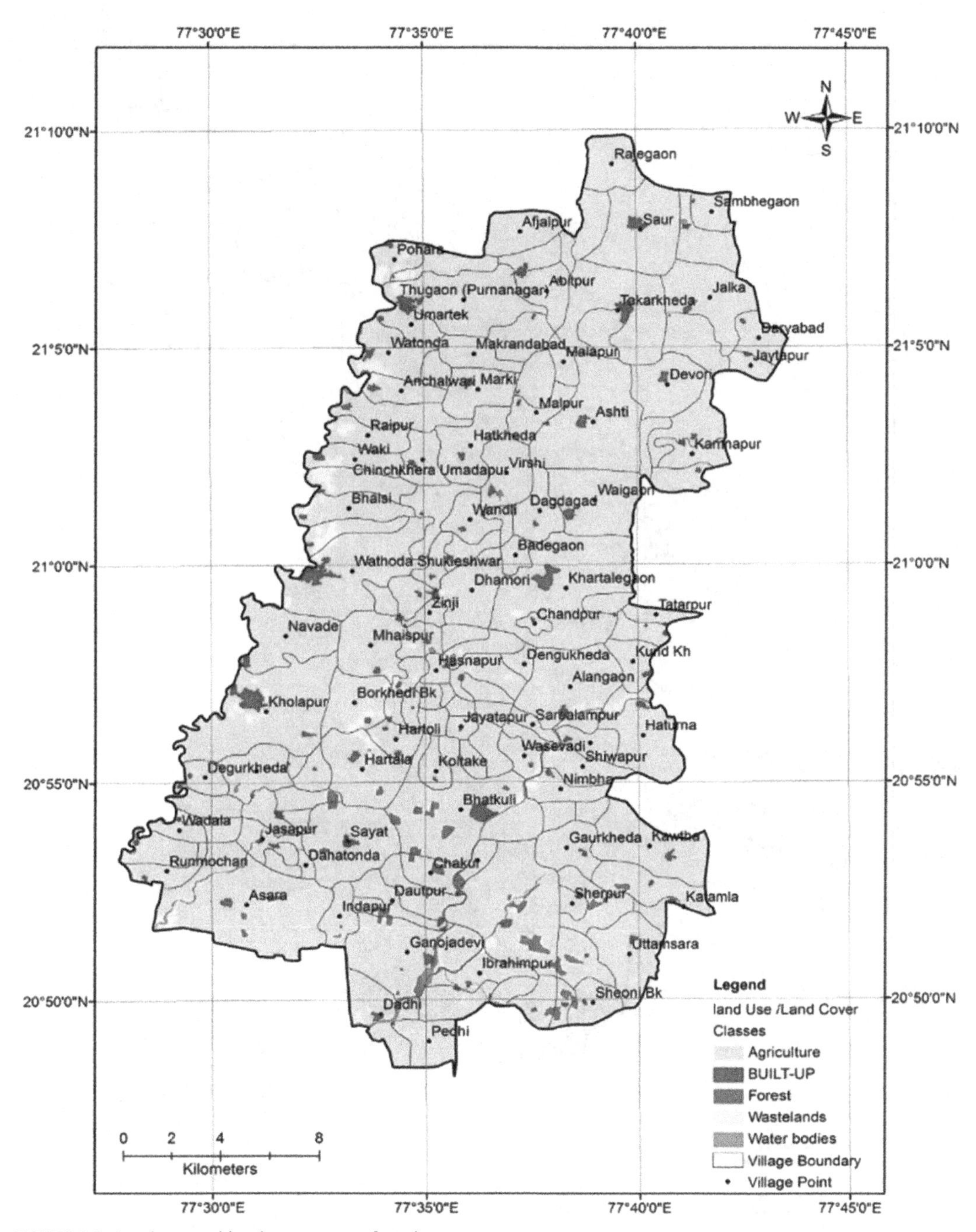

FIGURE 5.9 Land use and land cover map of study area.

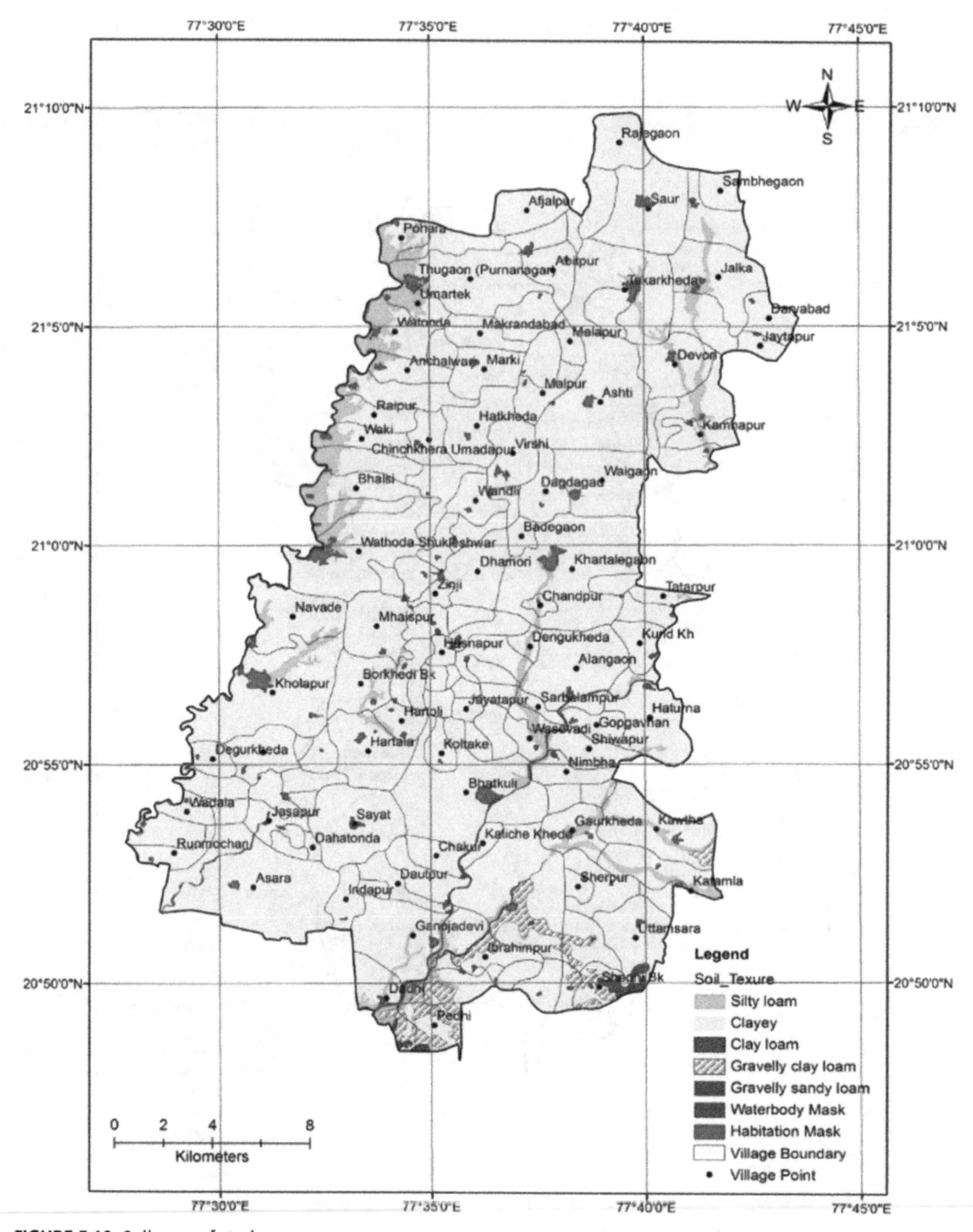

FIGURE 5.10 Soil map of study area.

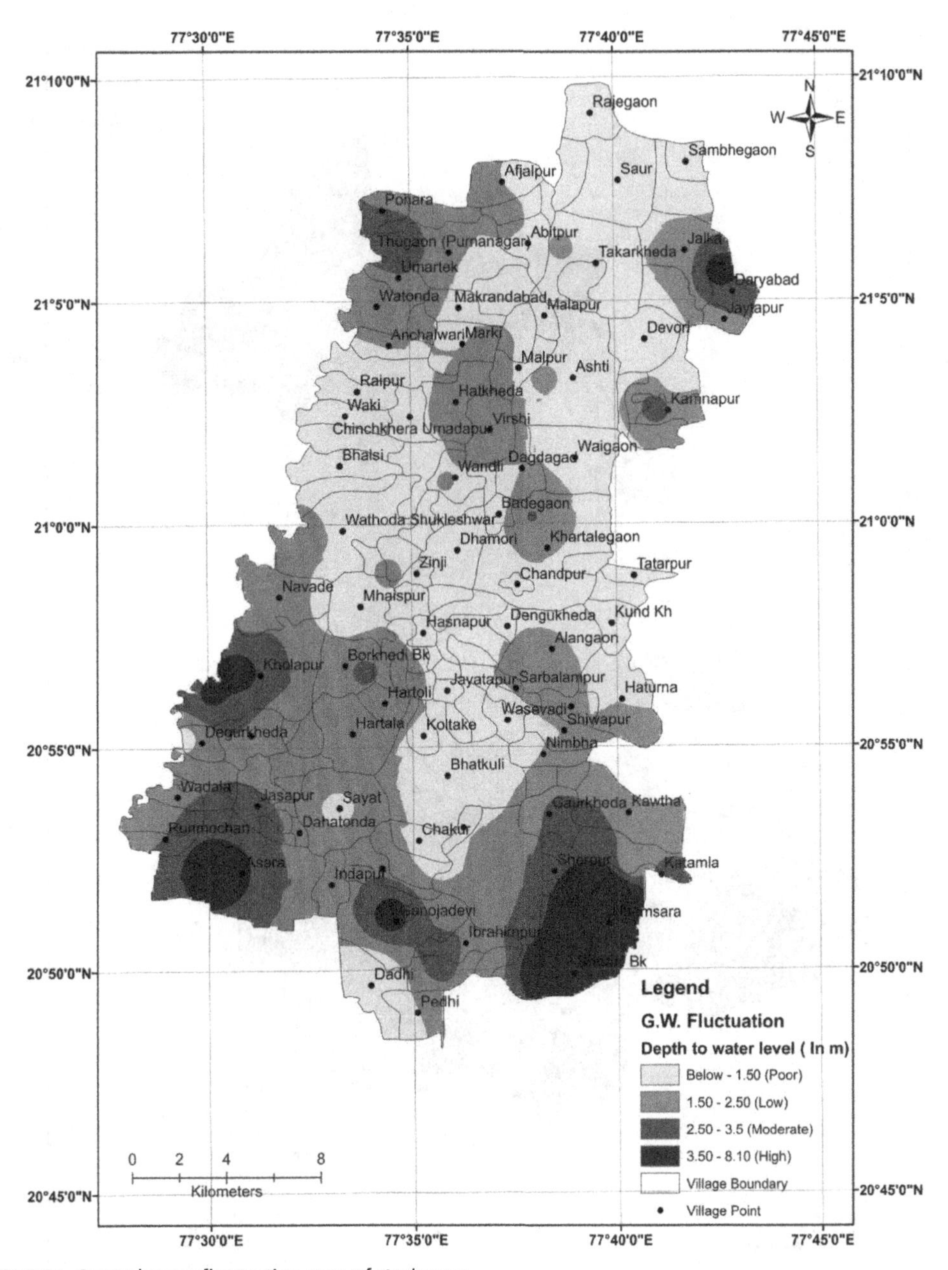

FIGURE 5.11 Groundwater fluctuation map of study area.

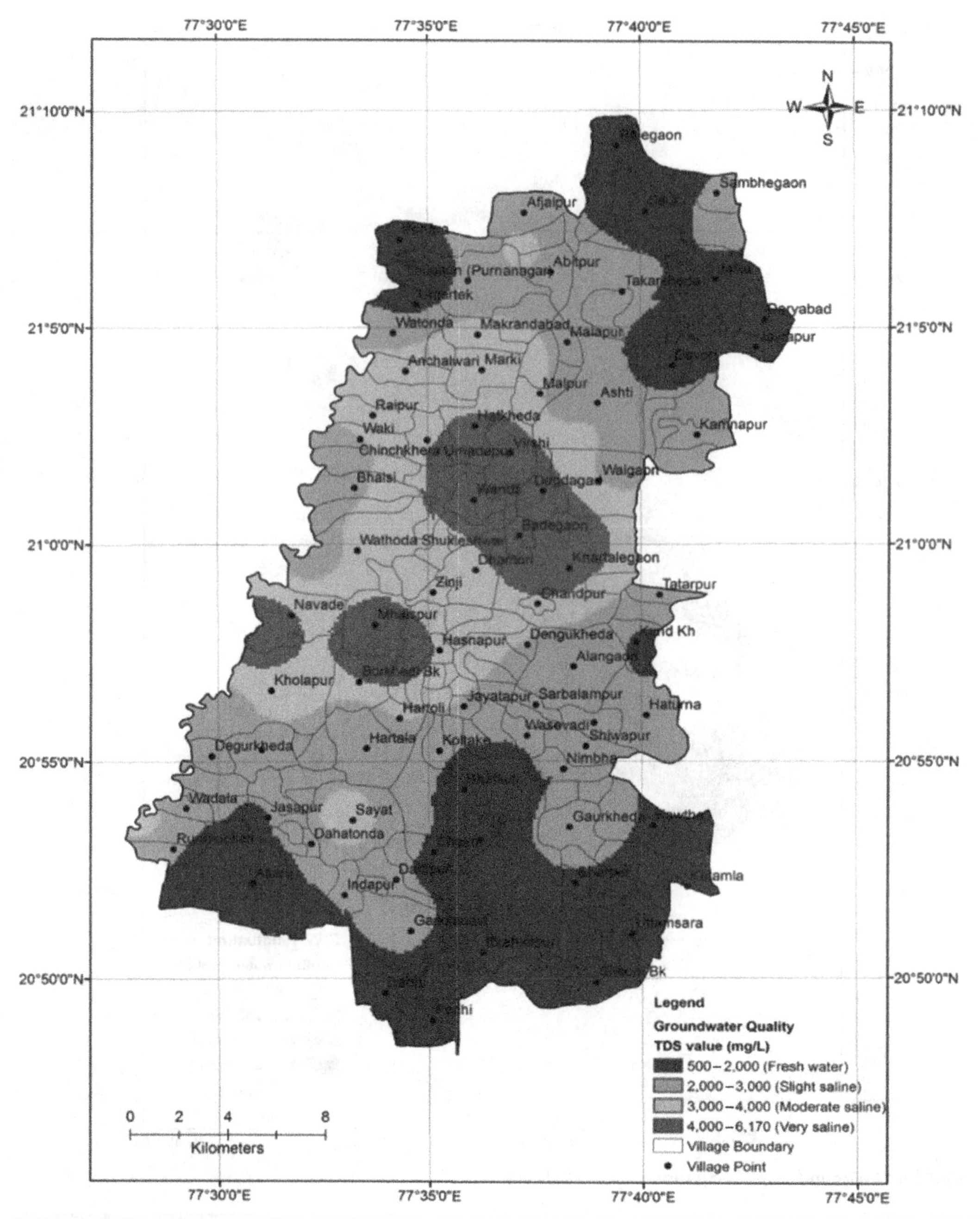

FIGURE 5.12 Groundwater quality map of study area.

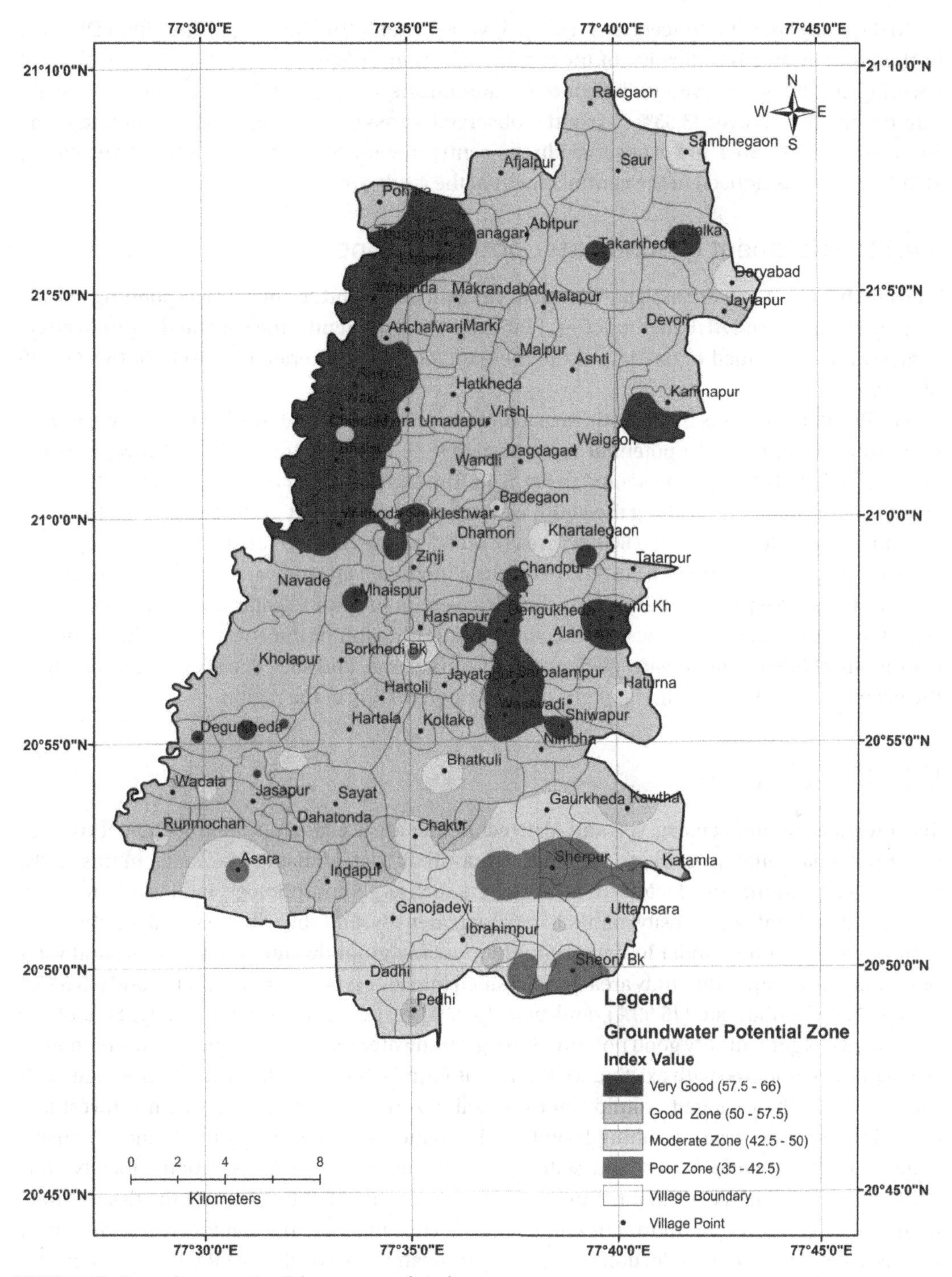

FIGURE 5.13 Groundwater potential zone map of study area.

In the study area, the concentration of TDSs varies from 570 to 6169 mg/L. The high TDS value in the groundwater is indicative of more mineralization in alluvial formation. The freshwater covering 38.89% is observed in the southern and northern region of the study area. The slight saline zone spreads over 33.33% area and is observed across the study area. The moderate saline zone cover 16.67% area and is observed in the central region. Similarly, very saline zone covers 11.11% area and is noticed in the central region of the study area.

5.4.11 Assessment groundwater potential zone

The suitable weights were assigned to each thematic layer and to their corresponding classes employing GIS-based AHP methodology (Table 5.1). Subsequently, the weighted index overlay analysis was performed to overlay all the thematic layers to generate the GWPI in the ArcGIS platform.

GWPI value expresses the quantities in a dimensionless form and divided the study area into four probable groundwater potential zones as high (57.5–66), moderate (50–57.5), low (42.5–50), and very poor (34–42.5), as presented in Fig. 5.13. The high potential zone contributes 78.77 km^2 area (13.54%) and spreads along the south western and central part of the study area. The good potential zone extends only 390.86 km^2 area (67.20%) and occurs overall in the central region of the area. About 92.98 km^2 area (15.99%) exhibits a moderate groundwater potential zone and is observed in southern part of the study area. The very poor groundwater potential zone covers 19.01 km^2 area (3.27%), observed in the south eastern region of the study area. The favorable conditions of lineament density, geology, landforms, slope, and soil type affect significantly in the occurrence and development of groundwater potentiality in the region.

5.5 Conclusion

The integrated remote sensing, GIS and AHP techniques are used for the identification of groundwater potential zonation in Bhatkuli taluka, Amravati district, Maharashtra, India. In this study, various geoenvironmental factors/thematic layers such as geomorphology, land use/land cover, geology, slope, drainage density, rainfall, lineament density, and soil with assigned weights were incorporated into the model to determine the possible groundwater zones. The groundwater potential zones map of the study area was classified into four groups, namely, very good (13.54%), good (67.20%), moderate (15.99%), and poor (3.27%) respectively. Approximately, 80% of the basin area has good to very good potentiality of groundwater resources. The groundwater potential map was compared with existing wells and was found to be basically in good agreement with the field data. The excellent groundwater possibility zone was identified at the northwest and central parts of the basin, resulting from the distribution of high lineament, drainage density, suitable rainfall, gentle slopes, and sedimentary plains with high penetration capacity. This study concludes that geospatial technology integrated with AHP technique provides accurate quantitative groundwater information in a cost-effective manner. The significance of this study is to assist planners in the selection of appropriate locations for drilling wells. This groundwater potential map would support policymakers to formulate better planning and management of groundwater resources in the region.

References

Adeyeye, O., Ikpokonte, E., Arabi, S., 2018. GIS-based groundwater potential mapping within Dengi area, North Central Nigeria. Egyptian J. Remote Sensing Space Science.

Aggarwal, M., Subbarayan, S., Jacinth Jennifer, J., Abijith, D., 2019. Delineation of groundwater potential zones for hard rock region in Karnataka using AHP and GIS. In: El-Askary, H., Lee, S., Heggy, E., Pradhan, B. (Eds.), Advances in Remote Sensing and Geo Informatics Applications. CAJG 2018. Advances in Science, Technology & Innovation. Springer, Cham, pp. 315–317.

Ahmad, I., Dar, M.A., Teka, A.H., Teshome, M., Andualem, T.G., Tehsome, A., Shafi, T., 2020. GIS and fuzzy logic techniques-based demarcation of groundwater potential zones: a case study from Jemma River basin. Ethiopia. J. Afr. Earth Sci. 169, 103860.

Asoka, A., Gleeson, T., Wada, Y., Mishra, V., 2017. Relative contribution of monsoon precipitation and pumping to changes in groundwater storage in India. Nat. Geosci. 10 (2), 109–117.

Baskaran, R., Pothiraj, P., 2013. Mapping of lineaments for groundwater targeting and sustainable water resource management in hard rock hydrogeological environment using RS- GIS. Climate Change Regional/Local Responses 235–247.

BIS, 2003. Indian Standard Drinking Water Specifications IS 10500:1991, Edition 2.2 (2003-2009). Bureau of Indian Standards, New Delhi.

Chowdhury, A., Jha, M.K., Chowdary, V.M., 2010. Delineation of groundwater recharge zones and identification of artificial recharge sites in West Medinipur district, West Bengal using RS, GIS and MCDM techniques. Environmental Earth Sci. 59, 1209–1222.

Chang, F.J., Huang, C.W., Cheng, S.T., Chang, L.C., 2017. Conservation of groundwater from over-exploitation—scientific analyses for groundwater resources management. Sci. Total Environ. 598, 828–838.

Chepchumba, M.C., Raude, J.M., Sang, J.K., 2019. Geospatial delineation and mapping of groundwater potential in EMBU county, Kenya. Acque Sotterranee - Italian Journal of Groundwater.

Chezgi, J., Pourghasemi, H.R., Naghibi, S.A., Moradi, H.R., Kheirkhah Zarkesh, M., 2016. Assessment of a spatial multi-criteria evaluation to site selection underground dams in the Alborz Province, Iran. Geocarto Int. 31 (6), 628–646.

CGWB 2013. Ground water pollution by industrial clusters. Bhu-Jal news, 28 (1–4), Ministry of Water Resources, River Development and Ganga Rejuvenation, Government of India.

CWC and CGWB, 2016. A 21st century institutional architecture for India's water reforms. wrmin.nic.in/writereaddata/Report_on_Restructuring_CWC_CGWB.pdf. [Access date - 19.5.2021].

Deepa, S., Venkateswaran, S., Ayyandurai, R., Kannan, R., Vijay Prabhu, M., 2016. Groundwater recharge potential zones mapping in upper Manimuktha Sub basin Vellar river Tamil Nadu India using GIS and remote sensing techniques. Modeling Earth Syst. Environ. 2 (3), 1–13.

Dadgar, M.A., Zeaieanfirouzabadi, P., Dashti, M., Porhemmat, R., 2017. Extracting of prospective groundwater potential zones using remote sensing data, GIS, and a probabilistic approach in Bojnourd basin, NE of Iran. Arab J. Geosci. 10, 114.

Donselaar, M.E., Bhatt, A.G., Ghosh, A.K., 2017. On the relation between fluvio-deltaic flood basin geomorphology and the wide-spread occurrence of arsenic pollution in shallow aquifers. Sci. Total Environ. 574, 901–913.

Fienen, M.N., Arshad, M., 2016. The international scale of the groundwater issue. Concepts, approaches and challenges. In: Integrated Groundwater Management. Springer International Publishing, AG Switzerland, pp. 21–48.

Famiglietti, J.S., 2014. The global groundwater crisis. Nat. Clim. Change 4 (11), 945. https://doi.org/10.1038/nclimate2425.

Ghorbani Nejad, S., Falah, F., Daneshfar, M., Haghizadeh, A., Rahmati, O., 2016. Delineation of groundwater potential zones using remote sensing and GIS-based data-driven models. Geocarto Int. 32 (2), 167–187.

Ghosh, P.K., Bandyopadhyay, S., Jana, N.C., 2016. Mapping of groundwater potential zones in hard rock terrain using geoinformatics: a case of Kumari watershed in Western part of West Bengal. Modeling Earth Syst. Environ. 2 (1), 1–12.

Gleeson, T., Wada, Y., Bierkens, M.F., van Beek, L.P., 2012. Water balance of global aquifers revealed by groundwater footprint. Nature 488 (7410), 197. https://doi.org/10.1038/nature11295.

IPCC, 2001. Climate change 2001: The scientific basis. Contribution of Working Group I to the Third Assessment Report of the Intergovernmental Panel on Climate Change. Houghton, J.T., Ding, Y., Griggs, D.J., Noguer, M., Van der Linden, P.J., Dai, X., Maskell, K., and Johnson, C.A. (Eds.). Cambridge University Press, Cambridge and New York.

Jesiya, N.P., Gopinath, G., 2019. A Customized Fuzzy AHP-GIS based DRASTIC-L model for intrinsic groundwater vulnerability assessment of urban and peri urban phreatic aquifer clusters. Groundw. Sustain. Dev. 8, 654–666.

Jasrotia, A.S., Bhagat, B.D., Kumar, A., Kumar, R., 2013. Remote sensing and GIS approach for delineation of groundwater potential and groundwater quality zones of western doon valley, uttarakhand, India. J. Indian Soc. Remote Sens. 41 (2), 365–377.

Patra, S., Mishra, P., Mahapatra, S.C., 2018. Delineation of groundwater potential zone for sustainable development: a case study from Ganga Alluvial Plain covering Hooghly district of India using remote sensing, geographic information system and analytic hierarchy process. J. Clean. Prod. 172, 2485–2502.

Kaliraj, S., Chandrasekar, N., Magesh, N.S., 2014. Identification of potential groundwater recharge zones in Vaigai upper watershed, Tamil Nadu, using GIS-based analytical hierarchical process (AHP) technique. Arab. J. Geosci. 7, 1385–1401.

Kirubakaran, M., Johnny, J.C., Ashokraj, C., Arivazhagan, S., 2016. A geostatistical approach for delineating the potential groundwater recharge zones in the hard rock terrain of Tirunelvelitaluk, Tamil Nadu, India. Arabian J. Geosci. 9 (5).

Kumar, A., Pandey, A.C., 2016. Geoinformatics based groundwater potential assessment in hard rock terrain of Ranchi urban environment, Jharkhand state (India) using MCDM–AHP techniques. Groundw. Sustain. Dev. 2, 27–41.

Martin, S.L., Hayes, D.B., Kendall, A.D., Hyndman, D.W., 2017. The land-use legacy effect: towards a mechanistic understanding of time-lagged water quality responses to land use/cover. Sci. Total Environ. 579, 1794–1803.

Moges, D.M., Bhat, H.G., Thrivikramji, K.P., 2019. Investigation of groundwater resources in highland Ethiopia using a geospatial technology. Model. Earth Syst. Environ. 5, 1333–1345. https://doi.org/10.1007/s40808-019-00603-0.

Mogaji, K.A., Lim, H.S., Abdullah, K., 2014. Regional prediction of groundwater potential mapping in a multifaceted geology terrain using GIS-based Dempster–Shafer model. Arabian J. Geosci. 8, 3235–3258.

Mumtaz, R., Baig, S., Kazmi, S.S.A., Ahmad, F., Fatima, I., Ghauri, B., 2019. Delineation of groundwater prospective resources by exploiting geo-spatial decision-making techniques for the Kingdom of Saudi Arabia. Neural Comput. Appl. 31 (9), 5379–5399.

Nasir, M.J., Khan, S., Zahid, H., Khan, A., 2018. Delineation of groundwater potential zones using GIS and multi influence factor (MIF) techniques: a study of district swat, Khyber Pakhtunkhwa, Pakistan. Environ. Earth Sci. 77 (10), 1–11.

Nayyer, S., Huq, M., Nana Yaw Danquah, T., Akib, J., Asif, S., 2019. Parameters derived from and/or used with digital elevation models (DEMs) for landslide susceptibility mapping and landslide risk assessment: a review. ISPRS Int. J. Geo-Inf. 8 (12).

Nhamo, L., Yimer, G., Mabhaudhi, T., Mpandeli, S., Magombeyi, M., Chitakira, M., Magidi, J., 2019. An assessment of groundwater use in irrigated agriculture using multi-spectral remote sensing. Phys. Chem. Earth, 115, 102810.

Razandi, Y., Pourghasemi, H.R., Neisani, N.S., Rahmati, O., 2015. Application of analytical hierarchy process, frequency ratio, and certainty factor models for groundwater potential mapping using GIS. Earth Sci. Inf. 8 (4), 867–883.

Pinto, D., Shrestha, S., Babel, M.S., Ninsawat, S., 2017. Delineation of groundwater potential zones in the Comoro watershed, Timor Leste using GIS, remote sensing and analytic hierarchy process (AHP) technique. Appl. Water Sci. 7 (1), 503–519.

Rahman, M.A., Rusteberg, B., Gogu, R.C., Ferreira, J.L., Sauter, M., 2012. A new spatial multi-criteria decision support tool for site selection for implementation of managed aquifer recharge. J. Environ. Manag. 99, 61–75.

Roy, S., Hazra, S., Chanda, A., Das, S., 2020. Assessment of groundwater potential zones using multi-criteria decision-making technique: a micro-level case study from red and lateritic zone (RLZ) of West Bengal. India. Sustain. Water Resources Manage. 6 (1), 1–14.

Sandoval, J.A., Tiburan, C.L., 2019. Identification of potential artificial groundwater recharge sites in mount makiling forest reserve, Philippines using GIS and analytical hierarchy process. Appl. Geogr. 105, 73–85.

Saaty, T.L., 1980. The Analytical Hierarchy Process. McGraw Hill, New York, NY.

Sarker, M.N.I., Wu, M., Alam, G.M.M., Shouse, R.C., 2020. Life in riverine islands in Bangladesh: local adaptation strategies of climate vulnerable riverine island dwellers for livelihood resilience. Land Use Pol. 94, 104574.

Sander, P., 2007. Lineaments in groundwater exploration: a review of applications and limitations. Hydrogeol. J. 15 (1), 71–74.

Saravanan, S., 2012. Identification of artificial recharge sites in a hard rock terrain using remote sensing and GIS. Int. J. Earth Sci. Eng. 5 (6), 1590–1598.

Siebert, S., Burke, J., Faures, J.M., Frenken, K., Hoogeveen, J., Doll, P., Portmann, F.T., 2010. Groundwater use for irrigation—a global inventory. Hydrol. Earth Syst. Sci. 14 (10), 1863–1880.

Siddi Raju, R., Sudarsana Raju, G., Rajasekhar, M., 2019. Identification of groundwater potential zones in Mandavi River basin, Andhra Pradesh, India using remote sensing, GIS and MIF techniques. HydroRes. 2, 1–11.

Silwal, C.B., Pathak, D., 2018. Review on practices and state of the art methods on delineation of ground water potential using GIS and remote sensing. Bull. Dep. Geol. 7–20.

Snyder, Shannyn. 2019. Water In Crisis - Spotlight India. https://thewaterproject.org/water-crisis/water-in-crisis-india. [Access date - 25.8.2021].

Taweesin, K., Seeboonruang, U., Saraphirom, P., 2018. The influence of climate variability effectson groundwater time series in the lower central plains of Thailand. Water 10 (3), 290.

Venkatesan, G., Pitchaikani, S., Saravanan, S., 2019. Assessment of groundwater vulnerability using GIS and DRASTIC for upper Palar River basin, Tamil Nadu. J. Geol. Soc. India 94 (4), 387–394.

Yu, H.L., Lin, Y.C., 2015. Analysis of space-time non-stationary patterns of rainfall groundwater interactions by integrating empirical orthogonal function and cross wavelet transform methods. J. Hydrol. 525, 585–597.

6

GIS-based groundwater recharge potentiality analysis using frequency ratio and weights of evidence models

Suraj kumar Mallick, Biswajit Maity, Pritiranjan Das and Somnath Rudra

DEPARTMENT OF GEOGRAPHY, VIDYASAGAR UNIVERSITY, MIDNAPORE, WEST BENGAL, INDIA

6.1 Introduction

Water is an essential natural resource for life and an important controlling factor for the socio-economic development of a country. Our blue earth is covered by almost 70% of its surface by water, but only 1% of it is usable for us. Groundwater contributes nearly 30% of the total usable water. It has occurred in pore and fracture space below the surface. Groundwater is the most accessible and safest water use mode due to its freshness and being less vulnerable to pollution than surface water (Mallick and Rudra, 2021). Since the last century, the increasing demand for groundwater for agricultural and industrial revolution has increased the overexploitation of groundwater throughout the world. The nonuniform geographical distribution of groundwater and lack of microlevel information of potential groundwater sources creates a water crisis to the extensive population.

For this reason, it is necessary to explore the groundwater potential zone for increasing access and security to water (Wang et al., 2012). The infiltration of rain and snow melt water to the saturated area is a prerequisite for groundwater recharge and sustainable management (Freeze and Cherry, 1979; Yeh et al., 2016). Several factors are responsible for the occurrence and movement of groundwater like geological structure, slope, drainage pattern, lineaments characteristics, climatic condition, land use, land cover, etc. (Jiswal et al., 2003; Thapa et al., 2017). Consequently, it is necessary to consider the entire above factor for delineating potential groundwater zone for identifying site and suitability for future residential, agricultural, and industrial location of an area.

Several studies have been carried out that the remote sensing (RS) and geographic information system (GIS) act as a potent tool for delineating groundwater potential zone (Evans and Myers, 1990; Saraf and Choudhury, 1998; Rai et al., 2005; Singh and Singh, 2009;

Case Studies in Geospatial Applications to Groundwater Resources. DOI: https://doi.org/10.1016/B978-0-323-99963-2.00016-X

Yan et al., 2010; Hammouri et al., 2012; Kuria et al., 2012; Anbazhagan and Jothibasu, 2014; Chen et al., 2019; Gueretz et al., 2019; Mallick and Rudra, 2021). Many effective models have evolved this issue using the conditioning parameters and applying different efficient integrating techniques (Ghezelbash et al., 2019; Eraslan et al., 2019). Recently, multivariate geostatistical analysis and machine learning algorithms have been considered more efficient tools for model development. The geostatistical approaches like the novel hybrid intelligence approach (Miraki et al., 2019), boosted tree (Naghibi et al., 2016; Sameen et al., 2019), logistic regression model (Ozdemir, 2011; Pourtaghi and Pourghasemi, 2014; Douglas et al., 2018), random forest (Rahmati et al., 2016; Naghibi et al., 2016; Naghibi and Pourghasemi, 2015; Gayen et al., 2020), multicriteria decision evaluation (Murthy and Mamo, 2009; Rahmati et al., 2014; Razandi et al., 2015), maximum entropy model (Rahmati et al., 2016), analytic hierarchy process (AHP) (Shekhar and Pandey, 2015; Rahmati et al., 2015), frequency ratio (FR) (Oh et al., 2011; Elmahdy and Mohamed, 2015; Zeinivand and Nejad, 2018; Maity et al., 2022), fuzzy analytic hierarchy process (Aryafar et al., 2013; Şener et al., 2018; Mallick and Rudra, 2021), certainty factor (Razandi et al., 2015), and weights of evidence (WOE) (Al-Abadi, 2015; Chen et al., 2018; Kordestani et al., 2019) are used for groundwater recharge potentiality analysis. Besides, support vector machines, artificial neural networks (Lee et al., 2018), GIS-based trees, machine learning algorithms (Al-Abadi et al., 2019) etcetera have been used to delineate groundwater potentiality as means of machine learning algorithm approach.

The steadily increasing demand for food crops and industrial products has pressured India's agricultural and industrial sectors. However, India has received rainfall only from June to September, so Indian agriculture mostly depends on groundwater. It is reported that 89% of groundwater water is used in irrigation (CGWB, 2014). The application of multimodel has improved the prediction accuracy of models (Pourghasemi et al., 2017). The present work is based on FR and WOE for delineating groundwater recharge potentiality zone (GRPZ) in the drier most of South Bengal, namely the Jhargram District. Jhargram consists of alluvium surface, pediment, dissected hills, and valleys. Jhargram has high groundwater stress during summer and winter, both the season. So, it is essential to analyze the groundwater recharge potentiality in this kind of drier region.

6.2 Study area

Jhargram in West Bengal, India lies between 21°45′ N to 22° 42′ N latitude and 86°42′ E to 87°16′ E (Fig. 6.1) Longitude covered with 3,037.64 sq. km area. This area is famous for its wooded beauty, popularly known as "Jungalmahal" and topography culminating in the hill ranges of Belpahari, Kankrajhor to the north, and Subarnarekha to the south. This district experiences an extremely humid and tropical climate with an average 46°C temperature in the May–June months, chilly nights in December–January, and an annual average rainfall of 1428 mm. The entire area is situated on the lower part of the Chota Nagpur plateau with undulating, lateritic covered, hard rock upland, flat alluvial, and deltaic plains. The highly rugged topography in the western part is gradually sloping toward the east. The heterogeneous terrain condition creates a problematic situation for groundwater restoration in this region. It is one of the drought-prone areas of South

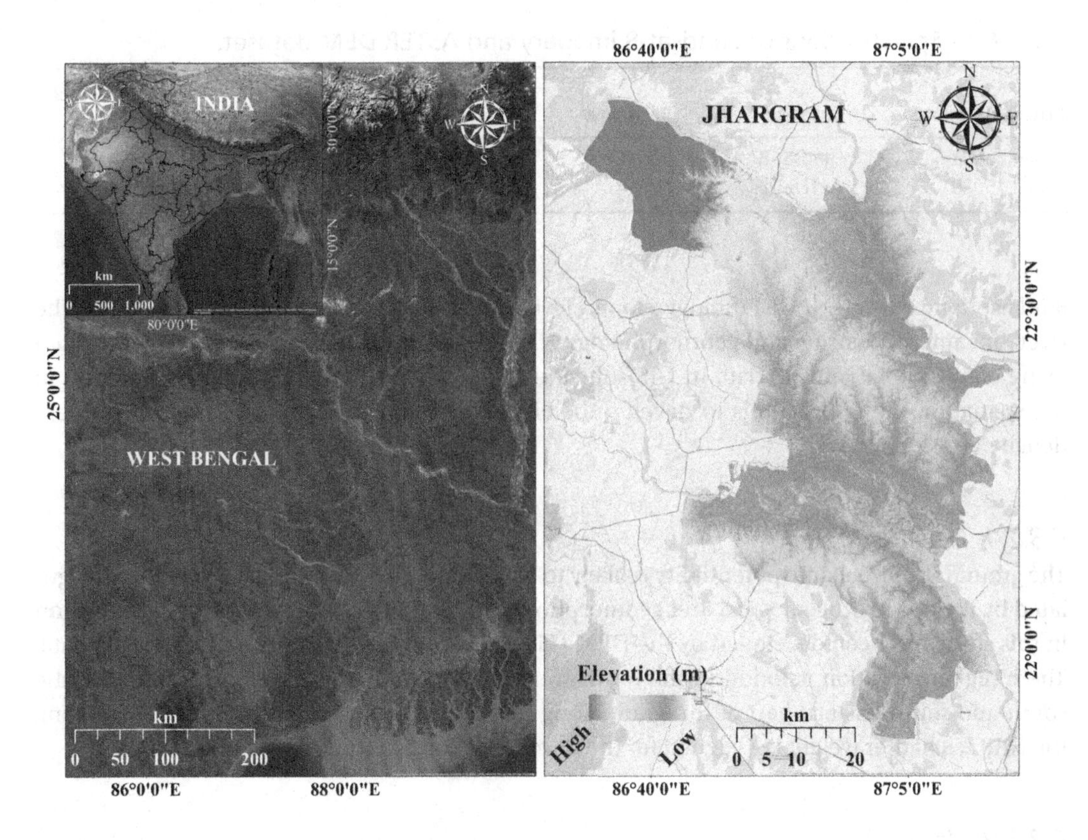

FIGURE 6.1 Location map of Jhargram district along with digital elevation model.

Bengal with a particularly severe drought situation. Agriculture is the main occupation of the rural part of this region. So, groundwater and surface water are both being harnessed by installing various irrigation systems in this region. Hence, this study tried to identify the GRPZ for planning and distribution of water resources to the entire area using geospatial techniques.

6.3 Database and methodology

6.3.1 Data used

This study used multispectral satellite imagery from Landsat-8 (OLI/ TIRS) and digital elevation model (DEM) that obtained from advanced space-borne thermal emission and reflection radiometer (ASTER) satellite as the primary satellite datasets. Table 6.1 provided detailed information about the data acquisition. Both datasets were derived from the United States Geological Survey (USGS) Earth Explorer web portal (http://earthexplorer.usgs.gov).

6.3.2 Data processing and generation of thematic Maps using GIS

The methodology adopted for the present study is to develop the GRPZ, which is given in Fig. 6.3. To evaluate the GRPZ of the Jhargram district, eight thematic maps have been prepared those are

Table 6.1 Specifications of Landsat-8 imagery and ASTER DEM dataset.

Acquisition date	Satellite	Path/row	Resolution	Referencing system
November 23, 2020	Landsat-8 OLI	139/43	30 m, 100 m (b-10, 11), 15 m (b-8)	UTM 45°N
March 15, 2011	ASTER DEM	NA	30 m	and WGS 84

land use/land cover (LULC), lithology, slope, elevation, drainage density (DD), distance from the river, lineaments density, and geomorphology. These conditioning layers have been generated using conventional and RS data in GIS software (Fig. 6.2). Further use of density function in the spatial toolset of ArcGIS is to develop the elevation map, slope map, DD, and lineament density.

6.3.2.1 Elevation (E)

The groundwater recharge potentiality is likely to be affected by topographic elevation and regulated by various hydrogeological and geomorphological processes. The topographical elevation in this study has been developed by the DEM (Fig. 6.2A) and extracted from ASTER DEM data. There higher elevation is found in the north-western part, and a lower height is found in the south and south-eastern part of Jhragram along the Subarnarekha Riverbank. For delineating the GRPZ, a higher weight is assigned for the lower elevated area and vice versa.

6.3.2.2 Slope (S)

The availability and flow of groundwater are highly controlled by the slope (Yeh et al., 2016). The slope was derived from ASTER-DEM data of Jhargram (Fig. 6.2B). In this study, the slope has been reclassed into five classes according to the value of slope and represented by percentage value. The slope area of 0%–6% covers the almost entire area, which refers to the gentle slope in the favor of highest rate of infiltration and 6%–12%, 12%–18%, 18%–24%, and >24% covering the remaining part of the study area. The infiltration rate of shallow water into subsurface water is high in the gentle slope area, allowing more time to percolate and vice versa.

6.3.2.3 Drainage density (DD)

The DD is the sum of all stream lengths in a river basin divided by the total volume of the river basin. The structural analysis of a drainage network helps identify the groundwater recharge potentiality and the quality of the drainage system relies on a significant index on percolation rate (Yeh et al., 2016). The DD is significantly correlated with the groundwater recharge; higher the DD refers to the higher degree of groundwater recharge. The DD map (Fig. 6.2C) is extracted from the ASTER-DEM in ArcGIS. In this study, higher the weight attributed to high DD and vice-versa.

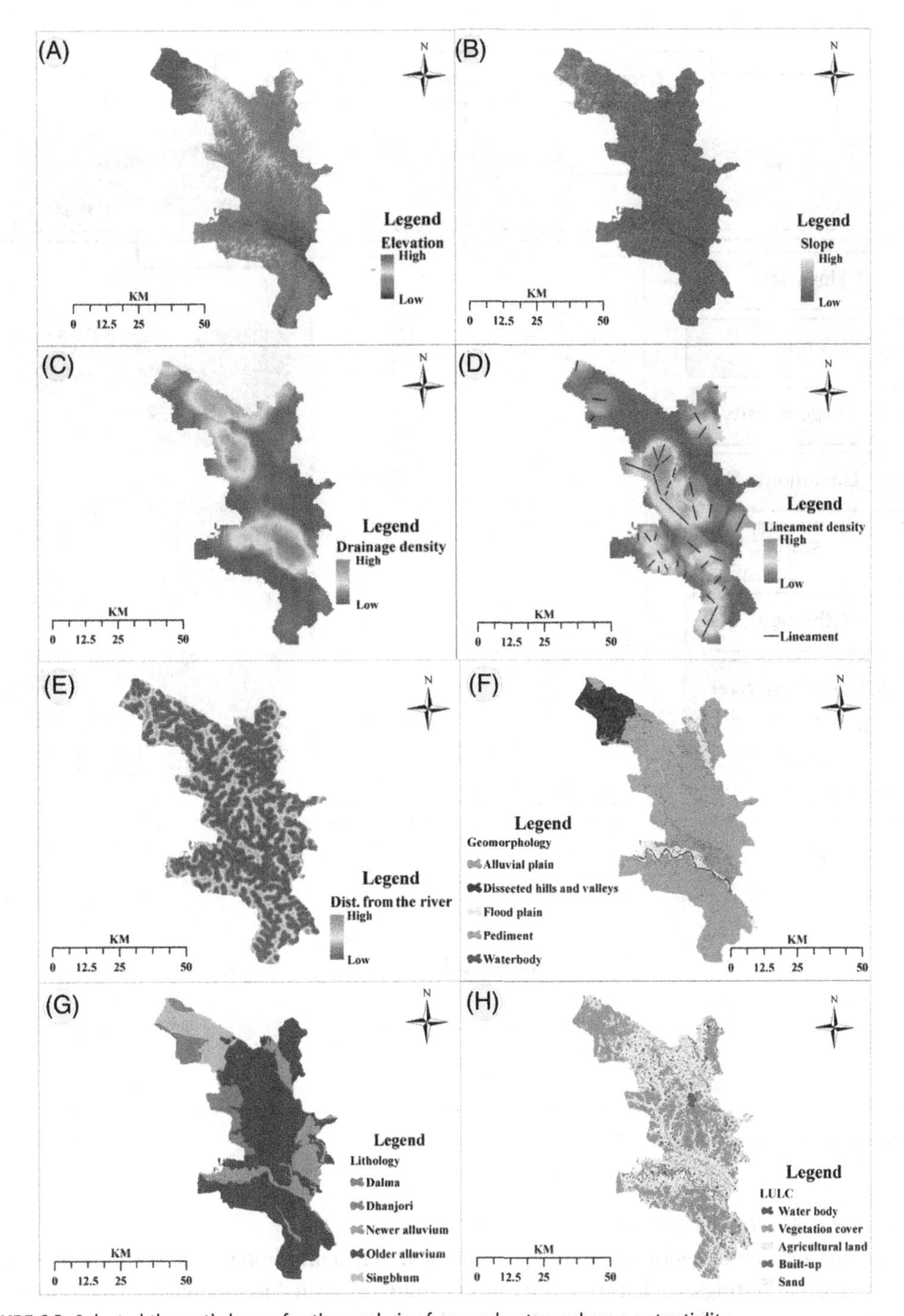

FIGURE 6.2 Selected thematic layers for the analysis of groundwater recharge potentiality.

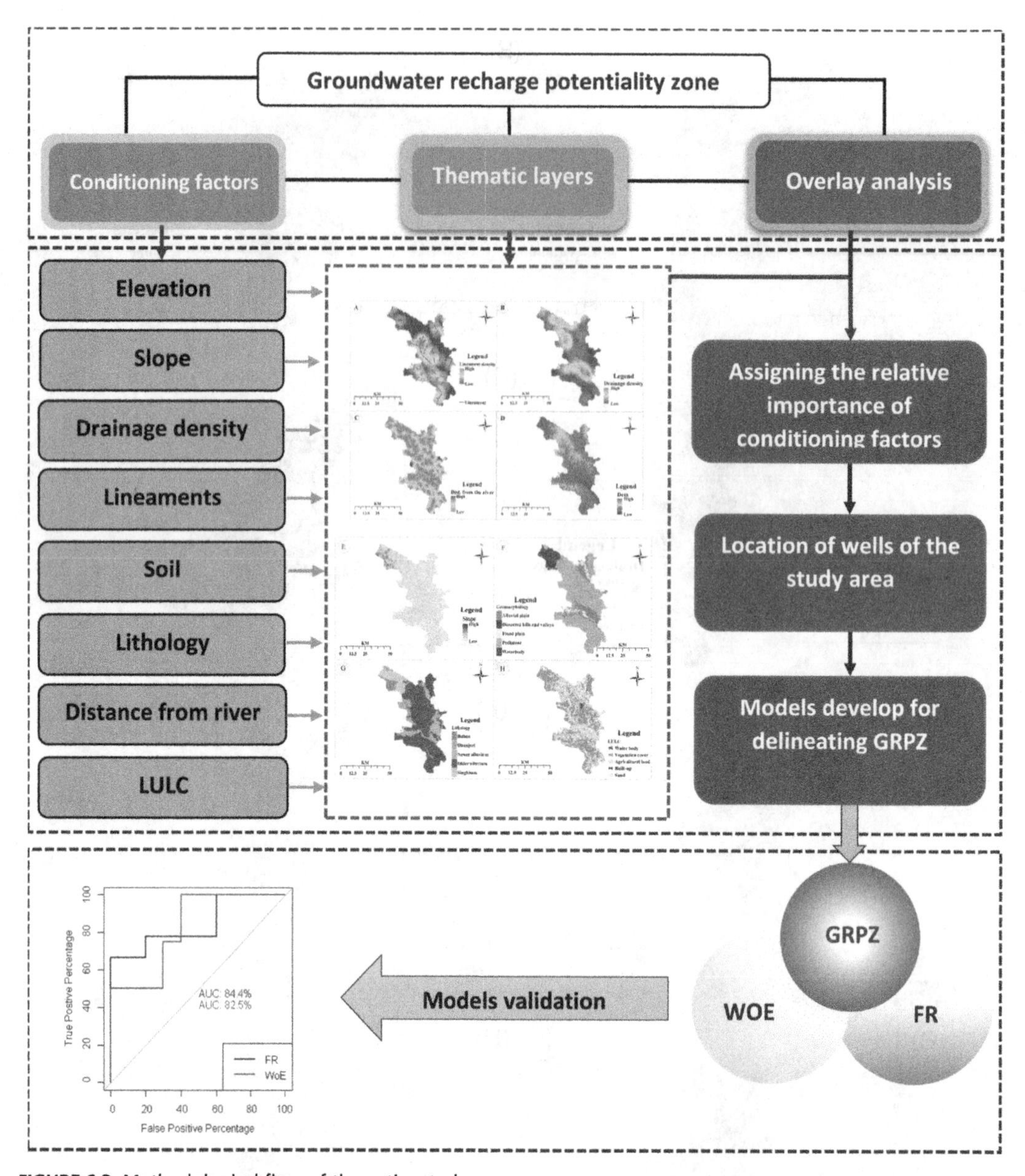

FIGURE 6.3 Methodological flow of the entire study.

6.3.2.4 Lineament density (LD)

The lineaments are the region of dimness surface with particular linear to curved topographies in the geological settings, such as fractures, faults, joints, etc. The lineaments density influences groundwater intensity (Yeh et al., 2016). Used the geological map to extract the lineaments density in this study. The high weight assigned for high lineament density represents the high recharge zone and vice versa. In this study, five lineament density zones (Fig. 6.2D) were identified which are 0–0.47 (very low), 0.47–0.94 (low), 0.94–1.41 (moderate), 1.41–1.88 (high), >1.88 km/km^2 (very high).

6.3.2.5 Distance from the river (DR)

Distance from the river also influences the groundwater table (Mallick and Rudra, 2021). Land of the proximity of river has high recharge potentiality, and the furthest one has an inverse relation. Although DR is not considered a primary indicator, it is an essential element that helps increase recharge possibilities.

6.3.2.6 Geomorphology (G)

Geomorphological structures have played a significant character in assessing the groundwater recharge and percolation into the subsurface layers (Balamurugan et al., 2017). The geomorphological map is prepared based on the topographical map of the Geological Survey of India (GSI) and satellite images. The map has been categorized into five classes: alluvial plain, dissected hills and valleys, flood plain, pediment, and water body (Fig. 6.2F). Alluvial plain, flood plain, and water body have given maximum weights.

6.3.2.7 Lithology (L)

Lithological rock types exposed to the surface significantly impact the groundwater recharge (Fig. 6.2G) has been collected from the GSI. This map illustrates five classes of lithological units, for example, Dalma, Dhanjori, newer alluvium, older alluvium, and Singhbhum. The newer alluvium land of the Subarnarekha river has the highest water holding capacity as compared to other lithological units.

6.3.2.8 Land use/land cover (LULC)

Globally, anthropogenic activities (directly or indirectly) are responsible for environmental damage, such as groundwater depletion, deforestation, soil erosion, and loss of soil quality. LULC change is a crucial factor for groundwater change (Mallick and Rudra, 2021). This study has applied a supervised classification method using the Maximum Likelihood Classifier Algorithm in ArcGIS software. The classification result has been validated using the Kappa coefficient. As a result, the six different LULC classes (Fig. 6.2H) have been found. The proportionate share of each class is about agricultural land (24.75%), built-up area (5.20%), dry/barren land (28.20%), mining and industrial area (4.91%), vegetation cover (31.29%), and water bodies (5.65%).

6.3.3 Statistical models

6.3.3.1 Frequency ratio (FR)

FR model is one of the bivariate statistical method which helps to measure the probability between dependent variables and independent variables of the groundwater potential zones (Balamurugan et al., 2017; Boughariou et al., 2021). In this study, groundwater conditioning factors are considered as independent variables and bore well data is considered as dependent variable. The FR can be calculated as follows (Eq. 6.1):

$$\mathrm{FR} = \frac{W/\mathrm{BW}}{P/\mathrm{TP}} \tag{6.1}$$

where W denotes the number of pixels of the bore well locations for each conditioning factor; BW denotes the number of total bore well pixels in the study area; P represents the number of pixels of each class of the conditioning factor; TP denotes the number of total pixels in the study area. In FR calculation, the FR model value is obtained from the thematic layers of the conditioning factors that determine the groundwater potentiality (Balamurugan et al., 2017).

6.3.3.2 Weights of evidence (WOE)

WOE model is a widely accepted method to calculate the groundwater potentiality (Lee et al., 2018; Kordestani et al., 2019; Boughariou et al., 2021; Maity et al., 2022). Based on relative weightage of the conditioning factors, WOE model is calculated on a study area. WOE can be expressed as follows (Eqs. 6.2 and 6.3):

$$W^{+} = \mathrm{Ln}\frac{P\left(\frac{B}{A}\right)}{P\left(\frac{B'}{A'}\right)} \tag{6.2}$$

$$W^{-} = \mathrm{Ln}\ \frac{P\left(\frac{B}{A}\right)}{P\left(\frac{B'}{A'}\right)} \tag{6.3}$$

where W^{+} and W^{-} denotes the positive and negative weights of the conditioning factors; P is the probability, B stands for conditioning factor, stands for the absence of conditioning factors, A stands for bore well data, and stands for the absence of borewell data.

Then, the standard deviation of the contrast value can be calculated as follows (Eq. 6.4):

$$\mathrm{SD}_c = \sqrt{((S^2(W^{+}) + S^2(W^{-})))} \tag{6.4}$$

where S^2 denotes the influencing weight of the conditioning factors (Zeinivand and Nejad, 2018).

The standardized contrast τ is considered as measure of confidence and can be calculated as follows (Eq. 6.5):

$$\tau = \left(\frac{C}{SD_c}\right) \tag{6.5}$$

After assigning all the conditioning parameters with the bore well data, the value of W^{+},W^{-}, SD_c, and τ has been calculated.

6.3.4 Delineation of the GRPZ through groundwater potentiality index (GPI)

6.3.4.1 Delineation of GRPZ using FR

In FR model, the weightage of the each class is not put based on the properties of the conditioning factor but given in the form of spatial occurrence of the wells in each class. Similarly, the FR is calculated for all the conditioning factors. Finally, the groundwater potential index (GPI) has been computed using Eq. (6.6):

$$GPI = Fr1 + Fr2...... + Frn \tag{6.6}$$

where FR is the final weight of the conditioning factors. Then, GRPZ is calculated based on FR-based GPI value of each conditioning factors as follows (Eq. 6.7):

$$GRPZ = \sum (E_{FR} + S_{FR} + L_{FR} + DD_{FR} + LULC_{FR} + G_{FR} + LD_{FR} + DR_{FR}) \tag{6.7}$$

where FR represents the weights of FR of the individual classes. *E* is the elevation, *S* denotes slope, *L* represents lithological condition of the study area, DD represents drainage density, LULC denotes land use the land cover of the study area, *G* indicates geomorphology of the study area, LD is the lineament density, and DR represents distance from the river.

6.3.4.2 Delineation of GRPZ using WOE

WOE is quite similar like FR; it is calculated and mapped according to the τ values using Eq. (6.8):

$$GPI = \tau 1 + \tau 2 .... + \tau n \tag{6.8}$$

where τ is the final weight that comes through the calculation of W^{+}, W^{-}, C, and SD_C. Then, the GRPZ is calculated according to this GPI value as follows (Eq. 6.9):

$$GRPZ = \sum (E\tau + S\tau + L\tau + DD|\tau + LULC\tau + G\tau + LD\tau + DR\tau) \tag{6.9}$$

where τ represents the final weights of the individual classes. *E* is the elevation, *S* denotes slope, *L* represents lithological condition of the study area, DD represents drainage density, LULC denotes land use land cover of the study area, *G* indicates geomorphology of the study area, LD is the lineament density, and DR represents the distance from the river.

6.3.5 Validation method

The FR and WOE models have been validated by receiver operating characteristics (ROCs) curve. The area under the ROC curve (AUC) defines the prediction or classification accuracy (BougHariou et al., 2021). In this study, we have run and validated two models based on their classification results of the GRPZ. AUC values is ranged from 0 to 1. If the value is given below 0.5, then it indicates that the result of the model is not suitable, and it needs to classify once again, and close to 1 implies that the result is well delineated (Mallick and Rudra, 2021).

6.4 Results and discussion

6.4.1 Groundwater recharge potentiality zone (GRPZ) analysis

After normalization of different fuzzy layers, researchers put the weightage to find out the suitable GRPZs for the plateau region using two geostatistical methods, that is, FR and WOE model.

6.4.1.1 Assessment of GRPZ using FR model

The groundwater potentiality zone estimation using the FR model is very significant. This model executed the GRPZ through the correlation between conditioning factors and the location of bore wells. Moreover, the higher correlation value indicated the more significant groundwater potential, while the lower correlation value specified low groundwater potential and vice versa. In this study, eight conditioning factors (Fig. 6.2) and 12 bore wells (WRIS) have been used to create GRPZ map. The results (Table 6.2) of the FR model, slope, and wells occurrence depicted that lower the slope percent (10-20) is indicated by the highest value of FR (4.11). In contrast, the absence of wells in higher the slope percent (2-10) area which is indicated the lowest FR (0.00) and reflects the low groundwater potential. Meanwhile, the higher the value of lineaments density (1.41-1.88) is displayed shows the high FR (5.04) value which indicates the greater chance to groundwater potential, while was found low FR value (0.0) for the lineaments class 0.94-1.41. For lithological conditions, the result shows that the alluvium plain and flood plain has the more remarkable ability to groundwater potential with FR value (6.83). In the case of DD, moderate DD has highest FR value (1.88) followed by low DD. So that, less DD has more water holding capacity. The geomorphological characteristics are also significant for delineating GRPZ. According to the FR model, the alluvium plain (FR = 6.83) has more groundwater recharge potentiality than other geomorphological settings found in this region. Finally, the impact of urbanization on groundwater potentiality is highly influenced by LULC classes. In this study, the water body is contributed a high potential ability followed by agriculture and vegetation cover. At the same time, mining/industrial area and built-up area have found insignificant FR values because FR is analyzed using conditioning factor and bore wells data. Due to these classes' lack of bore wells data, the result indicated the low FR value while vegetation cover has always greatly influenced the infiltration rate. A similar result is found by Balamurugan et al. (2017) .The GRPZ map (Fig. 6.4) obtained using the FR model specifies four potential groundwater zones. Highly GRPZ covering 26.02% of the total area of Jhargram along the Subarnarekha river bank (south-eastern part of the study area), low recharge potential zone (31.80%) is found in the hard rock areas and very low recharge potential zone (29.10%) is found in the built-up and industrial areas of Jhargram (Table 6.2).

6.4.1.2 Assessment of GRPZ using WOE model

The groundwater potentiality zone estimation using the WOE model is also helpful for this study. According to WOE (τ) values, it is noted that for the lineament, the highest τ value (11.88) is given which indicates a high groundwater potentiality (Table 6.2). Then, low elevated areas have a high τ value (6.42). Geomorphology and lithology thematic maps are given the next level of τ value of 10.36 and 8.01, respectively. Similarly, the LULC (specifically vegetation and agricultural land) has been given a positive standard value. For DEM, the highest weight of τ is given to the lowest altitude in the Subarnarekha riverbank area and slope also influences significantly ($\tau = 8.76$). The lowest weight of DD also has a higher value of τ about 4.31, while the highest value of this parameter is detected in the highest class of lineament. Similar result is found by Boughariou et al. (2021). The quintile method was also adopted for the WOE model to classify

Table 6.2 Spatial relationship between wells locations and conditioning factors using FR and WOE models.

Factors	Class	No. of pixel in a class	% of pixel in a class	No. of pixel of wells	% of pixel of wells	FR	C	S^2 (W^+)	S^2 (W^-)	S (C)	τ
Lineament density	Very high	636	3.41	0	0	0	−0.04	0	0.01	0.10	−0.35
	High	2280	12.21	8	61.54	5.04	2.44	0.016	0.03	0.21	11.88
	Moderate	4731	25.34	4	30.77	1.21	0.27	0.032	0.01	0.22	1.25
	Low	5479	29.34	1	7.69	0.26	−1.61	0.13	0.01	0.38	−4.28
	Very low	5546	29.7	0	0	0	−0.35	0	0.01	0.10	−3.52
DEM	Very high	1426	5.8	0	0	0	−0.06	0	0.01	0.10	−0.60
	High	7366	29.96	0	0	0	−0.36	0	0.01	0.10	−3.56
	Moderate	7005	28.49	5	38.46	1.35	0.45	0.026	0.02	0.21	2.19
	Low	7366	29.96	8	61.54	2.05	1.32	0.016	0.03	0.21	6.42
	Very low	1426	5.8	0	0	0	−0.06	0	0.01	0.10	−0.60
Geomorphology	Dissected hills and valleys	1887	9.95	0	0	0	−0.11	0	0.01	0.10	−1.05
	Dissected plateau	1	0.01	0	0	0	0.00	0	0.01	0.10	0.00
	Alluvial plain	854	4.5	4	30.77	6.83	2.24	0.032	0.01	0.22	10.36
	Pediment pediplain	14768	77.86	9	69.23	0.89	−0.45	0.014	0.03	0.22	−2.06
	Flood plain	728	3.84	0	0	0	−0.04	0	0.01	0.10	−0.39
	Water body	728	3.84	0	0	0	−0.04	0	0.01	0.10	−0.39
	Quarry and mine dump	1	0.01	0	0	0	0.00	0	0.01	0.10	0.00
Lithology	Newer alluvium	1370	7.27	4	30.77	4.23	1.74	0.032	0.01	0.22	8.01
	Older alluvium	8746	46.4	3	23.08	0.5	−1.06	0.043	0.01	0.24	−4.46
	Singhbhum	3854	20.44	0	0	0	−0.23	0	0.01	0.10	−2.29
	Dalma	327	1.73	0	0	0	−0.02	0	0.01	0.10	−0.18
	Chhotanagpur gneiss	4093	21.71	4	30.77	1.42	0.47	0.032	0.01	0.22	2.18
	Unclassified metamorphics	460	2.44	2	15.38	6.3	1.98	0.065	0.01	0.28	7.16
	Dhanjori	1	0.01	0	0	0	0.00	0	0.01	0.10	0.00

(continued on next page)

Table 6.2 Spatial relationship between wells locations and conditioning factors using FR and WOE models—cont'd

Factors	Class	No. of pixel in a class	% of pixel in a class	No. of pixel of wells	% of pixel of wells	FR	C	S^2 (W^+)	S^2 (W^-)	S (C)	τ
Drainage density	Very high	1274	6.25	0	0	0	−0.07	0	0.01	0.10	−0.65
	High	2760	13.55	0	0	0	−0.15	0	0.01	0.10	−1.46
	Moderate	4174	20.49	5	38.46	1.88	0.89	0.026	0.02	0.21	4.31
	Low	5261	25.82	3	23.08	0.89	−0.15	0.043	0.01	0.24	−0.63
	Very low	6905	33.89	5	38.46	1.13	0.20	0.026	0.02	0.21	0.96
Slope	Very high	1658177	2.63	0	0	0	−0.03	0	0.01	0.10	−0.27
	High	5902083	9.36	5	38.46	4.11	1.80	0.026	0.02	0.21	8.76
	Moderate	12443763	19.74	3	23.08	1.17	0.20	0.043	0.01	0.24	0.84
	Low	21198807	33.63	5	38.46	1.14	0.21	0.026	0.02	0.21	1.02
	Very low	21827560	34.63	0	0	0	−0.43	0	0.01	0.10	−4.25
Distance from river	Very high	419	2.06	0	0	0	−0.02	0	0.01	0.10	−0.21
	High	1279	6.28	2	15.38	2.45	1.00	0.065	0.01	0.28	3.60
	Moderate	4660	22.87	4	30.77	1.35	0.41	0.032	0.01	0.22	1.87
	Low	6520	32	3	23.08	0.72	−0.45	0.043	0.01	0.24	−1.90
	Very low	7496	36.79	4	30.77	0.84	−0.27	0.032	0.01	0.22	−1.25
Land use land cover	Water body	2040637	2.06	0	0	0	−0.02	0	0.01	0.10	−0.21
	Vegetation cover	34422487	34.76	2	66.67	1.92	1.32	0.015	0.03	0.21	6.24
	Agricultural land	55495935	56.04	11	84.62	1.51	1.46	0.012	0.07	0.28	5.28
	Built-up	6625728	6.69	0	0	0	−0.07	0	0.01	0.10	−0.69
	Open land	442997	0.45	0	0	0	−0.01	0	0.01	0.10	−0.05

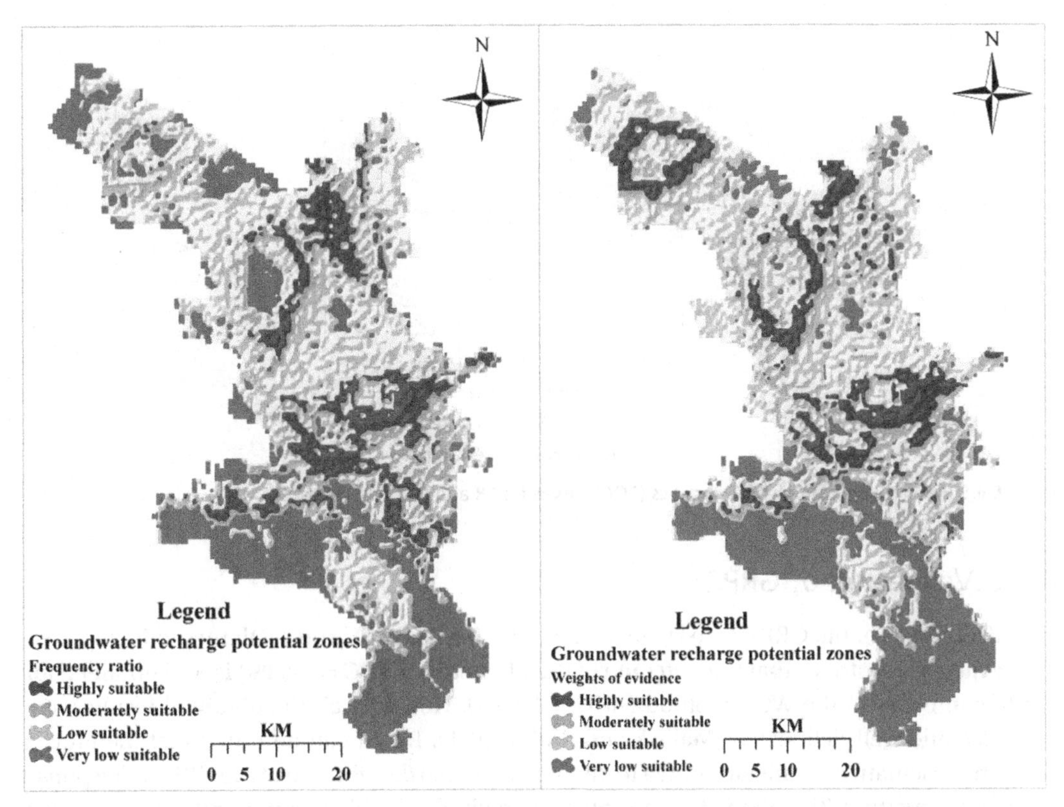

FIGURE 6.4 Groundwater recharge potential zone analysis using FR and WOE models.

Table 6.3 Groundwater recharge potential zonation derived from FR and WOE models.

GRPZ	WOE model		FR model	
	Area (km^2)	% of Area	Area (km^2)	% of area
Highly potential	745.97	24.55	790.54	26.02
Moderately potential	936.44	30.82	397.39	13.08
Low potential	589.99	19.42	966.15	31.80
Very low potential	765.60	25.20	883.92	29.10

the groundwater potential map into four classes: very low, low, moderate, and high (Fig. 6.4). According to the final GRPZ map, the low GRPZ is mostly found on the Subarnarekha river bank. The delineation of the WOE-based GRPZ map shows a similar surface cover for the high and moderate groundwater potential. The very low GRPZ class is covered a slightly higher area (25.20%) followed by low recharge potential zone (19.42%), moderate recharge potential zones (30.82%), and highly recharge potentiality (24.55%) (Table 6.3).

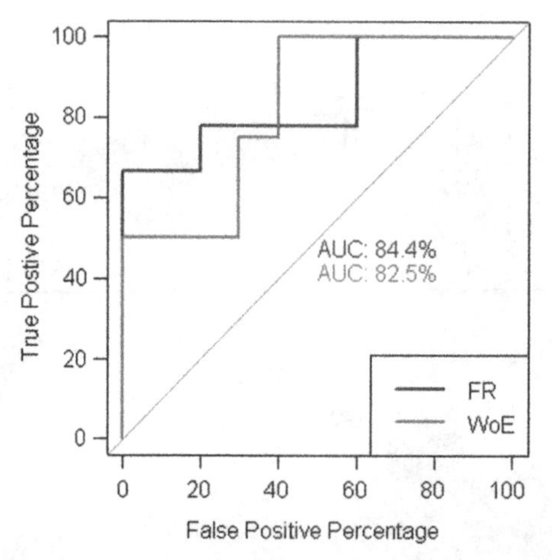

FIGURE 6.5 Receiver operating characteristics (ROC) curves for FR and WOE methods.

6.4.2 Validation of GRPZ

The ROC curves of the GRPZ maps were generated to validate FR and WOE methods (Fig. 6.5). The result demonstrates that the outcome of the FR method (AUC = 84.4%) is well obtained for GRPZ compared to the WOE method (AUC = 82.5%). However, all the obtained results were validated and well-delineated (Mallick and Rudra, 2021). But, it can be stated that the FR is a better representative for this study area to denote the spatial distribution of the GRPZ in response to the WOE method. Therefore, this geostatistical method is highly recommended for assessing the GRPZ for this study area.

6.5 Conclusion

In this study, FR and WOE were used for groundwater potential zone assessment based on geospatial techniques. Eight conditioning factors of thematic maps were used as indispensible input parameters to calculate the GRPZ. The GRPZ values of the two models were calculated based on GPI values. The FR and WOE were imposed the distribution of wells that helps to estimate the weights of each class of the conditioning factors. The output maps were classified into four zones, that is, very low potential, low potential, moderately potential, and highly potential zones. However, the maps using FR and WOE methods revealed better representation where 26.06% and 24.55% of high groundwater recharge potentiality includes the high potential area of 790.54 km^2 and 745.97 km^2, respectively. But, most of the areas are given as low recharge potential abilities. For these two adopted methods, the high groundwater potential area was basically located on the riverine area on the southern part of the Jhargram. Moreover, the areas under the ROC curve (AUC) were calculated and validated.

This study will manage the sustainable groundwater recharge potentiality including regional land use activities, well construction, and groundwater security. Accordingly, the application of the geostatistical techniques is depicted as a significant tool for decision-making process in groundwater exploitation and groundwater management in confined aquifer. Moreover, several advanced techniques might be incorporated to assess the groundwater resources especially in the drier region like Jhargram but, the outcome is well enough for determining the restoration and conservation process of groundwater.

Conflict of interest

The authors declare that there is no such conflict of interest regarding data and diagrams.

References

Al-Abadi, A.M., 2015. Groundwater potential mapping at northeastern Wasit and Missan governorates, Iraq using a data-driven weight of evidence (WOE) technique in framework of GIS. Environ. Earth Sci. 74 (2), 1109–1124.

Al-Abadi, A.M., Handhal, A.M., Al-Ginamy, M.A., 2019. Evaluating the Dibdibba aquifer productivity at the Karbalae Najaf plateau (Central Iraq) using GIS-based tree machine learning algorithms. Nat. Resour. Res. 29 (3), 1989–2009. https://doi.org/10.1007/s11053-019-09561-x.

Anbazhagan, S., Jothibasu, A., 2014. Geoinformatics in groundwater potential mapping and sustainable development: a case study from southern India. Hydrol. Sci. J. doi:10.1080/02626667.2014.990966.

Aryafar, A., Yousefi, S., Ardejani, F.D., 2013. The weight of interaction of mining activities: groundwater in environmental impact assessment using fuzzy analytical hierarchy process (FAHP). Environ. Earth Sci. 68 (8), 2313–2324.

Balamurugan, G., Seshan, K., Bera, S., 2017. Frequency ratio model for groundwater potential mapping and its sustainable management in cold desert, India. J. King Saud Univ.—Sci. 29, 333–347. http://dx.doi.org/10.1016/j.jksus.2016.08.003.

Boughariou, E., Allouche, N., Brahim, F.B., Nasri, G., & Bouri, S., (2021). Delineation of groundwater potentials of Sfax region, Tunisia, using fuzzy analytical hierarchy process, frequency ratio, and weights of evidence models. Environ. Develop. Sustain. http://dx.doi.org/10.1007/s10668-021-01270-x.

Chen, W., Li, H., Hou, E., Wang, S., Wang, G., Panahi, M., Xiao, L., 2018. GIS-based groundwater potential analysis using novel ensemble weights-of-evidence with logistic regression and functional tree models. Sci. Total Environ. 634, 853–867.

Chen, W., Panahi, M., Khosravi, K., Pourghasemi, H.R., Rezaie, F., Parvinnezhad, D., 2019. Spatial prediction of groundwater potentiality using Anfis ensemble with teaching-learning-based and biogeography-based optimization. J. Hydrol. 572, 435e448.

CGWB, 2014. Ground Water Year Book, 2013–14. Ministry of Water Resources, Govt. of India. Ministry of Water Resources, Govt. of India. http://cgwb.gov.in/Documents/Ground%20Water%20Year%20Book%202013-14.pdf.

Douglas, S.H., Dixon, B., Griffin, D., 2018. Assessing the abilities of intrinsic and specific vulnerability models to indicate groundwater vulnerability to groups of similar pesticides: a comparative study. Phys. Geog. 39 (6), 487–505.

Elmahdy, S.I., Mohamed, M.M., 2015. Probabilistic frequency ratio model for groundwater potential mapping in Al Jaww plain, UAE. Arab. J. Geosci. 8 (4), 2405–2416.

Eraslan, G., Avsec, Z., Gagneur, J., Theis, F.J., 2019. Deep learning: new computational modelling techniques for genomics. Nat. Rev. Genet. 20 (7), 389e403.

Evans, B.M., Myers, W.L., 1990. A GIS-based approach to evaluating regional groundwater pollution potential with drastic. J. Soil Water Conserv. 45 (2), 242–245.

Freeze, R.A., Cherry, J.A., 1979. Groundwater. Prentice-Hall, Englewood Cliffs, NJ, p. 604.

Gayen, A., Haque, S.M., Saha, S., 2020. Modeling of gully erosion based on random forest using GIS and R. In: Gully Erosion Studies from India and Surrounding Regions. Springer, Cham, p. 35e44.

Ghezelbash, R., Maghsoudi, A., Carranza, E.J.M., 2019. Performance evaluation of RBF-and SVM-based machine learning algorithms for predictive mineral prospectivity modeling: integration of SA multifractal model and mineralization controls. Earth Sci. Inform. 12 (3), 277–293. https://doi.org/10.1007/s12145-018-00377-6.

Gueretz, J.S., Da Silva, F.A., Simionatto, E.L., Ferard, J.F., Radetski, C.M., Somensi, C.A., 2019. A multi-parametric study of the interaction between the parati river and babitonga bay in terms of water quality. J. Environ. Sci. Health, Part B 55 (3), 257–264. https://doi.org/10.1080/03601234.2019.1685813.

Hammouri, N., El-Naqa, A., Barakat, M., 2012. An integrated approach to groundwater exploration using remote sensing and geographic information system. J. Water Resour. Prot. 4, 717e724. https://doi.org/10.4236/jwarp.2012.49081.

Jaiswal, R., Mukherjee, S., Krishnamurthy, J., Saxena, R., 2003. Role of remote sensing and GIS techniques for generation of groundwater prospect zones towards rural development-an approach. Int. J. Remote Sens. 24 (5), 993e1008.

Kordestani, M.D., Naghibi, S.A., Hashemi, H., Ahmadi, K., Kalantar, B., Pradhan, B., 2019. Groundwater potential mapping using a novel data-mining ensemble model. Hydrogeol. J. 27 (1), 211e224.

Kuria, D.N., Gachari, M.K., Macharia, M.W., Mungai, E., 2012. Mapping groundwater potential in Kitui district, Kenya using geospatial technologies. Int. J. Water Resour. Environ. Eng. 4 (1), 15–22.

Lee, S., Hong, S.M., Jung, H.S., 2018. GIS-based groundwater potential mapping using artificial neural network and support vector machine models: the case of Boryeong city in Korea. Geocarto. Int. 33 (8), 847e861.

Maity, B., Mallick, S.K., Das, P., Rudra, S., 2022. Comparative analysis of groundwater potentiality zone using fuzzy AHP, frequency ratio and Bayesian weights of evidence methods. App. Water Sci. 12, 63. https://doi.org/10.1007/s13201-022-01591-w.

Mallick, S.K., Rudra, S., 2021. Analysis of groundwater potentiality zones of Siliguri urban agglomeration using GIS-Based fuzzy-AHP approach. In: Shit, P.K., et al. (Eds.), Groundwater and Society. Springer, Cham., pp. 141–160.

Miraki, S., Zanganeh, S.H., Chapi, K., Singh, V.P., Shirzadi, A., Shahabi, H., Pham, B.T., 2019. Mapping groundwater potential using a novel hybrid intelligence approach. Water Resour. Manage. 33 (1), 281e302.

Murthy, K.S.R., Mamo, A.G., 2009. Multi-criteria decision evaluation in groundwater zones identification in moyale-teltele subbasin, South Ethiopia. Int. J. Remote Sens. 30, 2729–2740.

Naghibi, S.A., Pourghasemi, H.R., Dixon, B., 2016. Groundwater spring potential using boosted regression tree, classification and regression tree, and random forest machine learning models in Iran. Environ. Monit. Assess. 188, 44. http://dx.doi.org/10.1007/s10661-015-50496.

Naghibi, S.A., Pourghasemi, H.R., Pourtaghi, Z.S., Rezaei, A., 2015. Groundwater qanat potential mapping using frequency ratio and Shannon's entropy models in the moghan watershed, Iran. Earth Sci. Inf. 1 (8), 171–186.

Oh, H.J., Kim, Y.S., Choi, J.K., Park, E., Lee, S., 2011. GIS mapping of regional probabilistic groundwater potential in the area of Pohang city, Korea. J. Hydrol. 399 (3–4), 158–172.

Ozdemir, A., 2011. GIS-based groundwater spring potential mapping in the Sultan mountains (Konya, Turkey) using frequency ratio, weights of evidence and logistic regression methods and their comparison. J. Hydrol. 411 (3–4), 290–308.

Pourghasemi, H.R., Yousefi, S., Kornejady, A., Cerd, A., 2017. Performance assessment of individual and ensemble data-mining techniques for gully erosion modeling. Sci. Total Environ. 609, 764e775.

Pourtaghi, Z.S., Pourghasemi, H.R., 2014. GIS-based groundwater spring potential assessment and mapping in the Birjand Township, southern Khorasan province, Iran. Hydrogeology 22, 643–662. http://dx.doi.org/10.1007/s10040-013-1089-6.

Rahmati, O., Nazari Samani, A., Mahdavi, M., Pourghasemi, H.R., Zeinivand, H., 2014. Groundwater potential mapping at Kurdistan region of Iran using analytic hierarchy process and GIS, Arab. J. Geosci. doi:10.1007/s12517-014-1668-4.

Rahmati, O., Pourghasemi, H.R., Melesse, A., 2016. Application of GIS-based data driven random forest and maximum entropy models for groundwater potential mapping: a case study at Mehran region, Iran. Catena 137, 360–372. http://dx.doi.org/10.1016/j.catena.2015.10.010.

Rahmati, O., Samani, A.N., Mahdavi, M., Pourghasemi, H.R., Zeinivand, H., 2015. Groundwater potential mapping at Kurdistan region of Iran using analytic hierarchy process and GIS. Arab. J. Geosci. 8 (9), 7059–7071.

Rai, B., Tiwari, A., Dubey, V.S., 2005. Identification of groundwater prospective zones by using remote sensing and geo-electrical methods in Jharia and Raniganj coalfields, Dhanbad district, Jharkhand state. J. Earth Syst. Sci. 114 (5), 515–522.

Razandi, Y., Pourghasemi, H.R., Samani Neisani, N., Rahmati, O., 2015. Application of analytical hierarchy process, frequency ratio, and certainty factor models for groundwater potential mapping using GIS. Earth Sci. Inform. 8 (4), 867–883. http://dx.doi.org/10.1007/s12145-015-0220-8.

Sameen, M.I., Pradhan, B., Lee, S., 2019. Self-learning random forests model for mapping groundwater yield in data-scarce areas. Nat. Resour. Res. 28 (3), 757e775.

Saraf, A.K., Choudhury, P.R., 1998. Integrated remote sensing and GIS for groundwater exploration and identification of artificial recharge sites. Int. J. Remote Sens. 19 (10), 1825–1841.

Şener, E., Şener, Ş., Davraz, A., 2018. Groundwater potential mapping by combining fuzzy analytic hierarchy process and GIS in Beyşehir lake Basin, Turkey. Arab. J. Geosci. 11, 1–21.

Shekhar, S., Pandey, A.C., 2015. Delineation of groundwater potential zone in hard rock terrain of India using remote sensing, geographical information system (GIS) and analytic hierarchy process (AHP) techniques. Geocarto. Int. 30 (4), 402–421.

Singh, P.K., Singh, U.C., 2009. Water resource evaluation and management for Morar river basin, Gwalior district, Madhya Pradesh, using GIS e-journal. Earth Sci. Ind. 2 (III), 174–186.

Thapa, R., Gupta, S., Guin, S., Kaur, H., 2017. Assessment of groundwater potential zones using multi-influencing factor (MIF) and GIS: a case study from Birbhum district, West Bengal. Appl. Water Sci. 7 (7), 4117e4131. https://doi.org/10.1007/s13201-017-0571-z.

Wang, J., He, J., Chen, H., 2012. Assessment of groundwater contamination risk using hazard quantification, a modified drastic model and groundwater value, Beijing plain, China. Sci. Total Environ. 432, 216e226.

Yan, E., Milewski, A., Sultan, M., Abdeldayem, A., Soliman, F., Gelil, K.A., 2010. Remote-sensing based approach to improve regional estimation of renewable water resources for sustainable development. In: US-Egypt Workshop On Space Technology and Geoinformation for Sustainable Development. US-Egypt Workshop on Space Technology and Geo-information for Sustainable Development, Cairo, Egypt, pp. 1–7.

Yeh, H.-F., Cheng, Y.-S., Lin, H.-I., Lee, C.-H., 2016. Mapping groundwater recharge potential zone using a GIS approach in Hualian river, Taiwan. Sustain. Environ. Res. 26 (1), 33e43.

Zeinivand, H., Ghorbani Nejad, S., 2018. Application of GIS-based data-driven models for groundwater potential mapping in Kuhdasht region of Iran. Geocarto. Int. 33 (6), 651-666.

Delineation of groundwater potential zones in the hard rock terrain of an extended part of Chhotanagpur plateau applying frequency ratio (FR) model

Arijit Ghosh and Biswajit Bera

DEPARTMENT OF GEOGRAPHY, SIDHO-KANHO-BIRSHA UNIVERSITY, PURULIA, INDIA

7.1 Introduction

Groundwater is a precious natural resource for its multipurpose uses such as drinking, agricultural, and industrial purposes (Andualem and Demeke, 2019). Groundwater is formed by the melting of snow or rain that percolates through the porous media such as soil and structural imprints (e.g., fracture, bedding, lineation, and joint) of rocks (Banks and Robins, 2002; Bera, 2008, 2010b). In recent years, the population is gradually increasing from a regional to a global level and hence, the demand for drinking and industrial water is also significantly amplifying (Oh et al., 2011). In the last few decades, the crisis of freshwater has been extensively augmented particularly in tropical regions of the world. In India, the principal source of drinking water is groundwater. Today, the groundwater crisis is a burning issue in India because of rapid population growth, speedy urbanization, irregularity of monsoonal rainfall, and overexploitation of groundwater (Chatterjee et al., 2020). The western part of Bengal is under the semiarid climatic region and extended part of the Chhotanagpur plateau which is facing recurrent drought hazards along with severe water crisis mostly during premonsoonal months (Bera et al., 2021; Bera and Bandyopadhyay, 2017; Chakraborty et al., 2021b; Saha, 2015; Sahoo et al., 2021).

The remote sensing technique is well organized to handling different features of the earth in dissimilar electromagnetic radiation (Machiwal et al., 2011) while the geographical information system (GIS) technique has an important role in the analysis of multithematic layers to delineate potential and prospect zones of groundwater (Carver, 1991). Groundwater availability in any region depends on several hydro-geomorphological factors like lineament density, land

Case Studies in Geospatial Applications to Groundwater Resources. DOI: https://doi.org/10.1016/B978-0-323-99963-2.00020-1

use/land cover (LULC) pattern, slope, drainage density, etc. These parameters are considered to delineate the groundwater potential zones of any region, particularly in hard rock terrain (Todd, 2005). Similarly, there are several models to delineate groundwater potential zones such as the frequency ratio (FR) model (Guru et al., 2017; Moghaddam et al., 2015; Nampak et al., 2014; Oh et al., 2011; Shaban et al., 2006), multicriteria analysis and analytic hierarchy process (Abijith et al., 2020; Andualem and Demeke, 2019; Nag and Kundu, 2018; Razandi et al., 2015; Thapa et al., 2017), logistic regression technique (Ozdemir, 2011; Rizeei et al., 2019), novel data mining ensemble model (Kordestani et al., 2019), weights-of evidence model (Chen et al., 2018; Tahmassebipoor et al., 2015), boosted regression tree (Choubin et al., 2019; S. Naghibi et al., 2015), random forest model (Zabihi et al., 2016), evidential belief functional model (Pourghasemi and Beheshtirad, 2015). These statistical models have been applied to demarcate groundwater potential zones in different regions worldwide. The FR model has high precision and accuracy to detect the groundwater potential and prospects zones in arid and semiarid regions (Arshad et al., 2020; Elmahdy and Mohamed, 2015; Guru et al., 2017; Manap et al., 2014; Razandi et al., 2015).

Some previous studies have been also applied to identify the groundwater potential and prospects zone using remote sensing and GIS techniques like the multi-influencing factor model, an analytic hierarchy process model in the western part of West Bengal (Acharya and Nag, 2013; Ghosh et al., 2016; Nag, 2016; Nag and Kundu, 2018; Pal et al., 2020). Although, all the above-mentioned techniques have been applied in this region but FR model has yet not been applied. This is the main research gap in this present context and this model has also been applied to comprehend the dynamic status of the groundwater in this semiarid region. The major priority of the population of this region is the fulfillment of water needs in different sectors. According to the 2011 census, the total population of this district (Purulia) is 29,30,115 and the groundwater of this region is highly vulnerable due to fluoride contamination (Bera et al.,2021; Chakraborty et al., 2021a). The people of this region will be benefitted if it is possible to accurately identify the dynamic groundwater status (Moench, 2003). The main objective of this study is to investigate the groundwater potential zones using the FR model.

7.2 Methods and materials

7.2.1 Study area

Purulia is the westernmost district of West Bengal (India). It is situated in the extended part of the Eastern province of Chhotanagpur plateau. It is located between 22°42′35 N to 23° 42′00″ N and 85°49′25″ E to 86°54′37″ E. According to census 2011, about 29,30,115 human habitants live within a 6259 km^2 area. This district is bounded by the Jharkhand state of India in the West, Northwest and Southwest, Bankura district in the east, Paschim Bardhaman district in the Northeast, and Jhargram district in the southeast (Fig. 7.1). Out of the total population of the Purulia district, around 87.26% populations reside in the rural pocket and about 12.74% populations live in the urban area. This region lies in an arid hot climate with an annual average

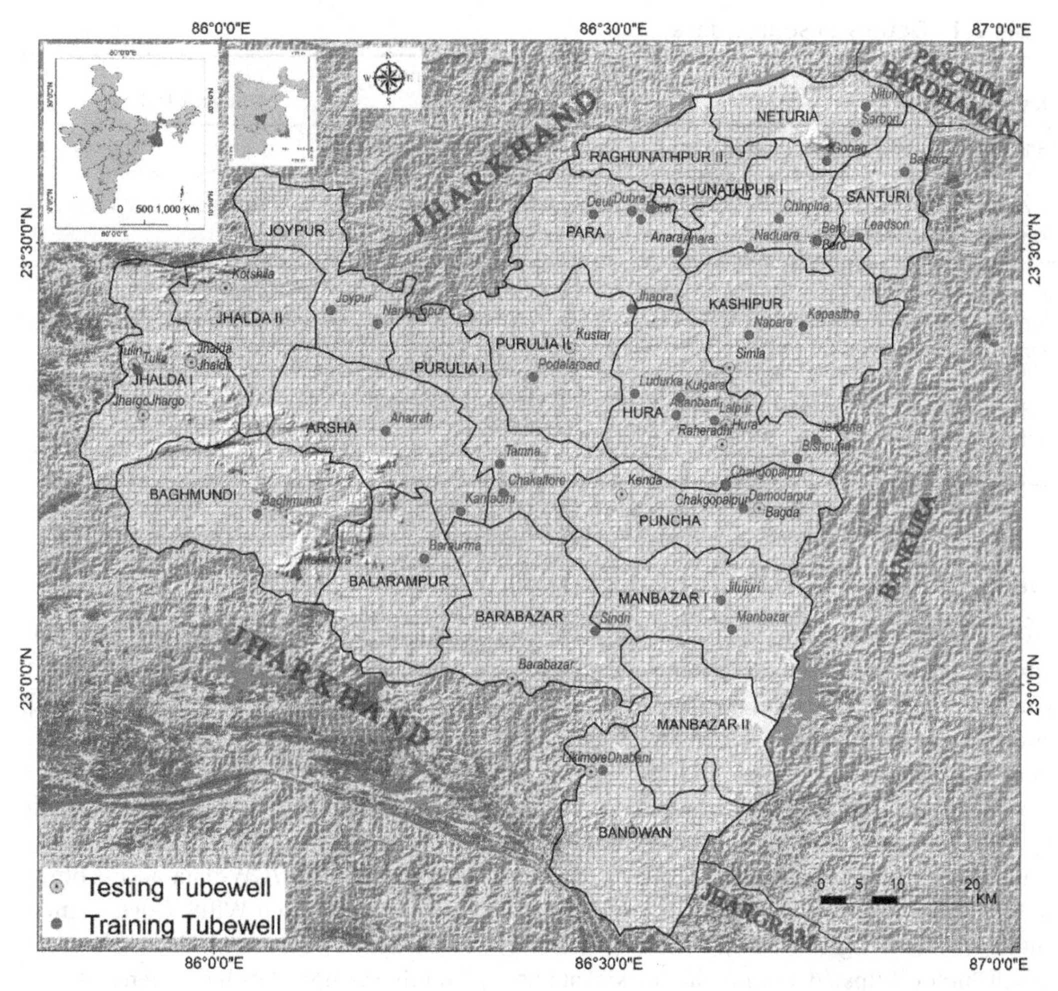

FIGURE 7.1 Study area.

rainfall of 1280 mm. This Precambrian landscape is drained by numerous nonperennial rivers and their tributaries such as Subarnarekha, Kangsabati, Kumari, Darakeshar, Silabati, etc. The significant rivers Damodar (in the northern flank) and Subarnarekha (in the southern flank) are flowing as state boundaries. This extended part of the Chhotanagpur plateau is mostly composed of Precambrian granite-gneiss and pegmatite. Various first- and second-generation structural units like bedding, thrust, fault, joint, cleavage, and schistosity have been exposed on lithological diversity of this region. According to Central Ground Water Board (CGWB), groundwater mostly occurs in unconfined and semiconfined conditions. Previously, most of the rural people practice monsoonal paddy cultivation but in recent years, nonmonsoonal cultivation is being practiced through the huge extraction of groundwater or groundwater irrigation methods.

Table 7.1 Details of source data.

Data and sources	Resolution and scale of data	Thematic layers
Landsat 8 OLI Satellite Image Date of acquisition on December 15, 2020 www.glovis.usgs.gov	30 m and PAN 15 m	Land use and land cover Lineament density Hydro-geomorphology
Aster DEM dated December 9, 2011, www.earthexplorer.usgs.gov	30 m	Slope Drainage density Geology
Geology and geomorphology Geological Survey of India	1:200000	Geology Hydro-geomorphology Geomorphological unit
WIRS, Govt. of India and Public Health Engineering Department, Govt. of West Bengal,2019–2020	–	Groundwater fluctuation Rainfall distribution zone Well locations

7.2.2 Dataset and data acquisition

To formulate thematic layers for this study, various satellite imageries, hydro-geomorphic data and meteorological data have been collected from different consistent sources (Table 7.1). Landsat 8 OLI satellite imageries (www.glovis.usgs.gov) were used to prepare LULC map and extraction of lineament density. A geological map along with structures was collected from the Geological Survey of India (www.gsi.gov.in). Aster DEM with 30 m resolution (www.earthexplorer.usgs.gov) has been used to prepare the slope map and validation for the hydro-geomorphological map. Groundwater level data were collected from Central Groundwater Board, Ministry of Jalsakti, Govt. of India (http://cgwb.gov.in/GW-data-access.html). Rainfall data of different stations of Purulia district were obtained from WRIS, Govt. of India (https://indiawris.gov.in/wris). The data regarding soil were collected from the National Bureau of soil Survey (https://www.nbsslup.in/kolkata.html). Training sample data have been collected from Central Ground Water Board (https://indiawris.gov.in/wris).

7.2.3 Preparation of input database/factors influencing groundwater potential zone

Thematic layers of geology, hydro-geomorphology, lineament, lineament density, slope, LULC, groundwater fluctuation, and drainage density have been designed to process FR model using ArcGIS 10.8, PCI Geomatica and ERDAS Imagine 2015 software (Fig. 7.2). Kappa statistics have been applied for the measurement of the accuracy of LULC maps. Finally, area under curve (AUC) method has been used to validate the result of groundwater potential by FR model (in the Arc SDM Master tool in ArcGIS 10.8 Software) (Moghaddam et al., 2015).

Revealing to the literature, several scholars carefully chose the influences constructed on their knowledge and availability of data from the study area. Thematic layers of LULC, lineament

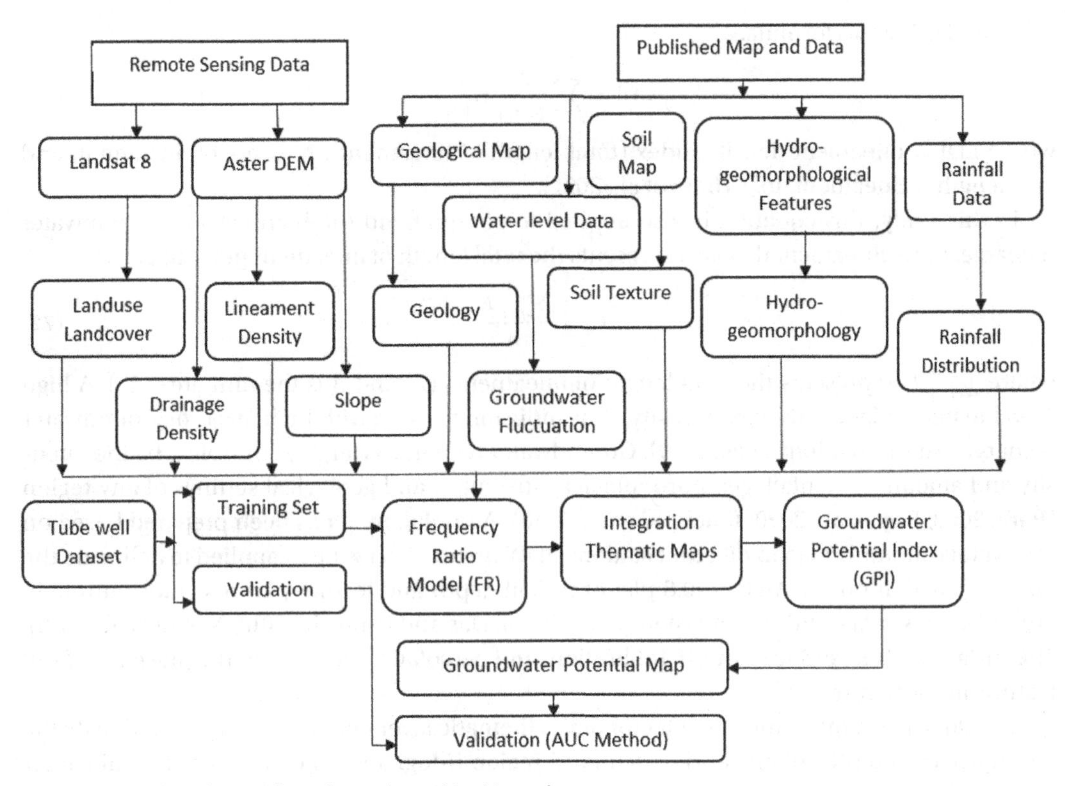

FIGURE 7.2 Methodological flow chart adopted in this study.

density (Ld), rainfall (Rf), drainage density (Dd), slope (S), soil texture, geology, hydrogeomorphology, and groundwater fluctuation were selected for this study. LULC is a crucial determining factor for the volume of surface runoff, amount of groundwater recharge, groundwater fluctuation, depletion of groundwater and groundwater storage and level (Elmahdy et al., 2020; Owuor et al., 2016; Prabhakar & Tiwari, 2015; Scanlon et al., 2005; Zhang et al., 2014). LULC map has been prepared using supervised classification methods with the help of ArcGIS 10.8 and ERDAS Imagine software into six categories (Chamling and Bera, 2020; Oberthur and Warnat, 2009). Kappa statistics method has been applied to validate or ground verification of LULC (Chamling and Bera, 2020; Duan et al., 2016; Esaid et al., 2018; Verma et al., 2020) with the help of 124 collected samples in different parts of our study area. Finally, the accuracy of the LULC map was 84.56%. Area of the different classes of LULC was also calculated using the field calculator tool in ArcGIS 10.8 platform. A lineament is a structural unit like fault, fracture, joint, etc., on images and it is also an expression of fundamental geological structures of surface and subsurface (Muthumaniraja et al., 2019). In the present study, lineament or horizontal and vertical curvature (Florinsky, 2016) has been extracted from Landsat 8 and Aster DEM using the line module tool (Bera and Ghosh, 2019) in PCI Geomatica Software. Finally, lineament density layers were prepared in ArcGIS 10.8 software (Zakir et al., 1999). Lineament density index (LDI) has been measured using this formula.

$$\text{LDI} = \sum \frac{N}{i=1} xi \frac{\text{km}}{\text{km}^2} \tag{7.1}$$

where LDI = lineament density index (total length of lineament), N = no. of lineament, and xi = length of lineament no. i. (Bera et al., 2021).

In this study, this equation is also applied to comprehend the high level of groundwater recharge. Here, lineament density represents the total length of lineament per unit area.

$$\text{L}_d = \frac{\sum_{i=1}^{i=n} L_i}{A} \tag{7.2}$$

where $\sum_{i=1}^{i=n} L_i$ represents the total length of lineaments (L) and A is the unit area (L^2). A high L_d value infers high secondary porosity, thus indicating a zone with high levels of groundwater recharge (Muthumaniraja et al., 2019). Groundwater recharge is largely controlled by the intensity and amount of rainfall, geomorphological attributes, and geological settings of any region (Bera, 2008; Dey et al., 2020; Kotchoni et al., 2018). A rainfall map has been prepared based on the average monthly data in different stations. IDW methods have been applied to delineate the rainfall zonation on the ArcGIS 10.8 platform. Soil depth and texture play a significant role in groundwater storage and transmission (Bera, 2010a; Das and Pardeshi, 2018; Mehra et al., 2016). The surface and subsurface run-off, infiltration, and percolation depend on the presence of soil texture and structure.

The slope is the most important factor of the thematic layer and it is applied to delineate the groundwater potential of any arid or semiarid region (Moghaddam et al., 2020; Senthilkumar et al., 2019). Therefore, a thematic layer of the degree of the slope has been prepared for this scientific study or investigation using ArcGIS 10.8 software. The degree of the slope has been extracted from the Aster DEM (www.earthexplorer.usgs.gov) for this study (Toutin, 2008). Z factor has been adjusted by using this formula (https://desktop.ArcGIS10.8.com).

$$\text{Adjusted } Z \text{ Factor} = (Z \text{ Factor}) + (\text{Pixel Size}) \text{ Pixel Size Power} \times (\text{Pixel Size Factor}) \tag{7.3}$$

Groundwater is largely depending on the nature of lithology as well as the structure and it acts as a geologic agent (Toth, 1999). The geological condition of any region is a key factor in analyzing the groundwater status most specifically groundwater potential mapping (Altafi Dadgar et al., 2017; Guru et al., 2017; Maity and Mandal, 2019; Rizeei et al., 2019). Therefore, geological features have been collected to prepare thematic layers to delineate groundwater potential zone using remote sensing and GIS technique (Arulbalaji et al., 2019; Choubin et al., 2019; Duraiswami and Patankar, 2011; Elmahdy and Mohamed, 2015; Oh et al., 2011; Pinto et al., 2017; Tahmassebipoor et al., 2015). As an example, unconsolidated sedimentary and fractured rock has a positive role in groundwater movement and storage (Guru et al., 2017). In this study, the geological factor is more important to assess groundwater-related issues. Here, geological structures have been digitized from the base map and the outcrop location has been mapped using GPS and assembled geospatial data layers to make the final geological map. Groundwater resources are correctly evaluated by the hydro-geomorphic condition of any area (Teixeira et al., 2013). Hydro-geomorphological map (Dinesan et al., 2015) is a very much necessary thematic layer for the estimation of groundwater resources (Das et al., 1997; Shah and Lone, 2019; Sikakwe, 2018; Thomas et al., 1999). Hydro-geomorphic features have been identified in the field and it is verified with the help of some published articles (www.gsi.gov.in). The magnitude of drainage density is defined as the ratio of the sum of lengths of streams to the size of the area of the grid under consideration (Adiat et al., 2012; Mogaji et al., 2015). The drainage density has been calculated using this formula.

$$\mathrm{Dd} = \sum_{i=1}^{i=n} \frac{Di}{A} \frac{\mathrm{Km}}{\mathrm{km}^2} \tag{7.4}$$

where ΣDi = sum of lengths of streams in an area i km and A = area of the grid (km^2) (Razandi et al., 2015). Groundwater fluctuation is a salient parameter to identify groundwater potential zone as it is the chief variable in the FR model (Bera, 2009; Das et al., 2019; Guru et al., 2017). Therefore, the groundwater fluctuation layer was prepared to delineate the potential zone of groundwater using the FR model. In this study, average data for the past 10 years (2010–2020) have been used and geocoding methods of geospatial analysis have been applied to prepare the perfect thematic layers for groundwater fluctuation. Training and testing samples among 56 tube wells have been considered for the verification and smooth model. Around 70% of total tube well data has been used as a training sample and 30% of data as a testing sample in this study. It is the standard percentage of training and testing sample to run this model, particularly for groundwater-related analysis (Guru et al., 2017). Finally, the area under the curvature method (Chen et al., 2018; Shasan et al., 2017; Tahmassebipoor et al., 2015) has been applied based on 30% of testing samples among the total tube-well dataset.

7.3 Frequency ratio model

The FR model is an effective bivariate statistical method that is determined the probability of groundwater potential of an area and shows the relationship between dependent and independent variables (Graeme, 1994; Oh et al., 2011; Shasan et al., 2017). The equation is following as:

$$\mathrm{FR} = \frac{W f\mathrm{TW}}{\mathrm{CP} f\mathrm{TP}} \tag{7.5}$$

where FR represents the frequency ratio of the class of factor, W denotes the number of pixels of tube well locations for each class of thematic layers; TW means the number of total pixels of a tube well in the study area. CP signifies the number of pixels in each thematic layer, and TP stands for the total number of pixels in the study area. FR value of each class was considered as the weightage of a particular class in thematic parameters to determine the potentiality of groundwater. FR value has been calculated from the formula based on the tube-well training dataset (Guru et al., 2017; Naghibi et al., 2015). After getting GWPZ, area under curvature method (AUC) has been applied to the verification of groundwater potential based on the testing tube well data and the final thematic layer of groundwater potential (Altafi Dadgar et al., 2017; Guru et al., 2017; Moghaddam et al., 2020; Pradhan, 2009).

7.4 Result and discussion

7.4.1 Land use/land cover

LULC has an impact on the development of groundwater in an area. In this study, LULCs thematic layer was classified into six categories using the supervised classification method (Oberthür and Warnat, 2009), namely, water body, forest area, built-up land, agricultural land, agricultural fallow, and other vegetation. Bagmundi, Nituria, and Bundwan block of Purulia district are vastly covered by forest compared to other blocks (Fig 7.3A). Among the total area 30.87% under forest cover, 22.98% area under agricultural land, 24% under agricultural fallow land, and 5.93% covered with waterbodies (Table 7.2).

Built-up-land has a low permeable surface that's why the infiltration rate is low. On other hand, the infiltration rate is high in the forest and agricultural land area. The water body is also allowed infiltration greatly. The relatively highland area of Ajodhya hill or southwestern and southeastern part of the study area is dominated by forest cover and also good groundwater potential than the other lands.

7.4.2 Lineament density

In this study, lineament density map was classified into five categories, namely, very low (<0.10 km/km^2), low (0.10–0.20 km/km^2), moderate (0.21–0.30 km/km^2), high (0.31–0.40 km/km^2), and very high (>0.40 km/km^2) (Fig. 7.3B). Lineament density is calculated in km/km^2 for this study and the highest lineament density was found in the southwestern part of the Purulia district. 37.72% (<0.10 km/km^2) area of this district is covered very low lineament density and 13.17% (0.30–0.40 km/km^2) area with high and very high (>40 km/km^2) respectively. Very low lineament density is found in the extreme western part of the study area.

Lineament controls the rate of recharge and storage of groundwater in any region (Aluko and Igwe, 2018). High lineament density indicates good secondary porosity but 37% area is dominated by the low lineament density of the study area (Table 7.2). Although, the south western part is a good recharge zone because high density of the lineaments than the other area of this region.

7.4.3 Rainfall

In this study, the rainfall zonation contour map was classified into four categories, namely, low (<1556.15 mm), moderate (1556.16–2386.47 mm), high (2386.48–3216.79 mm), and very high (>3216.80 mm) (Fig. 7.3C). Around 52.24% of the total area received moderate rainfall followed by 20.49% area under high (2386.48–3216.79 mm) and 21.68% area received low (<1556.15 mm) rainfall (Table 7.2).

Infiltration, as well as runoff of any area, is partially depending on the intensity and duration of rainfall. According to the spatial interpolation map, very high rainfall in the northwestern part and southeastern part of the study area indicates good infiltration. The probability of groundwater recharge is high in the south and south western parts of the region. On other hand, the

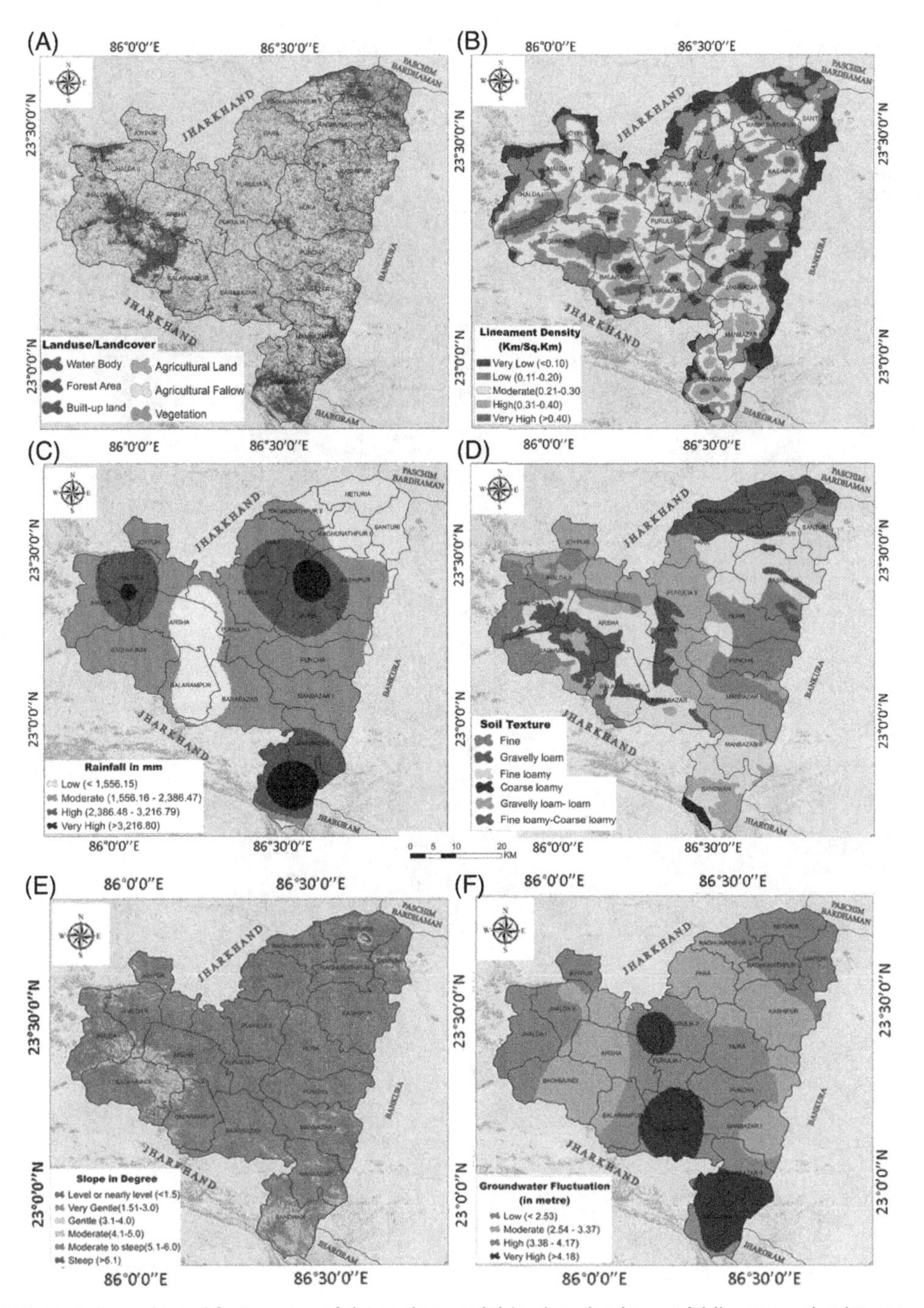

FIGURE 7.3 Hydrogeological factors map of the study area: (A) land use/land cover, (B) lineament density map, (C) Rainfall map, and (D) soil texture map of the study area, (E) slope map, (F) groundwater fluctuation, (G) geology and (H) drainage density map of the study area, (I) hydro-geomorphological map of the study area.

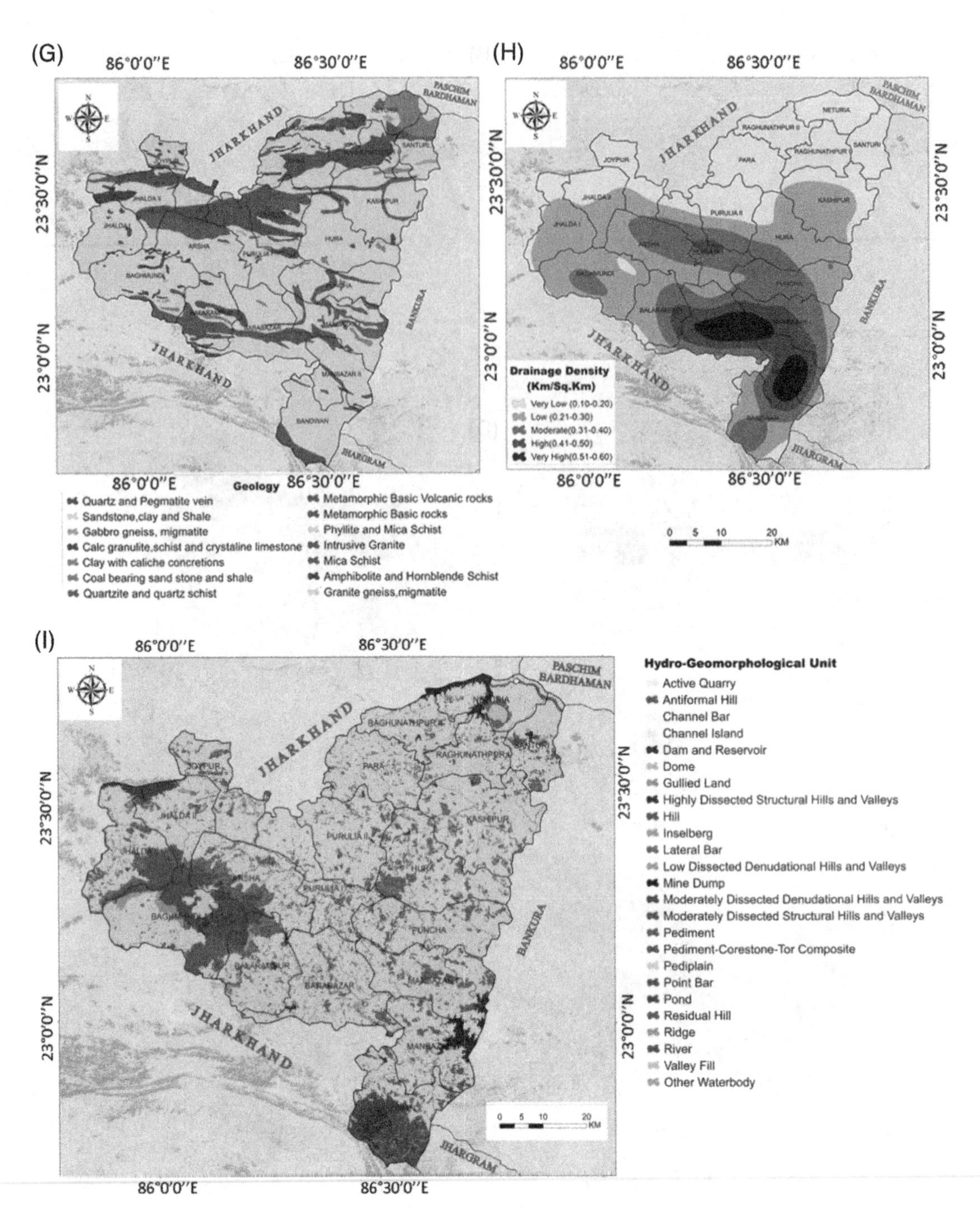

FIGURE 7.3, cont'd.

Table 7.2 Area and percentage of different hydrological factors of groundwater potential.

Factor	Subclass	Area (km^2)	Percentage (%)
Land use & land cover	Water body	370.83	5.93
	Forest area	1932.16	30.87
	Built-up land	452.87	7.23
	Agricultural land	1438.36	22.98
	Agricultural fallow	1539.66	24.61
	Other vegetation	524.76	8.38
Lineament density	Very low	2361.12	37.72
	Low	1853.83	29.62
	Moderate	1218.76	19.47
	High	518.51	8.28
	Very high	306.42	4.89
Rainfall	Low	1356.74	21.68
	Moderate	3269.43	52.24
	High	1282.24	20.49
	Very high	350.23	5.59
Soil texture	Coarse loam	33.89	0.54
	Gravelly loam	1709.5	27.31
	Fine loamy, Coarse loamy	814.91	13.02
	Fine	1315.19	21.02
	Gravelly loam	442.39	7.07
	Fine loamy	1942.76	31.04
Slope	Level or nearly level	5824.72	93.07
	Very gentle	291.8	4.66
	Gentle	89.05	1.42
	Moderate	42.32	0.68
	Moderate to steep	8.07	0.13
	Steep	2.68	0.04
Groundwater fluctuation	Low	571.92	9.14
	Moderate	2165.86	34.61
	High	2895.08	46.26
	Very high	625.78	9.99
Geology	Quartz and pegmatite vein	3.93	0.06
	Sandstone, clay, and shale	37.63	0.6
	Gabbro gneiss, migmatite	25.6	0.41
	Calc granulite, schist, and crystalline limestone	45.52	0.73
	Clay with caliche concretions	48.72	0.78
	Coal bearing sandstone and shale	132.54	2.12
	Quartzite and quartz schist	22.76	0.36
	Metamorphic basic volcanic rocks	51.76	0.83
	Metamorphic basic rocks	165.67	2.65
	Phyllite and mica schist	724.92	11.58
	Intrusive granites	823.24	13.15
	Mica schist	483.57	7.73
	Amphibolite and hornblende schist	74.16	1.18
	Granite gneiss, migmatite	3618.62	57.82

(continued on next page)

Table 7.2 Area and percentage of different hydrological factors of groundwater potential—cont'd

Factor	Subclass	Area (km²)	Percentage (%)
Drainage density	Very low	2052.76	32.8
	Low	478.65	7.65
	Moderate	1270.35	20.3
	High	246.09	3.93
	Very high	2210.79	35.32
Hydro-geomorphological unit	Residual hill	33.83	0.54
	Pediplain	4043.212	64.6
	Inselberg	6.66	0.11
	Pediment	786.56	12.56
	Other waterbody	35.96	0.57
	Moderately dissected structural hills and valleys	385.39	6.16
	Valley fill	533.43	8.52
	Highly dissected structural hills and valleys	60.41	0.95
	Ridge	14.17	0.23
	Pond	40.57	0.65
	Low dissected denudational hills and valleys	7.94	0.13
	Antiformal hill	36.68	0.58
	Moderately dissected denudational hills and valleys	31.41	0.5
	Gullied land	31.01	0.49
	Hill	3.5	0.05
	Point bar	5.35	0.07
	Channel bar	0.85	0.01
	Channel island	0.82	0.01
	Dam and reservoir	111.71	1.78
	River	68.08	1.09
	Lateral bar	3.72	0.06
	Dome	13.68	0.29
	Mine dump	0.008	0.003
	Pediment–corestone–tor composite	3.4	0.05
	Active quarry	0.29	0.001

northern part experienced low rainfall, hence groundwater recharge is low in this location. The rainfall pattern and intensity of this region has a significant role in groundwater storage, although other factors are responsible for the overall groundwater recharge scenario.

7.4.4 Soil texture

Soil texture map was classified into six categories, namely, coarse loamy, gravelly loam-loam, fine loamy-coarse loamy, fine, gravelly loam, and fine loamy. Fine loamy soil is covered greatly among 31.04% area of the study area followed by gravelly loam (27.31%) and fine soil (21.02%) (Table 7.2). Different parts of Kashipur, Para, Manbazar, Balarampur, Arsha, Hura, Nituria, Raghunathpur block out of 20 development blocks are covered by fine loamy soil (Fig. 7.3D). Sandy soil is responsible for good groundwater potential in this region.

7.4.5 Slope

Slope map also classified into six categories, namely, level or nearly level (<1.5°), very gentle (1.51°-3.0°), gentle (3.1°-4.0°), moderate (4.1°-5.0°), moderate to steep (5.1°-6.0°) and steep (>6.1°). In respect of the degree of slopes, 93.07% of the total area covers under level and nearly level, 4.66% area under very gentle, 0.04% area of this district under steep slope class (Table 7.2). Most of the steep slope is found in the southwestern part of the study area, mainly Ajodhya Hilly area in the Purulia district (Fig. 7.3E). The runoff of a specific area depends on the variation of elevation and slope (Bera, 2008; Bera et al., 2019; Grinevskii, 2014). Runoff is higher in south western part of the study area due to the steep slope.

7.4.6 Groundwater fluctuation

Groundwater fluctuation map was prepared based on the premonsoon, postmonsoon water level datasets, and arranged them the groundwater contour map designed for zonation of the water level fluctuation. Groundwater fluctuation zone was classified into four categories, namely, low (<2.53 m), moderate (2.54-3.37m), high (3.38-4.17m), and very high (>4.18m) (Fig. 7.3F). The highest (>4.8m) groundwater fluctuation that occurred in Brabazar, Bandwan, and Purulia I block indicates good groundwater potentiality and lowest (<2.53) groundwater fluctuation is found in the west and northeastern part of the study area which indicates poor groundwater potentiality. In total, 46.26% of areas of this district are under high groundwater fluctuation, and 9.99% area under very highest groundwater fluctuations with good groundwater potential (Table 7.2).

7.4.7 Geology

In respect of geological features, intrusive granite is mostly found in the middle portion (Hura, Kashipur, Arsha, Jhalda, Bagmundi, Barabazar block, etc.) and 13.15% area is followed by phyllite and mica Schist (11.58%) (Table 7.2). Maximum portion of Bandwan and Manbazar II block covers by phyllite and mica schist. The study area is situated under the Sighhum craton which is dominated by the Pre Cambrian metamorphic rocks and Gondwana deposits. Mica Schist and phyllite have more cracks than the granitic gneiss in this region. According to CGWB, Pre Cambrian granite, pegmatite, and some metasedimentary rocks like crystalline limestones, sillimanite schist is found in this region. Northeastern part of this district is covered by Gondwana rocks such as sandstone, shale, hornblende, etc. Despite Gondwana formation, some types of rocks of Panchet and Raniganj formations are present in different parts of the district (Fig. 7.3G). Unconsolidated sediments or sand, silt, clay are distributed in some areas which play a crucial role in groundwater storage.

7.4.8 Drainage density

Low drainage density is found in 32.8% area of the total study area followed by 35.32% of very high drainage density (Table 7.2). The south-eastern part of the Purulia district has the highest

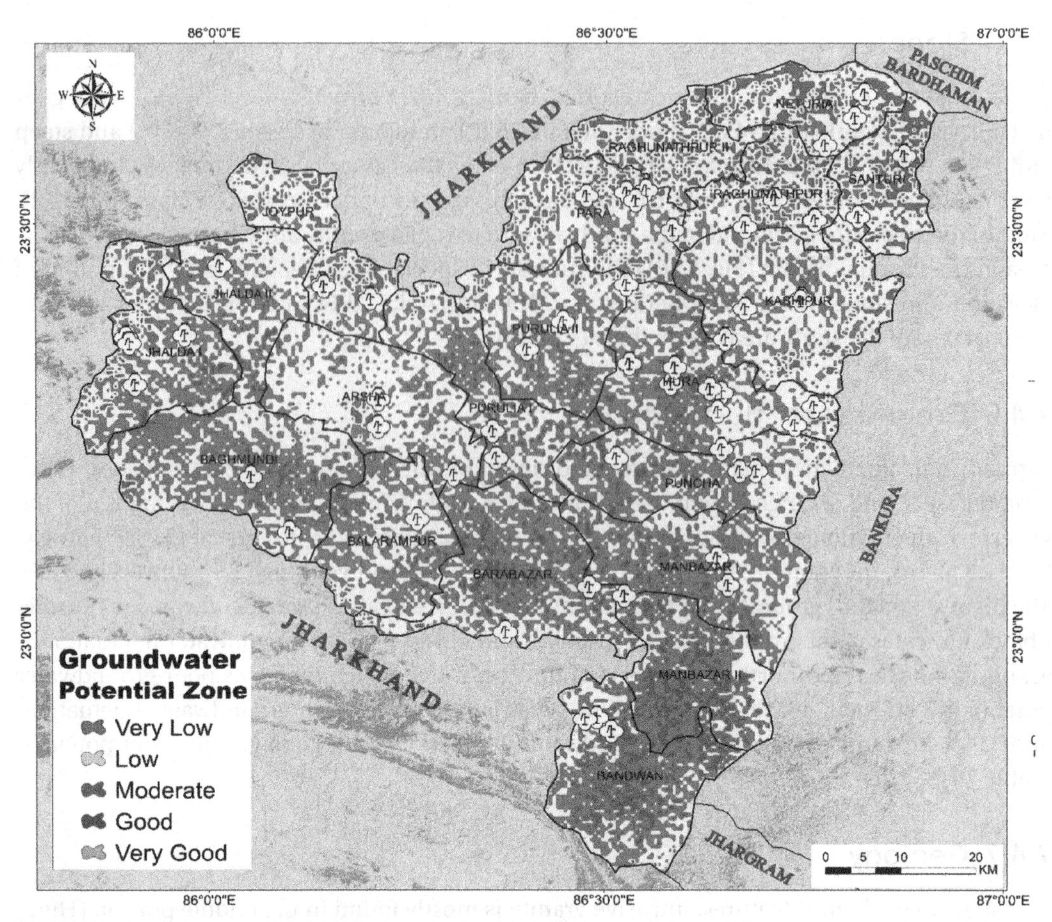

FIGURE 7.4 Groundwater potential zonation map based on frequency ratio model.

drainage density in respect of the northern portion. Manbazar I, Manbazar II, and Barabazar experienced the highest drainage density (Fig. 7.3H). Drainage density was calculated per km^2 by the raster statistics tool in ArcGIS 10.8 software and classified into five categories such as very low (0.10-0.20 km/km^2), low (0.21-0.30 km/km^2), moderate (0.31-0.40 km/km^2), high (0.41-0.50 km/km^2), and very high (0.51-0.60 km/km^2).

7.4.9 Hydro-geomorphology

Among the all hydro-geomorphic features 64.6% of the total area of the Purulia district is covered by pediplain and 12.56% of the total area is dominated by pediment. Pond, river, dam-reservoir, and other water body cover 0.65%, 1.09%, 1.78%, and 0.57%, respectively. In this study, most of the portion of the Purulia district except the extreme south and south-western part is covered by Pediplain (Fig. 7.3I).

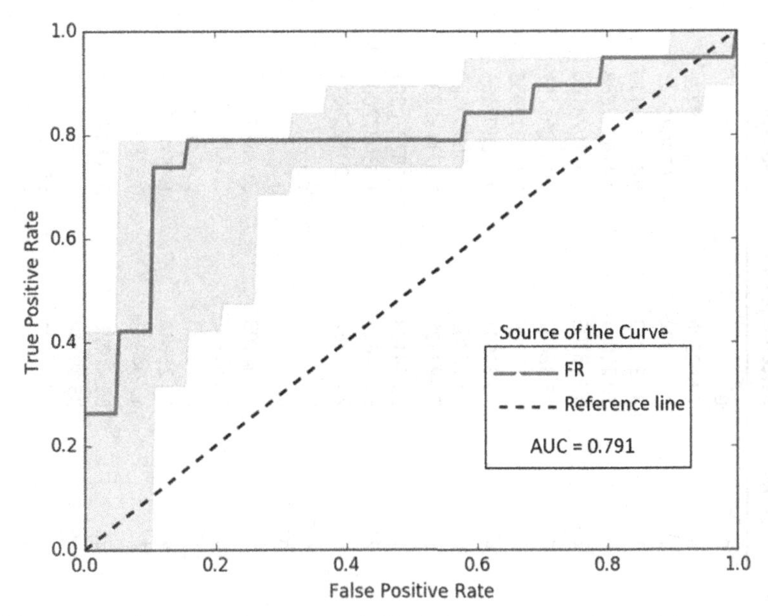

FIGURE 7.5 Validation result from groundwater potential index by frequency ratio model.

7.5 Groundwater potential zone

FR value of less than 1 indicates a high probability of groundwater accumulation (Chen et al., 2016). Five zones, namely, very low, low, moderate, good, very good potential zone of groundwater potential are delineated using FR Model-based GWPI. Based on the relationship among different factors (Table 7.3). In the south-western part of the study area mainly Arsha, Bagmundi, and Jhalda blocks are not suitable for groundwater storage and groundwater potential. Similarly, the southern portion of this district namely the Bandwan and Manbazar II block is much potential for groundwater due to some suitable factors according to GWPI values. In the northern part of this district, namely, Saturi and Nituria block is moderate to low groundwater potential. The middle portion of the district considers as a moderate groundwater potential (Fig. 7.4).

7.6 Validation of FR model

The area under curvature (AUC) method is used for the prediction of the accuracy of a model in addition to the value always ranging from 0.5 to 1.0. Value closed to 1.0 indicates the degree of accuracy is highest and close to 0.5 indicates the inaccuracy in the model (Fawcett, 2006). Validation of GWPI with the AUC model is important to know the accuracy of groundwater potentiality (Jothibasu and Anbazhagan, 2016). On this method, it can be divided into five classes, namely, 0.9–1.0 (excellent), 0.8–0.9 (very good) 0.7–0.8 (good), 0.6–0.7 (average), and 0.5–0.6 (poor) (Naghibi et al., 2015). If the value ranges from 7.0 to 0.8, the accuracy of potential mapping

Table 7.3 Frequency ratio values of the hydrological factors with respect to the groundwater occurrences.

Factor	Subclass	Total number of pixel		Groundwater occurrence of pixel		Frequency ratio
		Number	Percentage (%)	Number	Percentage (%)	
Land use & land cover	Water body	3,33,747	5.925089	0	0	0
	Forest area	1738944	30.87188	3	7.692308	0.249168726
	Built-up land	407583	7.235917	17	43.58974	6.024080041
	Agricultural land	1294524	22.98199	13	33.33333	1.450411116
	Agricultural fallow	1385694	24.60055	4	10.25641	0.416917888
	Other vegetation	472284	8.384569	2	5.128205	0.611624166
Lineament density	Very low	2125008	37.72577	6	15.38462	0.407801252
	Low	1668447	29.62033	20	51.28205	1.731312458
	Moderate	1096884	19.47324	4	10.25641	0.526692536
	High	466659	8.284707	6	15.38462	1.856989629
	Very high	275778	4.895952	3	7.692308	1.571156733
Rainfall	Low	1221066	21.67787	10	25.64103	1.182820207
	Moderate	2942487	52.23867	21	53.84615	1.030772007
	High	1154016	20.48752	8	20.51282	1.001235018
	Very high	315207	5.595944	0	0	0
Soil texture	Coarse loam	30501	0.541491	0	0	0
	Gravelly loam	1538550	27.31424	10	25.64103	0.938742022
	Fine loamy, Coarse loamy	733419	13.02056	6	15.38462	1.181563231
	Fine	1183671	21.01399	12	30.76923	1.464225994
	Gravelly loam	398151	7.068469	4	10.25641	1.451008827
	Fine loamy	1748484	31.04125	7	17.94872	0.57822152
Slope	Level or nearly level	5242248	93.06686	22	56.41026	0.606126109
	Very gentle	262620	4.662355	14	35.89744	7.699421803
	Gentle	80145	1.422833	2	5.128205	3.604221195
	Moderate	38088	0.676185	0	0	0
	Moderate to Steep	7263	0.128942	0	0	0
	Steep	2412	0.042821	1	2.564103	59.87983161

(continued on next page)

Table 7.3 Frequency ratio values of the hydrological factors with respect to the groundwater occurrences—cont'd

Factor	Subclass	Total number of pixel		Groundwater occurrence of pixel		Frequency ratio
		Number	Percentage (%)	Number	Percentage (%)	
Groundwater fluctuation	Low	514728	9.138088	1	2.564103	0.280595098
	Moderate	1949274	34.60592	23	58.97436	1.704169623
	High	2605572	46.25733	15	38.46154	0.831468986
	Very high	563202	9.998658	0	0	0
Geology	Quartz and pegmatite vein	3537	0.062793	0	0	0
	Sandstone, clay, and shale	33867	0.601249	0	0	0
	Gabbro gneiss, migmatite	23040	0.409035	0	0	0
	Calc granulite, schist, and crystalline limestone	40968	0.727315	0	0	0
	Clay with caliche concretions	43848	0.778444	1	2.564103	3.293882363
	Coal bearing sandstone and shale	119286	2.117712	3	7.692308	3.632366426
	Quartzite and quartz schist	20484	0.363657	0	0	0
	Metamorphic basic volcanic rocks	46584	0.827017	0	0	0
	Metamorphic basic rocks	149103	2.647061	1	2.564103	0.968660281
	Phyllite and mica schist	652428	11.58271	1	2.564103	0.221373322
	Intrusive granites	740916	13.15366	6	15.38462	1.169607517
	Mica schist	435213	7.726439	4	10.25641	1.327443379
	Amphibolite and hornblende schist	66744	1.184922	0	0	0
	Granite gneiss, migmatite	3256758	57.81799	23	58.97436	1.020000116
Drainage density	Very low	1847484	32.79882	15	38.46154	1.172650106
	Low	430785	7.647828	16	41.02564	5.364352198
	Moderate	1143315	20.29754	6	15.38462	0.757954652
	High	221481	3.932004	1	2.564103	0.652110808
	Very high	1989711	35.32381	1	2.564103	0.072588509

(continued on next page)

Hydro-geomorphological unit	Residual hill	30447	0.540533	0	0	0
	Pediplain	3638891	64.60209	31	79.48718	1.230411962
	Inselberg	5994	0.106413	0	0	0
	Pediment	707904	12.56759	4	10.25641	0.816100227
	Others water body	32364	0.574905	0	0	0
	Moderately dissected structural hills and valleys	346851	6.157728	0	0	0
	Valley Fill	480087	8.523098	2	5.128205	0.601683253
	Highly dissected structural hills and valleys	54369	0.965226	0	0	0
	Ridge	12753	0.226407	0	0	0
	Pond	36513	0.648224	0	0	0
	Low dissected denudational hills and valleys	7146	0.126865	0	0	0
	Antiformal hill	33012	0.58607	1	2.564103	4.37508039
	Moderately dissected denudational hills and valleys	28269	0.501866	0	0	0
	Gullied land	27909	0.495475	1	2.564103	5.175038656
	Hill	3150	0.055923	0	0	0
	Point bar	4815	0.085482	0	0	0
	Channel bar	765	0.013581	0	0	0
	Channel island	738	0.013102	0	0	0
	Dam and reservoir	100539	1.784893	0	0	0
	River	61272	1.087776	0	0	0
	Lateral bar	3348	0.059438	0	0	0
	Dome	12312	0.218578	0	0	0
	Mine dump	7.2	0.000128	0	0	0
	Pediment–core stone–tor composite	3060	0.054325	0	0	0
	Active quarry	261	0.004634	0	0	0

is equitably accurate (Moghaddam et al., 2015). In this study, validation is good because of AUC value is 0.791 in this study. AUC is calculated based on testing samples (30% of the total sample) in our study area. So this groundwater potential map using FR Model has a good prediction accuracy of 79.1% (Fig. 7.5).

7.7 Conclusion

This model-based study identifies groundwater potential pockets along with relevant factors which bring significant results. Many studies considered different models and successfully applied in this hard rock terrain but FR model gives very good accuracy to delineate potential pockets for this Precambrian granitic terrain. Groundwater potential is much significant to planning the water-related difficulties of any area. Several studies also assist different benefits of groundwater-related conservational planning and management. This study is also accommodating to future planners and investigators for district-level water resources management. Finally, based on the 30% of testing data, the overall accuracy of the groundwater potential zone is 79.1%. This region is situated in a semiarid climatic region and it is facing a severe water crisis during the summer months. So this study will definitely assist to develop proper water resource management and irrigation facility in this region. In the recent years, the land use has been significantly modified in this semiarid climatic region of India.

References

Abijith, D., Saravanan, S., Singh, L., Jennifer, J.J., Saranya, T., Parthasarathy, K.S.S., 2020. GIS-based multi-criteria analysis for identification of potential groundwater recharge zones - a case study from Ponnaniyaru watershed, Tamil Nadu, India. Hydro Res. 3, 1–14. https://doi.org/10.1016/j.hydres.2020.02.002.

Acharya, T., Nag, S., 2013. Study of groundwater prospects of the crystalline rocks in Purulia district, West Bengal, India using remote sensing data. Earth Res. 1, 54–59. https://doi.org/10.12966/er.07.03.2013.

Adiat, K.A.N., Nawawi, M.N.M., Abdullah, K., 2012. Assessing the accuracy of GIS-based elementary multi criteria decision analysis as a spatial prediction tool - a case of predicting potential zones of sustainable groundwater resources. J. Hydrol. 440–441, 75–89. https://doi.org/10.1016/j.jhydrol.2012.03.028.

Altafi Dadgar, M., Zeaieanfirouzabadi, P., Dashti, M., Porhemmat, R., 2017. Extracting of prospective groundwater potential zones using remote sensing data, GIS, and a probabilistic approach in Bojnourd basin, Ne of Iran. Arab. J. Geosci. 10 (5). https://doi.org/10.1007/s12517-017-2910-7.

Aluko, O.E., Igwe, O., 2018. Automated geological lineaments mapping for groundwater exploration in the basement complex terrain of akoko-edo area, edo-state Nigeria using remote sensing techniques. Modeling Earth Syst. Environ. 4 (4), 1527–1536. https://doi.org/10.1007/s40808-018-0511-4.

Andualem, T.G., Demeke, G.G., 2019. Groundwater potential assessment using GIS and remote sensing: a case study of Guna tana landscape, upper blue Nile basin, Ethiopia. J. Hydrol.: Regional Studies 24 (May), 100610. https://doi.org/10.1016/j.ejrh.2019.100610.

Arshad, A., Zhang, Z., Zhang, W., Dilawar, A., 2020. Mapping favorable groundwater potential recharge zones using a GIS-based analytical hierarchical process and probability frequency ratio model: a case study from an agro-urban region of Pakistan. Geosci. Front. 11 (5), 1805–1819. https://doi.org/10.1016/j.gsf.2019.12.013.

Arulbalaji, P., Padmalal, D., Sreelash, K., 2019. GIS and AHP techniques based delineation of groundwater potential zones: a case study from Southern Western Ghats, India. Sci. Rep. 9 (1), 1–17. https://doi.org/10.1038/s41598-019-38567-x.

Banks, D., Robins, N., 2002. An introduction to groundwater in crystalline bedrock. Norges Geologiske Undersøkelse 1–64, ISBN 8273861001.

Bera, B., 2008. Study of some hydrological parameters like rainfall, infiltration and runoff in the Rorachu and Ranikhola water divide regions in Sikkim Himalayas. https://www.researchgate.net/publication/327578736_Study_of_some_hydrological_parameters_like_rainfall_infiltration_and_runoff_in_the_Rorachu_and_Ranikhola_water_divide_regions_in_Sikkim_Himalayas. (Accessed date: 22.12.2020).

Bera, B., 2009. Application of GPS and ERS techniques for the mapping of out-crops vs soil cover and its thickness, in some mountainous part of the Sikkim Himalayas. https://www.researchgate.net/publication/327578573_Application_of_GPS_and_ERS_Techniques_for_the_Mapping_of_Out-Crops_vs_Soil_Cover_and_Its_thickness_In_the_Mountainous_part_of_the_Sikkim_Himalayas. (Accessed date: 22.12.2020).

Bera, B., 2010a. Ground water potential zones in the hardrock area, applying RS, GIS & ERS techniques, Gangtok, Sikkim Himalayas. https://www.researchgate.net/publication/326059495_Ground_Water_Potential_Zones_in_the_Hardrock_Area_Applying_RS_GIS_ERS_Techniques_Gangtok_Sikkim_Himalayas. (Accessed date: 22.12.2020).

Bera, B., 2010b. Jalobigyan (Text book of hydrology). https://www.researchgate.net/publication/327578595_Jalobigyan_Text_Book_of_Hydrology. (Accessed date: 22.12.2020).

Bera, B., Bhattacharjee, S., Chamling, M., Ghosh, A., Sengupta, N., 2021. Fluoride hazard and risk enumeration of hard rock unconfined aquifers in fluoride hazard and risk enumeration of hard rock unconfined aquifers in the extended part of Chhota Nagpur Gneissic complex. https://doi.org/10.1007/s12594-021-1651-0.

Bera, B., Bhattacharjee, S., Chamling, M., Ghosh, A., Sengutpa, N., Ghosh, S., 2021. High fluoride in groundwater and fluorosis related health hazard in rarh bengal, india: a socio-environmental study. Curr. Sci. 120, 1225–1233. https://doi.org/10.18520/cs/v120/i7/1225-1233.

Bera, B., Bhattacharjee, S., Ghosh, A., Ghosh, S., Chamling, M., 2019. Dynamic of channel potholes on precambrian geological sites of Chhota Nagpur plateau, Indian peninsula: applying fluvio-hydrological and geospatial techniques. SN Applied Sci. 1, 494. https://doi.org/10.1007/s42452-019-0516-2.

Bera, B., Ghosh, A., 2019. Fluoride dynamics in hydrogeological diversity and fluoride contamination index mapping: a correlation study of north Singbhum Craton, India. Arab. J. Geosci. (24) 12. https://doi.org/10.1007/s12517-019-4994-8.

Bera, K., Bandyopadhyay, J., 2017. Drought analysis for agricultural impact through geoinformatic based indices, a case study of Bankur district, West Bengal, India. J. Remote Sens. GIS 06 (03). https://doi.org/10.4172/2469-4134.1000209.

Carver, S.J., 1991. Integrating multi-criteria evaluation with geographical information systems. Int. J. Geogr. Inf. Syst. 5 (3), 321–339. https://doi.org/10.1080/02693799108927858.

Chakraborty, B., Roy, S., Bera, A., Adhikary, P.P., Bera, B., Sengupta, D., Bhunia, G., & Shit, P. (2021b). Geospatial assessment of groundwater quality for drinking through water quality index and human health risk index in an upland area of Chota Nagpur plateau of West Bengal, India (pp. 327–358). https://doi.org/10.1007/978-3-030-63422-3_19.

Chakraborty, B., Roy, S., Bera, A., Adhikary, P.P., Bera, B., Sengupta, D., Bhunia, G.S., Shit, P.K., 2021a. Groundwater vulnerability assessment using GIS-based drastic model in the upper catchment of Dwarakeshwar river basin, West Bengal, India. Environ. Earth Sci. 81 (1), 2. https://doi.org/10.1007/s12665-021-10002-3.

Chamling, M., Bera, B., 2020. Spatio-temporal patterns of land use/land cover change in the Bhutan–Bengal foothill region between 1987 and 2019: study towards geospatial applications and policy making. Earth Syst. Environ. 4 (1), 117–130. https://doi.org/10.1007/s41748-020-00150-0.

Chatterjee, R.S., Pranjal, P., Jally, S., Kumar, B., Dadhwal, V.K., Srivastav, S.K., Kumar, D., 2020. Potential groundwater recharge in north-western India vs spaceborne grace gravity anomaly based monsoonal groundwater storage change for evaluation of groundwater potential and sustainability. Groundw. Sustain. Develop. 10, 100307. https://doi.org/10.1016/j.gsd.2019.100307.

Chen, W., Chai, H., Sun, X., Wang, Q., Ding, X., Hong, H., 2016. A GIS-based comparative study of frequency ratio, statistical index and weights-of-evidence models in landslide susceptibility mapping. Arab. J. Geosci. 9 (3), 204. https://doi.org/10.1007/s12517-015-2150-7.

Chen, W., Li, H., Hou, E., Wang, S., Wang, G., Panahi, M., Li, T., Peng, T., Guo, C., Niu, C., Xiao, L., Wang, J., Xie, X., Ahmad, B.B, 2018. GIS-based groundwater potential analysis using novel ensemble weights-of-evidence with logistic regression and functional tree models. Sci. Total Environ. 634, 853–867. https://doi.org/10.1016/j.scitotenv.2018.04.055.

Choubin, B., Rahmati, O., Soleimani, F., Alilou, H., Moradi, E., Alamdari, N., 2019. Regional groundwater potential analysis using classification and regression trees (pp. 485–498). 10.1016/B978-0-12-815226-3.00022-3.

Das, B., Pal, S.C., Malik, S., Chakrabortty, R., 2019. Modeling groundwater potential zones of Puruliya district, West Bengal, India using remote sensing and GIS techniques. Geol. Ecol. Landsc. 3 (3), 223–237. https://doi.org/10.1080/24749508.2018.1555740.

Das, S., Behera, S.C., Kar, A., Narendra, P., Guha, S., 1997. Hydrogeomorphological mapping in ground water exploration using remotely sensed data—a case study in Keonjhar district, Orissa. J. Ind. Soc. Remote Sens. 25 (4), 247–259. https://doi.org/10.1007/BF03019366.

Das, S., Pardeshi, S.D., 2018. Integration of different influencing factors in gis to delineate groundwater potential areas using IF and FR techniques: a study of Pravara basin, Maharashtra, India. Appl. Water Sci. 8 (7), 1–16. https://doi.org/10.1007/s13201-018-0848-x.

Dey, S., Bhatt, D., Haq, S., Mall, R.K., 2020. Potential impact of rainfall variability on groundwater resources: a case study in Uttar Pradesh, India. Arab. J. Geosci. 13 (3). https://doi.org/10.1007/s12517-020-5083-8.

Dinesan, V.P., Gopinath, G., Ashitha, M.K., 2015. Application of geoinformatics for the delineation of groundwater prospects Zones—a case study for melattur grama panchayat in Kerala, India. Aquat. Procedia 4, 1389–1396. https://doi.org/10.1016/j.aqpro.2015.02.180.

Duan, H., Deng, Z., Deng, F., Wang, D., 2016. Assessment of groundwater potential based on multicriteria decision making model and decision tree algorithms. Math. Probl. Eng. 2016, 1–12. https://doi.org/10.1155/2016/2064575.

Duraiswami, R.A., Patankar, U., 2011. Occurrence of fluoride in the drinking water sources from Gad river basin, Maharashtra. J. Geol. Soc. Ind. 77 (2), 167–174. https://doi.org/10.1007/s12594-011-0020-9.

Elmahdy, S., Mohamed, M., Ali, T., 2020. Land use/land cover changes impact on groundwater level and quality in the northern part of the United Arab Emirates. In: Remote Sensing, 12 (11), p. 1715. https://doi.org/10.3390/rs12111715.

Elmahdy, S.I., Mohamed, M.M., 2015. Probabilistic frequency ratio model for groundwater potential mapping in Al Jaww plain, UAE. Arabian J. Geosci. 8 (4), 2405–2416. https://doi.org/10.1007/s12517-014-1327-9.

Esaid, O., Abdelkareem, O., Mohamed, H., Elamin, A., Elyas, M., Eltahir, S., Adam, H., Elhaja, M., Rahamtallah Abualgasim, M., Osunmadewa, B., Elmar, C., 2018. Accuracy assessment of land use land cover in Umabdalla natural reserved forest. 3, 5–9.

Fawcett, T., 2006. An introduction to ROC analysis. Pattern Recog. Lett. 27 (8), 861–874. https://doi.org/10.1016/j.patrec.2005.10.010.

Florinsky, I.v., 2016. Lineaments and Faults. Digital Terrain Analysis in Soil Science and Geology 2(14), pp. 353–376). Academic Press, Elsevier. https://doi.org/10.1016/B978-0-12-804632-6.00014-6

Ghosh, P.K., Bandyopadhyay, S., Jana, N.C., 2016. Mapping of groundwater potential zones in hard rock terrain using geoinformatics: a case of Kumari watershed in western part of West Bengal. Modeling Earth Syst. Environ. 2 (1), 1–12. https://doi.org/10.1007/s40808-015-0044-z.

Graeme F. Bonham-Carter, 1994. Geographic Information Systems For Geoscientists: Modelling with GIS, 1st Ed. Elsevier. (No-13).

Grinevskii, S.O., 2014. The effect of topography on the formation of groundwater recharge. Moscow Univ. Geol. Bull. 69 (1), 47–52. https://doi.org/10.3103/S0145875214010025.

Guru, B., Seshan, K., Bera, S., 2017. Frequency ratio model for groundwater potential mapping and its sustainable management in cold desert, India. J. King Saud Univ.—Sci. 29 (3), 333–347. https://doi.org/10.1016/j.jksus.2016.08.003.

Jothibasu, A., Anbazhagan, S., 2016. Modeling groundwater probability index in Ponnaiyar river basin of South India using analytic hierarchy process. Modeling Earth Syst. Environ. 2 (3), 1–14. https://doi.org/10.1007/s40808-016-0174-y.

Kordestani, M.D., Naghibi, S.A., Hashemi, H., Ahmadi, K., Kalantar, B., Pradhan, B., 2019. Groundwater potential mapping using a novel data-mining ensemble model. Hydrol. J. 27 (1), 211–224. https://doi.org/10.1007/s10040-018-1848-5.

Kotchoni, D., Vouillamoz, J.-M., Lawson, F.M., Adjomayi, P., Boukari, M., Taylor, R., 2018. Relationships between rainfall and groundwater recharge in seasonally humid benin: a comparative analysis of long-term hydrographs in sedimentary and crystalline aquifers. Hydrol. J. https://doi.org/10.1007/s10040-018-1806-2.

Machiwal, D., Jha, M.K., Mal, B.C., 2011. Assessment of groundwater potential in a semi-arid region of India using remote sensing, GIS and MCDM techniques. Water Resour. Manage. 25 (5), 1359–1386. https://doi.org/10.1007/s11269-010-9749-y.

Maity, D.K., Mandal, S., 2019. Identification of groundwater potential zones of the Kumari river basin, India: an RS & GIS based semi-quantitative approach. Environ., Develop. Sustain. 21 (2), 1013–1034. https://doi.org/10.1007/s10668-017-0072-0.

Manap, M.A., Nampak, H., Pradhan, B., Lee, S., Sulaiman, W.N.A., Ramli, M.F., 2014. Application of probabilistic-based frequency ratio model in groundwater potential mapping using remote sensing data and GIS. Arab. J. Geosci. 7 (2), 711–724. https://doi.org/10.1007/s12517-012-0795-z.

Mehra, M., Oinam, B., Singh, C.K., 2016. Integrated assessment of groundwater for agricultural use in Mewat district of Haryana, India using geographical information system (GIS). J. Ind. Soc. Remote Sens. 44 (5), 747–758. https://doi.org/10.1007/s12524-015-0541-6.

Moench, M., 2003. Groundwater and poverty: exploring the connections. Intensive use of groundwater: challenges and opportunities, 441–456, ISBN 9058093905.

Mogaji, K.A., Lim, H.S., Abdullah, K., 2015. Regional prediction of groundwater potential mapping in a multifaceted geology terrain using GIS-based Dempster–Shafer model. Arab. J. Geosci. 8 (5), 3235–3258. https://doi.org/10.1007/s12517-014-1391-1.

Moghaddam, D.D., Rahmati, O., Haghizadeh, A., Kalantari, Z., 2020. A modeling comparison of groundwater potential mapping in a mountain bedrock aquifer: QUEST, GARP, and RF models. Water (Switzerland) 12 (3). https://doi.org/10.3390/w12030679.

Moghaddam, D.D., Rezaei, M., Pourghasemi, H.R., Pourtaghie, Z.S., Pradhan, B., 2015. Groundwater spring potential mapping using bivariate statistical model and GIS in the Taleghan watershed, Iran. Arab. J. Geosci. 8 (2), 913–929. https://doi.org/10.1007/s12517-013-1161-5.

Muthumaniraja, C.K., Anbazhagan, S., Jothibasu, A., Chinnamuthu, M., 2019. Remote Sensing and fuzzy logic approach for artificial recharge studies in Hard Rock Terrain of South India. GIS and Geostatistical Technique for Groundwater Studies. Elsevier, 91–112.

Nag, S., 2016. Delineation of groundwater potential zones in hard rock terrain in kashipur block, purulia district, west bengal, using geospatial techniques. Int. J. Waste Res. 06. https://doi.org/10.4172/2252-5211.1000201.

Nag, S.K., Kundu, A., 2018. Application of remote sensing, GIS and MCA techniques for delineating groundwater prospect zones in Kashipur block, Purulia district, West Bengal. Appl. Water Sci. 8 (1), 1–13. https://doi.org/10.1007/s13201-018-0679-9.

Naghibi, S., Pourghasemi, H.R., Dixon, B., 2015. GIS-based groundwater potential mapping using boosted regression tree, classification and regression tree, and random forest machine learning models in Iran. Environ. Monit. Assess. 188. https://doi.org/10.1007/s10661-015-5049-6.

Naghibi, S.A., Pourghasemi, H.R., Pourtaghi, Z.S., Rezaei, A., 2015. Groundwater qanat potential mapping using frequency ratio and Shannon's entropy models in the Moghan watershed, Iran. Earth Sci. Inf. 8 (1), 171–186. https://doi.org/10.1007/s12145-014-0145-7.

Nampak, H., Pradhan, B., Manap, M.A., 2014. Application of GIS based data driven evidential belief function model to predict groundwater potential zonation. J. Hydrol. 513, 283–300. https://doi.org/10.1016/j.jhydrol.2014.02.053.

Oberthür, A., Warnat, P., 2009. Supervised Classification BT - Encyclopedia of Cancer (M. Schwab, Ed.; pp. 2862–2864). Springer Berlin and Heidelberg. https://doi.org/10.1007/978-3-540-47648-1_5581

Oh, H.J., Kim, Y.S., Choi, J.K., Park, E., Lee, S., 2011. GIS mapping of regional probabilistic groundwater potential in the area of Pohang city, Korea. J. Hydrol. 399 (3–4), 158–172. https://doi.org/10.1016/j.jhydrol.2010.12.027.

Owuor, S.O., Butterbach-Bahl, K., Guzha, A.C., Rufino, M.C., Pelster, D.E., Díaz-Pinés, E., Breuer, L., 2016. Groundwater recharge rates and surface runoff response to land use and land cover changes in semi-arid environments. Ecol. Process. 5 (1), 16. https://doi.org/10.1186/s13717-016-0060-6.

Ozdemir, A., 2011. Using a binary logistic regression method and GIS for evaluating and mapping the groundwater spring potential in the Sultan Mountains (Aksehir, turkey). J. Hydrol. 405 (1), 123–136. https://doi.org/10.1016/j.jhydrol.2011.05.015.

Pal, S.C., Ghosh, C., Chowdhuri, I., 2020. Assessment of groundwater potentiality using geospatial techniques in Purba Bardhaman district, West Bengal. Appl. Water Sci. 10 (10), 1–13. https://doi.org/10.1007/s13201-020-01302-3.

Pinto, D., Shrestha, S., Babel, M.S., Ninsawat, S., 2017. Delineation of groundwater potential zones in the Comoro watershed, Timor Leste using GIS, remote sensing and analytic hierarchy process (AHP) technique. Appl. Water Sci. 7 (1), 503–519. https://doi.org/10.1007/s13201-015-0270-6.

Pourghasemi, H.R., Beheshtirad, M., 2015. Assessment of a data-driven evidential belief function model and GIS for groundwater potential mapping in the Koohrang watershed, Iran. Geocarto. Int. 30 (6), 662–685. https://doi.org/10.1080/10106049.2014.966161.

Prabhakar, A., Tiwari, H., 2015. Land use and land cover effect on groundwater storage. Modeling Earth Syst. Environ. 1, 45. https://doi.org/10.1007/s40808-015-0053-y.

Pradhan, B., 2009. Groundwater potential zonation for basaltic watersheds using satellite remote sensing data and GIS techniques. Central Eur. J. Geosci. 1 (1), 120–129. https://doi.org/10.2478/v10085-009-0008-5.

Razandi, Y., Pourghasemi, H.R., Neisani, N.S., 2015. Application of analytical hierarchy process, frequency ratio, and certainty factor models for groundwater potential mapping using GIS. https://doi.org/10.1007/s12145-015-0220-8.

Rizeei, H.M., Pradhan, B., Saharkhiz, M.A., Lee, S., 2019. Groundwater aquifer potential modeling using an ensemble multi-adoptive boosting logistic regression technique. J. Hydrol. 579, 124172. https://doi.org/10.1016/j.jhydrol.2019.124172.

Saha, P., 2015. Identifying the causes of water scarcity in Purulia, West Bengal, India—A geographical perspective. IOSR J. Environ. Sci. Ver. I 9 (8), 2319–2399. https://doi.org/10.9790/2402-09814151.

Sahoo, S., Chakraborty, S., Pham, Q.B., Sharifi, E., Sammen, S.S., Vojtek, M., Vojteková, J., Elkhrachy, I., Costache, R., Linh, N.T.T., 2021. Recognition of district-wise groundwater stress zones using the GLDAS-2 catchment land surface model during lean season in the Indian state of West Bengal. Acta Geophys. 0123456789. https://doi.org/10.1007/s11600-020-00509-x.

Scanlon, B.R., Reedy, R.C., Stonestrom, D.A., Prudic, D.E., Dennehy, K.F., 2005. Impact of land use and land cover change on groundwater recharge and quality in the southwestern US. Glob. Change Biol. 11 (10), 1577–1593. https://doi.org/10.1111/j.1365-2486.2005.01026.x.

Senthilkumar, M., Gnanasundar, D., Arumugam, R., 2019. Identifying groundwater recharge zones using remote sensing & GIS techniques in Amaravathi aquifer system, Tamil Nadu, South India. Sustain. Environ. Res. 29 (1), 15. https://doi.org/10.1186/s42834-019-0014-7.

Shaban, A., Khawlie, M., Abdallah, C., 2006. Use of remote sensing and GIS to determine recharge potential zones: the case of Occidental Lebanon. Hydrol. J. 14 (4), 433–443. https://doi.org/10.1007/s10040-005-0437-6.

Shah, R.A., Lone, S.A., 2019. Hydrogeomorphological mapping using geospatial techniques for assessing the groundwater potential of Rambiara river basin, Western Himalayas. Appl. Water Sci. 9 (3), 64. https://doi.org/10.1007/s13201-019-0941-9.

Shasan AL-Zuhairy, M., Alauldeen Abdulrahman Hasan, A., Mezher Shnewer, F., 2017. GIS-Based frequency ratio model for mapping the potential zoning of groundwater in the western desert of Iraq. Int. J. Sci. Eng. Res. 8 (7), 52–65.

Sikakwe, G.U., 2018. GIS-based model of groundwater occurrence using geological and hydrogeological data in Precambrian Oban Massif southeastern Nigeria. Appl. Water Sci. 8 (3), 79. https://doi.org/10.1007/s13201-018-0700-3.

Tahmassebipoor, N., Rahmati, O., Noormohamadi, F., Lee, S., 2015. Spatial analysis of groundwater potential using weights-of-evidence and evidential belief function models and remote sensing. Arab. J. Geosci. 9 (1), 79. https://doi.org/10.1007/s12517-015-2166-z.

Teixeira, J., Chaminé, H.I., Carvalho, J.M., Pérez-Alberti, A., Rocha, F., 2013. Hydrogeomorphological mapping as a tool in groundwater exploration. J. Maps 9 (2), 263–273. https://doi.org/10.1080/17445647.2013.776506.

Thapa, R., Gupta, S., Guin, S., Kaur, H., 2017. Assessment of groundwater potential zones using multi-influencing factor (MIF) and GIS: a case study from Birbhum district, West Bengal. Appl. Water Sci. 7 (7), 4117–4131. https://doi.org/10.1007/s13201-017-0571-z.

Thomas, A., Sharma, P.K., Sharma, M.K., Sood, A., 1999. Hydrogeomorphological mapping in assessing ground water by using remote sensing data—a case study in lehra gaga block, Sangrur district, Punjab. J. Ind. Soc. Remote Sens. 27 (1), 31. https://doi.org/10.1007/BF02990773.

Todd, D.K., M., L., 2005. Groundwater Hydrology, 3rd Ed. Wiley, New York, NY.

Tóth, J., 1999. Groundwater as a geologic agent: an overview of the causes, processes, and manifestations. Hydrol. J. 7, 1–14. https://doi.org/10.1007/s100400050176.

Toutin, T., 2008. ASTER DEMs for geomatic and geoscientific applications: a review. Int. J. Remote Sens. 29 (7), 1855–1875. https://doi.org/10.1080/01431160701408477.

Verma, P., Raghubanshi, A., Srivastava, P.K., Raghubanshi, A.S., 2020. Appraisal of kappa-based metrics and disagreement indices of accuracy assessment for parametric and nonparametric techniques used in LULC classification and change detection. Model. Earth Syst. Environ. 6 (2), 1045–1059. https://doi.org/10.1007/s40808-020-00740-x.

Zabihi, M., Pourghasemi, H.R., Pourtaghi, Z.S., Behzadfar, M., 2016. GIS-based multivariate adaptive regression spline and random forest models for groundwater potential mapping in Iran. Environ. Earth Sci. 75 (8), 665. https://doi.org/10.1007/s12665-016-5424-9.

Zakir, F.A., Qari, M.H.T., Mostafa, M.E., 1999. Technical note a new optimizing technique for preparing lineament density maps. Int. J. Remote Sens. 20 (6), 1073–1085. https://doi.org/10.1080/014311699212858.

Zhang, X., Zhang, L., He, C., Li, J., Jiang, Y., Ma, L., 2014. Quantifying the impacts of land use/land cover change on groundwater depletion in Northwestern China—a case study of the Dunhuang oasis. Agric. Water Manage. 146, 270–279. https://doi.org/10.1016/j.agwat.2014.08.017.

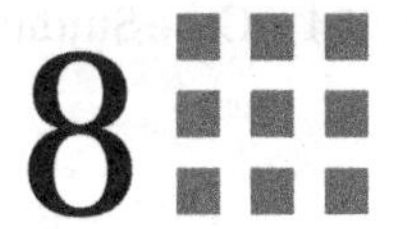

Assessment of groundwater salinity risk in coastal belt of Odisha using ordinary kriging and its management

Ch. Jyotiprava Dash[a], Partha Pratim Adhikary[b] and Jotirmayee Lenka[a]

[a]ICAR-INDIAN INSTITUTE OF SOIL AND WATER CONSERVATION, RESEARCH CENTRE, KORAPUT, ODISHA, INDIA [b]ICAR-INDIAN INSTITUTE OF WATER MANAGEMENT, BHUBANESWAR, ODISHA, INDIA

8.1 Introduction

Groundwater is considered as one of the vital natural resources and has been played an important role in the survival of human being. In the context of rapid urbanization, population growth, and changing climate, the importance of groundwater has been increased manifold. It plays an important role for both irrigation and drinking. Groundwater irrigation is considered as a kind of silent revolution and also a source of social conflicts in many parts of the world (Llamas and Santos, 2005). Groundwater has played a significant role during the green revolution in India and helped in ensuring food security to more than 1 billion people. Presently, India stands first in terms of groundwater use in the world (Margat and Gun, 2013). The excess withdrawal of groundwater leads to a lowering of water table and groundwater quality deterioration. A report published by Central Ground Water Board (CGWB) explains the severity of the overexploitation of groundwater resources across India (Central Ground Water Board, 2019), and its subsequent impact on the aquifer system. The water table depletion reduces the volume of water stored in aquifers, which results in saline water intrusion in the case of coastal aquifers. In the coastal aquifers, the groundwater is generally influenced by sea and brackish water estuaries, making the groundwater saline in nature (Fig. 8.1), which leads to a reduction in crop yield, and creates social and economic issues.

The groundwater salinity is usually monitored through a network of monitoring wells, which is very expensive and time consuming. As the estimation of groundwater salinity is very important at unsampled locations, various interpolation techniques are used to quantify the salinity

Case Studies in Geospatial Applications to Groundwater Resources. DOI: https://doi.org/10.1016/B978-0-323-99963-2.00017-1

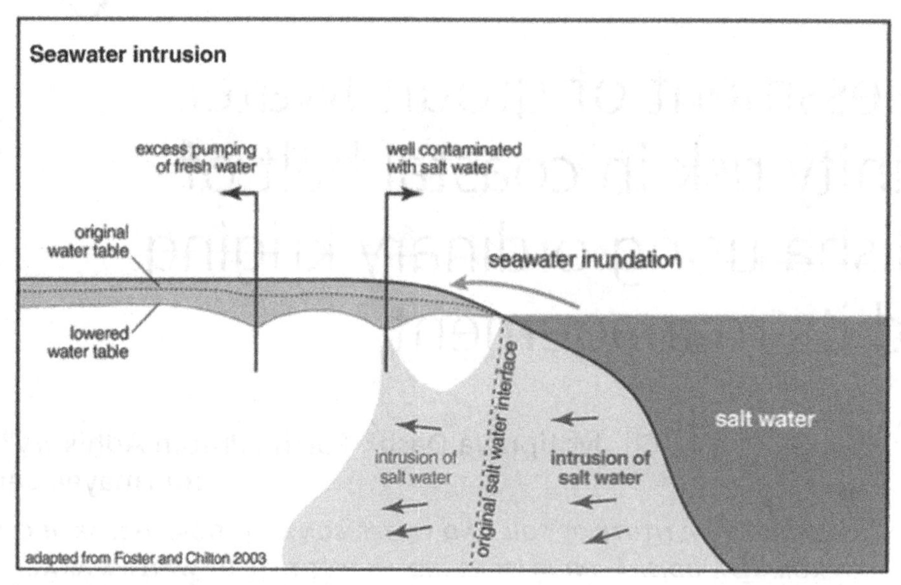

FIGURE 8.1 Seawater intrusion to fresh groundwater.

at unsampled location. Kriging is one of the most widely used geostatistical interpolation techniques (Dash et al., 2010). In this method, the value of a spatial random variable is estimated for an unsampled location by incorporating spatial autocorrelation among the measured values. There are different types of kriging methods such as simple kriging, ordinary kriging, disjunctive kriging, universal kriging and indicator kriging. In recent years, several researchers applied ordinary kriging for spatial analysis of groundwater depth and salinity for its accurate prediction and simplicity.

Spatio-temporal analysis of groundwater salinity was performed considering 97 wells monitored over a period of 7 years (2004–2010) in Bafra Plain, Turkey using ordinary kriging, and reported decrease in groundwater salinity over the years (Arslan, 2012). Jeihouni et al. (2015) used 120 observation wells located in Tabriz plain, north-west of Iran for spatial analysis of groundwater salinity by using ordinary kriging. In a study, Ashrafzadeh et al. (2016) reported the suitability of groundwater salinity for paddy cultivation in the alluvial plains of Guilan Province, northern Iran, using ordinary kriging and cokriging. Similarly, spatio-temporal behavior of groundwater depth and groundwater salinity was reported for Nain experimental farm, Panipat, Haryana using ordinary kriging and emphasized on the suitability of ordinary kriging in prioritizing the area for implementing the optimum groundwater use plan in salt-affected area (Narjary et al., 2017). Spatial variation and risk zonation map of groundwater salinity was developed for Mazandaran Plain, Iran, using ordinary kriging (Amiri-Bourkhani et al., 2017). Shahraki et al. (2021) compared ordinary kriging and inverse distance weighting (IDW) interpolation techniques for estimating groundwater level and groundwater salinity in Salman Farsi Sugarcane Plantation, Khouzestan province, western Iran, and reported higher prediction accuracy of the ordinary kriging over

IDW. The coastal part of Odisha has been suffering from groundwater salinity, seawater intrusion, and other extreme events. In Odisha, saline groundwater zones have a width of 2–3 km in south-east, 1.5–5 km in north and 15 km in north-east (Central Ground Water Board, 2014). Some researchers reported the increased concentration of major ions in coastal aquifer indicating the reason as mixing of seawater with fresh groundwater (Mohapatra et al., 2011; Prusty et al., 2018; Mohanty and Rao, 2019; Prusty et al., 2020). In an earlier report, Radhakrishna (2001) reported the deterioration of groundwater quality in Mahanadi delta area due to mixing of seawater with freshwater. The previous studies related to salinization of coastal aquifer are mostly confined to some specific locations. Therefore, the present study aims to assess the spatial variation of groundwater salinity in the entire coastal belt of Odisha using ordinary kriging.

8.2 Material and method

8.2.1 Study area

The state of Odisha has a geographical area of 155,707 km^2 with a population density of 270 per km^2 and stretches along the eastern coast of India. It is surrounded by Jharkhand and West Bengal in the North, the Bay of Bengal in the East, Chhattisgarh on the West, and Andhra Pradesh in the South. The state is classified into four zones such as the Northern Plateau, Coastal Plains, Central Table Lands, and Eastern Ghats region. The coastal plain stretches 480 km long and its width varies 10–50 km from the coastline into the mainland. This coastal plain is the combination of many deltas formed by the major rivers of Odisha (the Baitarani, the Brahmani, the Budhabalanga, the Mahanadi, the Rushikulya, and the Subarnarekha). Therefore, the coastal plain of Odisha has been termed as the "Hexadeltaic region" or the "Gift of six rivers." The coastal belt of Odisha is located in eastern part of the state and comes under agro-ecological subregion of 18.4. In this study, the coastal belts of Odisha comprising 10 districts such as Balasore, Bhadrak, Cuttack, Jajpur, Jagatsinghpur, Kendrapada, Khordha, Nayagarh, Puri, and Ganjam have taken in to consideration (Fig. 8.2). The coastal belt of Odisha is highly populated with a total population of 19.3 million and average population density of 570.1 per km^2 (Census, 2011; Environmental Systems Research Institute ESRI, 2010).

Coastal Odisha is characterized by hot and humid climate. The long-term mean monthly rainfall, temperature, and potential evaporation of study area are presented in Table 8.1. The annual average rainfall ranges from 1130.8 to 1533.2 mm, with a mean value as 1456.9 mm, of which more than 70% of rainfall is received during south-west monsoon (June–September). August is the rainiest month having average rainfall of 308.68 mm, whereas the study area receives only 5.15 mm of rainfall during December. The mean temperature ranges from 23.06 to 33.15°C. In May, evaporation is as high as 7.22 mm/day, whereas during August, it comes down to 5.30 mm/day. In the study area, the soil texture varies from coarse sand to clay and mostly depends on geomorphulogy of the flood plain and the type of alluvial material carried by river water. The structure may be granular or platy. Flooding occurs usually in rainy season. The presence of excess salinity in the soil is due to the presence of saline groundwater table at

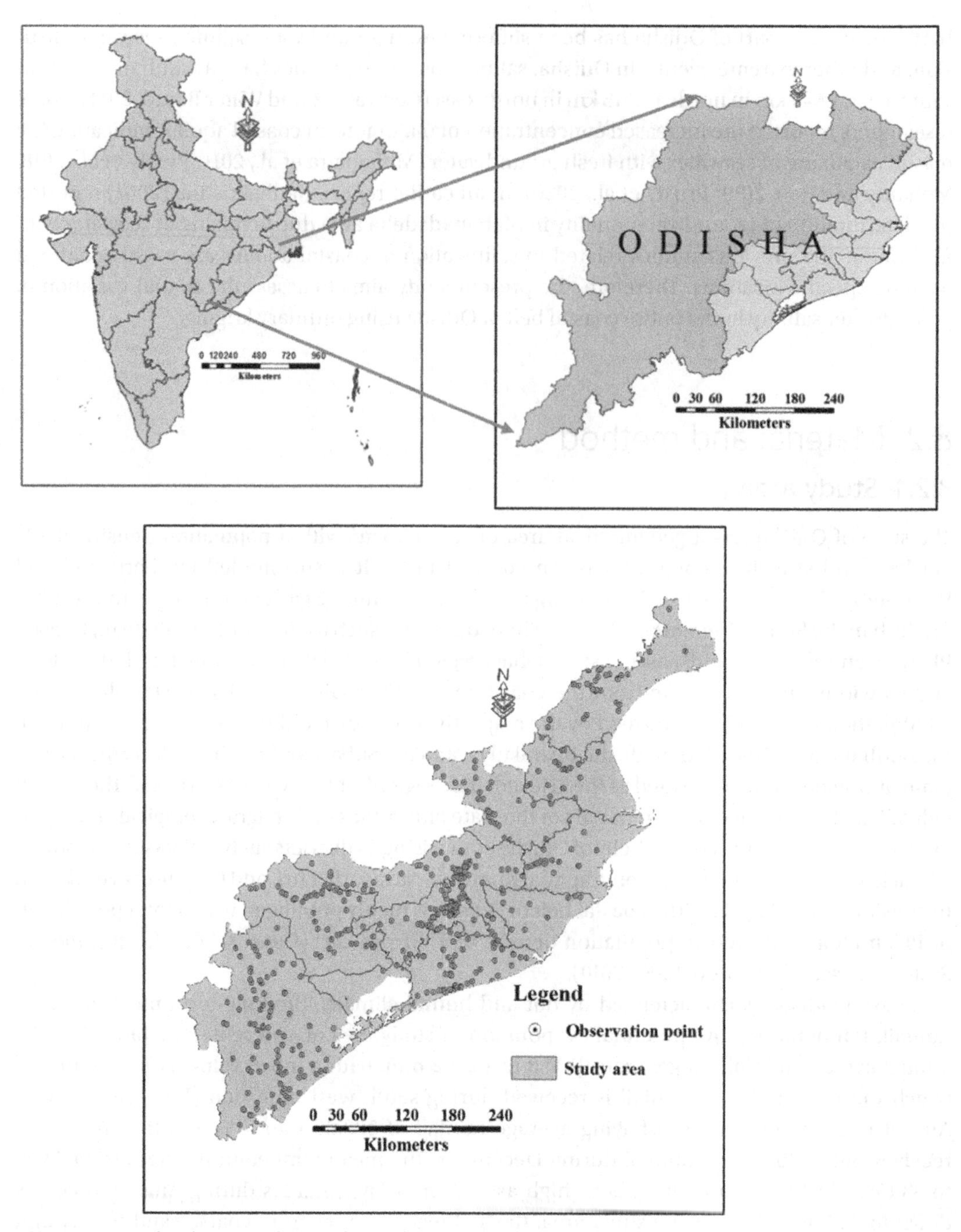

FIGURE 8.2 Map showing groundwater observation points in the study area.

Table 8.1 Average values for some of the long-term climatic parameters of the study area.

District	Parameters	Months											
		Jan.	**Feb.**	**Mar.**	**Apr.**	**May**	**Jun.**	**Jul.**	**Aug.**	**Sep.**	**Oct.**	**Nov.**	**Dec.**
Balasore	Rainfall (mm)	11.80	30.72	32.43	52.24	96.28	231.32	286.68	342.09	265.06	157.87	21.76	4.90
	Temperature (°C)	21.12	23.80	27.99	30.73	31.77	30.69	28.73	28.48	28.53	27.43	23.93	20.94
	Evaporation (mm/day)	5.20	5.86	6.72	7.41	7.41	6.33	5.26	4.92	4.95	5.29	5.47	5.15
Bhadrak	Rainfall (mm)	10.86	26.78	27.78	37.13	81.83	214.87	297.60	319.94	249.09	158.09	26.11	4.93
	Temperature (°C)	21.88	24.57	28.33	30.76	31.78	30.83	28.82	28.61	28.77	27.94	24.60	21.63
	Evaporation (mm/day)	5.19	5.82	6.57	7.14	7.17	6.21	5.20	4.91	4.95	5.30	5.45	5.17
Cuttack	Rainfall (mm)	11.36	22.80	20.79	26.74	52.35	184.42	275.53	284.44	227.44	173.17	40.38	3.60
	Temperature (°C)	21.37	23.74	27.02	29.21	30.53	29.89	28.06	27.94	28.13	27.22	23.96	21.21
	Evaporation (mm/day)	5.15	5.72	6.26	6.62	6.63	5.91	5.06	4.92	5.01	5.30	5.47	5.17
Jajpur	Rainfall (mm)	10.49	26.11	27.01	35.39	69.11	205.79	303.15	316.60	241.62	152.15	28.53	4.01
	Temperature (°C)	21.70	24.37	28.13	30.66	31.81	30.78	28.57	28.34	28.50	27.68	24.37	21.40
	Evaporation (mm/day)	5.22	5.86	6.61	7.17	7.26	6.29	5.27	5.01	5.05	5.38	5.51	5.20
Jagatsinghpur	Rainfall (mm)	11.48	18.53	17.72	16.54	53.63	185.33	274.51	290.00	236.12	185.79	44.15	5.79
	Temperature (°C)	21.84	23.98	27.00	28.74	29.76	29.56	28.28	28.36	28.61	27.77	24.68	21.84
	Evaporation (mm/day)	4.93	5.35	5.69	5.76	5.73	5.36	4.71	4.65	4.71	5.00	5.23	4.97
Kendrapada	Rainfall (mm)	10.68	21.10	23.06	25.69	74.06	214.64	307.75	316.70	255.17	177.26	38.49	6.52
	Temperature (°C)	22.07	24.50	27.91	29.99	30.97	30.35	28.74	28.65	28.84	28.10	24.90	21.91
	Evaporation (mm/day)	5.09	5.61	6.17	6.50	6.49	5.79	4.95	4.77	4.81	5.15	5.33	5.07
Khordha	Rainfall (mm)	11.52	20.93	16.84	25.65	45.49	166.98	239.87	247.79	216.04	196.03	51.56	4.09
	Temperature (°C)	21.52	23.75	26.71	28.63	29.99	29.65	28.10	28.00	28.21	27.25	24.03	21.38
	Evaporation (mm/day)	5.14	5.68	6.11	6.34	6.29	5.72	4.97	4.88	5.00	5.27	5.45	5.15
Nayagarh	Rainfall (mm)	11.32	22.11	19.28	32.19	46.49	173.82	255.80	256.64	208.17	162.01	41.58	2.75
	Temperature (°C)	20.29	22.56	25.77	28.01	29.56	28.77	26.81	26.66	26.90	25.78	22.52	20.01
	Evaporation (mm/day)	5.23	5.83	6.41	6.81	6.81	6.00	5.07	4.94	5.09	5.38	5.48	5.16
Puri	Rainfall (mm)	9.82	16.22	13.92	21.39	48.16	169.12	236.34	242.78	202.07	163.04	48.60	6.85
	Temperature (°C)	18.94	20.84	23.48	25.19	26.29	25.88	24.54	24.46	24.63	23.81	21.16	18.85
	Evaporation (mm/day)	4.43	4.85	5.25	5.43	5.41	4.94	4.28	4.18	4.26	4.49	4.61	4.40
Ganjam	Rainfall (mm)	10.28	17.17	15.28	29.94	44.72	152.69	212.45	222.07	202.81	169.07	50.32	3.95
	Temperature (°C)	20.37	22.42	25.49	27.56	29.07	28.23	26.48	26.38	26.58	25.51	22.40	20.03
	Evaporation (mm/day)	5.24	5.85	6.37	6.69	6.66	5.91	5.00	4.93	5.07	5.31	5.41	5.13
Coastal Odisha	**Rainfall (mm)**	**12.11**	**24.34**	**23.10**	**32.86**	**65.76**	**206.60**	**292.96**	**308.68**	**251.96**	**189.05**	**44.30**	**5.15**
	Temperature (°C)	**23.26**	**25.83**	**29.45**	**31.81**	**33.15**	**32.43**	**30.52**	**30.39**	**30.59**	**29.57**	**26.06**	**23.06**
	Evaporation (mm/day)	**5.60**	**6.21**	**6.83**	**7.22**	**7.22**	**6.42**	**5.47**	**5.30**	**5.39**	**5.71**	**5.89**	**5.57**

Table 8.2 Aquifer characteristics in coastal belts of Odisha (adopted from CGWB, 2014).

Aquifers	Quaternary alluvium	Tertiary sediments
Aquifer thickness	10–55	8–76
Static water level (m)	0.8–7.6	1.2 m agl–4.5m bgl
Dischage (lps)	18–75	8.5–45
Drawdown (m)	3.1–13.1	10.3–25.3
Yield factor (lpm/m/m^2)	2.9–42.8	0.73–11.1
Transmissivity (m^2/day)	1000–8198	151–1900
Hydraulic conductivity (m/day)	9–330	4–60
Storage coefficient	1.3×10^{-3}–7.5×10^{-5}	8.9×10^{-4}–1.9×10^{-6}

shallow depth and frequent brackish water inundation in the low-lying areas (Srinivasan et al., 2018).

The aquifers present in the coastal region of Odisha are mostly dominated by sand and gravel layers and these aquifers are shallow or phreatic in nature (Central Ground Water Board, 2014). Some of the aquifer characteristics in coastal belts of Odisha is presented in Table 8.2 (Central Ground Water Board, 2014). The average yield factor of quaternary or younger alluvium ranges 3.0–43 lpm/m/m^2, whereas that of tertiary sediments ranges 0.73–11 lpm/m/m^2 (Table 8.2). In the study area, people extract groundwater mainly through dug wells and hand pumps fitted with shallow filter point tube wells. Paddy is the dominating crop in this area. Apart from paddy farmers also cultivates pulses (green gram, black gram, horse gram), oilseeds, jute, and seasonal vegetables (Sethi et al., 2014).

8.2.2 Data and methodology used

The data used in this study were collected from CGWB. The groundwater depth and electrical conductivity (EC) data of 479 wells for the year 2018 were used for this study. Both premonsoon (month-April) and postmonsoon (month-January) water level data were used to assess the seasonal variation of groundwater in the study area. However, groundwater salinity data are a single-time observation. The location of the 479 observation wells in the study area is shown in Fig. 8.2. The spatial variation of groundwater depth, salinity in groundwater over the study area was mapped using the ordinary kriging in Arc GIS Software 10.0 (Environmental Systems Research Institute ESRI, 2010).

8.2.3 Ordinary kriging

Ordinary kriging is one of the most widely used geostatistical method, which is simple and have a remarkable flexibility (Dash et al., 2010; Adhikary and Biswas, 2011). It is the best linear unbiased estimation of the variables at the unsampled location. Ordinary kriging is used to reduce the estimation error of the variance. In this method, there used to be zero mean error (ME) of the estimate. Ordinary kriging linearly combines the sample means and the obtained coefficients from this linear combination are called weights. In practice, these weights are dependent on sample and estimated points and also the spatial structure of the data. Ordinary kriging also assumes that the mean of the process is constant and invariant within the spatial domain.

This is expressed as follows:

$$z(x) = \mu + \varepsilon(x) \tag{8.1}$$

where μ is an unknown constant and $z(x)$ is the measured value at any location x with stochastic residual $\varepsilon(x)$ with zero mean and unit variance.

Exploratory data analysis was carried out using groundwater depth and groundwater salinity (expressed as groundwater EC) data to identify outliers. Log normal and square root transformations of the data were carried out to ensure data normality. Further semivariogram parameters for different model such as exponential, Gaussian, linear, and spherical were generated, and best-fitted model was selected based on minimum residual sum square and maximum coefficient of determination (R^2). The semivariogram parameters such as nugget, range, and sill with respect to the best-fitted model were used for the generation of spatial variation map for groundwater depth and salinity.

8.2.4 Performance evaluation of ordinary kriging

The performance of ordinary kriging in predicting groundwater level and salinity was evaluated based on three commonly used statistics such as root mean square error (RMSE), ME, and mean relative error (MRE) (Nash and Sutcliffe, 1970). These measures can be determined using the following formulas:

$$\text{MAE} = \sum_{i=1}^{n} |(O_i - P_i)|/n \tag{8.2}$$

$$\text{RMSE} = \sqrt{\sum_{i=1}^{n} (O_i - P_i)^2/n} \tag{8.3}$$

$$\text{MRE} = \frac{\text{RMSE}}{\Delta} \tag{8.4}$$

where O_i and P_i are the observed and predicted groundwater table depth at ith location and n is the number of observations. Δ is the range and can be obtained by subtracting the minimum value from the maximum. The values of ME, and RMSE close to 0 indicate unbiased and accurate prediction.

Table 8.3 Descriptive statistics of premonsoon groundwater level in the study area.

District	Mean (m)	Minimum (m)	Maximum (m)	SD	CV (%)	SE
Balasore	3.88	1.45	6.55	1.43	36.84	0.30
Bhadrak	3.24	1.30	6.08	1.59	49.00	0.37
Cuttack	2.70	0.35	6.22	1.51	55.89	0.18
Jajpur	4.90	2.20	12.66	2.84	57.96	0.52
Jagatsinghpur	3.05	1.80	7.00	1.45	47.71	0.39
Kendrapada	3.12	1.60	9.63	1.72	55.05	0.38
Khordha	5.35	0.60	12.69	2.83	52.85	0.35
Nayagarh	4.94	0.76	9.90	1.98	39.96	0.31
Puri	3.00	0.48	10.89	1.85	61.59	0.20
Ganjam	4.01	0.35	8.58	1.86	46.43	0.20
Coastal Odisha	**4.04**	**0.35**	**12.69**	**2.23**	**55.25**	**0.10**

8.3 Results and discussion

8.3.1 Descriptive statistics, distribution of dataset, and semivariogram parameters

District-wise as well as for the whole study area, descriptive statistical analysis was carried out to provide a preliminary inference on groundwater level and EC. The average groundwater level during premonsoon varied from 2.70 to 5.35 m, with a mean value for coastal belts of 4.04 m (Table 8.3). The minimum depth to the groundwater table was observed in Cuttack district during premonsoon (0.35 m), and maximum depth to the groundwater was observed in Khordha district (12.69 m). The coefficient of variation (CV) was lowest in Balasore district (36.84%), whereas maximum variation in premonsoon groundwater level was observed in Puri district (61.59%). The average variation of premonsoon groundwater level in the study area was 55.25%. Similarly during postmonsoon, the groundwater level varied between 1.73 and 4.56 m (Table 8.4). The mean groundwater level across the study area was 3.23 m during postmonsoon. The rise in groundwater level by 0.81 m for the whole study area indicates groundwater recharge during rainy season. The desciptive statistics of EC in groundwater is presented in Table 8.5. The mean EC in the groundwater of coastal belt was found to be 0.79 dS/m. Minimum groundwater salinity was found in Cuttack district (0.06 dS/m) followed by Jajpur (0.07 dS/m) and Balasore (0.08 dS/m). Maximum groundwater salinity was observed in Bhadrak district (3.77 dS/m) followed by Puri district (3.66 dS/m), Jagatsinghapur district (3.14 dS/m), and Khordha district (3.10 dS/m), respectively. The co-efficient of variation was more than 80% in three districts such as Jagatsinghpur district (95.28%), Bhadrak district (91.52%), and Khordha district (88.83%) respectively.

The Kolmogorov and Smirnov test was carried out to verify normality of the groundwater level and salinity data of the study area, and it was observed that data were not normally distributed. Therefore log transformation was used to make the datasets normal distributed. After completion of log transformation, semivariogram parameters were generated. The nugget,

Table 8.4 Descriptive statistics of postmonsoon groundwater level in the study area.

District	Mean	Minimum (m)	Maximum (m)	SD	CV	SE
Balasore	4.12	1.43	6.99	1.68	40.74	0.35
Bhadrak	2.77	1.20	5.42	1.33	47.96	0.31
Cuttack	3.90	0.45	10.25	2.01	51.48	0.25
Jajpur	3.38	1.12	8.52	2.02	59.72	0.37
Jagatsinghpur	1.93	0.48	5.35	1.22	63.25	0.33
Kendrapada	1.73	0.39	3.0	0.70	40.52	0.16
Khordha	4.56	0.36	10.20	2.45	53.78	0.30
Nayagarh	3.63	0.11	7.73	1.71	47.08	0.27
Puri	2.64	0.14	8.89	1.80	68.03	0.19
Ganjam	3.38	0.21	8.41	1.79	53.03	0.19
Coastal Odisha	**3.23**	**0.11**	**10.20**	**1.95**	**60.32**	**0.09**

Table 8.5 Descriptive statistics of EC in groundwater in the study area.

District	Mean	Minimum	Maximum	SD	CV	SE
Balasore	0.66	0.08	1.31	0.34	51.82	0.06
Bhadrak	0.90	0.29	3.77	0.83	91.52	0.18
Cuttack	0.61	0.06	1.46	0.33	54.81	0.04
Jajpur	0.52	0.07	1.51	0.40	77.80	0.08
Jagatsinghpur	0.78	0.33	3.14	0.78	95.28	0.22
Kendrapada	1.15	0.50	2.51	0.51	44.61	0.12
Khordha	0.41	0.10	3.10	0.36	88.83	0.04
Nayagarh	0.83	0.20	2.50	0.53	63.05	0.10
Puri	1.03	0.16	3.36	0.64	62.54	0.07
Ganjam	1.00	0.21	2.95	0.58	57.63	0.06
Coastal Odisha	**0.79**	**0.06**	**3.77**	**0.57**	**72.36**	**0.03**

Table 8.6 Descriptive statistics of EC in groundwater in the study area.

Groundwater parameter	Best-fit model	Nugget	Sill	Range	Nugget/Sill	R^2	Residual sum square
Premonsoon water level	Linear	0.325	0.325	293.9	1.00	0.856	0.020
Postmonsoon water depth	Linear	0.451	0.451	303.5	1.00	0.755	0.043
Electrical conductivity	Spherical	0.271	0.617	75.6	0.439	0.981	0.002

sill, and range values (Fig. 8.3) of the best-fit models for premonsoon, postmonsoon groundwater level, and groundwater salinity are reported in Table 8.6. The spatial dependence of variables can be infered from nugget-to-sill ratio values (Cambardella et al., 1994; Adhikary and Dash, 2017). When the value of nugget-to-sill is less than 0.25, it indicates strong spatial dependence and moderate and weak spatial dependence are characterized by nugget-to-sill ratio as 0.25 to 0.75 and more than 0.75, respectively (Liu et al., 2006). Therefore, in this study, the water table

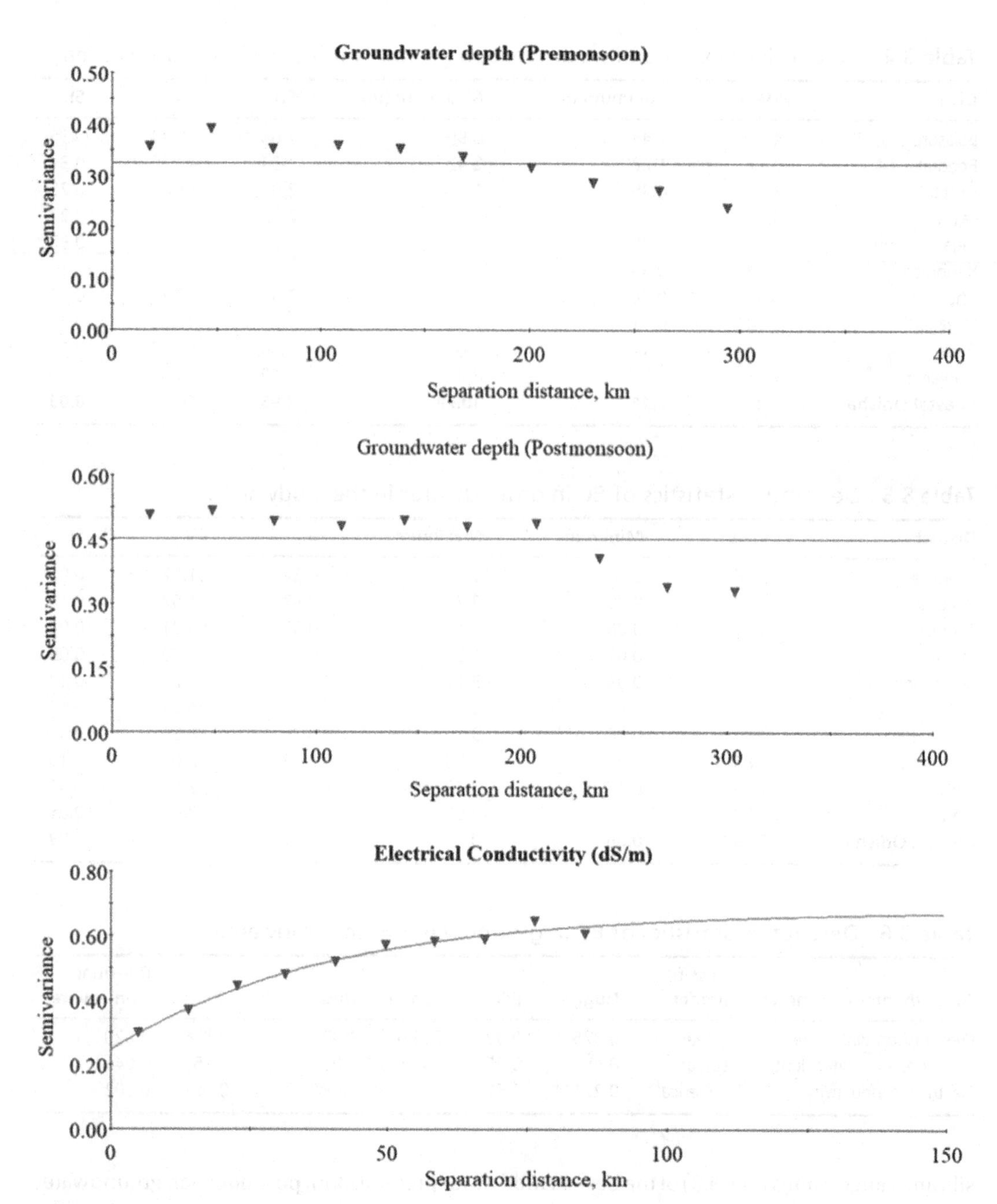

FIGURE 8.3 Best fit semivariograms of groundwater depth and EC in groundwater of coastal belt of Odisha.

Table 8.7 Performance accuracy of ordinary kriging in predicting groundwater level and salinity.

Groundwater parameter	ME	RMSE	MRE	R^2
Premonsoon water level	0.0193	0.216	0.0175	0.86
Postmonsoon water level	0.0254	0.379	0.0376	0.81
Electrical conductivity	0.0065	0.538	0.1450	0.83

depths were weakly spatially correlated, whereas groundwater salinity was moderately spatially correlated. The spatial correlation distances (range) of groundwater level (both premonsoon and postmonsoon) was more than groundwater salinity. The linear semivariogram model was observed to be the best-fit model for groundwater level, whereas the spherical model fit well for the groundwater salinity. Hu et al. (2005), Dash et al. (2010), Delgado et al. (2010), Yimit et al. (2011), Adhikary et al. (2011), and Nirjay et al. (2017) studied spatial variation of groundwater EC using geostatistics, respectively, in different parts of world and reported different best-fitted semivariogram models. Therefore, it can be stated that spatial patterns of groundwater EC vary from place to place depending on the climatic, soil, and geologic conditions of the area under consideration.

8.3.2 Performance accuracy of ordinary kriging

The performance accuracy of ordinary kriging in predicting groundwater level and salinity were computed and presented in Table 8.7. Both error terms ME and RMSE were close to 0 for all the three variables, which indicated that ordinary kriging's predictability capability for the coastal belt of Odisha was good and acceptable. The estimated MRE values for premonsoon, postmonsoon groundwater level, and salinity were 0.0175, 0.0376, and 0.1450, respectively (Table 8.7). These MRE values which indicate relative errors of the predicted data in comparison to the observed data were found to be very low. The standard error map of EC is presented in Fig. 8.4. The uncertainty related to prediction through ordinary kriging is very less as most of the areas come under uncertainty value less than 0.4 dS/m.

8.3.3 Spatial variability of groundwater depth

The spatial variability maps of groundwater depth during premonsoon and postmonsoon seasons of 2018 are shown in Fig. 8.5 with four classes, indicating areas having a water table within 1.5 m, between 1.5 and 3 m, between 3 and 5 m, and greater than 5 m from the ground level. The groundwater level during premonsoon ranged from 0.4 to 12.7 m. The water depth was below 3.0 m in areas near to coast during premonsoon, while other parts of the study area groundwater level is below 3.0 m. During postmonsoon, there was increase in groundwater level due to recharge because of monsoon, which was clearly shown in Fig. 8.5, as a significant part of area come under groundwater level below 5 m.

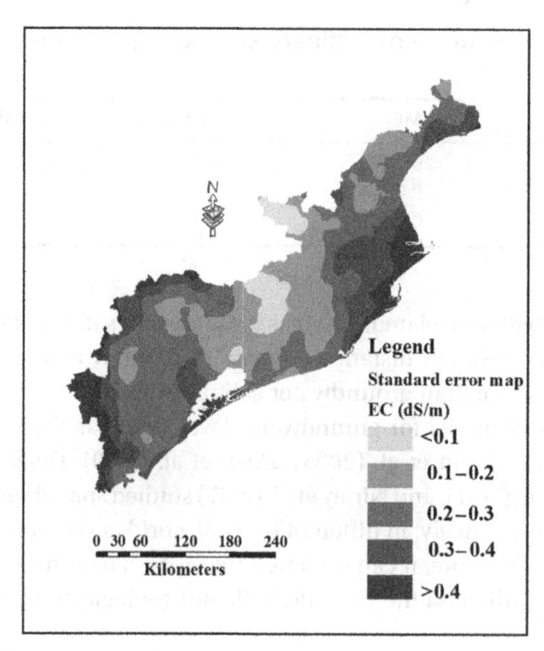

FIGURE 8.4 Prediction standard error map of EC in groundwater of coastal belts of Odisha.

Table 8.8 Classification of groundwater EC and its distribution in the study area.

Classification code	EC (dS/m)	Salinity level	Area (km^2)	Uses
C1	<0.25	Low salinity water	301.7 (0.84%)	Can be used for irrigation for all soil and most crops
C2	0.25–0.75	Medium salinity water	17193.6 (47.97%)	Can be used for crops with moderate salt tolerant
C3	0.75–2.25	High salinity water	18172.3 (50.70%)	Can be used for salt tolerant crops by providing good drainage
C4	>2.25	Very high salinity water	177.4 (0.49%)	Not suitable for irrigation purpose.

8.3.4 Spatial variability of electrical conductivity and suitability of groundwater for irrigation

The spatial variation of groundwater salinity is presented in Fig. 8.6. The EC values ranged from 0.06 to 3.77 dS/m. Srinivasan et al. (2018) reported water salinity ranged from 0.6 to 10.3 dS/m with an mean value of 9.4 dS/m in Ganjam block of Ganjam district. The entire study area is categorized into four classes depending on the USA Salinity Laboratory classification (Table 8.8). Groundwater with low salinity (EC < 0.25 ds/m) and very high salinity (EC >2.25 ds/m) level was confined to only a few pockets in the study area (Fig. 8.6), whereas the groundwater salinity in the study area was mostly medium to high in nature. The area-wise distribution of groundwater salinity level within the study area is presented in Table 8.8. The area under low

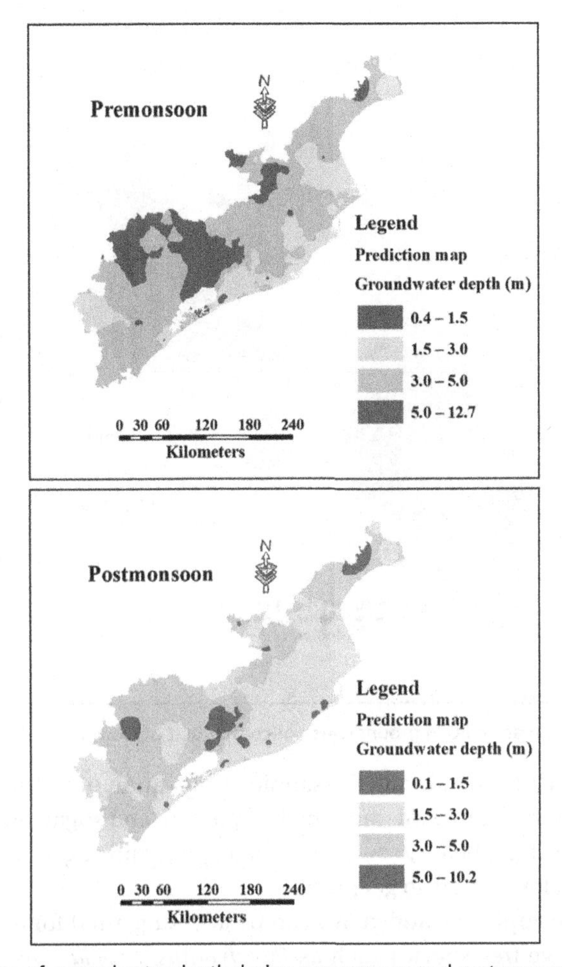

FIGURE 8.5 Spatial variation of groundwater depth during premonsoon and postmonsoon in coastal belt of Odisha.

salinity groundwater is 301.7 km^2 which is less than 1% of the study area. More than 98% of the study area is having groundwater salinity ranged medium to high. Therefore in the study area, moderate salt tolerant crops and salt tolerant crops can be grown. Researchers reported a 12% decrease in rice yield with every unit of increase in salinity from a certain threshold.

8.3.5 Groundwater salinity management

8.3.5.1 Groundwater salinity management in coastal belt of Odisha

The problems related to salinity and some management practices which can reduce salinity in coastal area are depicted in Fig. 8.7. Improvements in irrigation management such as appropriate irrigation scheduling, uniform water application and use of drip irrigation system can help in reducing salinity. Along with improved irrigation method, adoption of proper drainage techniques can solve the problem of salinity in many folds. Conjunctive use of fresh and saline

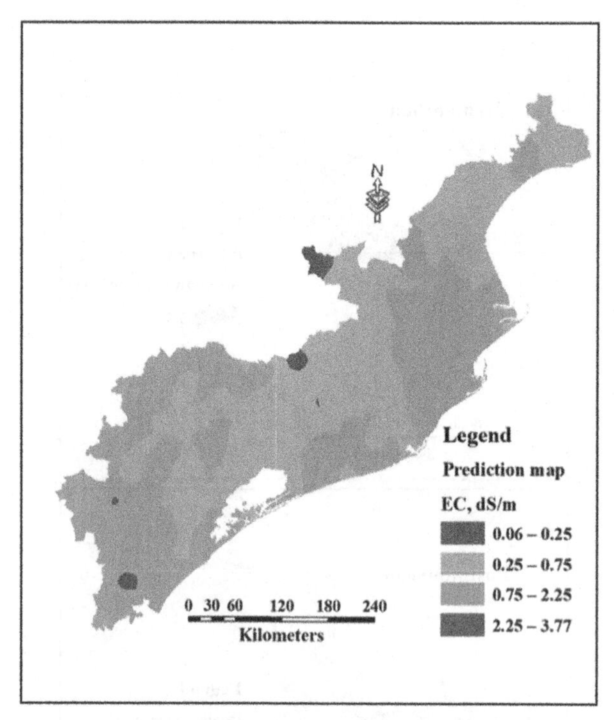

FIGURE 8.6 Spatial variability map of EC in groundwater of coastal belt of Odisha.

water is also a good practice to alleviate the salinity. Cultivation of salt-tolerant field crops like barley, wheat, sorghum, rice, mustard, rape seed, soyabean, and sugarcane, and vegetables like spinach, cucumber, zucchini, beans, cow pea, tomato, peas and sugar beet are the most viable and economical option for farmers to get profit.

Fruit crops like pine apple, fig, and guava can be also suggested for adoptation. Apart from this different biodrainage tree species such as *Eucalypatus, Acacia, Casurina, Ziziphus, Bamboo, Tamarix,* can be planted in large scale for removal of excess salts from the land surface (Dash et al., 2005). Construction of check dams, water harvesting structures like farm pond and bunding in large number can be taken up to impound fresh water. Techniques like farm pond, paddy-cum fish culture, and raised-sunken beds in coastal salt-affected area resulted in increasing cropping intensity from 114 to 186% in West Bengal (Mandal et al., 2018). Similarly Sharma and Chudhari (2012) emphasized on the importance of *rabi* cropping in mono-cropped coastal saline soils, adoption of salt-tolerant rice cultivars, rainwater harvesting through dugout farm ponds, integrated rice-fish culture, and efficient nutrient management to combat salinity problem in coastal areas.

8.4 Conclusion

Groundwater salinity in coastal belts of Odisha is a major concern with respect to crop production, as more than 95% of area falls under medium to high saline groundwater zone. In this

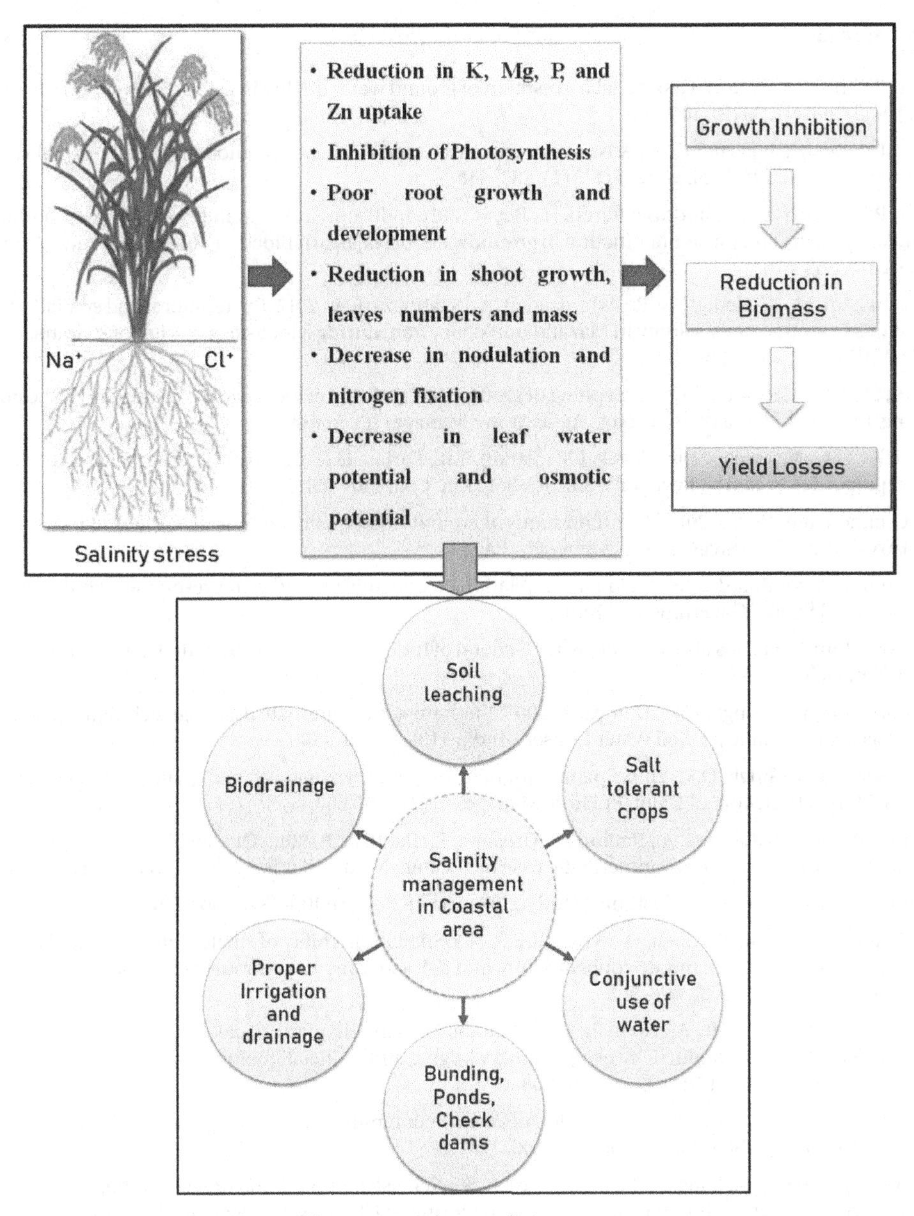

FIGURE 8.7 Management strategies to alleviate coastal area salinity.

context, execution of proper policy plan is necessary to compact the problem of groundwater salinity. The construction of water harvesting structures should be given a high priority which will not only enhance groundwater recharge, but also reduce pressure on freshwater aquifers. Apart from construction of water harvesting structures, salt-tolerant crops like rice, wheat, soyabean, mustard, and vegetables which include cucumber, tomato and peas may be grown in a large scale. Salt tolerant tree species may be planted to remove excess salts from the soil.

References

Adhikary, P.P., Biswas, H., 2011. Geospatial assessment of ground water quality in Datia district of Bundelkhand. Ind. J. Soil Conserv. 39 (2), 108–116.

Adhikary, P.P., Dash, Ch.J., 2017. Comparison of deterministic and stochastic methods to predict spatial variation of groundwater depth. Appl. Water Sci. 7 (1), 339–348.

Adhikary, P.P., Dash, Ch.J., Chandrasekharan, H., Bej, R., 2011. Indicator and probability kriging methods for delineating cu, fe, and mn contamination in groundwater of Najafgarh Block, Delhi, India. J. Env. Monit. Assess. 176, 663–676.

Amiri-Bourkhani, M., Khaledian, M.R., Ashrafzadeh, A., Shahnazari, A., 2017. The temporal and spatial variations in groundwater salinity in Mazandaran plain, Iran, during a long-term period of 26 years. GEOFIZIKA 34 (1), 119–139.

Arslan, H., 2012. Spatial and temporal mapping of groundwater salinity using ordinary kriging and indicator kriging: the case of Bafra plain, Turkey. Agric. Water Manage. 113, 57–63.

Cambardella, C.A., Moorman, T.B., Novak, J.M., Parkin, T.B., Karlen, D.L., Turco, R.F., 1994. Field scale variability of soil properties in central Iowa soils. Soil Sci. Soc. Am. J. 58, 1501–1511.

Central Ground Water Board, 2014. Report on status of ground water quality in coastal aquifers of India, 2014. Ministry of Water Resources, Government of India.

Central Ground Water Board, 2019. National compilation on dynamic ground water resources of India, 2017. Ministry of Jal Shakti, Government of India.

Census, 2011. Primary census abstracts, Registrar General of India, Ministry of Home Affairs. Government of India, New Delhi.

Dash, Ch.J., Sarangi, A., Singh, A.K., Dahiya, S., 2005. Biodrainage: an alternate drainage technique to control waterlogging and salinity. J. Soil Water Conserv. India 4 (3&4), 149–155.

Dash, J.P., Sarangi, A., Singh, D.K., 2010. Spatial variability of groundwater depth and quality parameters in the National Capital Territory of Delhi. Environ. Manage. 45 (3), 640–650.

Delgado, C., Pacheco, J., Cabrera, A., Batllori, E., Orellana, R., Bautista, F., 2010. Quality of groundwater for irrigation in tropical karst environment: the case of Yucatán, Mexico. Agric. Water Manag. 97, 1423–1433.

Environmental Systems Research Institute (ESRI). 2010. ArcGIS Release 10.1. Redlands, CA.

Hu, K., Huang, Y., Li, H., Li, B., Chen, D., White, R.E., 2005. Spatial variability of shallow groundwater level, electrical conductivity and nitrate concentration, and risk assessment of nitrate contamination in north China plain. Environ. Int. 31, 896–903.

Jeihouni, M., Delirhasannia, R., Alavipanah, S.K., Shahabi, M., Samadianfard, S., 2015. Spatial analysis of groundwater electrical conductivity using ordinary kriging and artificial intelligence methods (Case study: Tabriz plain, Iran). GEOFIZIKA 32 (2), 191–208.

Liu, C.W., Jang, C.S., Liao, C.M., 2004. Evaluation of arsenic contamination potential using indicator kriging in the Yun-Lin aquifer (Taiwan). Sci. Total Env. 321, 173–187.

Liu, D., Wang, Z., Zhang, B., Song, K., Li, X., Li, J., 2006. Spatial distribution of soil organic carbon and analysis of related factors in croplands of the black soil region, northeast China. Agric. Eco. Env. 113, 73–81.

Llamas, M.R., Santos, P.M. , 2005. Intensive groundwater use: silent revolution and potential source of social conflicts. J. Water Resour. Plann. Manage. 131 (5), 337–341.

Mandal, S., Raju, R., Kumar, A., Kumar, P., Sharma, P.C., 2018. Current status of research, technology response and policy needs of salt-affected soils in India—a review. Ind. Soc. Coastal Agric. Res. 36, 40–53.

Margat, J., van der Gun, J., 2013. Groundwater Around the world: a Geographical Synopsis. CRC Press, London.

Mohanty, A.K., Rao, V.V.S.G., 2019. Hydrogeochemical, seawater intrusion and oxygen isotope studies on a coastal region in the Puri district of Odisha, India. Catena 172, 558–571.

Mohapatra, P.K., Vijay, R., Pujari, P.R., Sundaray, S.K., Mohanty, B.P., 2011. Determination of processes affecting groundwater quality in the coastal aquifer beneath Puri city, India: a multivariate statistical approach. Water Sci. Tech. 64 (4), 809–817.

Narjary, B., Meena, M.L., Pathan, A.I., Kumar, N., Kumar, S., Kamra, S.K., Sharma, D.K., 2017. Assessing spatio-temporal variations in groundwater table depth and salinity using geostatistics and its relation with groundwater balance in a salt affected soil. Ind. J. Soil Conserv. 45 (3), 235–243.

Nash, J.E., Sutcliffe, L.V., 1970. River flow forecasting through conceptual models part I-a discussion of principles. J. Hydrol. 10 (3), 282–290.

Prusty, P., Farooq, S.H., Swain, D., Chandrasekharam, D., 2020. Association of geomorphic features with groundwater quality and freshwater availability in coastal regions. Int. J. Environ. Sci. Technol. 17, 3313–3328.

Prusty, P., Farooq, S.H., Zimik, H.V., Barik, S.S., 2018. Assessment of the factors controlling groundwater quality in a coastal aquifer adjacent to the Bay of Bengal, India. Environ. Earth Sci. 77 (22), 1–15.

Radhakrishna, I., 2001. Saline fresh water interface structure in Mahanadi Delta region, Orissa, India. Environ. Geol. 40, 369–380.

Sethi, R.R., Srivastava, R.C., Das, M., Anand, P.S.P., Tripathy, J.K., 2016. Comprehensive water resource management in coastal ecosystem of Odisha: a critical review. Nat. Environ. Poll. Technol. 15 (2), 589–594.

Shahraki, A.S., Nasab, S.B., Naseri, A.A., Mohammadi, A.S., 2021. Evaluation of interpolation techniques for estimating groundwater level and groundwater salinity in the Salman Farsi Sugarcane Plantation. Irrig. Sci. Eng. 44 (2), 67–78.

Sharma, D.K., Chaudhari, S.K., 2012. Agronomic research in salt-affected soils of India: an overview. Ind. J. Agronom 57, 175–185.

Srinivasan, R., Singh, S.K., Nayak, D.C., Dharumarajan, S., 2018. Assessment of soil and water salinity and alkalinity in coastal Odisha—a case study. J, Soil Salin. Water Qual. 10 (1), 14–23.

Yimit, H., Eziz, M., Mamat, M., Tohti, G., 2011. Variations in groundwater levels and salinity in the Ili river irrigation area, Xinjiang, northwest China: a geostatistical approach. Int. J. Sustain. Develop. World 18, 55–64.

Integrated GIS-based MCDA approach for suitability zoning of irrigation water quality in semiarid Kansai river basin, Purulia district, West Bengal

Amit Bera, Puja Chowdhury and Ananya Chakraborty

DEPARTMENT OF EARTH SCIENCES, INDIAN INSTITUTE OF ENGINEERING SCIENCE AND TECHNOLOGY, SHIBPUR, HOWRAH, WEST BENGAL, INDIA

9.1 Introduction

Water is the most basic requirement for survival. Water quality has deteriorated as a result of rapid population growth, industrialization, and human activities, and many employees from around the world have observed a scarcity of water resources (FAO, 1994; Kumari et al., 2019). Groundwater pollution from external contaminants such as industrial, urban, and agricultural practices is influenced by a variety of factors such as geology, soil, weathering, industry growth, pollutant emissions, sewage disposal, and other environmental conditions as it moves from point of entry to point of exit (Vasanthavigar et al., 2010; Barua et al., 2021a; Chowdhury et al., 2021). As the population continues to rise, so does the demand for food, resulting in an increase in agricultural goods, lands, and water for production. Irrigation uses fresh water for more than 70% of total withdrawals and 92% of the total water consumption, which is one of the key factors contributing to the degradation of surface and groundwater resources and their quality (FAO, 1994; Adikari et al., 2013; Narany et al., 2016). The criteria used to classify water for a certain purpose are incompatible with other standards; integrating the chemistry of all ions yields better results than using the individual or paired ionic character (Hem, 1985). Groundwater is the primary source of water for drinking and irrigation in the world's arid and semiarid regions (Adimalla and Qian, 2019). Because surface water supplies in those areas are mostly insufficient, reliance, on groundwater has risen considerably in recent years. CGWB (2010) estimates that 65% of groundwater is used for agriculture and roughly 85% is used for drinking purposes. The

Case Studies in Geospatial Applications to Groundwater Resources. DOI: https://doi.org/10.1016/B978-0-323-99963-2.00010-9

quality of groundwater is mostly determined by natural hydrogeological conditions as well as anthropogenic activity in a given area. As a result, regular water quality monitoring is critical for improved livelihood (Romanelli et al., 2012; Bera et al., 2021)

Crop yield and crop production have long-term effects on irrigated water, as well as soil structures and physical qualities, which are primarily influenced by ionic concentrations of sodium, bicarbonate, total dissolved solids (TDS), chloride (Cl^-), and carbonate in the water used. As a result, certain critical water quality metrics such as electrical conductivity (EC), sodium absorption ratio (SAR), residual sodium carbonate (RSC), permeability index (PI), Kelley's ratio (KR), magnesium adsorption ratio (MAR), and concentrations of specific elements such as nitrate, chlorine, and salt must be computed in order to assess the water appropriateness for irrigation purposes (Castellanos et al., 2002; Yidana et al., 2008).

GIS is a strong and user-friendly tool for analyzing and evaluating physical, geoenvironmental, hydrological, and landscape changes, as well as the generation of new spatio-temporal distribution maps. Multicriteria techniques have aided in the evaluation of various operations, such as disaster and water management, environmental studies, and so on (Üstün and Barbarosoğlu, 2015; Gunaalan et al., 2018). Using GIS and multiplecriteria decision analysis (MCDA) techniques, decision-making processes can be analyzed. In recent years, a tool named the analytic hierarchy process (AHP), developed by Saaty (1987), has been used in water resources management studies (Machiwal et al., 2011; Bera et al., 2020; Barua et al., 2021b). Many studies have employed GIS-based MCDA approaches in conjunction with various hydrochemical indicators such as SAR, RSC, or sodium percentage, as well as other parameters, to determine irrigational water quality suitability zones (Romanelli et al., 2012; Narany et al., 2016; Islam et al., 2018; Kumari et al., 2019; Singh et al., 2020).

Kansai river basin is a semiarid region with predominantly hard rock topography, with water mostly found in weathered and secondary cracked zones. Although there is a water shortage, irrigation is carried out in many portions of the district, mostly using surface and groundwater. In areas where there is a scarcity of water, proper water management planning is critical so that it can be used to the best of its ability without creating excessive exploitation and waste. Nag (2005) delineated the groundwater potential zones using lineament density and hydrogeomorphological parameters in the Bagmundi block of Purulia. Mandal et al. (2018) discussed about the land suitability for possible surface irrigation in the catchment regions, and found 51.82% of the slope, 68.62% of land cover/use are highly suitable and ideal for a surface irrigation system using the weighted overlay techniques. Ghosh et al. (2016) delineated the potential groundwater zones in the Kumari watershed using multi influencing factor (MIF) and the GIS technique. Kundu and Nag (2018) discussed on the water quality in the Kashipur block of Purulia. In areas where there is a scarcity of water, proper water management planning is vital so that it can be used to the best of its ability without creating excessive exploitation and waste (Bera and Das, 2021). Irrigational water suitability zones, as well as other irrigational water factors, were not studied in much detail in the Purulia district before. The aim of this research is to apply GIS-based MCDA methodologies to identify groundwater irrigation suitability zones (GISZ) based on water chemistry values and to explore the spatial distribution of water quality based on major cations

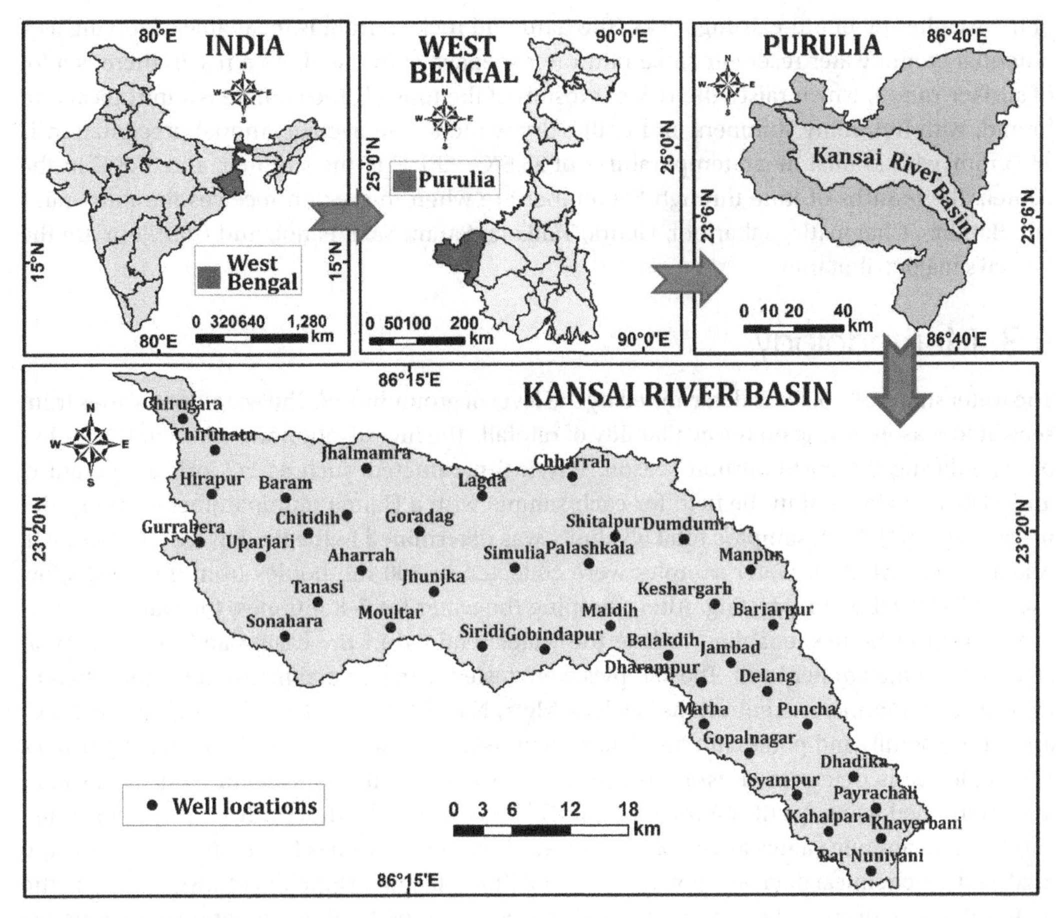

FIGURE 9.1 Location of the study area.

and anions in the Kansai river basin. The findings of this study will aid planners and policymakers in the future formulation of water management policies.

9.2 Study area

The research was carried out in the Kansai River basin of Purulia district, which covers an area of 1549.65 km^2. Geographically the area is bounded by latitudes 23°0′40″ N to 23°29′22″ N and the longitudes lie between 85°56′37″ E and 86°47′11″ E. The Kansai River rises in the Chota Nagpur plateau and flows in a twisted south-easterly direction. Purulia I, Purulia II, Jhalda I, Jhalda II, Bagmundi, Hura, Arsha, Joypur, Puncha, and Manbazar make up the research area (Fig. 9.1). The region lies at the crossroads of the Indo-Gangetic plain and Peninsular India. The Pre-Cambrian basement is overlain by recent alluvium, which is made up of metamorphic rocks, such as schists and gneisses. The district has a layer of residual soil that has formed as a result of the weathering of pre-existing rocks. The hard and rocky terrain is incapable of serving as a potential groundwater reservoir since rainwater seepage is minimal. As a result, there is a lot of surface runoff, which raises the risk of erosion of the topsoil. The climate is subtropical and humid, with hot, rainy summers and chilly, dry winters. The average annual precipitation is 1393 mm, with annual mean temperatures of 25.5°C, 29.1°C in the summer, and 21.3°C in the winter. The months of June through September are when the region receives the most rainfall. Bandhu, Chagmatia, Saharjhor, Goura, Talikota, Saranchaki, Patloi, and Gobrijhor are the Kansai's major tributaries.

9.3 Methodology

The water status of a place is determined by the level of groundwater. The water level varies from season to season, owing on the availability of rainfall. The investigation was place in December of 2020 during the postmonsoon season. Physical parameters such as EC, pH, temperature, and TDS are measured in the field for each sample with a Hanna multiparameter waterproof meter (HI98194). Each sample's total alkalinity was determined in the field using an AQUASOL Alkalinity kit (AE-214). Water samples were collected in 500 mL bottles from the 40 shallow tube wells for laboratory testing. After pumping the water for 7–8 minutes, the water was collected in fresh bottles, enabling to lose the water retained in the casing and obtain a fresh sample from the aquifer itself. The samples were tested using several instruments for different cationic and anionic concentrations, such as Mg^{+2}, Na^{+}, K^{+}, Ca^{+2}, HCO_3^{-}, Cl^{-}, SO_4^{2-}, pH, EC, and TDS. Sodium and potassium are determined using a digital flame photometer (Systronics 128), chloride is determined using the argentometric method, and calcium and magnesium are determined using the titrimetric method. Table 9.1 shows the descriptive statistics of varied cationic and anionic values as well as the physical parameter values for the research area. The analyzed geochemical data are shown on the Hill Piper diagram (Piper, 1944), which aids in the comprehension of groundwater facies. The analyzed values of the above-mentioned parameters were compared to WHO's (2011) recommended standardized values.

For validating the water quality, charge balance error (CBE) is calculated for the samples using the following equation:

$$\%\,\text{CBE} = \frac{\Sigma\,\text{Cations} - \Sigma\,\text{Anions}}{\Sigma\,\text{Cations} - \Sigma\,\text{Anions}} \times 100 \tag{9.1}$$

Here, all the cationic and anionic values are expressed in mg/L. The CBE for all the collected samples is found to be within the acceptable limit, that is, within $\pm 10\%$.

9.4 AHP technique

AHP is a multicriteria decision analysis approach that is widely used. In AHP, the several criteria that are taken into account are analyzed using a pair-wise comparison matrix, and then the weight analysis is calculated (Saaty, 1987). Nine groundwater quality parameters are used to determine irrigational suitability zones in this research. The scores are diagonally arranged in

Table 9.1 Descriptive statistics of the different physicochemical properties of the collected groundwater samples and irrigation water quality indices.

Parameters	Minimum	Maximum	Mean	Median	SD
EC (μs/cm)	402.00	1580.00	812.83	801.00	294.17
TDS (ppm)	257.28	1011.20	520.16	512.64	188.34
pH	6.30	8.10	7.10	7.10	0.41
TH (mg/L)	80.32	602.40	276.98	224.98	124.19
Ca^{+2} (mg/L)	21.60	198.60	66.19	57.90	36.69
Mg^{+2} (mg/L)	5.83	101.55	28.44	19.73	21.77
Na^{+} (mg/L)	11.55	140.84	46.04	35.91	31.01
K^{+} (mg/L)	0.69	34.52	5.36	2.58	8.00
Cl^{-} (mg/L)	19.74	508.45	105.46	70.80	94.25
SO_4^{2-} (mg/L)	10.00	174.00	63.10	49.50	43.07
HCO_3^{-} (mg/L)	102.00	678.00	257.93	251.50	109.76
SAR	2.03	23.12	6.91	5.93	4.49
RSC	-9.70	3.31	-1.42	-1.21	2.43
MAR	16.21	82.46	29.76	25.28	14.62
Na %	16.79	65.69	34.51	33.05	12.85
KR	0.12	1.90	0.54	0.46	0.38
PI	10.46	26.52	16.06	15.98	3.26

Table 9.2 Description of scales for pair comparison with analytic hierarchy process.

Scales	Degree of preferences	Descriptions
1	Equally important	The contribution of the two factors is equally important
3	Slightly important	Experiences and judgment slightly tend to certain factor
5	Quite important	Experiences and judgment strongly tend to certain factor
7	Extremely important	Experiences and judgment extremely strongly tend to certain factor
9	Absolutely important	There is sufficient evidence for absolutely tending to certain factor
2,4,6,8	Intermediate values	In between two judgments

Source: Saaty (1987).

Table 9.3 Random index value.

n	1	2	3	4	5	6	7	8	9	10
RI	0.0	0.0	0.58	0.90	1.12	1.24	1.32	1.41	1.45	1.49

Source: Saaty (1987).

an AHP comparison matrix, which has an equal number of rows and columns. Value 1 is set diagonally after passing through the center corner. The criteria are rated on a scale of 1 to 9 based on their relative relevance (Table 9.2). The principle Eigen value, which is determined from the random index scale (Table 9.3), is used to calculate the consistency ratio (CR). If the CR value is ≤ 0.1, then the consistency of the matrix is ok and it is acceptable for the weight analysis.

Table 9.4 Pairwise comparison matrix for all factors.

Factors	EC	SAR	Na %	Cl	MAR	RSC	KR	TH	PH	PI
EC	1	2	3	4	5	5	6	7	8	9
SAR	1/2	1	2	3	4	4	5	6	7	8
Na %	1/3	1/2	1	2	3	4	5	6	7	8
Cl	1/4	1/3	1/2	1	2	3	4	5	6	7
MAR	1/5	1/4	1/3	1/2	1	2	3	4	4	5
RSC	1/5	1/4	1/4	1/3	1/2	1	2	3	4	5
KR	1/6	1/5	1/5	1/4	1/3	1/2	1	2	3	3
TH	1/7	1/6	1/6	1/5	1/4	1/3	1/2	1	1	2
pH	1/8	1/7	1/7	1/6	1/4	1/4	1/3	1	1	2
PI	1/9	1/8	1/8	1/7	1/5	1/5	1/3	1/2	1/2	1

Table 9.5 Pairwise comparison matrix, weights, and consistency ratio of the data layers.

Factors	EC	SAR	Na%	Cl	MAR	RSC	KR	TH	pH	PI	Weight
EC	0.330	0.403	0.389	0.345	0.302	0.247	0.221	0.197	0.193	0.180	0.281
SAR	0.165	0.201	0.259	0.259	0.242	0.197	0.184	0.169	0.169	0.160	0.201
Na %	0.110	0.101	0.130	0.173	0.181	0.197	0.184	0.169	0.169	0.160	0.157
Cl	0.083	0.067	0.065	0.086	0.121	0.148	0.147	0.141	0.145	0.140	0.114
MAR	0.066	0.050	0.043	0.043	0.060	0.099	0.110	0.113	0.096	0.100	0.078
RSC	0.066	0.050	0.032	0.029	0.030	0.049	0.074	0.085	0.096	0.100	0.061
KR	0.055	0.040	0.026	0.022	0.020	0.025	0.037	0.056	0.072	0.060	0.041
TH	0.047	0.034	0.022	0.017	0.015	0.016	0.018	0.028	0.024	0.040	0.026
pH	0.041	0.029	0.019	0.014	0.015	0.012	0.012	0.028	0.024	0.040	0.023
PI	0.037	0.025	0.016	0.012	0.012	0.010	0.012	0.014	0.012	0.020	0.017

Principal eigen values: 10.796. Consistency ratio (CR): 0.059.

In this research by using comparison matrix (Tables 9.4 and 9.5), the weight calculation was also performed for the subclass of the thematic layers (Table 9.6). The thematic layers are then transformed to a 30 × 30 cell size and categorized based on their weight values (Fig. 9.1). The final groundwater irrigational suitability zone (GISZ) map was prepared by the using the overlay analysis technique.

$$\text{Consistency ratio (CR)} = \frac{\text{Consistency index (CI)}}{\text{Random consistency index (RI)}} \tag{9.2}$$

$$\text{Consistency index (CI)} = (\lambda_{max}\text{-n})/\text{n} - 1 \tag{9.3}$$

where λ_{max} is the principal Eigen value; *n* stands for the number of considering factors. For this present study, the CR value is 0.086 indicating the consistency of AHP matrix is good for weight analysis and appropriate for delineating the GISZ.

Table 9.6 Weights assignment of subclass of all groundwater quality parameters.

Parameters	Range	Suitability classification	Comparison matrix				CR	Weight
			1	2	3	4		
EC (μS/cm)	250–750	Good	1				0.000	0.750
	750–2250	Doubtful	1/3	1				0.250
SAR	<3	Excellent	1				0.050	0.483
	3–10	Good	1/2	1				0.272
	10–18	Moderate	1/3	1/2	1			0.157
	18–26	Doubtful	1/5	1/3	1/2	1		0.088
Na %	<20	Excellent	1				0.019	0.473
	20–40	Good	1/2	1				0.285
	40–50	Less suitable	1/3	1/2	1			0.170
	50–80	Unsuitable	1/5	1/4	1/3	1		0.073
Cl (mg/L)	<140	Highly suitable	1				0.056	0.528
	140–350	Suitable	1/2	1				0.333
	>350	Doubtful	1/3	1/3	1			0.140
MAR	<50	Suitable	1				0.000	0.800
	>50	Unsuitable	1/4	1				0.200
RSC	<1.25	Good	1				0.019	0.558
	1.25–2.5	Doubtful	1/2	1				0.320
	>2.5	Unsuitable	1/4	1/3	1			0.122
KR	<1	Suitable	1				0.000	0.750
	>1	Unsuitable	1/3	1				0.250
TH (mg/L)	75–150	Moderate	1				0.010	0.540
	150–300	Hard	1/2	1				0.297
	>300	Very hard	1/3	1/2	1			0.163
pH	7–8	Excellent	1				0.019	0.558
	6.5–7	Good	1/2	1				0.320
	<6.5, >8	Unsuitable	1/4	1/3	1			0.122
PI	<25	Excellent	1				0.000	0.667
	25–75	Good	1/2	1				0.333

9.5 Results and discussion

9.5.1 Hydrochemical facies

The piper plot was done depending upon the concentration of the major cations and anions as found in the water samples. The (Piper, 1944) plot helps in identifying the various hydrochemical facies type present within a given groundwater basin. The samples contain the cations and anions in decreasing order of dominance as $Ca^{+2} > Na^{+} > Mg^{+2} > K^{+}$ and $HCO^{3-} > Cl^{-} > SO^{4-}$, respectively. From Fig. 9.2, it is observed that most of the groundwater samples are Ca-Mg-SO_4^{2-} (calcium-magnesium-sulfate) water type which is moderately suitable for irrigation. The rest are within the Ca-Mg-HCO_3 (calcium-magnesium-bicarbonate) water type which is highly suitable for agricultural as well as drinking purposes (Kundu and Nag, 2018).

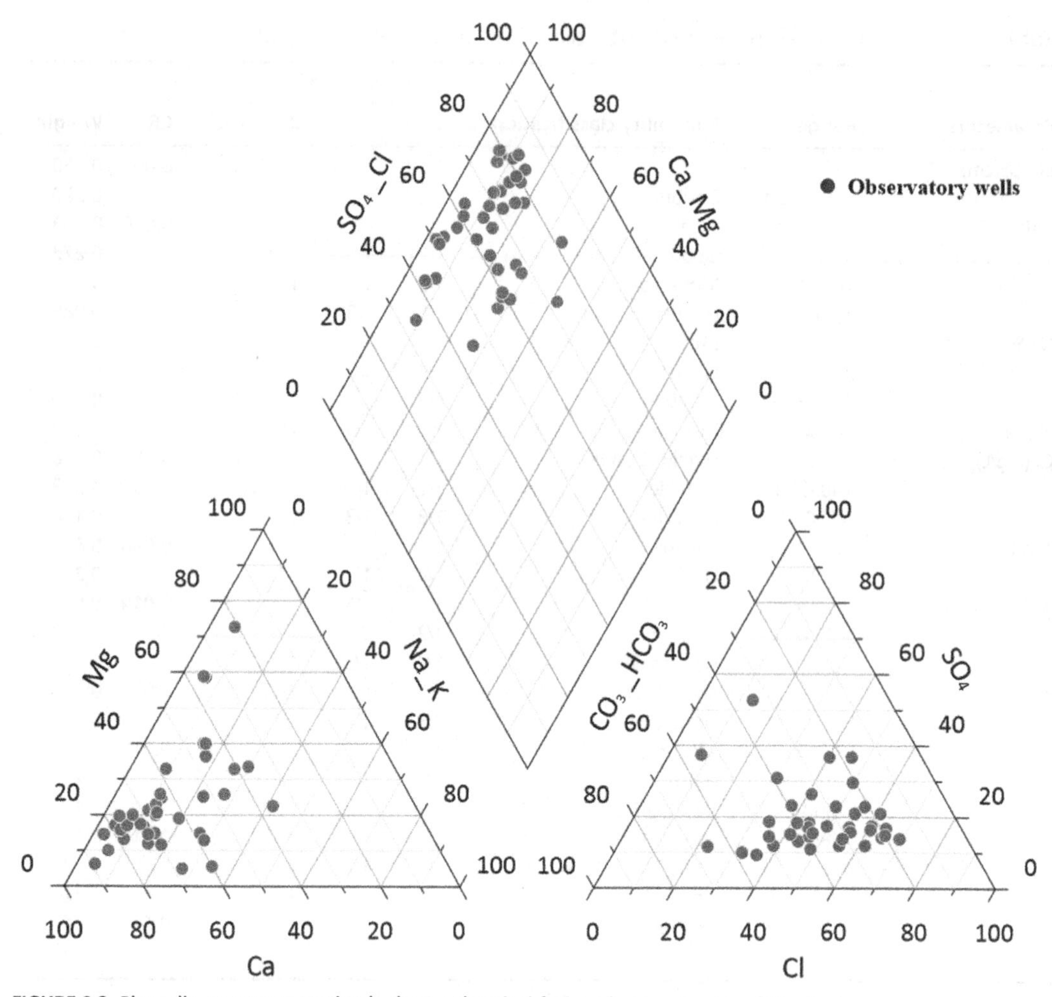

FIGURE 9.2 Piper diagram representing hydrogeochemical facies of groundwater of the study area.

9.6 Study of major cations and anions in the area

Due to the restricted attainability of surface water in the study region, the groundwater quality plays a vital role for irrigation purpose here, especially during the dry seasons (Aravinthasamy et al., 2021). However, the water quality solely depends upon the type of water and the quantity of the various ions dissolved in it. The descriptive statistics of the major physicochemical parameters of water from the study area are shown in (Table 9.1) which have been analyzed for a better understanding of the overall water quality. The analyzed data have been further compared with the standard values of WHO (2011) to better evaluate the condition of the groundwater and its appropriateness for drinking purposes.

From the mean value of the cationic and anionic concentrations in the collected samples from the study area, the cations and anions have been arranged in the following decreasing order of occurrences Ca^{+2} > Na^{+} > Mg^{+2} > K^{+} and HCO^{3-} > Cl^{-} > SO^{4-}, respectively. In the study area, Ca^{+2} ions vary from a lowest concentration of 21.60 mg/L at Chiruhatu to a maximum concentration of 198.60 mg/L at Sonahara block with a mean value of 66.19 mg/L. As per WHO (2011), the desirable range for Ca^{2+} ions in drinking water is 75 mg/L and 40% of the samples have values in the desirable limit while all the samples show Ca^{+2} ion values well within the maximum allowable limit of 200 mg/L (WHO, 2011). The Na^{+} ion concentration varies from a lowest concentration of 11.55 mg/L at Jhalmamra to a maximum concentration of 140.84 mg/L at Dumdumi with a mean value of 46.04 mg/L which is within the maximum allowable limit of 200 mg/L for the drinking purposes. The principal sources of sodium ions can be attributed to the presence of Archean gneisses and the weathering of the crystalline rocks. The concentration of potassium ions (K^{+}) range from a lowest value of 0.69 mg/L to a highest value of 34.52 mg/L with an average of 5.36 mg/L. Goradag shows the lowest value of K^{+} while Chiruhatu shows the highest. In total, 80% of the samples have value within the desirable limits of 12 mg/L while 20% of the samples exceed it. The K^{+} ions present in the groundwater occur as a required trace element to maintain a stability in the human body (He and Macgregor, 2008). Mg^{+2} ions range from a minimum of 5.83 mg/L at Siridi to a maximum of 101.55 mg/L at Chiruhatu with a mean of 28.44 mg/L. The maximum allowable limit is 150 mg/L and all the samples show values that are within the limit. Ca^{+2} and Mg^{+2} are very much required for human body but when taken in excess it can cause adverse effects on human health (Adimalla and Qian, 2019). Bicarbonate (HCO^{3-}) value varies from a lowest concentration of 102 mg/L at Bar Nuniyani to a maximum concentration of 678 mg/L at Chiruhatu with an overall average value of 257.93 mg/L. Cl^{-} concentration varies from 19.74 to 508.45 mg/L with an average value of 105.46 mg/L. Jhalmamra shows the lowest value of Cl^{-} while Aharrah shows the highest. Too much of chloride is known to cause damage to the human body (Marghade et al., 2011; Adimalla et al., 2018). Here it is the second most dominating anion, where 75% of the samples are within the desirable limit of 200 mg/L and all of the samples are within the maximum desirable limit of 600 mg/L. The higher chloride (Cl^{-}) concentration is primarily from effluents, wastes from leaked septic tanks and breakdown of the chloride-bearing minerals (Rao and Latha, 2019). Sulfate (SO_4^{2-}) in the study region ranges from 10 mg/L at Dharampur to 174 mg/L at Dhadika, with an average of 63.10 mg/L. The maximum allowable limit is 250 mg/L (WHO, 2011) and all the samples have values within the allowable limit.

9.7 Groundwater quality for irrigation based on the physicochemical parameters

Groundwater is evaluated for agricultural purpose based upon certain parameters which are calculated from the various irrigation suitability indicators. These parameters include salinity hazard (EC), sodium percentage (Na%), SAR, RSC, PI, MAR, and KR.

9.8 pH

The pH value of water indicates its strength, whether acidic, basic, or neutral. pH together with the alkalinity value constitutes the two prime characteristics of water that determines its suitability for irrigational purpose. Usually, the pH for irrigational water ranges from 6.5 to 8.4 (Bauder et al., 2011; Riaz et al., 2018). The pH value in the study area ranges from a minimum of 6.3 to a maximum of 8.10 with a mean value of 7.10 indicating that most of the samples in the area are neutral to weakly alkaline in nature. According to the guidelines of WHO (2011), all the samples, therefore, have pH values well within the acceptable limit for drinking purpose.

9.9 Total dissolved solids (TDS)

TDS in the Kansai basin area varies from a minimum value of 257.28 mg/L in Gobindapur block to a maximum of 1011.20 mg/L in the Gurrabera block with a mean value of 539.12 mg/L. Based upon WHO's guidelines, the highest permissible limit is 500 mg/L but 1500 mg/L is accepted as the maximum extended desirable limits for drinking purposes. In the present study area, nearly 50% of the samples have TDS value within 500 mg/L while all the samples show TDS value well within the maximum desirable limit as laid down by WHO (2011). Todd (1980) classified the groundwater as fresh water when the TDS value is less than 1000 mg/L; brackish water when TDS value ranged from 1000 to 10,000 mg/L; saline water when TDS ranged from 10,000 to 1,000,000 mg/L. In total, 95% of the samples in the area are freshwater whereas only 5% are brackish in nature.

9.10 Total hardness (TH)

The high value of total hardness (TH) can be attributed to high Ca and Mg concentration. As per WHO (2011), the desirable limit of TH in drinking water is 100 mg/L and the maximum allowable limit is 500 mg/L. The TH is determined using the equation by Todd (1980):

$$TH(CaCO_3)mg/L = (2.497)Ca + (4.115)\,Mg. \tag{9.4}$$

In the present area, water samples show TH ranging from a lowest value of 80.32 mg/L in Siridi to a highest value of 602.40 mg/L in Sonahara block with a mean value of 276.98 mg/L. From the analyzed results, 90% of the samples have values within the maximum allowable limits, while only 10% of the samples show high hardness values within 1500 mg/L.

9.11 Salinity hazard

The conductivity value of groundwater is a measure of the electric current that can be transported by it. The salinity hazard is a measure of the EC. It is usually dependent on the quantity of the sodium ions present and the total concentration of salts present in the given water. A higher EC of water causes saline soil formation leading to breakage of the soil structure and affecting the plant growth and crop yield adversely. Continued usage of groundwater with very high EC values directly affects the yield of the crops (Kumar et al., 2017). In the concerned area, all the samples have EC values ranging from 402 to 1580 μs/cm with a mean value of 812.83 μs/cm which is within the WHO desirable limits of 1500 μs/cm. From the EC values, salinity hazard can be categorized as—low EC value (EC < 250 μs/cm); medium EC value (EC—250-750 μs/cm); high EC value (EC—750-2250 μs/cm) and very high EC value (EC > 2250 μs/cm). Very highly saline waters are not suitable for usage on the soil. In the Kansai river basin area, 48% of the samples show medium EC values while the rest possess medium EC values.

9.12 Sodium absorption ratio (SAR)

As per WHO (2011), irrigation water is commonly graded based upon the SAR value. Thus, while deciding the suitability of water for irrigation purpose, evaluating the SAR value is absolutely necessary. SAR is one of the most important parameter while determining the sodium hazard (Mays and Todd, 2005). The SAR is calculated by the formula given by Richards (1954) as follows:

$$SAR = \frac{Na^{+}}{\frac{1}{2}\sqrt{Ca^{2+} + Mg^{2+}}} \tag{9.5}$$

Overall the SAR value varies from 2.03 at Matha to 23.12 at Dumdumi with an average value of 6.91. A SAR value below 10 is considered to be excellent water and are best to use for irrigation purpose (Richards, 1954; Bouwer, 1978). In the Kansai river basin, 70% of the samples show excellent water while 30% are graded as good water and can be used for irrigation purpose. To check the suitability of the groundwater for agricultural purpose more thoroughly, the United States Salinity Laboratory (USSL) diagram has been plotted for the concerned area (Fig. 9.3).

Here, the values of SAR and EC are reciprocally plotted against each other in a graphical manner. The samples are then graded into five major zones namely—C1S1, C2S1, C3S1, C2S2, and C3S2. The samples falling in the C1S1, C2S2, and C2S2 zone are grouped as good water for irrigation purpose while the samples falling in the C3S1 and C3S2 zone are graded as marginally suitable for irrigation due to higher salinity hazard or EC values. SAR further determines the relative movement of the Na^{+} (sodium) ions within the soil's exchange reactions. This ratio further develops the relative concentration of sodium ions to magnesium and calcium (Wang, 2013). When groundwater having a high SAR value is added to a soil, the Na^{+} ions present in the water can immediately dislocate the Mg^{+2} and Ca^{+2} ions present in the soil. Due to high SAR values, when the soil is dry, it tends to be compact, hard, and reduces the infiltration rates of air and water into the soil. This problem is related to several other factors like the soil type and salinity ratio so, the calculation of SAR value is very much necessary before using the water for agricultural purpose.

9.13 Sodium percentage (Na %)

Yet another critical parameter for understanding the water suitability for irrigation purpose is sodium percentage. An excess of sodium in water might combine with the carbonates leading to

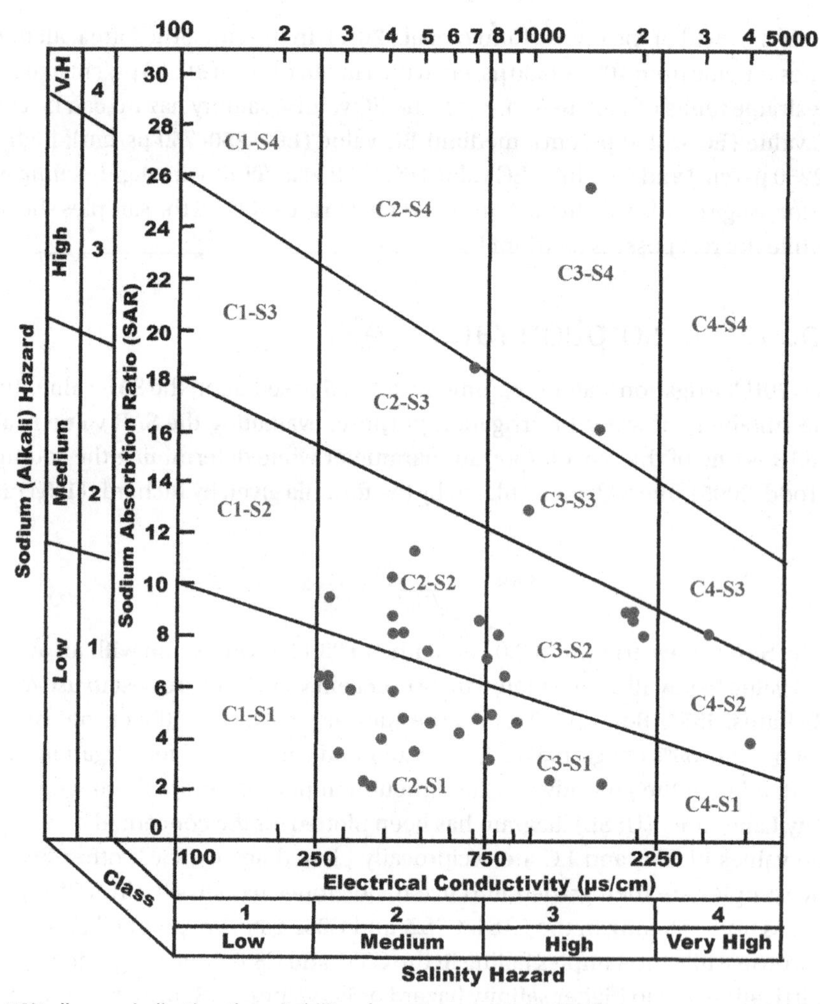

FIGURE 9.3 USSL diagram indicating the suitability of groundwater for irrigation.

an increase in alkaline soils. The excessive sodium with chlorides further increases soil salinity and reduces the soil permeability. The sodium percentage is calculated using (Hem, 1985):

$$\text{Na }\% = \frac{(\text{Na} + \text{K}) \times 100}{(\text{Ca} + \text{Mg} + \text{Na} + \text{K})} \tag{9.6}$$

The value of Na% in the study area lies from 16.79 to 65.69 with an average of 34.51. A Na % value within 50 is suited for agricultural purpose. In total, 90% of the samples in the Kansai river basin are within the permissible limits. The sodium percentage (Na %) and the EC values of the locations are plotted in Wilcox (1948) diagram (Fig. 9.4), which demonstrates a better pictorial representation of the samples best suited for irrigation. As per Wilcox's plot, irrigation water can be grouped into five different classes such as: excellent to good, good to permissible,

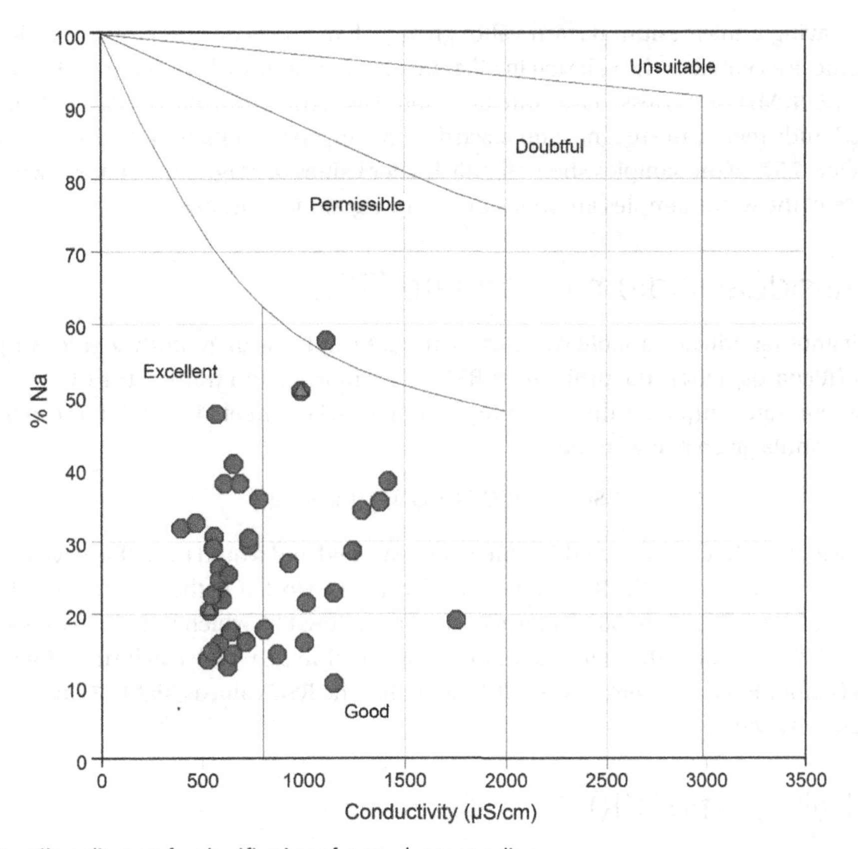

FIGURE 9.4 Wilcox diagram for classification of groundwater quality.

permissible to doubtful, doubtful to unsuitable and unsuitable. The result showed that 65% of the groundwater samples (from 26 locations) fall within the excellent category while only 32.5% (13 locations) have values that fall in the good category. Only 2.5% of the data falls in the permissible region.

9.14 Magnesium adsorption ratio (MAR)

Magnesium is a much-required nutrient for healthy plant growth. Insufficient quantity of magnesium might therefore pose a barrier in plant growth which would adversely affect the crop yield. Therefore, the magnesium concentration in water plays a vital role in determination of water suitability for the irrigation purpose. It is calculated by using the formula (Szabolcs and Darab, 1964):

$$\text{MAR} = \frac{\text{Mg}}{(\text{Mg} + \text{Ca})} * 100 \tag{9.7}$$

Water having a magnesium hazard value greater than 50 is considered unsuitable for irrigation purpose as it makes the soil rich in alkaline concentrations which reduces the crop yield. The waters with MH values less than 50 are best suited for agricultural purposes (Fig. 9.5E). In the concerned study region, the magnesium hazard value ranged from 16.21 to 82.46 with an average of 29.76. Only 7.5% of the samples show slightly higher values of magnesium hazard whereas the rest 92.50% of the water samples are well suited for irrigational purposes.

9.15 Residual sodium carbonate (RSC)

RSC is yet another critical variable while estimating the aptness of groundwater for irrigational purposes (Richards, 1954). The problem of RSC arises in irrigation water if the $CO_3^{2-} + HCO_3^{-}$ content in the water surpasses the $Ca^{2+} + Mg^{2+}$ content (Naseem et al., 2010). It is calculated by using the formula given below in meq/L:

$$RSC = [(HCO_3 + CO_3) - (Ca + Mg)] \tag{9.8}$$

Water with high RSC values would reduce the crop yield and would burn the leaves if used for a longer period of time pan. The RSC range varied from -9.70 to 3.31 in the study area with a mean value of -1.42 (Fig. 9.5F). Thus, 38% of the samples are below 1.25 which is defined as good water and the rest 62% are within the marginal values, that is, within 2.5 which can be used for irrigation purposes (Eaton, 1950). Naseem et al. (2010) stated that the RSC value is highly influenced by the EC, pH, and SAR values.

9.16 Kelly's ratio (KR)

KR is a significant parameter while testing the water suitability for agricultural purposes (Kelley, 1951). The formulae is expressed as:

$$KR = \frac{Na^+}{Ca^{2+} + Mg^{2+}} \tag{9.9}$$

$KR > 1$ indicates an excess quantity of Na in the water while water with $KR < 1$ is best suited for irrigation (Fig. 9.5G). The calculated KR varies from a minimum of 0.12 at Chirugara to a maximum of 1.90 at Dumdumi with a mean value of 0.54 in the concerned area. In total, 90% of the samples are within the permissible limits which means that they are suitable for irrigation purposes based upon the KR value while the rest 10% are unsuitable (Fig. 9.6).

9.17 Permeability index (PI)

Excessive quantity of certain ions such as Na^+, Ca^{+2}, Mg^{+2}, and HCO_3^- damages the soil structure thereby lowering the permeability. Doneen (1964) grouped the PI-based upon the relation between cations and alkalies, into three main categories—Class I, Class II, and Class III, to determine the usability of water for irrigational purposes. Classes I and II are good for irrigation purpose and Class III is unsuitable or irrigation or agricultural work. It is expressed in meq/L and

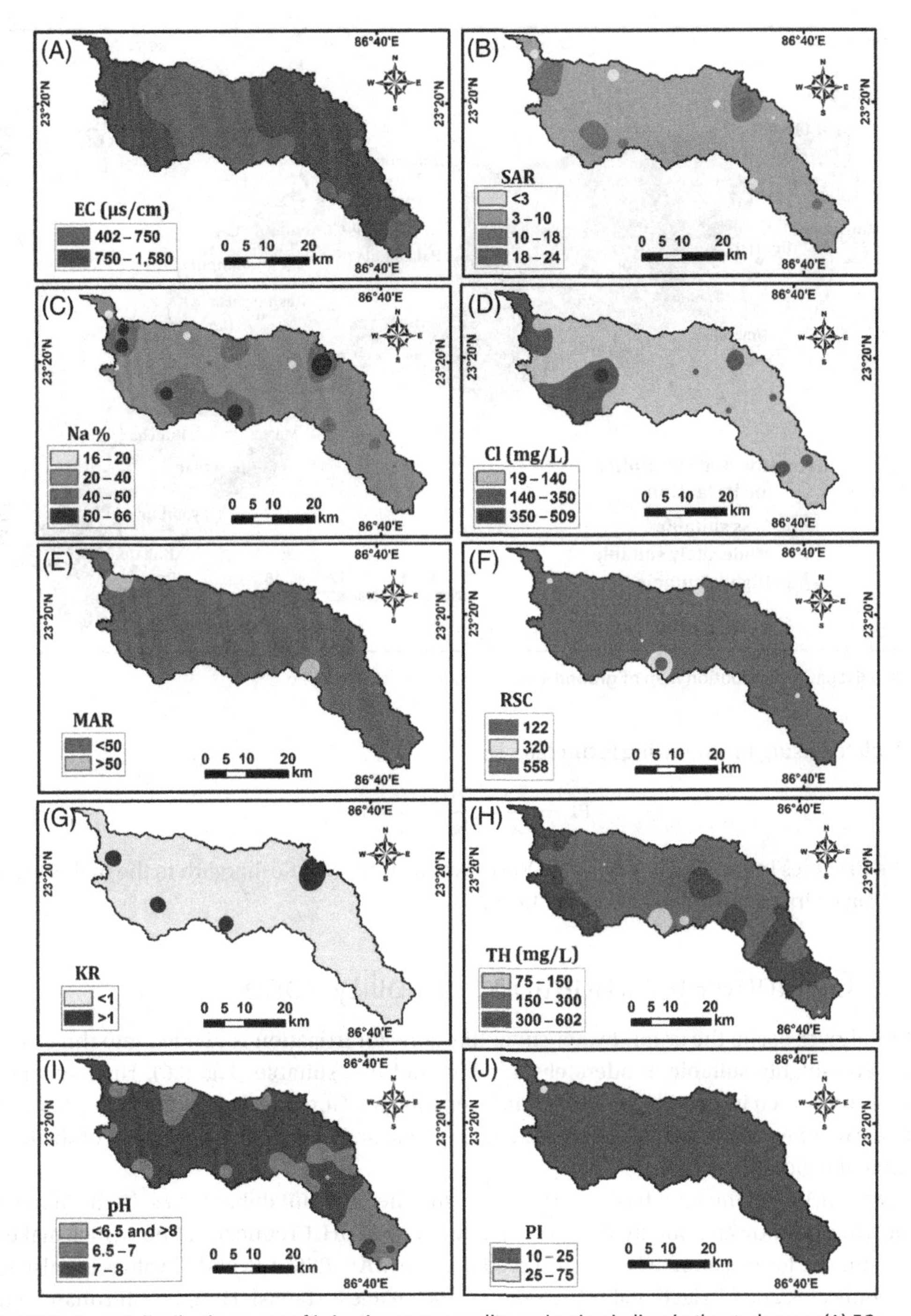

FIGURE 9.5 Spatial distribution maps of irrigation water quality evaluation indices in the study area; (A) EC; (B) SAR; (C) Na %; (D) Cl; (E) MAR; (F) RSC; (G) KR; (H) TH; (I) pH; (J) PI.

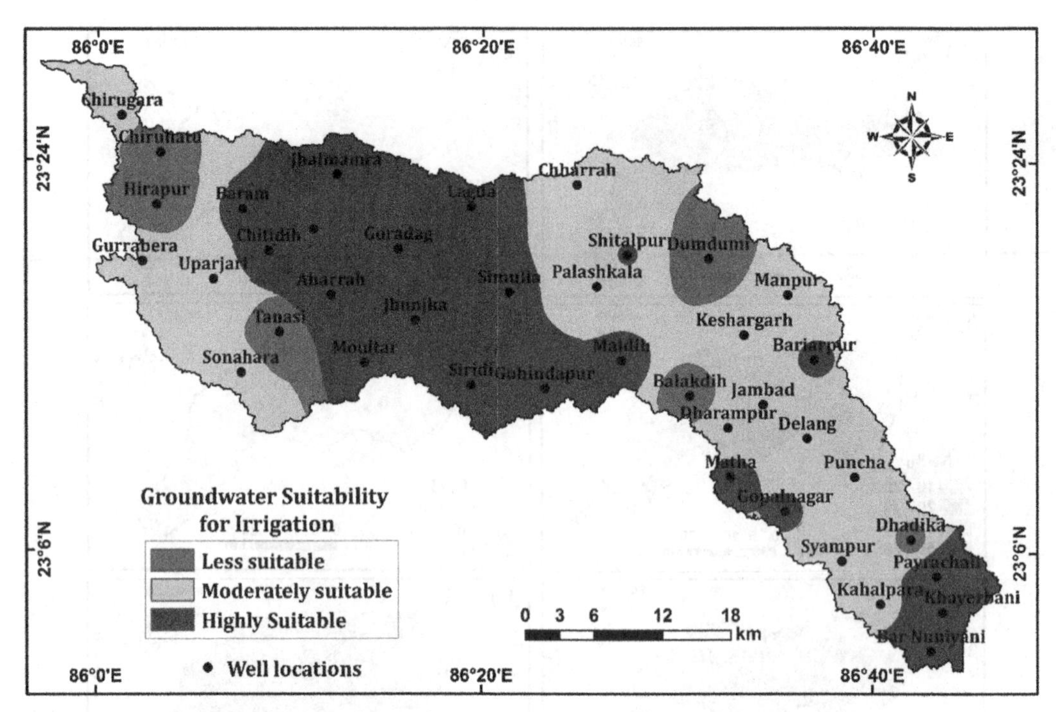

FIGURE 9.6 Spatial distribution map of groundwater irrigation suitability zone.

is calculated using the following formulae.

$$PI = \frac{Na + \sqrt{HCO_3}}{Ca + Mg + Na} \times 100 \tag{9.10}$$

The PI values in the area lies from the lowest value of 10.46 at Keshargarh to the highest value of 26.52 at Chiruhatu with an average of 16.06.

9.18 Groundwater irrigation suitability zone

In Kansai river basin, the groundwater suitability zone for irrigation is divided into three zones which are—highly suitable, moderately suitable, and less suitable (Fig. 9.6). High suitability zones are observed in the villages—Moultar, Sirdi, Jhunjka, Goradag, Lagda, Chitidih, etc., of the middle catchment area and villages—Matha, Gopalnagar, Payrachali, Khayerbani of the lower catchment region of the basin.

Nearly 41.18% of the total basinal area are within the high suitability zones. In the aforesaid regions, the groundwater quality show low to moderate EC, pH, Cl concentrations which makes it highly suitable for irrigation it is also observed that the MAR, RSC, KR, and PI values are also low which indicates good irrigational water quality. The villages—Tanasi, Hirapur, Chiruhatu of the upper catchment region and villages—Dumdumi, Balakdih, and Dhadika of the lower catchment regions are observed within the lower groundwater suitability zones which covers nearly 11.19%

of the total area. This region shows slightly higher concentrations of KR, TH, sodium %, SAR, and EC which are less suitable for irrigation. The rest 47.83% of the total basinal area shows moderate suitability zonation. Therefore, it can be inferred that moderately suitable zone is dominating in the Kansai river basin.

9.19 Conclusion

The present work mainly focusses upon understanding the overall water quality for both drinking and agricultural practices in the study area. From the calculated statistics, the mean values of the major cations are found to be 66.19 mg/L for Ca^{2+}, 46.04 mg/L for Na^{+}, 28.44 mg/L for Mg^{2+}, 5.36 mg/L for K^{+} and for the anions, the mean values are 257.93 mg/L for HCO^{3-}, 105.46 mg/L for Cl^{-}, 63.10 mg/L for $SO_4{}^{2-}$. The hydrochemical facies are inferred using the piper plot, which suggests that the dominant water type is Ca-Mg-$SO_4{}^{2-}$ (calcium-magnesium-sulfate) facies in the Kansai river basin. The USSL diagram was also plotted which shows that 62% of the samples are good water for irrigation purpose while the rest 38% are graded as marginally suitable for irrigation due to higher salinity hazard or EC values. The AHP technique was then executed by using nine of the irrigational water quality parameters and the final map delineating the groundwater suitability zones for irrigation practice was prepared using the weighted analysis. On the whole, the Kansai river basin has been categorized into three major groundwater suitability zones namely—(1) highly suitable zones; (2) moderately suitable zones, and (3) less suitable zones. Around 41.18% of the area falls under the highly suitable zone while 47.63% of it comes under moderately suitable zone. The rest 11.19% of the area falls in the less suitability zone. Therefore, the moderately suitable zone, covering an area of nearly 738.12 km^2 is the most dominating area in the region. These results can assist the policy makers while taking decisions regarding agricultural production, environmental planning, and water management programs in the near future.

Acknowledgment

The authors are grateful to the State Water Investigation Directorate (SWID), Kolkata and Agro-irrigation officers, Belguma for providing information's related to Purulia district. The authors also extend their thanks to anonymous reviewers for the valuable constructive comments and suggestion.

Competing interests

The authors declare that they have no competing interests.

References

Adhikari, K., Chakraborty, B., Gangopadhyay, A., 2013. Assessment of irrigation potential of ground water using water quality index tool. Asian J. Water Environ. Pollut. 10 (3), 11–21.

Adimalla, N., Li, P., Venkatayogi, S., 2018. Hydrogeochemical evaluation of groundwater quality for drinking and irrigation purposes and integrated interpretation with water quality index studies. Environ. Process. 5, 363–383. https://doi.org/10.1007/s40710-018-0297-4.

Adimalla, N., Qian, H., 2019. Hydrogeochemistry and fluoride contamination in the hard rock terrain of central Telangana, India: analyses of its spatial distribution and health risk. SN. Appl. Sci. 1, 202. https://doi.org/10.1007/s42452-019-0219-8.

Aravinthasamy, P., Karunanidhi, D., Subramani, T., Roy, P.D., 2021. Demarcation of groundwater quality domains using GIS for best agricultural practices in the drought-prone Shanmuganadhi river basin of South India. Environ. Sci. Pollut. Res. 28 (15), 18423–18435. https://doi.org/10.1007/s11356-020-08518-5.

Barua, S., Mukhopadhyay, B.P., Bera, A., 2021. Integrated assessment of groundwater potential zone under agricultural dominated areas in the western part of Dakshin Dinajpur district, West Bengal, India. Arab. J. Geosci. 14, 1042. https://doi.org/10.1007/s12517-021-07312-y.

Barua, S., Mukhopadhyay, B.P., Bera, A., 2021. Hydrochemical assessment of groundwater for irrigation suitability in the alluvial aquifers of Dakshin Dinajpur district, West Bengal, India. Environ. Earth Sci. 80, 514. https://doi.org/10.1007/s12665-021-09832-y.

Bauder, T.A., Waskom, R., Sutherland, P., Davis, J., Follett, R., Soltanpour, P., 2011. Irrigation water quality criteria service in action; No 0506. Colorado State University Extension, Fort Collins, Colorado, pp. 1–4.

Bera, A., Das, S., 2021. Water resource management in semi-arid Purulia district of West Bengal, in the context of sustainable development goals. In: Shit, P.K., Bhunia, G.S., Adhikary, P.P., Dash, C.J. (Eds.), Groundwater and Society. Springer, Cham https://doi.org/10.1007/978-3-030-64136-8_23.

Bera, A., Mukhopadhyay, B.P., Barua, S., 2020. Delineation of groundwater potential zones in Karha river basin, Maharashtra, India, using AHP and geospatial techniques. Arab. J. Geosci. 13 (15), 1–21. https://doi.org/10.1007/s12517-020-05702-2.

Bera, A., Mukhopadhyay, B.P., Chowdhury, P., Ghosh, A., Biswas, S., 2021. Groundwater vulnerability assessment using GIS-based drastic model in Nangasai river basin, India with special emphasis on agricultural contamination. Ecotoxicol. Environ. Saf. 214, 112085. https://doi.org/10.1016/j.ecoenv.2021.112085.

Bouwer, H., 1978. Groundwater hydrology. New York: McGraw-Hill, USA. p. 480.

Castellanos, J.Z., Ortega Guerrero, A., Grajeda, O.A., Vázquez Alarcón, A., Villalobos, S., Muñoz Ramos, J.J., Zamudio, B., Martínez, J.G., Hurtado, B., Vargas, P., Enríquez, S.A., 2002. Changes in the quality of groundwater for agricultural use in Guanajuato. Terra Latinoamericana 20 (2), 161.

CGWB, 2010. Ground water quality in shallow aquifers of India. Central Ground Water Board, Faridabad, India, pp. 1–117. http://cgwb.gov.in/documents/waterquality/gw_quality_in_shallow_aquifers.pdf. (Accessed 28 October 2021).

Chowdhury, P., Mukhopadhyay, B.P., Nayak, S., Bera, A., 2021. Hydro-chemical characterization of groundwater and evaluation of health risk assessment for fluoride contamination areas in the eastern blocks of Purulia district, India. Environ. Dev. Sustain. 1–28. doi:10.1007/s10668-021-01911-1.

Doneen, L.D., 1964. Notes on water quality in agriculture published as a water science and engineering, paper 4001. Department of Water Science and Engineering, University of California, Los Angeles, CA, USA.

Eaton, F.M., 1950. Significance of carbonate in irrigation water. Soil Sci. 62 (2), 123–133.

FAO, 1994. Water Quality For Agriculture. In: Irrigation and Drainage Paper-29, 3rd ed. Food and Agriculture Organization of the United Nations, Rome, p. 174.

Ghosh, P.K., Bandyopadhyay, S., Jana, N.C., 2016. Mapping of groundwater potential zones in hard rock terrain using geoinformatics: a case of Kumari watershed in western part of West Bengal. Model. Earth Syst. Environ. 2, 1–12. https://doi.org/10.1007/s40808-015-0044-z.

Gunaalan, K., Ranagalage, M., Gunarathna, M.H.J.P., Kumari, M.K.N., Vithanage, M., Srivaratharasan, T., Saravanan, S., Warnasuriya, T.W.S., 2018. Application of geospatial techniques for groundwater quality and availability assessment: a case study in Jaffna Peninsula, Sri Lanka. ISPRS Int. J. Geo Inf. 7 (1), 20. https://doi.org/10.3390/ijgi7010020.

He, F.J., MacGregor, G.A., 2008. Beneficial effects of potassium on human health. Physiol. Plant. 133 (4), 725–735. https://doi.org/10.1111/j.1399-3054.2007.01033.x.

Hem, J.D., 1985. Study and interpretation of the chemical characteristics of natural water, 3rd edn. United States Geological Survey Water-Supply Paper 2254, USA. p. 263.

Islam, M.A., Rahman, M.M., Bodrud-Doza, M., Muhib, M.I., Shammi, M., Zahid, A., Akter, Y., Kurasaki, M., 2018. A study of groundwater irrigation water quality in south-central Bangladesh: a geo-statistical model approach using GIS and multivariate statistics. Acta Geochim. 37 (2), 193–214. https://doi.org/10.1007/s11631-017-0201-3.

Kelley, W.P., 1951. Alkali soils; their formation, properties, and reclamation. Reinhold Publishing Corporation, New York.

Kumar, B., Gangwar, V., Parihar, S.K.S., 2017. Effect of saline water irrigation on germination and yield of wheat (Triticumaestivum L.) genotypes. Agrotechnology 6 (1), 156.

Kumari, M.K.N., Sakai, K., Kimura, S., Yuge, K., Gunarathna, M.H.J.P., 2019. Classification of groundwater suitability for irrigation in the Ulagalla tank cascade landscape by GIS and the analytic hierarchy process. Agronomy 9 (7), 351. https://doi.org/10.3390/agronomy9070351.

Kundu, A., Nag, S.K., 2018. Assessment of groundwater quality in Kashipur block, Purulia district, West Bengal. Appl. Water Sci. 8 (1), 33. https://doi.org/10.1007/s13201-018-0675-0.

Machiwal, D., Jha, M.K., Mal, B.C., 2011. Assessment of groundwater potential in a semi-arid region of India using remote sensing, GIS and MCDM techniques. Water Resour. Manage. 25, 1359–1386. https://doi.org/10.1007/s11269-010-9749-y.

Mandal, B., Dolui, G., Satpathy, S., 2018. Land suitability assessment for potential surface irrigation of river catchment for irrigation development in Kansai watershed, Purulia, West Bengal, India. Sustain. Water Resour. Manag. 4, 699–714. https://doi.org/10.1007/s40899-017-0155-y.

Marghade, D., Malpe, D.B., Zade, A.B., 2011. Geochemical characterization of groundwater from northeastern part of Nagpur urban, Central India. Environ. Earth. Sci. 62, 1419–1430. https://doi.org/10.1007/s12665-010-0627-y.

Mays, L.W., Todd, D.K., 2005. Groundwater Hydrology, 3rd ed. John Wily and Sons, Inc., Arizona State University, New York.

Nag, S.K., 2005. Application of lineament density and hydrogeomorphology to delineate groundwater potential zones of Baghmundi block in Purulia district, West Bengal. J. Indian Soc. Remote Sens. 33, 521. https://doi.org/10.1007/BF02990737.

Narany, T.S., Ramli, M.F., Fakharian, K., Aris, A.Z., 2016. A GIS-index integration approach to groundwater suitability zoning for irrigation purposes. Arab. J. Geosci. 9, 502. https://doi.org/10.1007/s12517-016-2520-9.

Naseem, S., Hamza, S., Bashir, E., 2010. Groundwater geochemistry of winder agricultural farms, Balochistan, Pakistan and assessment for irrigation water quality. Eur. Water. 31, 21–32.

Piper, A.M., 1944. A graphic procedure in the geochemical interpretation of water-analyses. Eos Trans. AGU. 25 (6), 914–928. https://doi.org/10.1029/TR025i006p00914.

Rao, K.N., Latha, P.S., 2019. Groundwater quality assessment using water quality index with a special focus on vulnerable tribal region of Eastern Ghats hard rock terrain, Southern India. Arab. J. Geosci. 12, 267. https://doi.org/10.1007/s12517-019-4440-y.

Riaz, U., Abbas, Z., Mubashir, M., Jabeen, M., Zulqadar, S.A., Javeed, Z., Rehman, S., Ashraf, M., Qamar, M.J., 2018. Evaluation of ground water quality for irrigation purposes and effect on crop yields: a GIS based study of Bahawalpur. Pak. J. Agric. Res. 31 (1). http://dx.doi.org/10.17582/journal.pjar/2018/31.1.29.36.

Richards, L.A., 1954. In: Diagnosis and Improvement of Saline and Alkaline Soils, 78. US Department of Agriculture Hand Book, Washington, DC, p. 154 p 60.

Romanelli, A., Lima, M.L., Londono, O.M.Q., Martínez, D.E., Massone, H.E., 2012. A GIS-based assessment of groundwater suitability for irrigation purposes in flat areas of the wet Pampa plain, Argentina. Environ. Manage. 50, 490–503. https://doi.org/10.1007/s00267-012-9891-9.

Saaty, R.W., 1987. The analytic hierarchy process-what it is and how it is used. Math. Model. 9 (3-5), 161–176.

Singh, G., Rishi, M.S., Herojeet, R., Kaur, L., Sharma, K., 2020. Multivariate analysis and geochemical signatures of groundwater in the agricultural dominated taluks of Jalandhar district, Punjab, India. J. Geochem. Explor. 208, 106395. https://doi.org/10.1016/j.gexplo.2019.106395.

Szabolcs, I., Darab, C., 1964. The influence of irrigation water of high sodium carbonate content of soils. In: Proceedings of the 8th International Congress of ISSS, Tsukuba, pp. 803–812.

Todd., D.K., Mays, L.W., 1980. Groundwater Hydrology. John Willey & Sons. Inc., New York, NY, p. 535.

Üstün, A.K., Barbarosoğlu, G., 2015. Performance evaluation of Turkish disaster relief management system in 1999 earthquakes using data envelopment analysis. Nat. Hazards 75, 1977–1996. https://doi.org/10.1007/s11069-014-1407-x.

Vasanthavigar, M., Srinivasamoorthy, K., Vijayaragavan, K., Ganthi, R.R., Chidambaram, S., Anandhan, P., Vasudevan, S., 2010. Application of water quality index for groundwater quality assessment: Thirumanimuttar sub-basin, Tamilnadu, India. Environ. Monit. Assess. 171, 595–609. https://doi.org/10.1007/s10661-009-1302-1.

Wang, S., 2013. Groundwater quality and its suitability for drinking and agricultural use in the Yanqi basin of Xinjiang province, Northwest China. Environ. Monit. Assess. 185, 7469–7484. https://doi.org/10.1007/s10661-013-3113-7.

WEF, 1998. Standard Methods for the Examination of Water and Wastewater, 20th ed. American Public Health Association, American Water Works Association, and Water Environmental Federation, Washington, DC.

WHO, 2011. Guidelines for drinking-water quality. Ed. F. chronicle 38 (4), 104–108.

Wilcox, L.V., 1948. The quality of water for irrigation use. US Department of Agriculture, technical bulletins-962, Washington DC. pp 1–40.

Yidana, S., Ophori, D., Banoeng-Yakubo, B., 2008. Groundwater availability in the shallow aquifers of the southern voltaian system: a simulation and chemical analysis. Environ. Geol. 55, 1647–1657. https://doi.org/10.1007/s00254-007-1114-y.

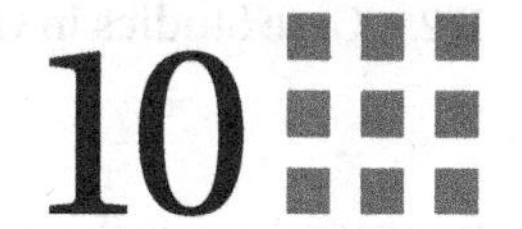

Field-based spatio-temporal monitoring of hydrograph network stations to predict the long-term behavioral pattern of groundwater regime and its implications in India: A review

Anadi Gayen

CENTRAL GROUND WATER BOARD, EASTERN REGION, KOLKATA, DEPARTMENT OF WATER RESOURCES, RIVER DEVELOPMENT AND GANGA REJUVENATION, MINISTRY OF JAL SHAKTI, GOVERNMENT OF INDIA, INDIA

10.1 Introduction

India consists of diverse hydrogeological units with complex spatial distribution and groundwater potential characteristics. Unconsolidated formations - alluvial are present below the extra-peninsular region of Himalaya in the east-west directions covering the states like parts of Gujrat and Rajasthan, Coastal areas of India, and Indo-Gangetic plains. Out of which Indo-Gangetic plains contain enormous groundwater reserves down to 600 m bgl depth. The Indo-Gangetic plains receive high rainfall and hence recharge is ensured and therefore it supports large-scale development through deep tube wells. Consolidated/semiconsolidated formations like sedimentaries, basalts, and crystalline rocks in peninsular areas, the availability of groundwater depend on secondary porosity developed due to weathering, fracturing, etc. Scope for availability of groundwater at shallow depths (within 20–40 m) in few areas and deeper depths (within 100–200 m) in other areas. In the hilly areas, storage capacity is low due to quick runoff.

The behavior of groundwater regimes in the Indian subcontinent is very complex owing to the occurrence of diversified geological formations with significant lithological and sequential variations, multifaceted tectonic framework, climatological variations, and different hydrochemical conditions. Studies since many years have shown that aquifer groups in alluvial/soft rocks even

Case Studies in Geospatial Applications to Groundwater Resources. DOI: https://doi.org/10.1016/B978-0-323-99963-2.00009-2

transcend the surface basin boundaries (CGWB, 2019). Generally, two groups of rock formations have been recognized on the basis typically different hydraulics of groundwater, namely, Porous Formations and Fissured Formations.

Three optimization models are proposed to select the best subset of stations from a large groundwater monitoring network: (1) one that maximizes spatial accuracy; (2) one that minimizes temporal redundancy; and (3) a model that both maximizes spatial accuracy and minimizes temporal redundancy (Nunes et al., 2004). It is essential that the sampling techniques utilized in groundwater monitoring provide data that accurately depicts the water quality of the sampled aquifer in the vicinity of the well. It would be desirable to minimize the requirements of sampling time, equipment, and quantity of contaminated waters pumped to the surface, without loss of data integrity (Powell and Puls, 1993). Using the criteria of maximizing information and minimizing cost, a methodology is developed for design of an optimal groundwater-monitoring network for water resources management. A monitoring system is essentially an information collection system. Therefore, its technical design requires a quantifiable measure of information which can be achieved through application of the information (or entropy) theory. The theory also provides information-based statistical measures to evaluate the efficiency of the monitoring network. The methodology is applied to groundwater monitoring wells in a portion of Gaza Strip in Palestine (Mogheir et al., 2002). Groundwater monitoring network plan in Taiwan includes basic groundwater data, including water-level and water-quality data are being collected, and a reliable database is being established for the purpose of managing total water resources (Hsu, 1998).

10.1.1 Study area

The study area covers the areas where the groundwater monitoring points are located throughout India. The network hydrograph monitoring stations in India are shown in Fig. 10.1.

State as well as Union Territory (UT) wise allocation the monitoring well and type of hydrograph network monitoring station are shown in Table 10.1.

10.1.2 Objectives of the study

Groundwater monitoring is the collection of data generally at set locations at regular time intervals in order to provide information pertaining to the long-term behavior of groundwater regime and the generated data would be useful for the following purposes:

- To determine the quantity, quality, and temperature of groundwater both in space and time.
- To provide the basis for detecting the spatio-temporal trends of groundwater regime.
- Preparation of management plans.
- Ensuring the sustainability of the groundwater development programs.
- To study inter-relationship between ground water and climatic parameters such as rainfall.
- To study the hydrochemical behavior of groundwater.
- To assess the groundwater resources.
- To study the salinity ingress in coastal areas.
- To demarcate the drought-prone area.

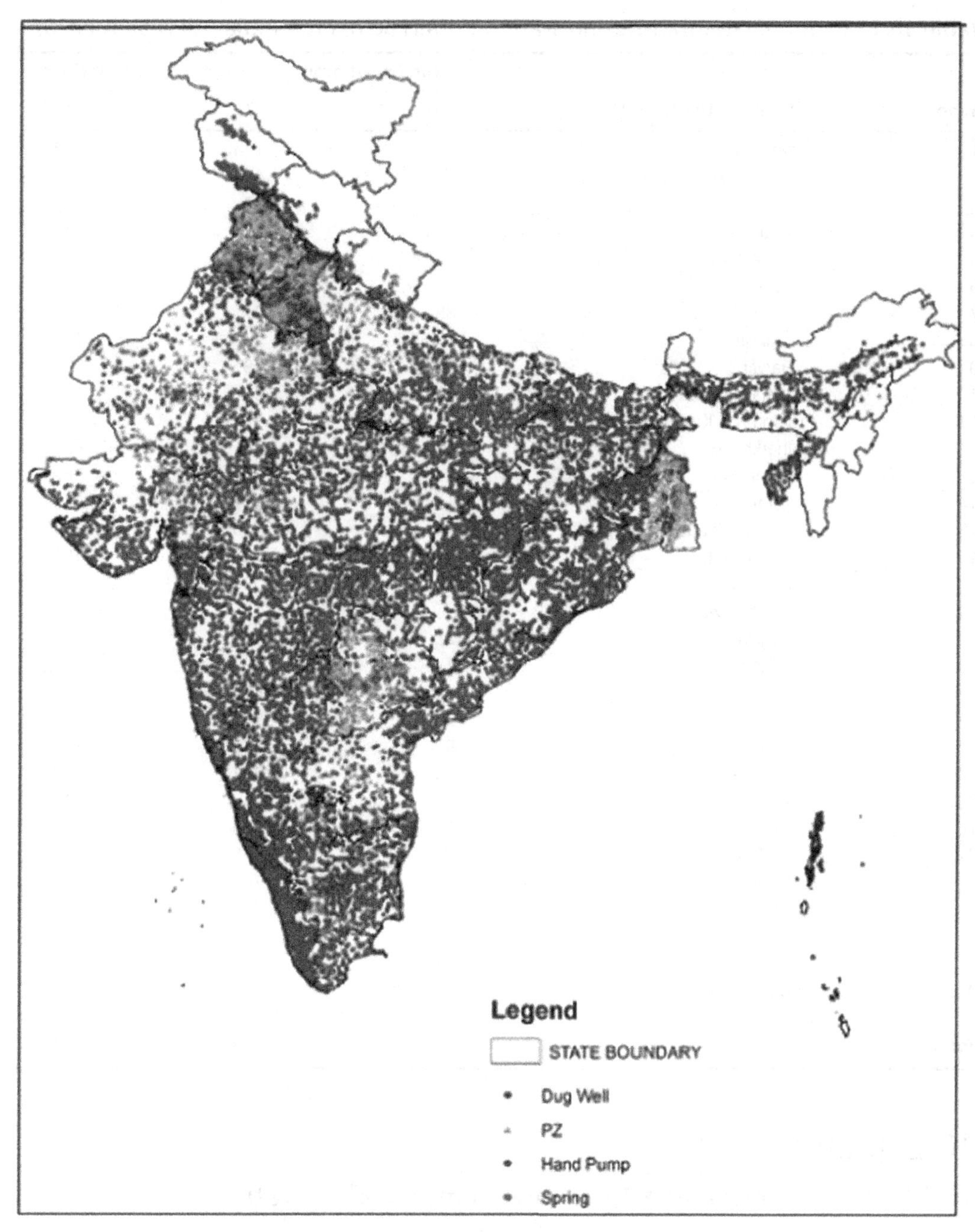

FIGURE 10.1 Location of the monitoring well throughout India (Source: CGWB).

Table 10.1 Groundwater monitoring station details across the India (till March 2020).

		Number of groundwater monitoring stations		
S. no.	**Name of the state/UTs**	**DW**	**PZ**	**Total**
1	Andhra Pradesh	674	164	838
2	Arunachal Pradesh	23	4	27
3	Assam	361	24	385
4	Bihar	734	23	757
5	Chhattisgarh	1157	268	1425
6	Delhi	17	84	101
7	Goa	88	44	132
8	Gujarat	638	266	904
9	Haryana	536	785	1321
10	Himachal Pradesh	128	0	128
11	Jammu & Kashmir	287	11	298
12	Jharkhand	447	20	467
13	Karnataka	1413	262	1675
14	Kerala	1381	221	1602
15	Madhya Pradesh	1203	316	1519
16	Maharashtra	1739	180	1919
17	Manipur	0	0	0
18	Meghalaya	56	11	67
19	Nagaland	22	8	30
20	Odisha	1510	92	1602
21	Punjab	205	850	1055
22	Rajasthan	709	556	1265
23	Tamil Nadu	814	472	1286
24	Telangana	306	442	748
25	Tripura	102	13	115
26	Uttar Pradesh	871	260	1131
27	Uttarakhand	39	166	205
28	West Bengal	766	765	1531
1	Andaman & Nicobar	111	2	113
2	Chandigarh	1	36	37
3	Dadra & Nagar Haveli	17	0	17
4	Daman & Diu	11	3	14
5	Pondicherry	9	7	16
Total		**16375**	**6355**	**22730**

10.2 Methodology

10.2.1 Design of network hydrograph monitoring stations

A groundwater monitoring network is a system of dedicated groundwater monitoring wells in a hydrogeologial unit at which groundwater levels and water quality are measured at pre-determined frequency (Uil et al., 1999). The design of an optimal network layout reflects the entire hydrogeological system of the area under consideration. The network provides long-term

FIGURE 10.2 Open well (dug well).

information on the different aquifers being developed. Depending on the need, the network stations can be categorized into three types: (1) basic network; (2) specific network; and (3) temporary network. Design of the monitoring network hydrograph stations may be undertaken based on the aim of the study, which includes:

a) Types of monitoring points.
b) The network density and location of measuring points.
c) Sampling frequency.

Hence, optimization of monitoring network stations is of prerequisite for the research. At a number of locations, no station will be available. Hence, the information is to be obtained from the network by interpolation. But if the interpolation is too large then an additional station or a redesign should be consider. The optimization process includes the coefficient of variation (Cv) and percentage of error (P) method:

$N = (\mathrm{Cv}/P)^2$

N = Optimum number of network station

Cv = Coefficient of variation

$N = (\mathrm{Cv}/E)^2$ where E = Percentage error (P) * range of variation of water table (R)/mean water table (X)

R = Maximum water level - minimum water level

In the optimization process, kriging technique holds good. The basic idea of kriging is to predict the value of a function at a given point by computing a weighted average of the known values of the function in the neighborhood of the point.

10.2.1.1 Types of monitoring points

i) Shallow wells/dug well (Fig. 10.2).
ii) Abandoned deep wells.

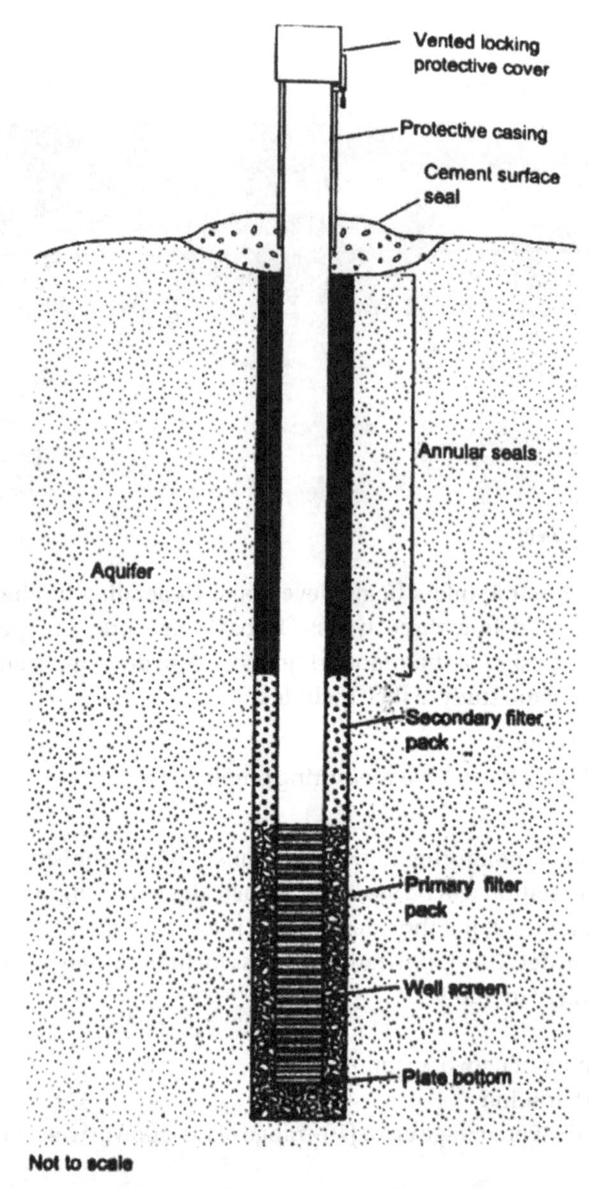

FIGURE 10.3 Piezometer: (A) nested piezometer, (B) monitoring wells with short screens, multiple monitoring wells with short screens, each installed in its own borehole installed in a single borehole (nested piezometers).

iii) Observation wells with one piezometer.

iv) Observation wells with nested piezometer (Fig. 10.3).

A piezometer (Fig. 10.3) is a simple structure, which is either a bore well/tube well-constructed to monitor a specifically identified aquifer/aquifers.

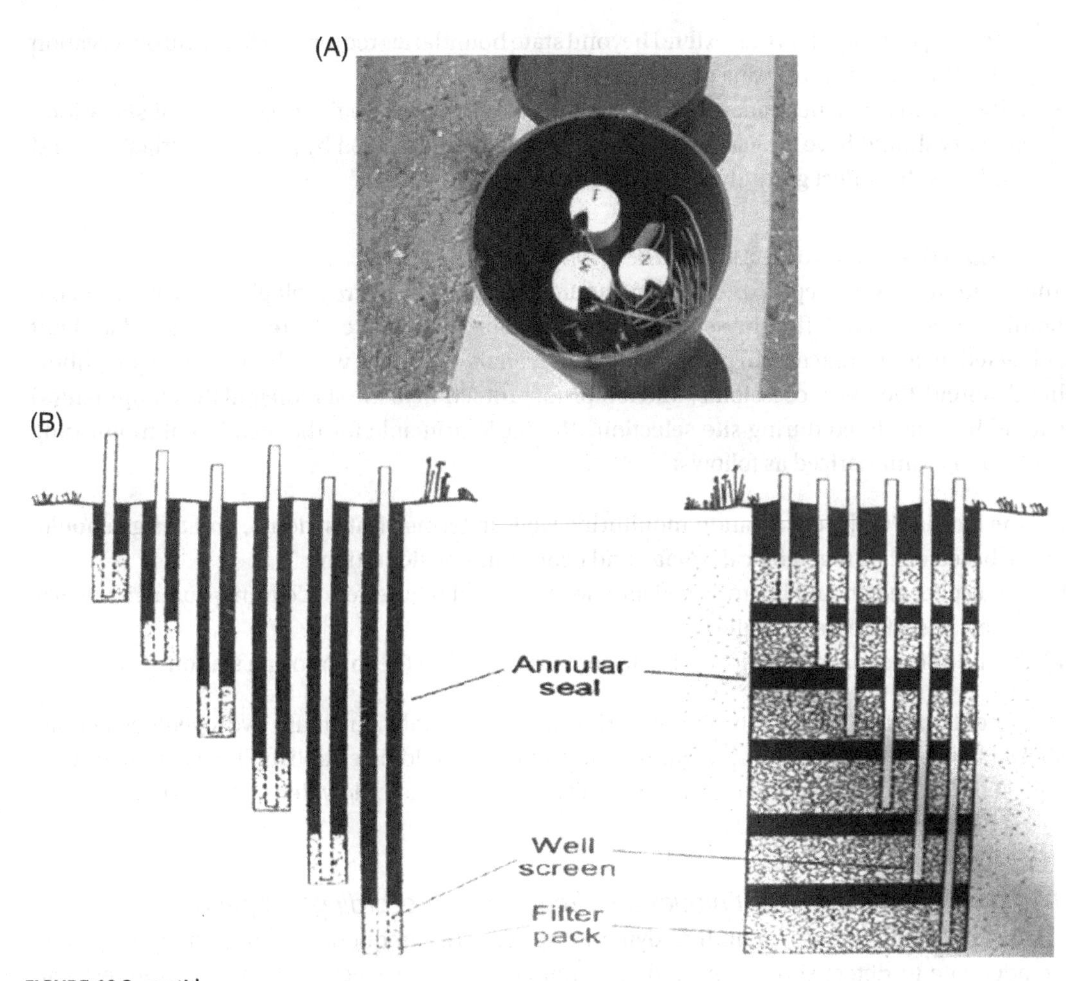

FIGURE 10.3, cont'd.

10.2.1.2 Dimensions of monitoring hydrograph network stations

- The dimensions of network have two aspects: (1) the density of network and (2) the sampling frequency.
- The number and locations of observation wells and piezometers are crucial to any water level or water quality data collection program.
- The sites chosen for an observation network should represent all topographic, geologic, climatic, and land-use environments.
- The areal distribution and depth of completion of piezometers depend on the physical boundaries and geologic complexity of aquifers under study.
- Groundwater monitoring program for complex and multilayer aquifer systems may require measurements in nested piezometers.

- Large, regional aquifers that extend beyond state boundaries require a network of observation wells distributed among one or more States.
- If the purpose of a network is to monitor the effects of hydrologic stresses, the observation network should have dedicated piezometers that are unaffected by pumping, irrigation, and land uses that affect groundwater recharge.

10.2.1.3 Location and density of the monitoring points

The location of sites, depth, and design should be based on hydrogeological, social, and economic considerations. The site selection should ensure that data collected is unbiased and not subjected to interference from production wells, canals or surface water bodies in the neighborhood. Round-the-year accessibility and foolproof protection to the station and the equipment(s) should be considered during site selection. The basic principle for the location of monitoring wells can be summarized as follows:

a) The location of the observation monitoring wells in terms of sites, depth, and design should be based on hydrogeological, social, and economic consideration.
b) Data collected should be unbiased and not subjected to interference from production wells canals or surface water bodies.
c) Round-the-year accessibility and foolproof protection to the monitoring station.

For determining the density of monitoring stations, initially, the main hydrogeological units should be identified and a limited number of stations should be established in each hydrogeological unit having homogeneous water quality, along the main flow lines. This forms the basic or the pilot network.

10.2.1.4 Measurement techniques and frequency of recording/sampling

The water level is a fast responding dynamic system. The frequency of measurement should be adequate to detect short-term and seasonal groundwater level fluctuation of interest and also to discriminate between the effects of short and long term hydrologic stresses. It also depends on factors like groundwater flow and recharge rate, aquifer development, and climatic conditions.

Attributes to be measured are the following: (1) peak and trough of the hydrograph; (2) time of shallow water level; (3) time of deep water level; and (4) rate of rise or decline. The current scenario of periodicity of groundwater monitoring is that groundwater levels are being measured four times in a year during January, April/May, August, and November by the Central Ground Water Board (CGWB) through a network of about 15,000 observation wells located all over the country. In general, the methods for monitoring of hydrograph stations two methods are adopted—(1) manual measurement (by measuring tapes) and (2) automatic measurement (by automatic recorders). However, continuous monitoring is usually carried out using digital water level recorders (DWLR), which is programed to make measurement at a specified frequency (Fig. 10.4).

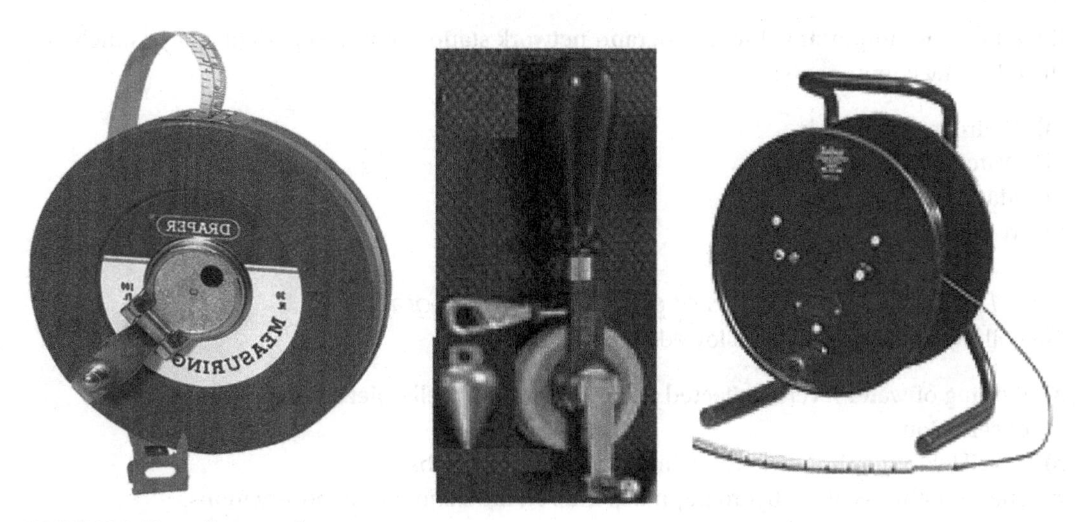

FIGURE 10.4 Manually measuring tapes.

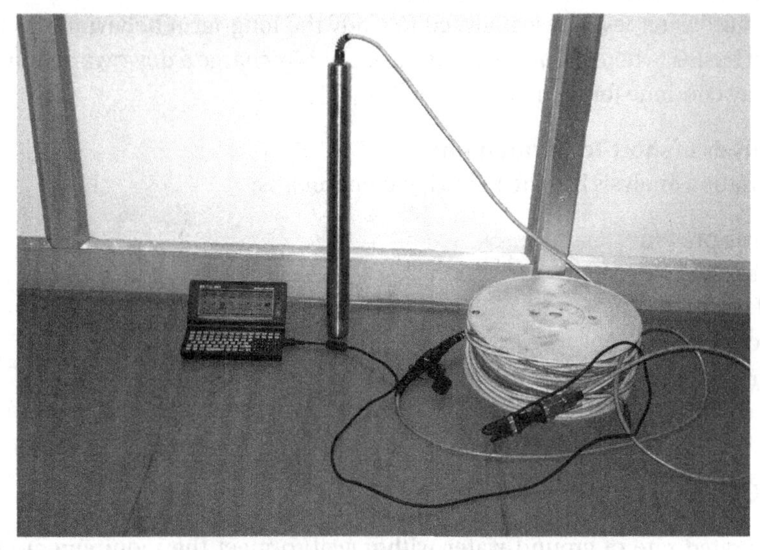

FIGURE 10.5 Digital water level recorders (DWLR).

10.2.1.4.1 Telemetry or remote monitoring

The perspectives of telemetric monitoring are as follows: (1) DWLRs with telemetry; (2) inaccessible locations can be monitored; (3) initial cost is more but data reliability and accuracy are also more; (4) high-frequency data can be collected; (5) needs maintenance; and (6) accessible through web (Fig. 10.5).

10.2.1.5 Analysis and presentation of data

The main objective of analyzing the groundwater regime data is to estimate groundwater resources. In the estimation of groundwater resources, the component of water level and its fluctuation are important. The hydrograph network station monitoring data analysis includes the following components:

i) Hydrograph analysis.
ii) Water level analysis.
iii) Maps.
iv) Water quality.

10.2.1.6 Plotting and analysis of groundwater hydrographs

The following steps are to be followed in the process:

a) Plotting of water levels, collected from observation wells/piezometers against time of observation.
b) A well hydrograph shows rising limbs and lowering limbs.
c) The smoothness of the hydrograph depends on the frequency of observations.

10.2.1.7 Long-term water level trend analysis

Sometimes groundwater levels are analyzed to study the long-term behavior of water table. In overdeveloped basins, where groundwater draft exceeds recharge a downward trend of groundwater levels may continue for many years.

10.2.1.7.1 Analysis of short-term variations

Short-term variation analysis has the following components:

i) Atmospheric pressure.
ii) Ocean tides.
iii) Earthquakes.
iv) External loads.
v) Evapotranspiration.

10.3 Discussion

There is accelarated rate of ground water withdrawal to meet the requirement of increasing population, urbanization and industrialisation. In the water nstressed state like Rajasthan, the Dug wells need to be replaced by construction of Piezometers, especially where the dug wells have gone dried on account of incessant declining of water level owing to more ground water extraction than its natural monsoon recharge. Hence, vigil on ground water regime in context of ground water development is the topmost priority for sustainable ground water management. There is a strong need for network strengthening with construction of purpose built stations for systematic monitoring of ground water level and water quality, especially in the vulnerable areas like the industrial zones, mining & smelting complexes and urban agglomerates. Alarming declines in ground water level are being witnessed both in hard rock and alluvial areas. Mostly contamination in ground water may be due to domestic wastes, industrial and sewage disposal activities. The ground water exploitation in such areas requires to be regulated and monitored through location specific management measures to achieve the sustainability and preservation of precious ground water resources. The premonsoon groundwater level data are collected from all the monitoring stations during the months of March/April/May, depending on the climatological conditions of the region. For north-eastern states, premonsoon data are collected during the month of March, since the onset of monsoon is normally observed in April. Similarly for Orissa, West Bengal, and Kerala, where monsoon appears early in May the monitoring is carried out during the month of April. Premonsoon monitoring month is fixed for May in case of the remaining states. Groundwater levels during August are monitored to access the impact of monsoon on the groundwater resources. Postmonsoon data collected during November reveal the cumulative effect of groundwater recharge and withdrawal of groundwater for many purposes. Groundwater level data for the month of January indicates the effect of extraction for *rabi* crops.

The data are analyzed to know about the frequency distribution of water levels during different periods in addition to the seasonal, annual, and decadal fluctuations in groundwater levels. The monitoring data are useful for preparing the groundwater table and groundwater table fluctuation maps for each monitoring period to study the spatial and temporal changes in groundwater regime in different seasons. In locations where the situations of over-extraction are observed through declining ground water levels, immediate measures need to be initiated to reduce the ground water draft by at least 20% and subsequently regular monitoring is mandatory to curb the ground water table depletion further. The impact of reduced draft should be monitored and documented for at least a period of over 2–3 years to understand the long term behaviour of ground water regime and that will facilitate to design a scientifically viable sustainable long-term management strategy.

10.4 Conclusions

- During network design location of well, network density, frequency of sampling is taking into consideration.
- Groundwater recharge can be estimated from groundwater hydrograph analysis.
- At a number of locations no station will be available. Hence, the information is to be obtained from the network by interpolation.
- One of the accurate methods which is generally adopted in estimating the groundwater resource is the water level fluctuation method through the analysis of hydrograph.
- Groundwater management should be in a systematic manner, create an integrated data base, analysis the data, and formulate strategies to develop and manage the resource.

Acknowledgments

The author would like to convey his earnest thanks and gratitude to the Chairman, CGWB and Member (East) for according kind permission to publish the paper as a book chapter in the edited book volume.

References

CGWB, 2019. National Compilation on Dynamic Ground Water Resources of India, 2017. Central Ground Water Board, Ministry of Water Resources, River Development and Ganga Rejuvenation, Government of India. http://cgwb.gov.in/GW-Assessment/GWRA2017-National-Compilation.pdf. National Compilation on Dynamic Ground Water Resources of India, 2017.

Hsu, S.K., 1998. Plan for a groundwater monitoring network in taiwan. Hydrol. J. 6 (3), 405–415.

Mogheir, Y., Singh, V.P., 2002. Application of information theory to groundwater quality monitoring networks. *Water Resources Management.* Water Resour. Manage. 16 (1), 37–49.

Nunes, L.M., Cunha, M.C., Ribeiro, L., 2004. Groundwater monitoring network optimization with redundancy reduction. J. Water Resour. Plan. Manage. 130 (1), 33–43.

Powell, R.M., Puls, R.W., 1993. Passive sampling of groundwater monitoring wells without purging: multilevel well chemistry and tracer disappearance. J. Contam. Hydrol. 12 (1–2), 51–77.

Uil, VanGeer, F.C., Gehrels, J.C., Kloosterman, F.H., 1999. State of the art on monitoring and assessment of groundwaters. UN/ECE Task Force on Monitoring & Assessment, Netherlands Institute of Applied Geoscience TNO. Lelystad. Netherlands Institute of Applied Geoscience TNO.

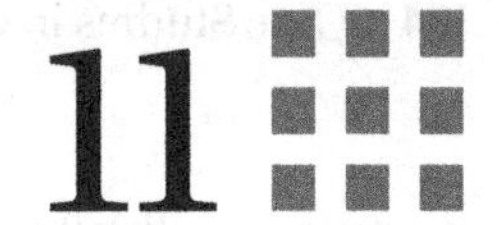

Groundwater resources in Nigeria: Case study of distribution and quality at a medium-size urban settlement-scale

Adebayo Oluwole Eludoyin and Adewole Abraham Fajiwe
DEPARTMENT OF GEOGRAPHY, OBAFEMI AWOLOWO UNIVERSITY, ILE-IFE, NIGERIA

11.1 Introduction

One of the themes of sustainable development goals is to have access to clean water and sanitation (World Bank, 2002). Twelve percent (12%) of the world population consumes 86% of the available water and a good percentage of world population have no access to adequate water supplies. Rise in the demand for clean water, climate variability and pollution have been reported to significantly cause reduction in potable water, giving rise to small-scale water projects, including personal boreholes and wells in many homes in most developing countries. Water is integrally linked to the provision and quality of ecosystem services. In homes, water is used for cooking and other domestic purposes but its availability is a subject of concern in many parts of developing countries, including Nigeria (Oloruntade et al., 2012). Hanidu (1990) reported that surface water resources in Nigeria yield up to 224 trillion units per year and groundwater resources are abstracted up to about 0–50 m depth for about 6 million km^2, still water resources are not evenly distributed. However, our focus which is groundwater, is the water body originally derived from percolation and it is contained in permeable rock known as aquifer and the saturated zone of the soil (Ayoade, 1988). In other words, it can be described as water body found beneath the earth's crust.

Groundwater resources (particularly, boreholes, and hand dug wells) are the ways to meeting goal number 6, of the sustainable development goals, that is, ensuring the availability of substantial and adequate quantity of water and sanitation for everyone. However, groundwater sources are often used conjunctively with other water sources (including rain and surface water)

Case Studies in Geospatial Applications to Groundwater Resources. DOI: https://doi.org/10.1016/B978-0-323-99963-2.00018-3

to ensure fair distribution of water supply in many regions. But despite the need for groundwater exploitation, it has been argued that oversinking of boreholes in an area can initiate seismic wave that may amplify to cause huge earth disturbance, tremor, and earthquake (Ameh et al., 2019). Also, groundwater sources (boreholes and hand dug wells) are known to vary in depth, in geological composition. Land use around them also varies. Other factors that can affect the chemistry of water derived from them are topography and probably seasons.

Two categories of groundwater pollution have been observed in different parts of the world; the first is from contaminated geology and the second from poor land management (see Eludoyin, 2014). Those that are related to contaminated geology includes rare chemical (e.g., arsenic) in wells, more than normal concentration of dissolved salts like calcium, carbonates, and fluorides in areas under which certain rocks occur, for example, regions where there are Malmesburyshales, granites, limestone, and coal deposits as contained in. Those attributed to poor land management are those due to or associated with latrines that are unlined and soak away system at homes, urbanization (markets in urban areas), land reclamation, chemicals preservative from burial sites and industrialization (chemicals for agricultural activities) as contained in (Mizumura, 2003; Eludoyin et al., 2004; Oyeku and Eludoyin, 2010; United Nations Environmental Protection, UNEP, 2002).

The World Health Organisation (WHO) reported that at least 75 L of water is needed to guide against sicknesses in homes, through appropriate sanitation practices in rural areas while residents in urban areas would use the average of 250 L of water per day (Francy et al., 1991). Unfortunately, only a privileged few have such access, especially in the urban areas, while the majority in the rural areas walk for more than 30 minutes to the nearest water sources. As a matter of fact, as population increases, water scarcity may move from bad to worse. Also, the United Nations World Water Assessment Programme predicted that by the year 2050, 7 billion people in over 60 countries will have to cope with water scarcity. The Nigerian documented efforts have shown that the governments have for many years attempted the development of water resources by constructing dams for both irrigation and domestic water supply but water supply to most communities in Nigeria is grossly inadequate. The National Water Supply and Sanitation Policy (2000) (report specified that the government recognized its capacity to provide water but more than half of the world population still cannot access good and steady water supply (Federal Ministry of Water Resources, 2000). The experience in the present study area, Ondo town in Ondo West Local Government area of Ondo State Nigeria, is typical of many advanced communities in Africa. The people living in the region are generally urbanized and being located in the tropics, the settlement water is expectedly not experiencing physical water scarcity. The study area is rather characterized by drainage systems including streams and underground water sources, but the main sources of domestic water supply are wells and boreholes (groundwater), albeit with different levels of water quality and quantity, on which no one had ever worked, hence, this study. Demand for quality water for drinking and washing at homes have increased in the town due to population increase, as a result of migration (immigration) and urbanization. Groundwater sources have become important water sources for the residents, especially as rainwater is seasonal and public supply of pipe-borne water is either inadequate or completely absent (Ezechukwu, 2016). The quality of groundwater depends on the geology and conditions that led to its formation, quantity of water available in the aquifer, and how it flows (Ramanathan et al., 2007). The nature and constituents of dissolved minerals are known factors that influence the quality of groundwater (Boyle and Okogbue, 1988). Groundwater quality may possess serious effects on human health when consumed (Spechler, 2010).

11.2 Research problem

Groundwater is generally conceptualized as an ecosystem source as well as a geologic agent (Toth, 1963, 1999). It is the major source of water for about 60% of the world population (United Nation Environmental Protection, UNEP, 2002) and also a preferred source of water for domestic use in most parts of Nigeria and other developing countries, where pipe-borne water supply is lacking and inadequate (Ayodele and Ajayi, 2015). Studies have however revealed that increase in demand for water is as a result of population increase (Eludoyin, 2020) and that the increased withdrawal for industrial and agricultural activities that is related to population increase and urbanization, this would eventually reduce the amount of available groundwater and will also contribute to the deterioration in groundwater quality (Piscopo, 2001).

Furthermore, the quality of groundwater is known to vary due to lithology, depth, and climate (Vrba, 2002). Consequently, there is no consensus on groundwater quality and associated geology, especially in Africa. This is particularly important as groundwater sources are now ubiquitous in centers where there is no effective government-controlled water supply. Also, studies have reported that there are health challenges associated with the consumption of water with abnormally high concentration of chemical parameters in boreholes and wells (US EPA, 2008). For example, the implication of excessive nitrates in drinking water has been death of both infants and adults. Both nitrates and nitrites have also been associated with diuresis, increase starchy deposits, and hemorrhaging of the spleen (US EPA, 2008).

The study area, Ondo town, is geologically underlain by basement complex rocks that are known to be selectively permeable, such that certain location, especially in the low topography are enriched by aquifer while other areas, particularly in the highlands generally experience water scarcity. Groundwater sources at lower topography are also known to be liable to sources of organic pollutants including leachate from dumpsites and commercial sites (Bayode et al., 2006). Many groundwater in the study area has been sited near sewage disposal units. It is, however, unclear if the quality of the groundwater sources vary with the location or altitude. Ondo town is a typical urban area where people are dependent on water from groundwater sources, especially boreholes, for domestic use. The area is also experiencing increase in population and urbanization with attendant pressure on water infrastructure, in recent time. Consequently, the number of boreholes has increased, and many residents have also complained about difference in taste of the water from the sources. There is no information about the nature of the distribution of the groundwater sources and there is no conscensus on whether it is location (topography), land-use around specific borehole or any other factor that causes the difference in the quality of the water. Existing studies have only focused on nearby Akure, Owo, Akoko and Okeigbo whose geology is different from that of the study area. Their studies have also focused on pathogens

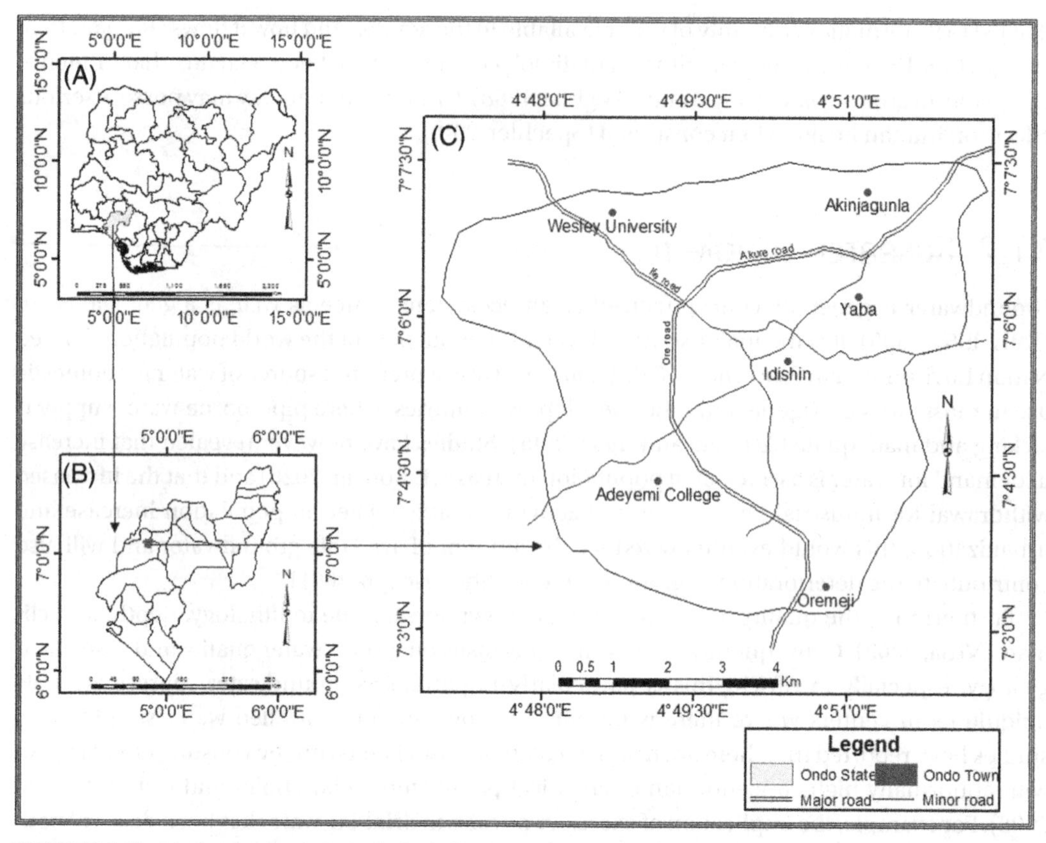

FIGURE 11.1 The study area, Ondo town in Ondo State, southwest Nigeria.

(Tekwa et al., 2006), basic groundwater geology (Ezeigbo, 1987) and general perceived impacts of human activities (Abimbola et al., 1999); hence this study.

11.2.1 Study objectives

The aim of this study is to contribute to existing knowledge on access to quality water in Africa, with focus on groundwater supply. Specific objectives are to assess the distribution of boreholes and wells in a typical urban community (Ondo town, Ondo state) in Nigeria and examine the chemical constituents and quality of selected groundwater sources for domestic uses. Water samples were taken and analyzed between August 2020 and February 2021.

11.3 Study area

The study area is the headquarters of Ondo West Local Government Area in Ondo State, Nigeria, about 970 km^2 (Fig. 11.1). It is characterized by tropical rain forest and wet and dry climate described in the revised Koppen's climate classification (Eludoyin et al., 2017) with two separate seasons. The rainy season lasts from April to October and the dry season from November and ends in March. Annual rainfall is about 2000mm with average monthly temperature ranging from 26°C to 30°C and the humidity is always high, above 50% in most times. Population is estimated (based on the National Population Commission figure of 288,868) 473,412 as at 2020.

The land is underlain by crystalline and sedimentary rocks, both occurring approximately in equal proportions (Woakes et al., 1987). The crystalline rocks are made up of Precambrian basement complex and the Phanerozoic rocks. The basement complex of Ondo is part of the West Africa craton of Precambrian early Paleozoic orogeny that has been affected by supracrustal plutonic (Oyawoye, 1972). Groundwater in the area occurs in both fractured aquifer and weathered aquifer (Olorunfemi and Fasuyi, 1993). Weathered aquifer is emanates from chemical modification processes while fractured aquifer is as a result of tectonic activities. At times, weathered layer aquifer may occur separately or along with the fractured aquifer (Olorunfemi and Fasuyi, 1993; Bayode et al., 2006). The direct exposure of the uppermost part of the vadose zone of the weathered layer aquifer system makes it liable to surface contaminants such as leachate that emanates from waste dump sites and flooding (Idowu et al., 2014). Many wells and boreholes drilled in the town have high tendency of failure and varied water chemistry which could be attributed to underlying complex geology and inadequate geophysical information before sitting the groundwater sources (Mogaji et al., 2011). Boreholes and hand dug well yield are frequently low in basement complex area during the dry season, so it is expedient to locate boreholes and wells on aquiferous zones or faulted basement rock.

The basement complex in Ondo differs in properties across the area due to type, origin, lithology, and structure of rocks that have great influence on the parameters of the water chemistry (Seidel and Lange, 2008; Oyinloye, 2011).

11.3.1 Groundwater supply in Nigeria: review

Groundwater is the most important source of freshwater for domestic use in most urban areas in Nigeria. The origin of groundwater in the world was probably subjected to speculation until the 17th century when Perrault (1608–1680) and Edme Mariote (1620–1684), through careful measurement and analysis demonstrated independently that the source of stream flow and groundwater was precipitation from the atmosphere also called atmospheric or meteoric water (Ayoade, 1975). Groundwater includes all water found beneath the earth's surface. It is the body of water derived primarily from percolation and contained in permeable rock formations known as aquifers. It may be found anywhere in the world and in all types of geological formations. However, its distribution in terms of quality and quantity varies from one geological formation to another. Edme Mariotte (1620–1684) demonstrated experimentally that the source of groundwater in an area could be the precipitation over that area. Groundwater is an integral part of the hydrological cycle. Marcus Vitruvius theorized that groundwater was mostly derived from rain and snow by infiltration from the surface (Ayoade, 1975). Groundwater, known to provide over 90% of the global freshwater supply, has been increasingly exploited in recent decades, probably to support the problem of the declining surface water sources in many regions (Eludoyin, 2014).

Groundwater sources are, however, known to vary in depth, geology, land use, and other factors that can affect the quality of water that is derived from them.

Cases of groundwater pollution observed in different parts of the world appear to be related to contaminated geology and poor land management. According to Eludoyin (2014), "water infrastructure in Nigeria is not centralized, and no standards are known to govern the exploitation of groundwater, except as driven by the need for water by individuals and communities. The implications of the decentralized infrastructure and poor enforcement of standard guidelines include individualized water quests with different quality of water from the groundwater sources. Vulnerability to geological and land use contamination is rarely stringently considered before water is exploited, especially in cases for household use." In their study on groundwater chemistry and quality of Nigeria, Edet et al. (2011) provided a characteristics of groundwater in Nigeria to include dominance of saline groundwater in parts of Benin and Niger Delta Basins due to intense dissolution of salts and seawater intrusion. Sokoto basin was characterized by high nitrate level why the chemistry of other parts was influenced by geology, chloride dissolution, weathering (silicate and carbonate) and ion exchange (Nganje et al., 2010). Also, Oyeku and Eludoyin (2010) illustrated that a major pollution source to groundwater is leachate from landfills. Clark (1991) described landfill practices as the disposal of solid wastes by infilling depressions on land. The depressions into which solid wastes are often dumped include valleys, (abandoned) sites of quarries, excavations, or sometimes a selected portion within the residential and commercial areas in many urban settlements where the capacity to collect, process, dispose of, or reuse solid waste in a cost-efficient, safe manner is often limited by available technological and managerial capacities.

11.3.2 Regulations on groundwater distribution in Nigeria

Nigeria Industrial Standard code of practice for water well and borehole construction (2010) controlled by the Standard Organisation of Nigeria (SON), in conjunction with The Technical Committee for Nigeria Code of Practice for Water and Well Construction. Itemized the basic legal requirement—for the constructing of water wells in Nigeria, the following are the conditions that should be met:

- **(i)** Individuals and drilling companies must possess drilling license.
- **(ii)** Permit shall be obtained before water well can be construction.
- **(iii)** Well-completion reports shall be signed by registered agents and should be deposited in the office of the minister in charge of water resources.
- **(iv)** Completion report shall be handled by registered hydro-geologists who has been appointed by the government.
- **(v)** Water quality analysis report shall be the responsibility of duly registered analyst from a tested and trusted laboratory.

No water well shall be constructed unless the owner a valid permit to do so. Permit shall be given by agency designated by the minister in charge of water resources.

11.4 Materials and methods

11.4.1 Data

Data used for the study were concentrations of selected ions (As, Pb, Mn, Ca^{2+}, K^+, Na^+, Al^{3+}, NO_3^-, Cl^-, and PO_4^{3-}) from systematically selected groundwater sources (hand dug wells and boreholes) in the study area. The following procedure was followed in the selection of the groundwater sources. First, the map of the study area was overlaid by a by 2 km^2 nest grid from which 20% of the total grid were randomly selected using the random number table, at both the upland and the lowlands. Groundwater sources that fall within the selected grids were considered for sampling. Selected groundwater sources were characterized in terms of:

a. Depth (m) and type (boreholes, pumped wells, and hand dug wells).
b. Presence or absence of concrete lining.
c. Conditions surrounding the location like farming or land use type, etc.
d. Presence or absence of cover as well as uses.

The coordinates (x, y, z) of the selected groundwater sources were determined with the use of global positioning system to a 5 m of accuracy. Water samples were collected in both wet (July–August, 2020) and dry (January–February, 2021) seasons. Water samples were collected in the morning and stored in a well labeled 2 L polyethylene plastic bottles which had been rinsed with deionized water. Water temperature, pH, conductivity and total dissolved solid (TDS) were determined at the site with handheld pH/temperature and conductivity/TDS meters. The water samples were then taken to the laboratory for chemical analysis.

In the laboratory, the water samples were stored in a refrigerator prior to analysis to minimize the activities of micro-organisms which might alter the water chemistry. The samples were first digested to determine the concentrations of the selected chemical parameters as described by Standard Laboratory Protocol. The concentration of systematically selected metals was determined with the use of atomic absorption spectrophotometer within their respective wavelength frequency while the anions were determined with colorimetric method. The parameters were selected based on their recognition as important ions in domestic water in Southwestern Nigeria (Ayoade, 1988). The chemical analysis was done in the Centre Science Laboratory of the Obafemi Awolowo University, Ile-Ife, Nigeria. Analysis was done using SPSS IBM 20 and PAST (version 3) software.

11.5 Results

11.5.1 Distribution of groundwater resources

Analysis of the distribution of the groundwater sources using their coordinates (Fig. 11.2) revealed that the groundwater sources clustered around the study area. Also, the result of the neighborhood analysis showed a mean distance of 1.5–3 m rather than the expected average distance of

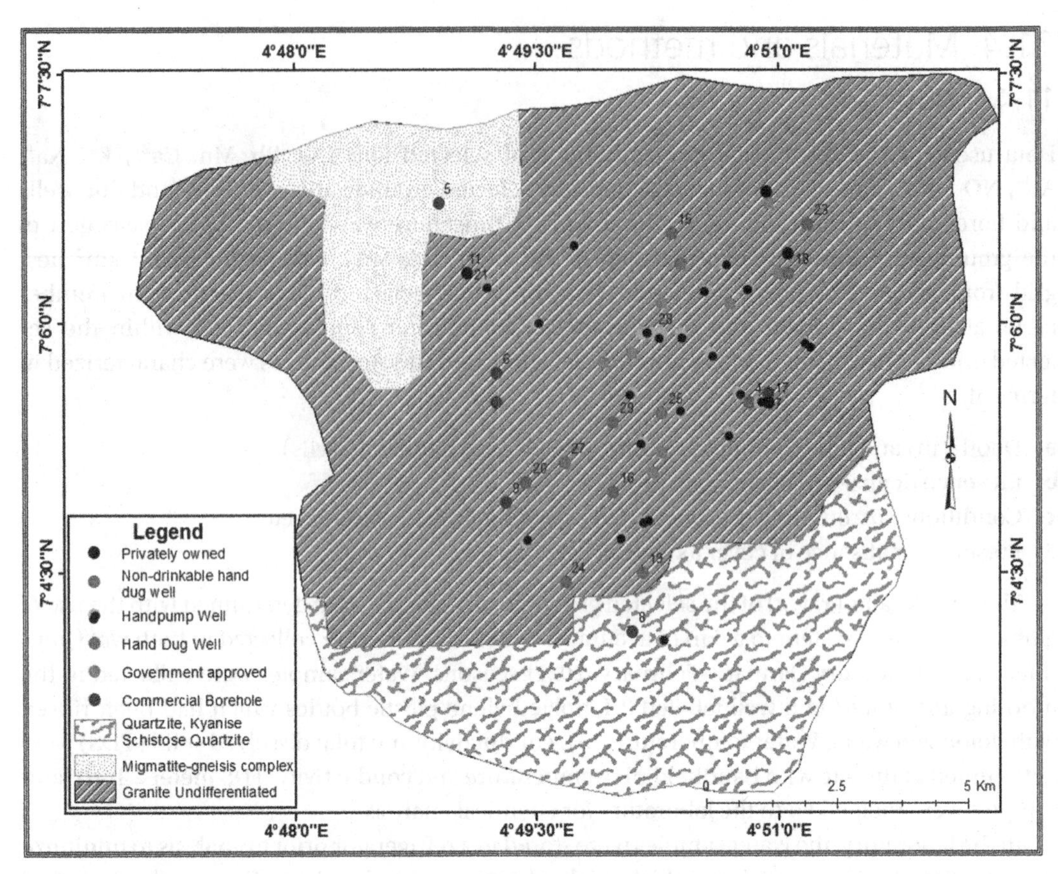

FIGURE 11.2 Distribution of groundwater sources across geological underlain.

4.5 m as stated in minimum groundwater source distance in Nigeria (Fig. 11.3). Majority of the wells were private hand dug wells and these were in good conditions as at the time of the study. The observed and expected neighboring distances revealed that most of the groundwater resources were within 1.5 m except three public hand dug wells (A4, A22, and A24). Evaluation of the distribution of the groundwater sources with respect to the topography showed that they mostly exist at the region between 236 and 269 m above the mean sea level, and around the northern part of the study area (Fig. 11.3). The outliers (outside the zone of cluster concentration) groundwater sources are public hand pump wells, which could have been provided with consideration for the population. In general, result of neighborhood analysis showed that the groundwater resources occurred in a statistically significant clustering pattern ($Z = -4.65$, $P = 0.006$).

11.5.2 Physical and chemical characteristics

The concentration or values of selected physiochemical variables examined in the study are presented in Table 11.1. The groundwater sources are distributed into hand dug wells, wells not

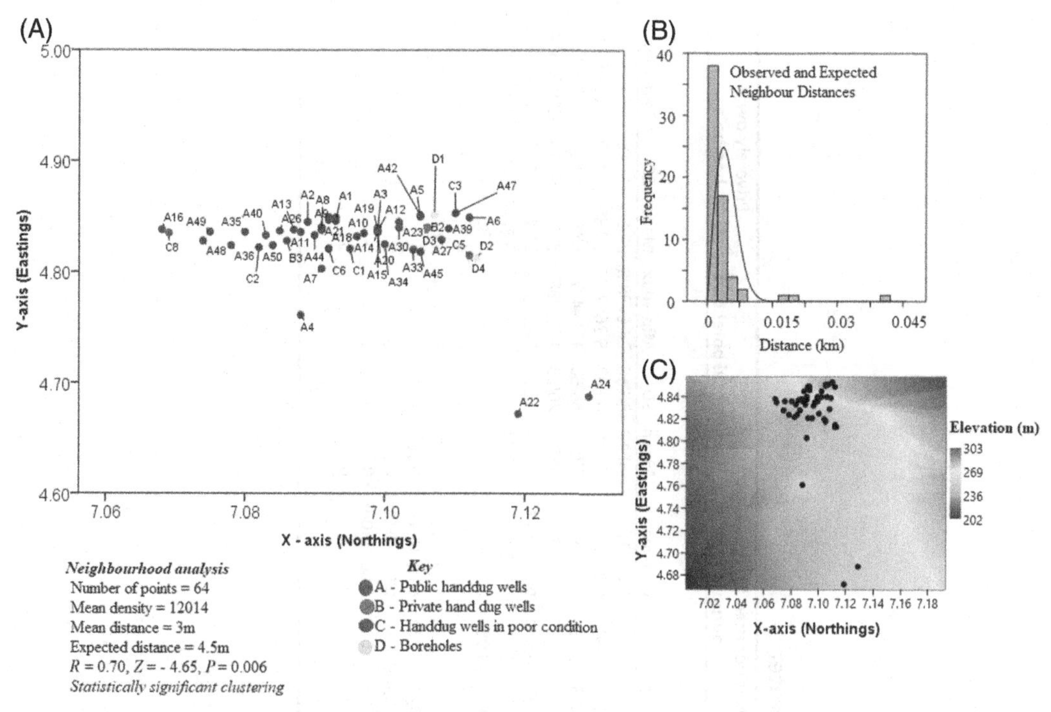

FIGURE 11.3 Distribution of groundwater sources across study area.

used for drinking purposes, commercial boreholes, privately owned boreholes and government provided hand pumped wells, for clarification on their uses and quality expectations. Water temperature generally ranged from 24.6 to 38.8°C with the warmest water being that of the commercial boreholes. The pH unit in the boreholes were also more acidic than those of the hand dug wells, varying 5.26–7.8 units while the hand dug wells were 6.5–8.9 units in range. In term of conductivity, the least value occurred at a privately owned hand dug well. Aluminum ion was detected in the boreholes (0–1.8 mg/L) but not in the hand dug wells while arsenic and lead mainly occurred in the hand dug wells (0–0.003, 0.003–0.001 mg/L) but were not detected in the boreholes (Table 11.1). When compared with recommended standards for drinking purposes, especially that of the World Health Organisation (WHO, 2011), the results show that based on pH (6.5–9.5 units), all the water sources are within the acceptable limits for drinking purpose. The water sources are also safe for conductivity, TDS, Na^+, Ca^{2+}, K^+, and Cl^- at 1000 µs/cm, 600 mg/L, 200 mg/L, and 250 mg/L, respectively. The water from a privately owned hand dug well was not safe from lead contamination; however, mean concentration of lead was 0.015 mg/L, slightly above WHO recommended limit of 0.01 mg/L. Also, phosphate concentration in the privately owned borehole was greater than the recommended limit of 10 mg/L. The level of arsenic concentration in the hand dug wells suggests that the hand dug wells were more vulnerable to arsenic concentration than the boreholes. The mean concentration of arsenic in privately owned hand dug wells in the area was exactly 0.1 mg/L, same level with recommended standards,

Table 11.1 Physical and chemical characteristics of groundwater.

Variables	Hand dug wells						Wells not used for drinking purpose			Commercial boreholes			Privately owned boreholes		
	Pubic owned Mean	SD	Min–max	Privately owned Mean	SD	Min–max	Mean	SD	Min–max	Mean	SD	Min–max	Mean	SD	Min–max
Water temperature	28.85	1.35	27.5–30.2	29.35	0.912	28–30.8	27.86	2.614	24.6–31	30.13	3.758	24.6–38.8	28.56	1.006	27.2–30.1
pH	6.9	0.1	6.80–7.00	7.061	0.811	6.48–8.94	7.41	0.849	6.81–8.61	6.398	0.832	5.26–7.77	6.235	0.503	5.62–7.24
Conductivity	229.5	56.5	173–286	240.5	62.91	1.51–335	257.3	32.31	212–285	275.2	85.50	151–407	331.5	51.50	256–392
TDS	160.9	39.3	121.6–200.2	224.7	71.52	118–350	250.3	73.3	151–326	350.5	205.3	158–788	300	99.10	200.2–504
Sodium	0.151	0.019	0.132–0.169	0.108	0.013	0.090–0.128	0.081	0.054	0.006 0.131	31.98	3.965	24.2–37.2	21.2	14.12	0.18–36.8
Calcium	0.124	0.004	0.120–0.128	1.043	3.063	0–11.20	0.09	0.06	0.005 0.137	5.404	12.66	0.16–38.9	6.848	8.973	0.27–24.8
Potassium	0.173	0.001	0.172–0.174	0.209	0.081	0.166–6.233	0.13	0.087	0.009–0.208	0.444	0.38	0.16–1.27	1.35	2.137	0.13–6.00
Phosphate	2.04	0.05	1.99–2.09	1.688	0.45	1.48–3.08	0.637	0.328	1.40–2.10	0	0	0	4.917	7.413	0–19.20
Chloride	33.1	1.1	32.0–34.2	36.17	8.055	28–56	38.26	12.54	29.3–56.0	48.87	22.90	6.8–96.4	39.83	14.64	8.00–52.4
Manganese	0.04	0	0–0.04	0.042	0.048	0.002–0.005	0.09	0.078	0.03–0.20	0.004	0.01	0–0.03	0	0	0
Lead	0.004	0.001	0.003–0.005	0.005	0.002	0.002–0.008	0.003	0.001	0.001–0.004	0	0	0	0	0	0
Arsenic	0.003	0	0–0.003	0.005	0.001	0.003–0.007	0.003	0.002	0.001 0.005	0	0	0	0	0	0
Aluminum	0	0	0	0	0	0	0	0	0	0.225	0.595	0–1.8	0.015	0.034	0–0.09
Nitrate	23.65	0.55	23.1–24.2	16.58	9.34	0.12–35.4	14.26	2.007	11.5–16.2	2.637	0.35	2.3–3.4	1.90	0.854	0–2.4

Table 11.2 Comparison of water chemistry from selected groundwater resources with the WHO (2017) recommended limit Standard for drinking water.

Parameters	WHO (2017)	Boreholes	Hand dug wells
pH	6.5–9.5	5.26–7.91	6.48–8.94
Ec (μs/cm)	1000	105–407	107–346
Temp (°C)	ND	24.6–38.8	24.6–32.2
TDS (mg/L)	600	103–788	64.5–350
Na^+ (mg/L)	200	0.18–37.2	0.00–0.99
Ca^{2+}(mg/L)	200	0.12–38.9	0.00–11.2
K^+ (mg/L)	200	0.13–6.00	0.01–0.23
PO_4^{3-}(mg/L)	ND	0.00–19.2	1.30–3.08
Cl^- (mg/L)	5	3.00–96.40	18.2–56.0
Mn (mg/L)	0.04	0.00–0.03	0.02–0.20
As (mg/L)	0.01	0.00–0.00	0.00–0.01
Pb (mg/L)	0.01	0.00–0.00	0.00–0.01
Al^{3+} (mg/L)	0.1	0.00–1.80	0.00–0.00
NO_3^- (mg/L)	50	0.00–3.40	0.03–35.4

ND, not determined.

Table 11.3a Distribution of factor scores in the determination of dominant physicochemical parameters in selected borehole wells in the study area.

	Factors			
Parameters	1	2	3	4
pH	0.197	0.088	0.728	0.09
EC (μs/cm)	−0.122	0.865	0.01	0.061
Temp (°C)	0.347	−0.478	−0.373	0.072
TDS (mg/L)	0.081	0.881	−0.014	0.036
Na^+ (mg/L)	0.45	−0.318	0.015	−0.095
Ca^{2+} (mg/L)	−0.037	0.095	−0.034	−0.864
K^+ (mg/L)	0.882	0.136	0.058	0.252
PO_4^{3-} (mg/L)	0.002	0.003	0.865	0.173
Cl^- (mg/L)	−0.785	0.263	−0.184	0.285
Mn (mg/L)	−0.638	−0.218	0.614	0.039

suggesting potential risk on the other hands, the commercial boreholes contain relatively more concentration of aluminum ions (0–1.8 mg/L) than the other water sources (Table 11.2).

Furthermore, the results of the factor analysis performed revealed that K^+, Cl^-, and Mn (factor 1) as well as conductivity and TDS (factor 2) explained 42.9% of the variations in the quality of water from hand dug wells (Tables 11.3a and b). The dominant variables often come from various anthropogenic sources including runoff from human wastes, (Cl^-), dump sites (Cl^- and Mn), vehicular wastes (Mn), and other sources (conductivity and TDS) as well as

Table 11.3b Total variance explained by each group of parameters.

	Initial eigenvalues			Rotation sums of squared loadings		
Factor	**Total**	**% of Variance**	**Cumulative %**	**Total**	**% of Variance**	**Cumulative %**
1	2.724	24.766	24.766	2.199	19.992	19.992
2	1.996	18.143	42.909	2.073	18.847	38.839
3	1.842	16.741	59.65	1.953	17.753	56.592
4	1.17	10.632	70.283	1.506	13.691	70.283
5	0.921	8.369	78.652			
6	0.77	7.003	85.655			
7	0.658	5.986	91.641			
8	0.4	3.634	95.275			
9	0.257	2.335	97.61			
10	0.177	1.612	99.222			
11	0.086	0.778	100			

Table 11.4a Distribution of factor scores in the determination of dominant physiochemical parameters in selected hand dug wells in the study area.

	Component					
Parameters	**1**	**2**	**3**	**4**	**5**	**6**
pH	0.1	−0.249	0.117	0.162	−0.89	0.118
EC (μs/cm)	0.112	−0.493	−0.226	0.694	0.102	0.121
Temp (°C)	0.117	−0.086	−0.185	−0.868	0.23	−0.025
TDS (mg/L)	0.347	−0.054	0.832	0.06	0.143	0.148
Na^+ (mg/L)	0.865	0.013	0.005	−0.042	−0.261	−0.263
Ca^{2+} (mg/L)	−0.181	0.926	−0.057	0.015	0.231	0.14
K^+ (mg/L)	−0.846	0.01	−0.055	0.257	0.272	−0.104
PO_4^{3-} (mg/L)	−0.045	0.007	−0.035	0.071	−0.089	0.976
Cl^- (mg/L)	0.705	0.021	0.339	−0.043	0.45	0.069
Mn (mg/L)	−0.064	−0.054	0.807	0.013	−0.197	−0.156
Al^{3+} (mg/L)	0.179	0.942	−0.085	−0.087	0.051	-0.1
NO_3^- (mg/L)	0.783	−0.043	0.055	0.33	0.312	0.025

natural sources (K^+). The dominance of the identified parameters suggests that the quality of water samples from the boreholes were affected by a mixture of earth minerals (Na^+, Al^{3+}, Ca^{2+}) and anthropogenic sources. In Table 11.3a and b, K^+ and Cl^- were important quality-explaining factors, suggesting the possible influence of geology on the water quality in the area. With regards to the boreholes, Na^+, Cl^-, and NO_3^- (factor 1) as well as Al^{3+} and Ca^{2+} (factor 2) were the dominant parameters among those which explained 45.6% of the total variance in the chemistry of the water samples obtained from the sources (Tables 11.4a and b).

Table 11.4b Total variance explained by each group of parameters in Table 11.4a.

	Initial eigen values			Rotation sums of squared loadings		
Component	**Total**	**% of Variance**	**Cumulative %**	**Total**	**% of Variance**	**Cumulative %**
1	3.029	25.239	25.239	2.801	23.34	23.34
2	2.442	20.35	45.589	2.065	17.207	40.547
3	1.409	11.745	57.334	1.575	13.128	53.676
4	1.381	11.507	68.841	1.456	12.13	65.806
5	1.183	9.856	78.697	1.42	11.833	77.639
6	1.016	8.463	87.161	1.143	9.521	87.161
7	0.696	5.802	92.963			
8	0.368	3.069	96.032			
9	0.239	1.989	98.021			
10	0.127	1.061	99.082			
11	0.096	0.804	99.886			
12	0.014	0.114	100			

Table 11.5 Different division of relationship among the variables with the topography.

Parameters	Intercept	Rate of recharge/slope	correlation	Coefficient of determination
	A	b	r	(R^2)
Mn (mg/L)	−0.05	0.00026	0.19	0.04
As (mg/L)	−0.004	0.000023	0.11	0.01
Na^+ (mg/L)	0.53	0.050	0.04	0.001
K^+ (mg/L)	6.54	−0.023	−0.25	0.06
Ca^{2+}(mg/L)	9.8	−0.03	−0.07	0.006
Al^{3+} (mg/L)	−0.02	0.00008	0.055	0.003
Pb (mg/L)	0.007	−0.00002	−0.09	0.008
PO_4^- (mg/L)	4.45	−0.010	−0.03	0.00089
Cl^- (mg/L)	−133.69	0.978	0.15	0.02
NO_3^-(mg/L)	18.065	−0.027	−0.03	0.001
pH	6.72	0.00013	0.007	0.00005
Temp (°C)	29.5	−0.001	−0.02	0.00055
Ec (μs/cm)	290.2	−0.08	−0.05	0.0027
TDS (mg/L)	256.74	0.07	0.023	0.00055

11.5.3 Relationship between topography and physiochemical characteristics of groundwater

The result of linear regression analysis that was performed to isolate the relationship between each of the investigated chemical parameters of groundwater sources and the elevation (topography) in the study area are presented in Table 11.5. Table 11.5 showed different divisions of relationship among the variables with the topography. Whereas, NO_3^-, Ca^{2+}, K^+, Pb, and PO_4^{3-} decreased with groundwater depth, Cl^-, Na^+, As, and Mn occurred more at higher water

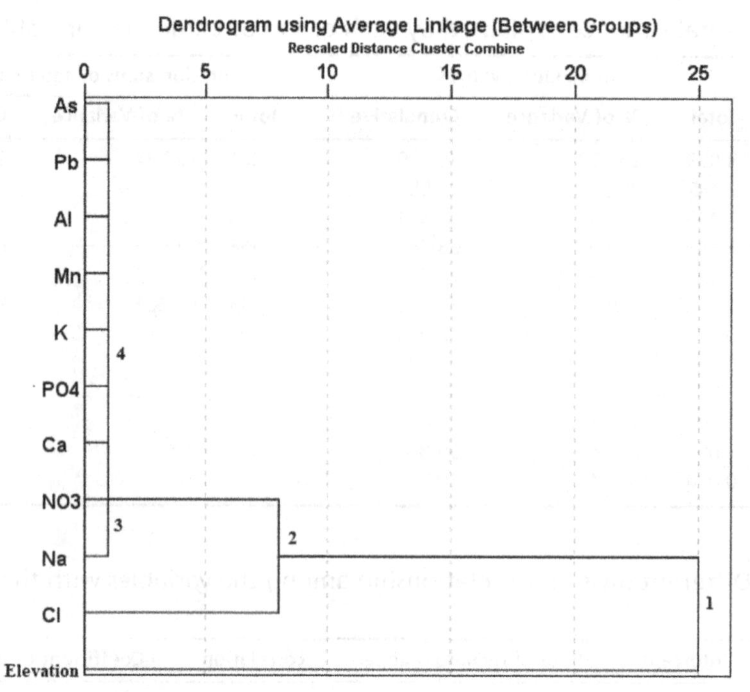

FIGURE 11.4 Dendogram from cluster analysis.

profile than at the lower profile. Water temperature and conductivity also declined with elevation (b = -0.001 and -0.08, respectively) while pH and TDS appeared to increase at 0.00013 and 0.07, respectively. In general, the result of Pearson correlation coefficient (r) indicated a positive relationship between each of all the chemical parameters and elevation ($r \leq 0.19$). The result of the coefficient of determination (R^2) also suggests that the influence of geology in the contamination of the groundwater resources may be low.

Furthermore, the dendrogram that was derived from the cluster analysis of the relationship of the chemical parameters and depth revealed that elevation was more influential to the concentration of chloride and nitrate than other parameters (Fig. 11.4). The association with other parameters (especially Al^{3+}, Mn, K^+, PO_4^{3-}, Ca^{2+}, and Na^+) were more connected to elevation through their association with Cl^- and NO_3^-. In order to distinguish the influence of difference in topography, depth was divided into two below 250 m and at 250 m above. The concentration of Pb, and No_3^- were significantly different between groundwater sources below 250 m and 250 m and above (Table 11.6), which was used to analyze the influence of specific parameter indicate that Na^+, Ca^{2+}, Pb, and Mn were the principal factors (-0.967, 0.977, 0.964, and 0.992, respectively) in groundwater sources at below 250 m while Cl^-, NO_3^- and Na^+ (0.928, 0.927 and -0.914) were the principal factors in groundwater sources at 250 m above elevation.

The parameter in factor 1 explained 64% and 45.7% of the total variance explained in the water quality of water samples from below and above 250 m elevation, respectively (Table 11.7).

Table 11.6 Analysis of variance of groundwater chemistry at different (below and above 250 m) landscape's heights.

Parameters	*F*-value	*P*-value	Interpretation
Mn (mg/L)	0.173	0.681	NS
As (mg/L)	2.19	0.151	NS
Na^+ (mg/L)	0.779	0.385	NS
K^+ (mg/L)	0.779	0.385	NS
Ca^{2+} (mg/L)	0.471	0.498	NS
Al^{3+} (mg/L)	0.376	0.545	NS
Pb (mg/L)	0.471	0.498	NS
Po_4^{3-} (mg/L)	0.202	0.656	NS
Cl^- (mg/L)	5.32	0.029	NS
NO_3^- (mg/L)	0.0067	0.798	NS
pH	1.976	0.171	NS
Temp (°C)	0.107	0.746	NS
EC (µs/cm)	0.477	0.496	NS

NS, difference is not significant at $P \leq 0.05$; *S*, difference is significant at $P \leq 0.05$.

Table 11.7 Results of rotated component matrix of investigated parameters in groundwater sources below and above 250 m elevation.

	Below 250 m			Above 250 m		
	Factor 1 (64.6%)	Factor 2 (18.6%)	Factor 3 (15.1%)	Factor 1 (45.7%)	Factor 2 (15.3%)	Factor 3 (13.9%)
Na^+ (mg/L)	−0.967	−0.214	−0.055	−0.914	−0.325	0.046
K^+ (mg/L)	−0.451	0.66	−0.598	−0.175	0.756	−0.099
Ca^{2+} (mg/L)	−0.977	−0.176	−0.068	−0.133	0.917	0.031
Pb (mg/L)	0.964	−0.019	0.153	0.854	−0.146	0.081
PO_4 (mg/L)	0.641	0.722	0.227	0.002	0.403	0.443
Cl^- (mg/L)	0.71	0.646	0.28	0.928	−0.208	−0.032
NO_3^- (mg/L)	0.716	0.577	0.385	0.927	−0.208	−0.037
pH (mg/L)	0.135	0.945	0.297	0.422	−0.317	0.716
EC (µs/cm)	0.23	0.123	0.965	−0.443	−0.006	0.665
Temp (°C)	0.041	−0.327	−0.944	−0.087	−0.021	−0.78
n (mg/L)	0.992	0.023	0.06	0.854	−0.21	0.152
As (mg/L)	0.784	0.235	0.477	0.884	−0.158	0.027

11.5.4 Seasonal variations

Results of the seasonal comparison and the seasonal averages (mean + standard deviation) of selected parameters are provided in Table 11.8. Table 11.8 shows that most of the parameters except electrical conductivity, occurred in significantly high concentration ($P \leq 0.05$) in the dry season than in the wet season, suggesting the influence of dilution in the wet season. On the other hand, more of the parameters in hand dug wells occurred in significantly high concentrations

Table 11.8 Seasonal variation.

	Boreholes		Hand dug wells		Overall difference between borehole and hand dug well
Variables	**Dry season**	**Wet season**	**Dry season**	**Wet season**	
pH	7.0±0.6	6.3±0.7	7.6±0.7	7.1±0.8	0.001
EC (μs/cm)	207.0±69.64	299.0±80.9	146.0±43.7	242.0±60.2	0.01
Temp (°C)	30.3±2.4	29.4±3.1	30.2±1.7	29.0±1.6	0.61 (NS)
TDS (mg/L)	251.0±17.0	328.0±176.0	130.0±54.9	221.0±75.2	0.001
Na^+ (mg/L)	24.8±10.9	27.3±11.4	0.2±0.3	108.0±0.03	0.002
Ca^{2+} (mg/L)	5.4±10.5	6.02±11.6	0.8±2.6	0.8±2.7	0.013
K^+ (mg/L)	0.8±1.56	0.8±1.6	0.2±0.1	0.2±0.05	0.02
PO_4^{3-} (mg/L)	2.1±5.6	2.1±5.6	1.7±0.4	1.7±0.4	0.69 (NS)
Cl^- (mg/L)	33.4±18.2	45.0±21.0	27.8±10.1	36.1±9.0	0.07 (NS)
NO_3^- (mg/L)	1.2±0.4	2.3±0.7	10.3±5.5	19.0±8.6	0.001
Mn (mg/L)	0.002±0.01	ND	0.05±0.05	0.05±0.05	0.000
As (mg/L)	ND	ND	0.004±0.001	0.04±0.01	0.000
Pb (mg/L)	ND	ND	0.42±0.001	0.42±0.001	0.000
Al^{3+} (mg/L)	0.13±0.47	0.13±0.47	ND	ND	0.000

Mean of Values with same alphabets (lower case) are not significantly ($P \leq 0.05$) different.
NS, not significant at $P \geq 0.05$.

in the wet season than they occurred in dry season. Since most of the ions and TDS could not have been more diluted in the dry season, the phenomenon suggests that they have been fed into the hand dug wells through runoff or seepages in the wet season. Water temperature was understandably lower in the wet season, due to more humidified weather effects in the wet season. When compared, Fig. 11.5, revealed varied level of spatial variability in the parameters. Parameters such as TDS, Na^+, Ca^{2+}, K^+, Al^{3+}, PO_4^{3-} varied conspicuously in the boreholes and well samples, probably owing to differences observed in ownership (refer to Table 11.1). On the other hand, Pb, As, and Mn appeared to vary conspicously in the hand dug wells. Averagely, significant differences occurred in the chemistry of the water samples obtained from the hand dug wells and those from the boreholes sources, except for water temperature, phosphate, and chloride.

11.5.5 Locational distribution of water chemistry parameters

The results of the spatial interpolation of the selected parameters over the study area are presented in Fig. 11.6. The figure showed that the values of pH and electrical conductivity were relatively different at the southeastern part of the study area than the other areas. Groundwater sources 5, 6, 10, and 12 (see Table 11.9 for interpretation of the codes) exhibited more conductivity than the other location. Also, the water resources in the northern part of the region were slightly more acidic along with pockets of other sources (13, 14, and 20) than the rest of the study area. In terms of water temperature, groundwater sources at the northern part of the study area appear warmer (32.5–36.3°C). For the major ions studied, the concentration of Na^+ was slightly more

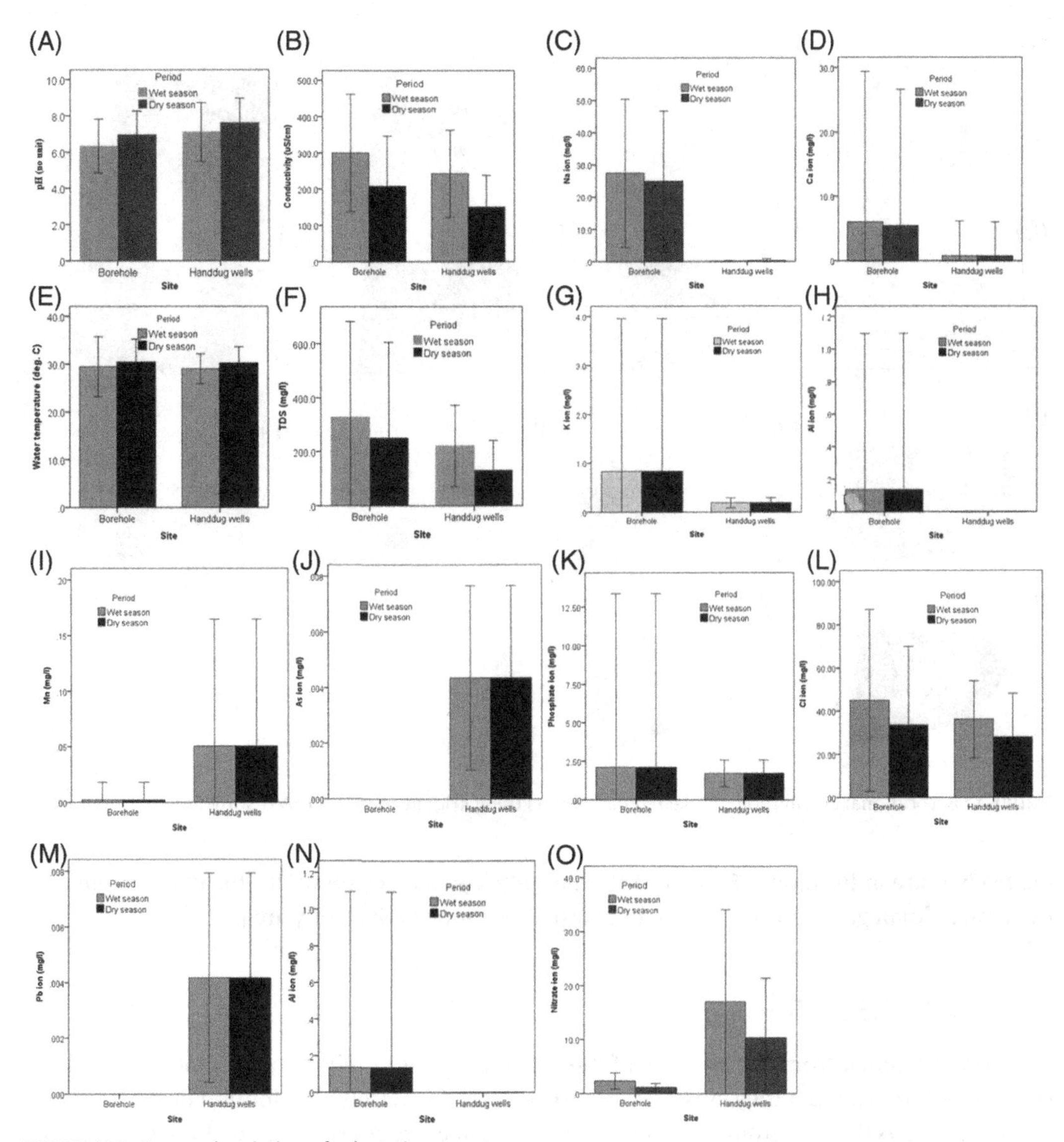

FIGURE 11.5 Seasonal variation of selected parameters.

at the southeastern region than the other part of the study area while K^+ concentration were more at the southwestern region. Ca^{2+} ion was more at a location at the center (13) of the study area. Cl^- and NO_3^- varied spatially across the study area at relatively higher than average concentration.

Phosphate, however, occurred more at low concentration in most of the sources except locations 13 and 10. Lastly, location 13 contained relatively more Al^{3+} while As concentration was

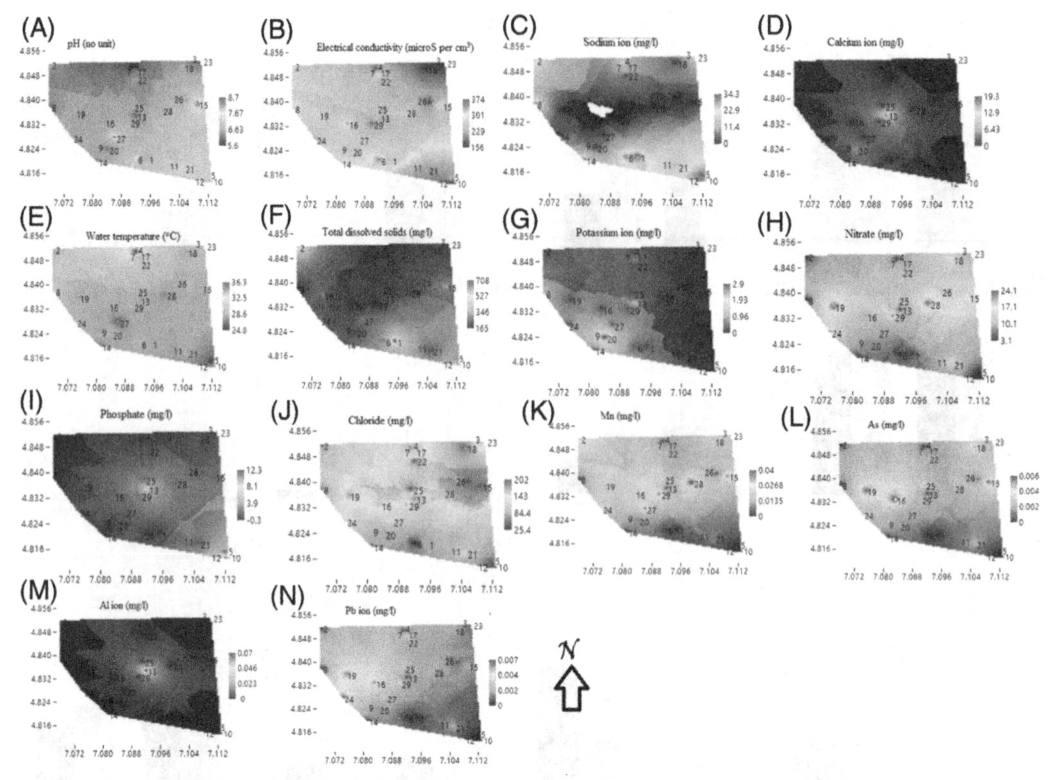

FIGURE 11.6 Locational distribution of selected parameters using kriging analysis in PAST software.

relatively more at locations 15, 16, and 19 than other areas. Locations 16, 19, and 24 were also richer in Pb than groundwater resources at other locations in the study area.

11.6 Discussion

The result obtained from the analysis of the distribution of boreholes and hand dug wells in Ondo Town showed that they clustered in distribution. Clustering distribution of groundwater sources suggests that the groundwater sources were not adequately planned. A review on water distribution and quality over the sub-Sahara Africa revealed that distribution of water sources may be may cluster because their availability are often dictated by age of the groundwater source (borehole came recently) and the socio-economic status of the residents. Naghibi et al. (2016) argued that groundwater though one of the most valuable freshwater resources is influenced by hydrological-geological-physiographical factors, including slope, degree, slope aspect among others. In the study area, however, majority of groundwater sources were sunk by individual house owners and the category of the groundwater sources tend to vary with the age of the well or borehole and the economic capacity or choice of house owners. In the past, up till 1990s, hand dug well was the only means of groundwater abstraction in the developing countries (in most

Table 11.9 Interpretation of the labels showing location of sampled groundwater resources in Fig. 11.5.

S/N	Address	Description
1	Adura Sabo	Commercial borehole
2	Austino, Laje Rd	
3	Bidmol, INEC	
4	Flourish GRA	
5	Klymax, Valentino	
6	Larry, Sabo	
7	MARPEC, Pele	
8	Oyem, Oka	
9	Laje Road	Privately owned borehole
10	Valentino	
11	Ife Road	
12	Ade Super	
13	Christland	
14	Fade-Olu	
15	Akinjagunla	
16	Okegbala	Residential hand dug well
17	Ahoyaya	
18	Omole	
19	Oke Odunwo	
20	Rainbow	
21	St Joseph	
22	GRA	
23	Akure Motor Park	
24	Adeyemi Gate	
25	Gear Weigh	Hand dug well in poor condition
26	Olorunsola	
27	Odosida	
28	Surulere	Public-owned hand dug well
29	Loro	

parts of Africa). Borehole is a recent development, around late 1990s (Ayoade, 1988) For example, hand dug wells are prominent in old buildings generally called "face-me-i-face-you" but machine drilled boreholes are more in relatively modern or separated apartment households. These probably became important in the study area because only two public government-owned water sources were functioning in the area as at the time this study was carried out. Consequently, one can assure that the resident of the area have decided on the need to seek water due to its importance to human life. Ameh et al. (2019), on the other hand, argued that concentration of groundwater sources in particular area may aggravate seismic activities and initiate significant huge disturbances over time.

With regards to the distribution of chemical constituents and quality, the results indicated that the water samples from boreholes varied from being slightly acidic to slight alkaline while the hand dug wells were generally alkaline. Both the hand dug wells and borehole sources contained higher than normal concentration of Chloride (Cl^-). It was also surprising that the water samples from borehole contained more TDS, Na^+, and Ca^{2+} than the samples from hand dug wells. These results may, however, require further investigation, although studies have indicated that location may exert significant impact on water chemistry in any area (WHO, 2017). In the study area, private boreholes at Adesuper street, Christland, and Akinjagunla streets possessed higher than normal concentration of TDS, sodium (Na^+) (at Adesuper), calcium (Ca^{2+}) and aluminum (Al^{3+}) at Christland and Chloride (Cl^-) (at Akinjagunla), respectively. Being boreholes privately owned, it is uncertain that investigations were carried out on the nature of the chemistry of the underlain rocks before the boreholes were drilled. Faniran and Omorinbola (1980) argued that groundwater sources especially shallow wells can also be contaminated if they are sunk in aquifer along water course, these groundwater sources are capable of producing large amount of water but are vulnerable to contamination. The study area is characterized by undulating topography with rock outcrops with interlock in some areas, which make groundwater sources to be sited mostly in the valleys and lowlands, for maximum water yield.

Furthermore, studies have revealed that some geology can contain high amount of contaminants, such as arsenic (As), calcium (Ca^{2+}), and chloride (Cl^-), which were detected in boreholes at the present study area. Consequently, further study is recommended to determine the chemistry of the geology of contaminated boreholes, it is essential that tests be carried out to determine the potential for contamination rather than focusing mainly on the quantity of water yield in the underlain geology. Results from this study is typical of occurrences in many parts of Nigeria, it is therefore recommended that efforts aimed at ensuring water quality be improved as the current efforts are apparently not effective as the present study has shown.

References

Abimbola, A.F., Tijani, M.N., Nurudeen, A., 1999. Some aspects of groundwater quality assessment of Abeokuta. J. Min. Geo. 32 (20), 127–130.

Adekunle, I.M., Adetunji, M.T., Gbadebo, A.M., Banjoko, O.B., 2007. Assessment of groundwater quality in a typical rural settlement in Southwest Nigeria. Int. J. Environ. Res. Public Health 4 (4), 307–318.

Adewumi, A.J., Ale, P.T., Oloye, B., 2004. Application of geographic information system and hydrochemistry in the analysis of groundwater across Okeluse Area of Ondo state Southwestern Nigeria. Int. J. Adv. Earth Environ. Sci. (IROSSS) (2) 1–8.

Adewumi, A.J., Ale, P.T., Oloye, B., 2014. Application of geographic information system and hydrogeochemistry in the analysis of groundwater across Okeluse area of Ondo state, southwestern Nigeria. Int. J. Earth Environ. Sci. 2 (1), 1–15.

Afuye, G.G., Oloruntade, A.J., Mogaji, K.O., 2015. Groundwater quality assessment in Akoko South East Area of Ondo State, Nigeria. Int. J. Sci. Technol. 5 (9), 1–5.

Akinro, A.O., Ologunagba, I.B., 2009. Bacteriological and physico-chemical analysis of some domestic water wells in peri-urban area of Akure, Nigeria. J. Appl. Irrig. Sci. 44 (2), 231–238.

Ameh, Y.C., Kogi, E., Eleojo, O.A., Cletus, A.T., Omoneke, Z.B., 2019. Challenges and Spatial Distribution of Water Infrastructures (Boreholes) in Okene Town, Kogi State, Nigeria. Int. J. Appl. Sci. 19 (1), 25–30.

Asiwaju-Bello, Y.A., Olabode, F.O., Duvbiama, O.A., Iyanu, J.O., Adeyemo, A.A., Onigbinde, M.T., 2013. Hydrochemical evaluation of groundwater in Akure area. Southwestern Nigeria, for irrigation purpose. Eur. Int. J. Sci. Technol. 2 (12), 22–26.

Ayoade, J.O., 1975. On water resources development in Nigeria. Niger. J. Econ. Soc. Stud. 17 (1), 35–48.

Ayoade, J.O, 1988. Tropical Hydrology and Water Resources. Macmillan Publishers Ltd, London and Basingstoke, pp. 230–235.

Ayodele, O.S., Ajayi, A.S., 2015. Water quality assessment of Otun and Ayetoro Area, Ekiti State, Southwestern Nigeria. Adv. Sci. Technol. Res. 2 (1), 8–18.

Bayode, S., OJO, J.S., Olorunfemi, M.O., 2006. Geoelectric characterization of aquifer types in the Basement Complex terrain of parh of Osun State, Nigeria. Global J. Pure Appl. Sci. 3 (1), 68–80.

Boyle, D.R., 1988. Application of Groundwater Geochemistry in Mineral Exploration, 1988. Geological Survey of Canada, Dartmouth, MA.

Clark, R., 1991. Water the International Crisis. Earth Scan Publications Ltd, London.

Clark, J., 2006. Rivers and their catchments: impact of landfill on waterquality. Earth Sci. Branch, Scottish Natural Heritage p. 2.

Edet, A., Ukpong, A.J., Ekwere, A.C., 2011. Impact of climate change on groundwater resources: An example from cross river state, Southeastern Nigeria. In: COLERM (Ed.), *COLERM Proceedings*, *1*, 2–22, Proceedings of the Environmental Management Conference, Federal University of Agriculture, Abeokuta, Nigeria, 1. Federal University of Agriculture, Abeokuta, Nigeria, pp. 2–22. https://publications.unaab.edu.ng/index.php/COLERM/article/view/233/216.

Eludoyin, A.O., Ofoezie, I.E., Ogunkoya, O.O., 2004. The effect of Oja-titun market effluent on the chemical quality of receiving OPA reservoir in Ile-Ife, Nigeria. J. Environ. Manage. 72 (4), 249–259.

Eludoyin, A.O., 2014. Environment and sustainability: How is the spatial information system relevant in groundwater quality assessment? Spring Newsletter 2014, West Afr. Res. Assn. 2014 (1), 16–17. doi:10.13140/RG.2.2.30891.34080, https://www.westafricanresearchassociation.org/wp-content/uploads/2019/11/WARA_Spring_NL.pdf.

Eludoyin, A.O., Abuloye, P.A., Nevo, O.A., Eludoyin, O.M., Awotoye, O.O., 2017. Climate events and impacts on cropping activities of small-scale farmers in a part of Southwest Nigeria. Weather Clim. Soc. (9) 235–237.

Eludoyin, A.O., 2020. Accessibility to safe drinking water in selected urban communities in southwest Nigeria. Water Prod. J. 1 (2), 1–10.

Ezeigbo, H.I., 1987. Quality of water resources in Anambra State, Nigeria. J. Min. Geol. 23, 97–103.

Ezechukwu, P.N., 2016. Water supply to cities through borehole sources. J. Environ. Urban. 2 (5), 17–20.

Federal Ministry of Water Resources, 2000. National water supply and sanitation policy. Federal Republic of Nigeria, Nigeria http://www.nwri.gov.ng/userfiles/file/National_Water_Supply_and_Sanitation_Policy.pd010.

Federal Republic of Nigeria (FRN), 2000. Water supply and sanitation interim strategy note, p.38 cited in Akinro, O.A., and Ologunagba, I.B., (2009): Bacteriological and physico- chemical analysis of some domestic water wells in peri-urban areas of Akure, Nigeria. J. Appl. Irrig. Sci. 44 (2) 231–238.

Faniran, A., Omorinbola, O.O., 1980. Trend surface analysis and practical implications of weathering depths in Basement Complex rocks of Nigeria. Nigerian Geogr. J. 23 (1), 113–126.

Francy, R., Pickford, J., Reed, R., 1991. A Guide to The Developing and Managing Community Water Supplies of On-Site Sanitation. World Health Organisation, Geneva.

Hanidu, J.A., 1990. National growth water demand and supply strategies in Nigeria in the 1990s. Water Resources J. Nigeria Assn. Hydrogeol. 2 (1), 1–6.

Idowu, N.A., Nardi, C., Long, H., Varslot, T., Øren, P.E., 2014. Effects of segmentation and skeletonization algorithms on pore networks and predicted multiphase-transport properties of reservoir-rock samples. SPE Reserv. Eval. Eng. 17 (04), 473–483.

Mizumura, K., 2003. Chloride ion in groundwater near disposal of solid wastes in landfills. J. Hydrol. Eng. 8 (4), 204–213.

Mogaji, K.A., Olagunju, G.M., Oladapo, M.I., 2011. Geophysical evaluation of rock type impact on aquifer characterization in the basement complex areas of Ondo State, Southwestern Nigeria: geo-electric assessment and geographic information systems (GIS) approach. Int. J. Water Resour. Environ. Eng. 3 (4), 77–86.

National Water Supply and Sanitation Policy (2000), Department of Water Supply and Quality Control, 41p, https://books.google.com.ng/books?id=mv4otAEACAAJ.

Nganje, T.N., Adamu, C.I., Ntekim, E.E.U., Ugbaja, A.N., Neji, P., Nfor, E.N., 2010. Influence of mine drainage on water quality along River Nyaba in Enugu South-Eastern Nigeria. Afr. J. Environ. Sci. Technol. 4 (3), 132–144.

Nyamboge, C., Karoli, N.N., George, V.L., Alfred, N.N., 2008. Hydrochemical characteristics of spatial distribution of groundwater quality in ausha fields, Northern Tanzania. Appl. Water Sci. 4 (2), 45–50.

Ojo, S.O., Olorunfemi, M.O., Aduwo, A.I., Sunday, B., Olaoluwa, J.A., Omosuyi, G.O., Akinleyi, F.O., 2014. Assessment of surface and groundwater quality of Akure Metropolis, Southwestern Nigeria. J. Environ. Health Sci. 2 (6), 987–992.

Okagbue, C.O. (1988). Hydrology and chemical characteristics of surface and groundwater resources of the Okigwi area and environs, Imo State, Nigeria. 3-6.

Okwere, A.O., Hart, L., Jackson, K.P., 2015. Mapping the spatial distribution of water borehole facilities in part of Rivers State using. Geogr. Inform. System (GIS) Techniq. 5 (4), 100–120.

Olagoke, E.A., Akeasa, O.S., 2017. GIS application for assessing spatial distribution of boreholes and hand dug wells in Boroboro Community, Atiba Local Government Oyo State. J. Remote Sens. GIS 5 (5), 34–36.

Olorunfemi, M.O., Fasuyi, S.A., 1993. Aquifer types and the geoelectric/hydrogeologic characteristics of part of the central basement terrain of Nigeria (Niger State). J. Afr. Earth Sci. (and the Middle East) 16 (3), 309–317.

Oloruntade, A.J., Mogaji, K.O., Alao, F., 2012. Quality of well water in Owo, Southwestern Nigeria. Int. J. Acad. Res. 3 (1), 445–447.

Oyawoye, M.O., 1972. The basement complex of Nigeria. Afr. Geol. 67–99.

Oyedotun, T.D.T., Obatoyinbo, O., 2012. Hydrogeological evaluation of groundwater quality in Akoko North West local government area of Ondo State, Nigeria. Ambi-aqua.net 7 (1), 67–68.

Oyeku, O.T., Eludoyin, A.O., 2010. Heavy metal contamination of groundwater resources in a Nigerian urban settlement. Afr. J. Environ. Sci. Technol. 4 (4), 201–214.

Oyinloye, A.O., 2011. Geology and Geotectonic Setting of the Basement Complex Rocks in SouthWestern Nigeria: Implications on Provenance and Evolution, Earth and Environmental Sciences, Dr. Imran Ahmad Dar (Ed.), ISBN: 978-953-307-468-9, InTech, Available from: http://www.intechopen.com/books/earth-and-environmental-sciences/geology-and-geotectonic-setting-of-the-basement-complex-rocks-in-south-western-nigeria-implications. (Accessed date 23 April 2020).

Piscopo, G., 2001. Groundwater Vulnerability Map, Explanatory Notes, Castlereagh Catchment, NSW. Department of Land and Water Conservation, Parramatta http://www.water.nsw.gov.au/__data/assets/pdf_file/0008/549377/quality_groundwater_castlereagh_map_notes.pdf.

Pritchad, M., Mkandawire, T., O'Neill, J.P, 2009. Assessment of groundwater in shallow wells within the southern districts of Malawi. Phys. Chem. Earth 120–125.

Ramanathan, A.L., Prasad, M.B.K., Chidambaram, S, et al., 2007. Ramanathan, AL Prasad MBK., and Chidambaram, S. 2007. Ground water Arsenic contamination and its health effects-Case studies from India and SE Asia. 22 (2) 371–384. Indian J. Geochem. 22 (2), 371–384.

Seidel, K., Lange, G. (2008). Direct Current Resistivity Methods in Environmental Geology Handbook of Field Methods and case Studies, Knodel K.G & H. Voigt (Eds), Springer. 205–237.

Spechler, R.M., 2010. Hydrogeology and Groundwater Quality of Highlands County, Florida: U.S. Geological Survey Scientific Investigations Report 1 (5097), 1–84. https://pubs.usgs.gov/sir/2010/5097/pdf/sir2010-5097.pdf (Accessed 23 April 2020).

Standard Organisation of Nigeria (SON), 2015. Nigerian Standard for Drinking Water Quality. NIGERIAN INDUSTRIAL STANDARD NIS-554-2015. https://rivwamis.riversstate.gov.ng/assets/files/Nigerian-Standard-for-Drinking-Water-Quality-NIS-554-2015.pdf.

Tekwa, I.J., Abba, M.U., Ray, H.H., 2006. An assessment of dug-well water quality and use in Mubi, Nigeria. J. Sus. Dev. Agric. Environ. 2 (1), 1–5.

Toth, J., 1963. A theoretical analysis of groundwater flow in small drainage basins. J. Geog. Dev. 68 (16), 4795–4812.

Toth, J., 1999. Groundwater as a geologic agent; an overview of the causes, processes and manifestation. Hydrol. J. 38 (7), 1–14.

United Nation Environmental Protection (UNEP), 2002. Past, present and future perspectives, Africa Environment Outlook. United Nations Environmental Programme.

United Nations, World Water Assessment Programme, 2006. Water: a shared responsibility, vol 2. Berghahn Books, New York.

U.S. EPA. EPA's Report on the Environment (ROE) (2008 Final Report). U.S. Environmental Protection Agency, Washington, D.C., EPA/600/R-07/045F (NTIS PB2008-112484), 2008.

Vrba, J., 2002. The impacts of aquifer intensive use on groundwater quality. In: Commission on Groundwater Protection of the International Association of Hydrogeologists (IAH), 19, Prague, The Czech Republic, pp. 113–132.

Woakes, M., Ajibade, C.A., Rahaman, M.A., 1987. Some metallogenic Features of the Nigerian Basement. African J. Sci. 5, 655–664.

World Bank, (2002). World Bank Global Report on population and access to quality water.

WHO, 2004. Water Sanitation and Hygiene Links to health.Water sanitation health publication, pp. 143–144.

WHO, 2011. Guidelines for Drinking-Water Quality, 1. World Health Organization, Geneva ed. ISBN 9789241548151.

World Health Organisation (WHO), (2017). Guidelines for Drinking Water Quality, 8th edition. Geneva, 2 (3), 210.

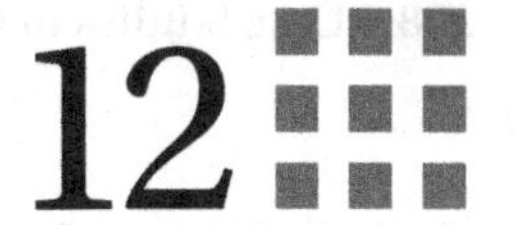

Assessing groundwater potential zone of Ong river basin using geospatial technology

Sanjoy Garai[a], Sk Mujibar Rahaman[a], Masjuda Khatun[a], Pulakesh Das[b] and Sharad Tiwari[a]

[a]INSTITUTE OF FOREST PRODUCTIVITY, LALGUTWA, RANCHI, INDIA [b]WORLD RESOURCES INSTITUTE INDIA, NEW DELHI, INDIA

12.1 Introduction

Sustainable use of water resources has become a national priority in countries where socio-economic development is primarily based on agriculture (Rao, 2002). The alteration in climate conditions and increased extreme weather events impose a greater threat to hydrological regime shift and availability of water resources (Gleick, 1989). Groundwater is a dynamic and recyclable natural resource that regulates economic development, human health, ecological diversity, etc., in a region (Chowdhury et al., 2009). As of April 2015, India's national water resource potential was estimated as 1869 billion cubic meters (BCM), while the net annual groundwater availability was 398 BCM (CGWB). About 68% of the groundwater recharge is dependent on rainfall, whereas 32% is contributed by canal seepage recharge due to irrigation and other water conservation structures like water tanks and ponds (Central Ground Water Board (CGWB), 2014). Overexploitation of groundwater is increasing rapidly due to population growth and intense irrigation (Singh and Singh, 2002; Kulkarni et al., 2018). Moreover, urbanization and increasing artificial structures significantly alter the infiltration process and reduce groundwater recharge (ODriscoll et al., 2010). Das et al. (2018) studied the impact of land use land cover (LULC) changes on the hydrological parameters in the Mahanadi river basin and reported significant changes in the surface runoff and base flow. Western India experienced a significant decrease in groundwater rejuvenation rate (5.81 ± 0.38 km^3/year) from 1996 to 2001, while replenishment (2.04 ± 0.20 km^3/year) occurred from 2002 to 2014 (Bhanja et al., 2017). The decreasing and replenishment rate was lower for southern India (-0.92 ± 0.12 km^3/year during 1996–2002) and (0.76 ± 0.08 km^3/year during 2003–2014) respectively. They concluded that the modification in the groundwater withdrawal and management policies has resulted in the aquifers' replenishment. Assessing the groundwater potential (GWP) and monitoring the water level is essential

Case Studies in Geospatial Applications to Groundwater Resources. DOI: https://doi.org/10.1016/B978-0-323-99963-2.00007-9

in developing improved management and conservation activities to enable sustained resource utilization.

For the last several decades remote sensing and geographic information system (GIS) technology have been widely used in water resource management and development activities (Kumar et al., 2014). The multi-temporal optical remote sensing imagery enables direct measurement of surface water resources and changes even in inaccessible areas. However, groundwater assessment using the remote sensing data-derived layers is primarily dependent proxy variables. In comparison to field investigations the GIS-based approach is cost-effective, less labor intensive, and facilitates rapid and periodic analysis. However, the GIS based analysis also requires sample ground observations for training and validation purposes. The optical and microwave remote sensing data is a reliable source to generate the layers on LULC, surface water availability, topographic parameters, drainage network, lineaments, and rainfall data. The remotely sensed rainfall data serves as an alternative data source providing daily data at high resolution freely.

The secondary data generated using the remote sensing data derived factors and GIS-based thematic layers are analyzed to delineate the groundwater potential zone (GWPZ). The commonly used factors include LULC and surface water, topography (elevation and slope), drainage and lineament density, surface and subsurface features (geology, geomorphology, soil) etc. (Agarwal et al., 2013; Chowdhury et al., 2009; Kumar et al., 2014; Magesh et al., 2012; Mallick et al., 2015; Prasad et al., 2008; Shekhar and Pandey, 2015; Abd Manap et al., 2014). The multi-criteria decision analysis (MCDA) is one of the effective approaches to integrating multiple thematic layers. Both the linear and nonlinear methods are applied in MCDA, where the differential importance of various thematic layers is assigned depending on their relative influence on GWPZ. Several techniques have been developed to estimate the weights of the input proxy variables, for example, the analytic hierarchy process (AHP) (Kumar et al., 2014), eigenvector technique (Mallick et al., 2015), frequency ratio model (Abd Manap et al., 2014; Garai and Das, 2021), regional probabilistic model (Oh et al., 2011), MIF technique (Magesh et al., 2012), random forest (RF) and maximum entropy models (Golkarian et al., 2018; Rahmati et al., 2016), support vector machine model (Lee et al., 2018), etc.

The AHP-based approach is the most widely used and reliable method in defining the weights in assessing the GWP (Chowdhury et al., 2009; Jha et al., 2010). In AHP, the hierarchical process is adopted based on expert opinion. Various studies have used the MCDA-AHP based approach to delineate the groundwater recharge zones and identifying the artificial recharge sites (Chowdhury et al., 2009), integrating the surface runoff with several other thematic layers (e.g., LULC, topography, geomorphology, stream, etc.) to prescribe suitable rainwater harvesting structures (Behera et al., 2019). The current research emphasizes the identification of groundwater potential in the Ong sub-basin using the MCDA-AHP based approach. The study aims to categorize the area under different groundwater levels and examine the important factors or variables.

12.2 Study area

Ong is an important tributary and a sub-basin of the Mahanadi river basin. The study area lies between 20°39′ to 21°29′ N latitude and 82°33′ to 83°55′ E longitude partially occupied by two

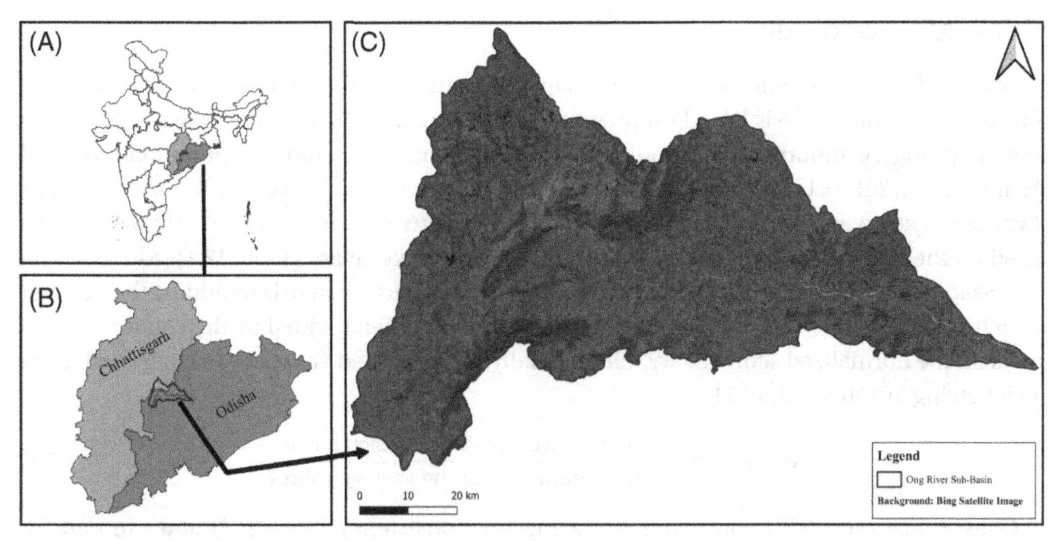

FIGURE 12.1 Location map of the study area showing (A) the country India map state boundary wise, (B) location of Ong river basin in the Odisha and Chhattishgarh state, and (C) map of the Ong river basin (a sub-basin of Mahanadi River) with Bing satellite image in the background.

states, that is, Odisha and Chhattisgarh (Fig. 12.1). The average elevation of the Ong sub-basin is 182 m, with a total area of 5128 km^2. The Ong sub-basin is occupied by the Nuapad, Bargarh, Balangir, and Subarnapur districts of Odisha state and Mahasamund district of Chhattisgarh state. The main tributaries of the Ong river are Surangi, Khira, ChiraNala, and Utalio.

12.3 Materials and methods

12.3.1 Data used

We used cloud free postmonsoon Sentinel-2 data for the year 2020, acquired from the Copernicus data portal (https://scihub.copernicus.eu/dhus/#/home). The ALOS PALSAR digital elevation model (DEM) data with a spatial resolution of 12.5 m was obtained from the Alaska Satellite Facility (ASF) site (https://search.asf.alaska.edu/#/). The maps on LULC, surface water, and lineament density were generated using the Sentinel-2 optical data. The DEM was used to generate the drainage density, slope, and elevation map. The vector layer of geology, geomorphology, and lithology features were acquired from the Geological Survey of India (GSI). The PERSIANN-CCS (PERSIANN-Cloud Classification System) high-resolution precipitation data (spatial resolution 4 km) was acquired from the CHRS portal (https://chrsdata.eng.uci.edu/). The annual rainfall data was accessed for the year 2020. The vector layer of soil type was acquired from the Food and Agriculture Organization (FAO) of the United Nations website (www.fao.org/geonetwork/srv/en/metadata.show?id=14116). To validate the results here also used the Dug Well data from Central Ground Water Board for the year of 2019–20 (http://www.indiaenvironmentportal.org.in/files/file/GROUND%20WATER%20YEAR%20BOOK%202019-2020%20Odisha.pdf).

12.3.2 Methodology

We applied MCDA to delineate the GWPZ map of the study area. A total of eleven factors were considered for this analysis based on previous studies includes LULC, surface water, drainage density, geology, geomorphology, lithology, lineament density, elevation, slope, rainfall, and soil (Agarwal et al., 2013; Chowdhury et al., 2009; Kumar et al., 2014; Magesh et al., 2012). The weights of various layers were derived using the AHP, where a pairwise comparison matrix was created based on the relative importance of the input thematic proxy layers (Table 12.1) (Kumar et al., 2014; Saaty, 1980, 2008). Then normalized the pairwise comparison matrix by adding the number in each column (Table 12.2). Each entry in the column was then divided by the column total to calculate the normalized score (Saaty, 1980). Finally, the weighted values were computed using the following formula (Eq. 12.1).

$$\text{Weight value} = \frac{100 * \text{Average value of each factor}}{\text{The summation of the average values}} \tag{12.1}$$

Consistency ratio (CR) was estimated using the consistency index (CI) and random CI. Wherein, the CI was computed using λ_{max} with the help of the consistency vector (Eq. 12.2–Eq. 12.4). Consistency vector (CV) was calculated by multiplying the pairwise matrix with the weighted vector followed by dividing the weighted sum vector by the average value of normalized pairwise comparison (Table 12.3). All the thematic layers were integrated based on the estimated weights using Quantum-GIS software to generate the groundwater potential zone map. Overall data processing flowchart is shown in Fig. 12.2. There are three phases to get the consistency ratio (CR), as follows:

$$\text{Consistency Ratio (CR)} = \frac{\text{CI}}{\text{RCI}} \tag{12.2}$$

$$\text{Consistency Index (CI)} = \frac{\lambda\text{max} - n}{n - 1} \tag{12.3}$$

where CI = consistency index; RCI = random consistency index; n = number of factors

$$\lambda_{\text{max}} = \frac{\sum CV}{n} \tag{12.4}$$

estimated values: $\lambda_{max} = 12.60$; CI = 0.16; CR = 0.10

The dug well data for the study area was accessed from the Central Ground Water Board (CGWB) from 2019 to 2020. The mean annual water level was derived from the dug well data to validate the GWP zones.

12.4 Result and discussion

12.4.1 Surface waterbody

Surface water is one of the main sources of infiltration, thus given high weightage in estimating the groundwater potential (Halder et al., 2020). Three major types of surface waterbody include

Table 12.1 Pairwise comparison matrix.

Factors	LULC	Lineament density	Drainage density	Elevation	Slope	Surface water	Rainfall	Soil	Geomorphology	Geology	Lithology
LULC	1	0.14	0.17	0.20	0.20	0.11	0.20	0.50	0.14	0.20	1.00
Lineament density	7	1	3.00	5.00	5.00	0.14	1.00	5.00	0.50	1.00	3.00
Drainage density	6	0.33	1	5.00	5.00	0.14	1.00	5.00	0.33	0.33	3.00
Elevation	5	0.20	0.20	1	0.33	0.14	0.33	3.00	0.33	0.33	3.00
Slope	5	0.20	0.20	3	1	0.20	0.33	3.00	0.33	0.33	3.00
Surface water	9	7	7	7	5	1	4.00	5.00	1.00	3.00	4.00
Rainfall	5	1	1	3	3	0.25	1	3.00	0.33	0.33	3.00
Soil	2	0.20	0.20	0.33	0.33	0.20	0.33	1	0.33	0.33	3.00
Geomorphology	7	2	3	3	3	1	3	3	1	3.00	5.00
Geology	5	1	3	3	3	0.33	3	3	0.33	1	3
Lithology	1	0.33	0.33	0.33	0.33	0.25	0.33	0.33	0.20	0.33	1
Total	53.00	13.41	19.1	30.87	26.20	3.77	14.53	31.83	4.84	10.20	32.00

Table 12.2 Normalized pairwise comparison matrix.

Factors	LULC	Lineament density	Drainage density	Elevation	Slope	Surface water	Rainfall	Soil	Geomorphology	Geology	Lithology	Average	Weighted value
LULC	0.02	0.01	0.01	0.01	0.01	0.03	0.01	0.02	0.03	0.02	0.03	0.02	1.76
Lineament density	0.13	0.07	0.16	0.16	0.19	0.04	0.07	0.16	0.1	0.1	0.09	0.12	11.59
Drainage density	0.11	0.02	0.05	0.16	0.19	0.04	0.07	0.16	0.07	0.03	0.09	0.09	9.11
Elevation	0.09	0.01	0.01	0.03	0.01	0.04	0.02	0.09	0.07	0.03	0.09	0.05	4.68
Slope	0.09	0.01	0.01	0.1	0.04	0.05	0.02	0.09	0.07	0.03	0.09	0.06	5.64
Surface water	0.17	0.52	0.37	0.23	0.19	0.27	0.28	0.16	0.21	0.29	0.13	0.25	25.44
Rainfall	0.09	0.07	0.05	0.1	0.11	0.07	0.07	0.09	0.07	0.03	0.09	0.08	7.8
Soil	0.04	0.01	0.01	0.01	0.01	0.05	0.02	0.03	0.07	0.03	0.09	0.04	3.56
Geomorphology	0.13	0.15	0.16	0.1	0.11	0.27	0.21	0.09	0.21	0.29	0.16	0.17	17.02
Geology	0.09	0.07	0.16	0.1	0.11	0.09	0.21	0.09	0.07	0.1	0.09	0.11	10.79
Lithology	0.02	0.02	0.02	0.01	0.01	0.07	0.02	0.01	0.04	0.03	0.03	0.03	2.63
Total												1	

Table 12.3 Estimation of CV.

Factors	LULC	Lineament density	Drainage density	Elevation	Slope	Surface water	Rainfall	Soil	Geomorphology	Geology	Lithology	Total	CV
LULC	0.02	0.02	0.02	0.01	0.01	0.03	0.02	0.02	0.02	0.02	0.03	0.2	11.687
Lineament density	0.12	0.12	0.27	0.23	0.28	0.04	0.08	0.18	0.09	0.11	0.08	1.59	13.72
Drainage density	0.1	0.04	0.09	0.23	0.28	0.04	0.08	0.18	0.06	0.04	0.08	1.21	13.318
Elevation	0.09	0.02	0.02	0.05	0.02	0.04	0.03	0.11	0.06	0.04	0.08	0.53	11.41
Slope	0.09	0.02	0.02	0.14	0.06	0.05	0.03	0.11	0.06	0.04	0.08	0.68	12.057
Surface water	0.16	0.81	0.64	0.33	0.28	0.25	0.31	0.18	0.17	0.32	0.11	3.56	13.986
Rainfall	0.09	0.12	0.09	0.14	0.17	0.06	0.08	0.11	0.06	0.04	0.08	1.02	13.127
Soil	0.03	0.02	0.02	0.02	0.02	0.05	0.03	0.04	0.06	0.04	0.08	0.39	11.152
Geomorphology	0.12	0.23	0.27	0.14	0.17	0.25	0.23	0.11	0.17	0.32	0.13	2.16	12.671
Geology	0.09	0.12	0.27	0.14	0.17	0.08	0.23	0.11	0.06	0.11	0.08	1.45	13.477
Lithology	0.02	0.04	0.03	0.02	0.02	0.06	0.03	0.01	0.03	0.04	0.03	0.32	12.102
Total													138.7

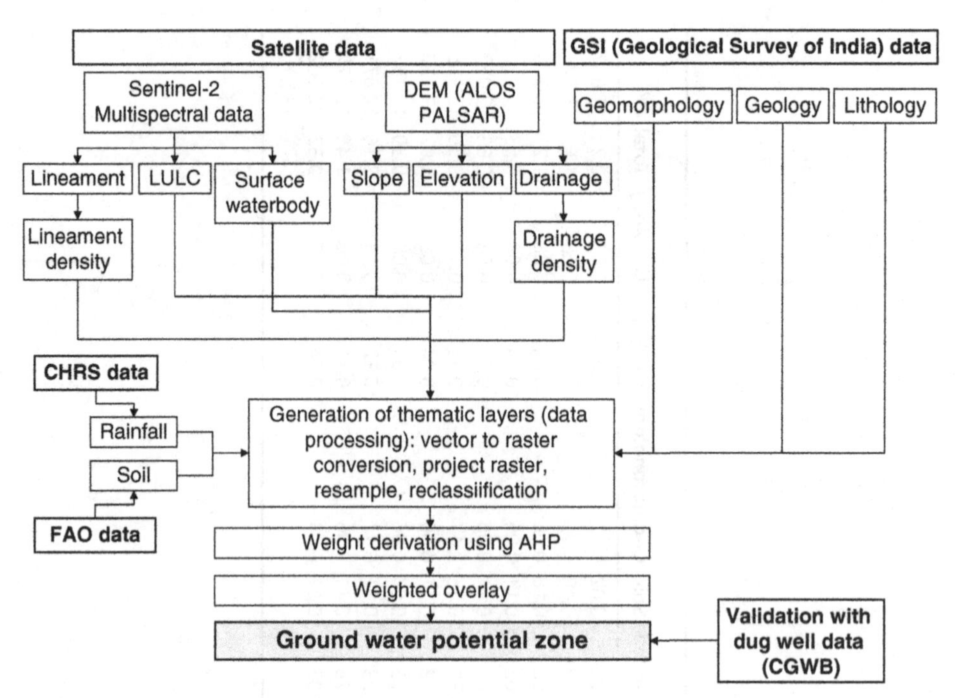

FIGURE 12.2 Overall data processing flowchart.

(1) permanent surface water (includes rivers, lakes, swamps, etc.), (2) semi-permanent surface water (includes creeks, waterholes, lagoons, etc.), and (3) artificial surface waterbody (includes dams, artificial swamps, lakes, etc.) (NGC, 2019). The surface water map was created by applying spectral enhancement on the Sentinel-2 optical data. The normalized difference water index (NDWI) $\left[= \frac{\text{Green}-\text{NIR}}{\text{Green}+\text{NIR}}\right]$ map was derived using the Sentinel-2 data, where the higher and lower values indicate water and nonwater land surface features in an image (Fig. 12.3).

12.4.2 Geomorphology

Geomorphology represents the earth surface features or the upper part of the crust. This layer contains water bodies, flood plain, pediment pediplain, plateau, hills, valley, etc. The spatial layer of geomorphology for the Ong sub-basin was acquired from the GSI. Geomorphology of any region is one of the essential criteria in regulating groundwater recharge (Arulbalaji et al., 2019). Pediplain is observed as the dominant feature, followed by hills and valleys in the study area (Fig. 12.4). Higher weightage was assigned to the waterbody and floodplain with the alluvium deposit as they allow the maximum water penetration. Moderate weight was assigned to pediment pediplain. These are flat areas primarily used for agriculture and an important source for water recharge (Chowdhury et al., 2009). On the contrary hills and valleys cause the maximum runoff provide the least time for infiltration and thus given lesser weight (Mallick et al., 2015).

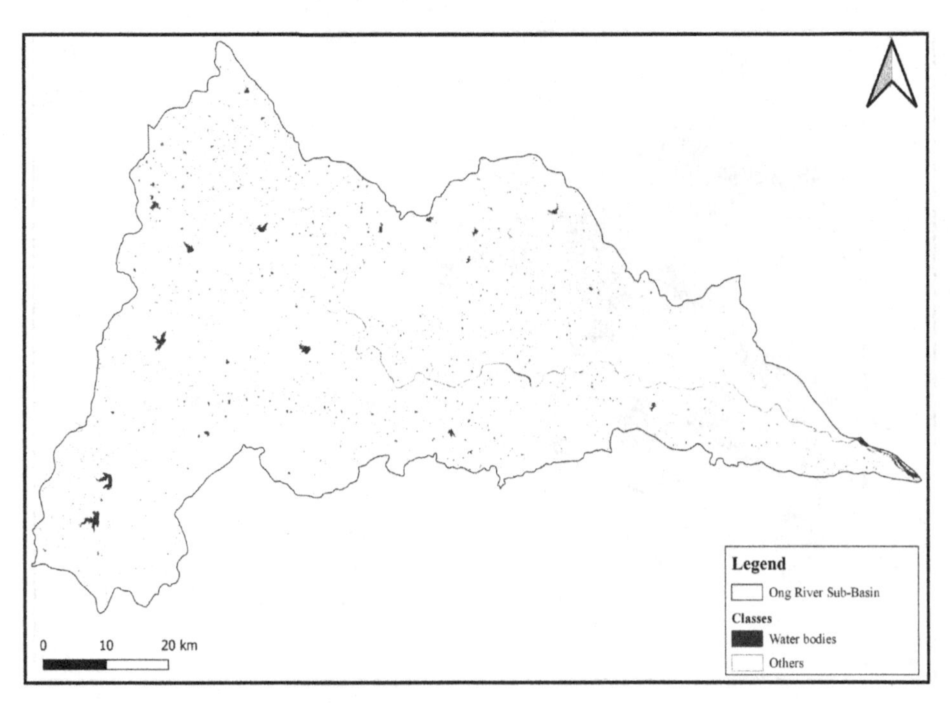

FIGURE 12.3 Surface waterbody map of Ong sub-basin.

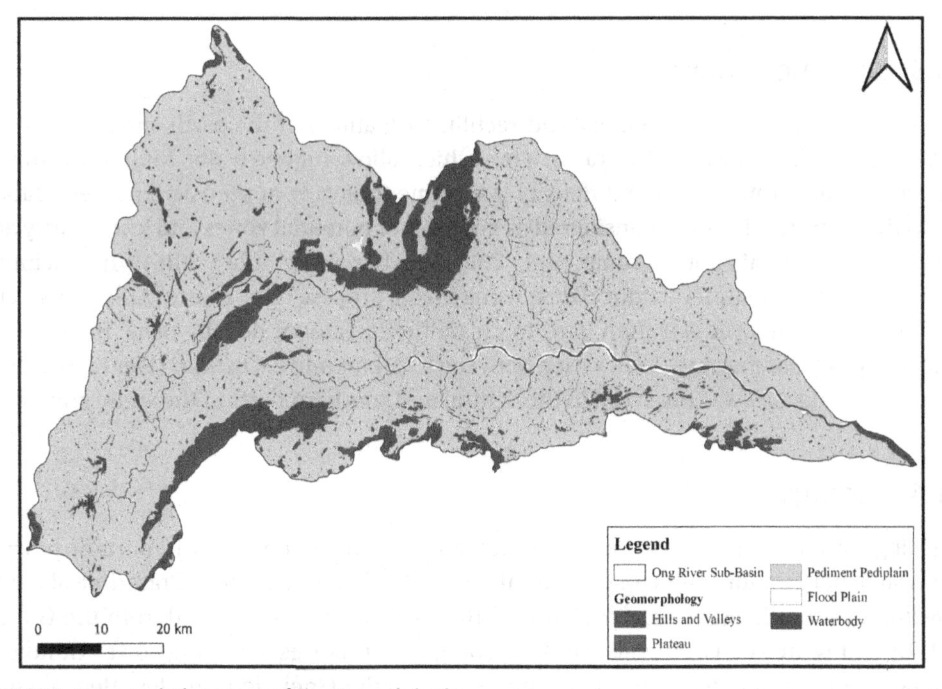

FIGURE 12.4 Geomorphology map of the Ong sub-basin.

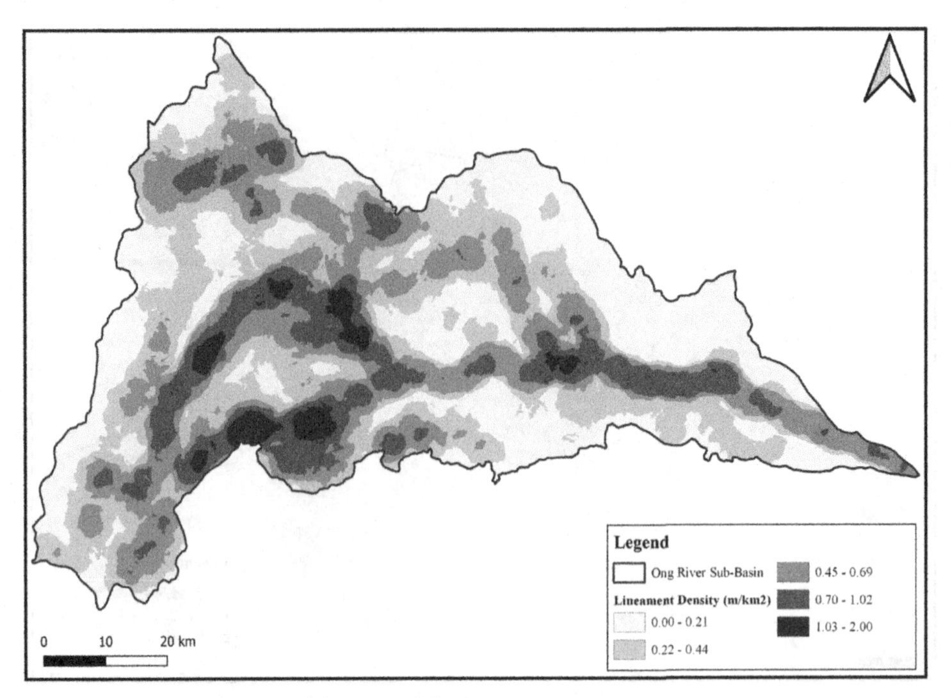

FIGURE 12.5 Lineament density map of the Ong sub-basin.

12.4.3 Lineament density

Lineaments are the linear, curvilinear and rectilinear features of the earth surface. This layer indicates the faults, folds, joints, cracks, etc., which allow higher water infiltration through the narrow breaks. Thus, lineament density is a prime indicator of porosity and permeability, wherein the high-density areas considered as good water potential zones and low-density areas as poor water potential zones (Kumar et al., 2014; Shekhar and Pandey, 2015). The lineaments were extracted from Sentinel-2 optical data using image processing software (Adb Manap et al., 2014). The generated layer was then used to create the lineament density map. The estimated lineament density for the study area varied between 0 m/km^2 and 2.0 m/km^2 (Fig. 12.5), wherein high lineament density was observed in the central and southern parts of the study area.

12.4.4 Geology

The geological layer represents the subsurface and is considered as an important proxy for groundwater penetration rate (Agarwal et al., 2013; Mallick et al., 2015; Prasad et al., 2008). The vector layer on the geological features for the study area was accessed from the GSI data portal. The study area consists of six types of geological structures including Lower Gondwana, Singhora Group, Khondalite Gneissic Complex, Migmatite Gneissic Complex, Bengal Gneiss Complex, and Dongargarh Granite (Fig. 12.6). The Lower Gondwana region is more suitable

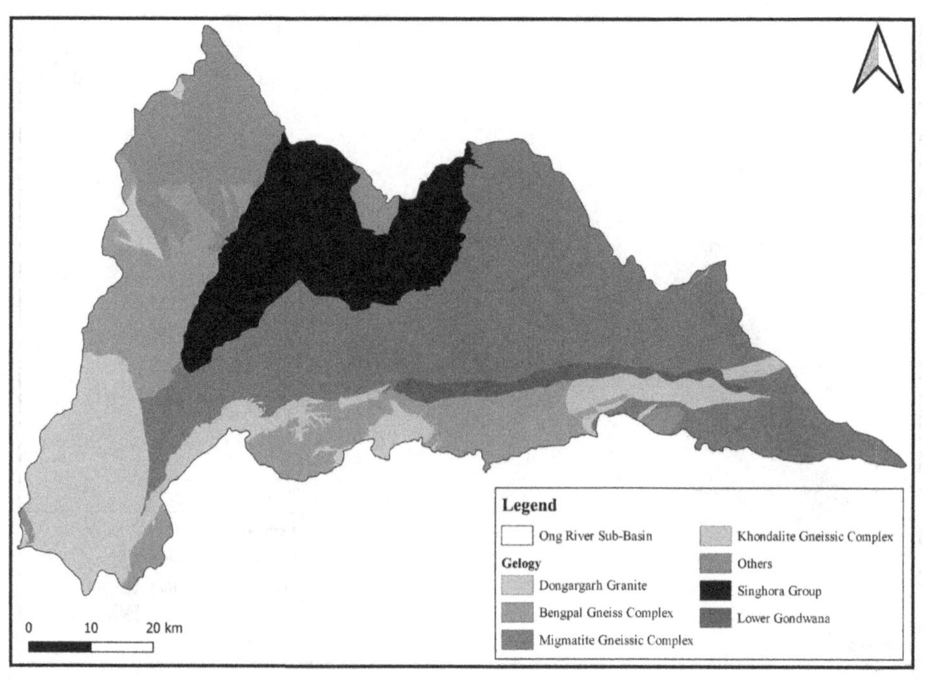

FIGURE 12.6 Geology map of the Ong sub-basin.

for the groundwater potential zonation due to faults than the Granite and Gneiss region as the Granite, and Gneiss rock interrupt the water penetration (Akanbi, 2018).

12.4.5 Drainage density

Drainage density is the ratio of the entire length of the drainage network to the total basin area. The layer on the drainage network of the study area was created using the DEM data. The drainage network layer was then used as input to create the drainage density layer. The groundwater potential is inversely related to drainage density where higher drainage density induces high surface runoff and poor potential zones for groundwater recharge. On the contrary, lower drainage density causes lower runoff and higher water infiltration (Shekhar and Pandey, 2015). Thus, lower values were assigned to less dense areas and higher values to higher drainage density areas. The estimated drainage density for the Ong sub-basin ranged between 0 m/km^2 and 31.68 m/km^2 (Fig. 12.7).

12.4.6 Rainfall

Rainfall is considered as one of the prime sources of groundwater restoration (Shekhar and Pandey, 2015). The satellite data-derived rainfall data for the year 2020 was downloaded from the Center for Hydrometeorology and Remote Sensing portal (CHRS). The annual rainfall in the sub-basin region varies between 1485 mm and 2189 mm, with an average value of 1851 mm

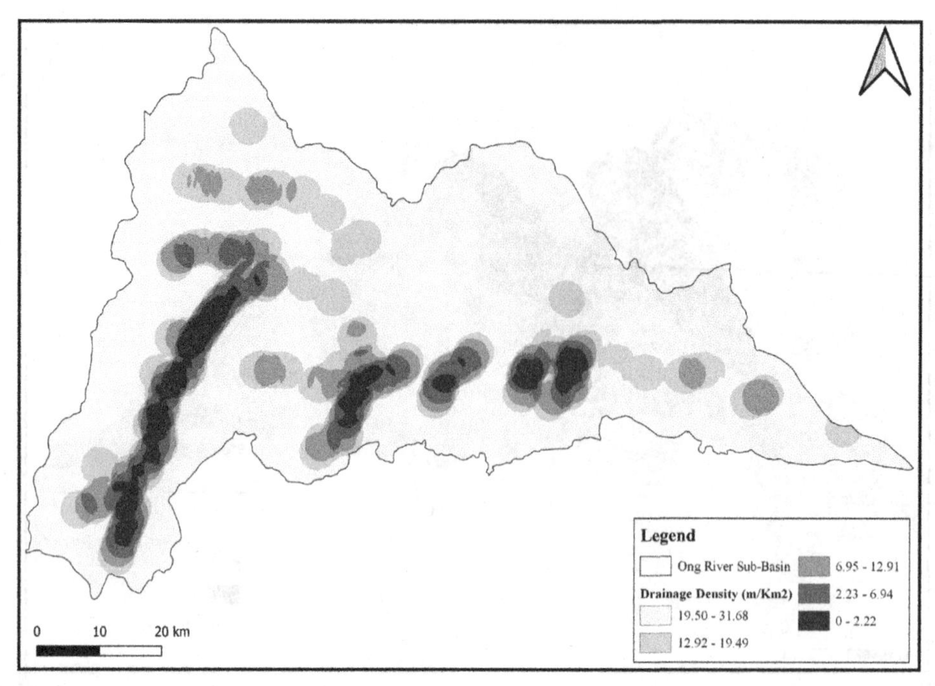

FIGURE 12.7 Drainage density map of the Ong sub-basin.

(Fig. 12.8). The dissemination of rainfall water in a region varies depending on the soil porosity, slope gradient, and elevation which regulates the penetration rate of surface runoff (Kumar et al., 2014). The areas with high annual rainfall were assigned higher weightage than the areas with lower annual rainfall. The analysis showed maximum rainfall in the eastern and southern regions, whereas lower rainfall was observed in the western and northern regions (Fig. 12.8).

12.4.7 Elevation and slope

The elevation and slope are vital terrain characteristics of the earth surface which regulates surface runoff and water infiltration (Thabile et al., 2020). The elevation map was reclassified into five categories wherein higher elevation sites that induce higher runoff were assigned higher weightage and lower weightage was given to low elevation sites indicating the plane landforms and more suitable for water infiltration (Abd Manap et al., 2014). The altitude of the study area ranges between 36 m and 944 m, where a relatively lower elevation was observed in the eastern region (Fig. 12.9). The DEM data was employed to generate the slope map. Higher weights were assigned to the lower slope (flat terrain), as these areas cause less surface runoff and enable higher water infiltration and groundwater recharge. On the contrary, a high slope causes higher surface runoff and lower water infiltration. The entire study area lies between 0° and 77.79° slope range wherein the majority of the study area falls under the lower slope range (<9°), and the steep slope range (>45°) was observed in hilly terrains in the southern region (Fig. 12.10).

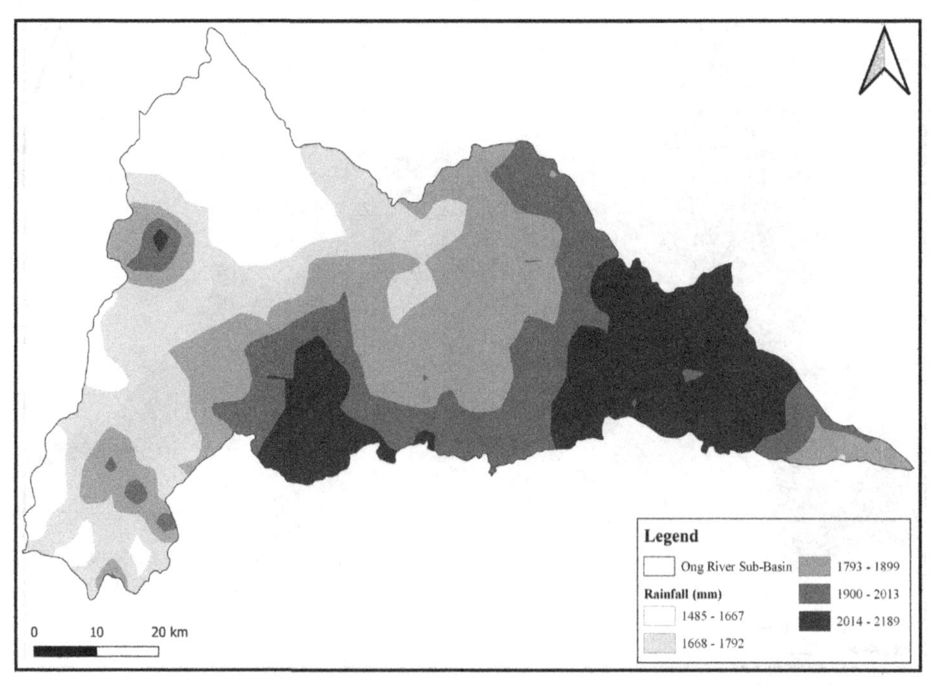

FIGURE 12.8 Annual rainfall distribution map of the Ong sub-basin.

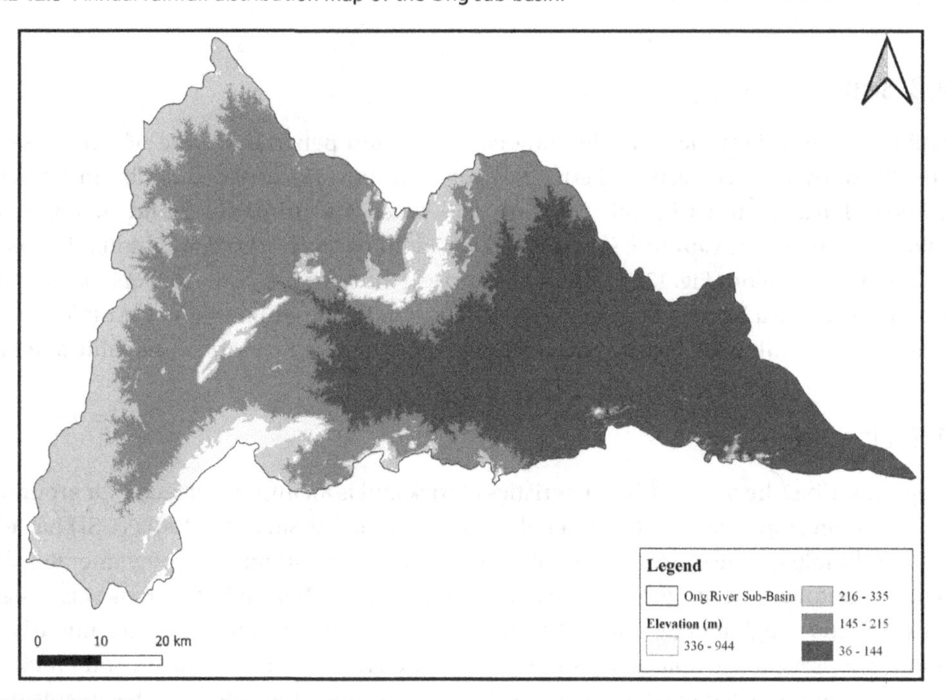

FIGURE 12.9 Elevation map of the Ong sub-basin.

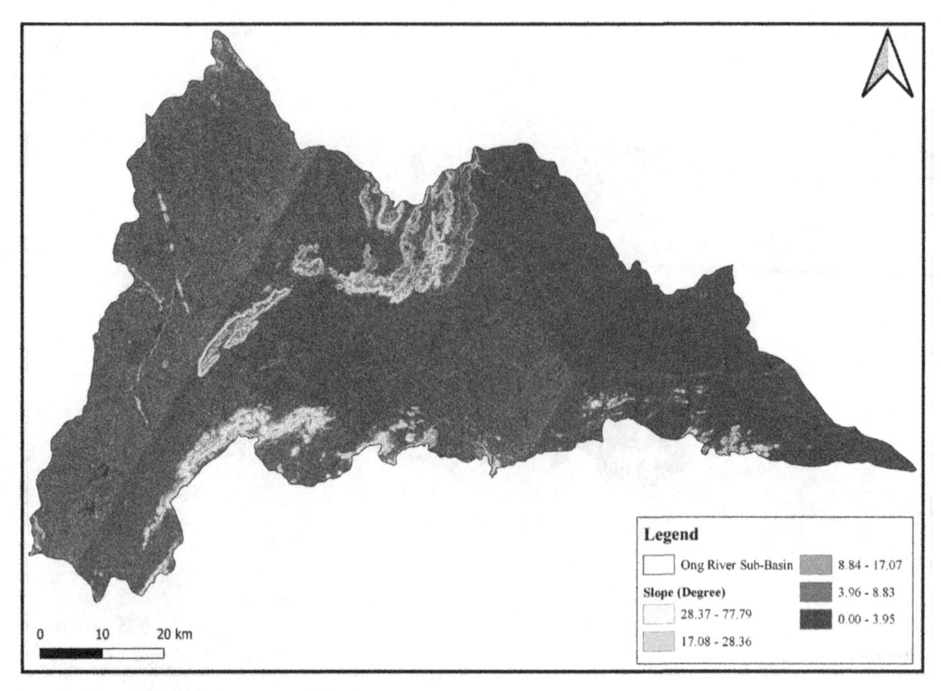

FIGURE 12.10 Slope map of the Ong sub-basin.

12.4.8 Soil

The soil type determines the water holding capability and penetration rate of water. The soil type of the study area consists of Ferric Luvisols, Chromic Vertisols, and Chromic Luvisols soils. Ferric Luvisols and Chromic Vertisols soil types are primarily found in flat terrain (<9° slope), whereas the Chromic Luvisols soil was dominantly observed in undulating terrain (>9° slope) in this region (Fig. 12.11). The soil texture of the Ferric Luvisols is coarse with a larger particle size causing a higher water penetration rate. On the other hand, the Chromic Vertisols are characterized by fine soil texture and smaller grains which reduces the water infiltration rate.

12.4.9 Lithology

Lithology describes the physical characteristics of rock and is an important factor for groundwater identification zone mapping (Bates et al., 1984). Geological Survey of India (GSI) has identified seven lithological features in this sub-basin, that is, sandstone and conglomerate, shale, quartzite, limestone, dolerite, granite and granite gneiss, and others. A higher rank was assigned to sandstone and conglomerate due to their higher porosity and permeability. In comparison, a lower rank was assigned to dolerite, granite, and granite gneiss which are mostly solid rocks and cause lower water infiltration (Barik et al., 2016). The majority of the Ong sub-basin falls under the granite and granite gneiss feature class (Fig. 12.12).

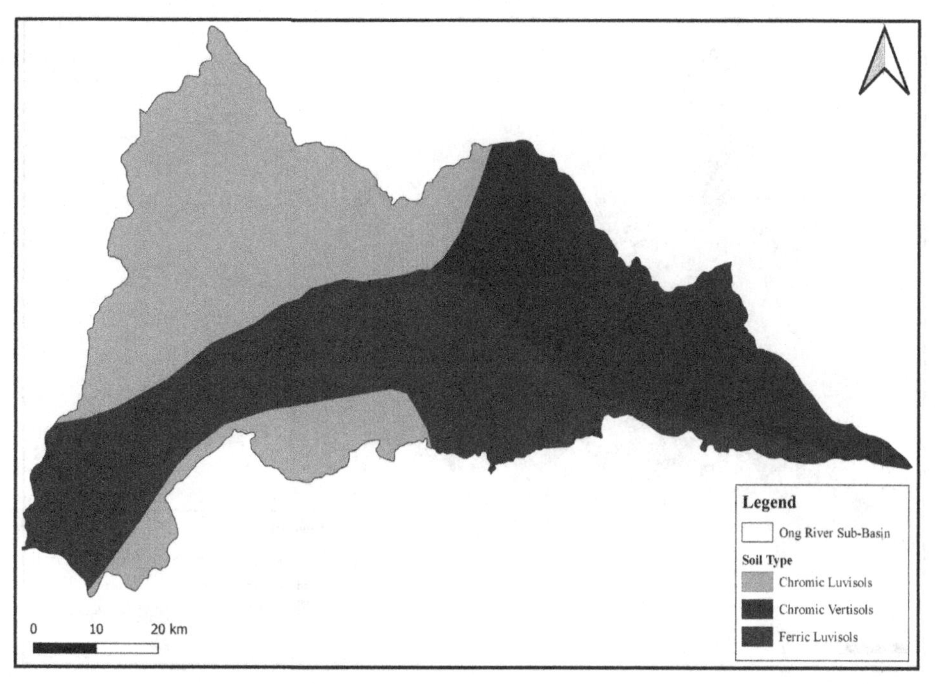

FIGURE 12.11 Soil map of the Ong sub-basin.

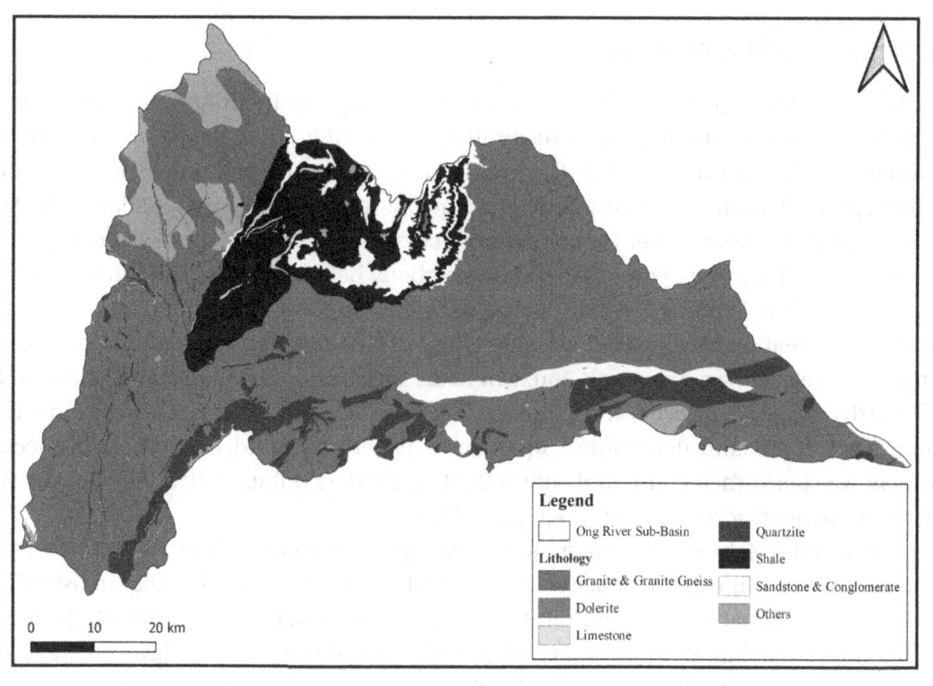

FIGURE 12.12 Lithology map of the Ong sub-basin.

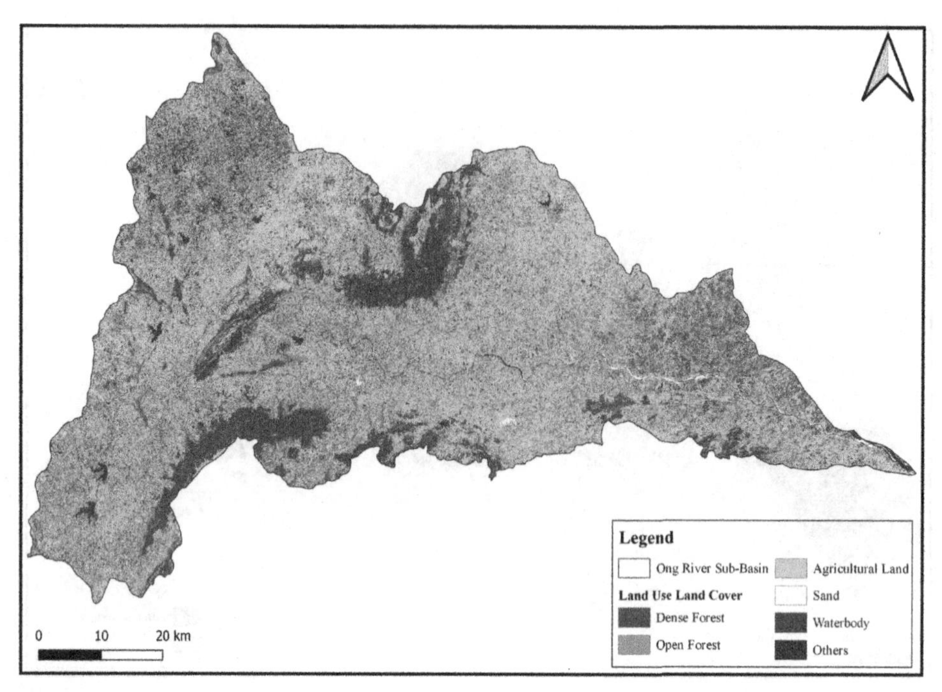

FIGURE 12.13 LULC map of the Ong sub-basin.

12.4.10 Land use land cover

The LULC distribution plays a dynamic role in regulating the hydrological parameters in a basin including surface runoff, evapotranspiration, water infiltration, baseflow, etc. (Das et al., 2018; Gupta and Srivastava, 2010). LULC controls many environmental processes, for example, infiltration, precipitation, surface runoff, evapotranspiration, etc. (Waikar and Nilawar, 2014). The agricultural land and forest cover reduce the surface runoff, prevent instant water evaporation, and help retain soil moisture where sand and waterbody help in water penetration (Gupta and Srivastava, 2010). The Sentinel-2 multispectral data was employed to generate the LULC using the maximum likelihood supervised classifier (Fig. 12.13). The training data for classification was created by visual image interpretation, where the high-resolution Quickbird (obtained from Google Earth) images were used for reference. Six LULC classes were classified in this region including agricultural land, dense forest, open forest, waterbody, sand, and others. The accuracy assessment was performed compared with 109 reference data points indicating high classification accuracy (overall accuracy: 89.9%; kappa: 0.853).

The maximum weight was estimated for the surface waterbody (25.44%), followed by geomorphology (17.02%), lineament density (11.59%), geology (10.79%), and drainage density (9.11%) (Table 12.4). The estimated weight for rainfall was 7.8%, followed by the slope (5.64%), elevation (4.68%), soil (3.54%), lithology (2.63%), and LULC (1.74%). The input proxy layers were used for weighted overlay applying the AHP-derived weights.

Table 12.4 Input thematic layers and computed weights.

Factors	Sub-classes	Category	Rank	Weight (%)
Surface water bodies	Water bodies	Very good	7	25.44
	Others	Very poor	1	
Geomorphology	Waterbody	Very good	7	17.02
	Flood plain	Good	6	
	Pediment pediplain	Moderate	4	
	Plateau	Less moderate	3	
	Hills and valleys	Very poor	1	
Lineament density (m/km^2)	1.03–2.00	Good	6	11.59
	0.70–1.02	Highly moderate	5	
	0.45–0.69	Moderate	4	
	0.22–0.44	Less moderate	3	
	0.00–0.21	Poor	2	
	Lower Gondwana	Very good	7	10.79
	Singhora Group	Good	6	
Geology	Others	Highly moderate	5	
	Khondalite Gneissic Complex	Moderate	4	
	Migmatite Gneissic Complex	Less moderate	3	
	Bengpal Gneiss Complex	Poor	2	
	Dongargarh Granite	Very poor	1	
Drainage density (m/km^2)	19.50–31.68	Very poor	1	9.11
	12.92–19.49	Poor	2	
	6.95–12.91	Moderate	4	
	2.23–6.94	Good	6	
	0–2.22	Very good	7	
Rainfall (mm)	2014–2189	Very good	7	7.8
	1900–2013	Good	6	
	1793–1899	Highly moderate	5	
	1668–1792	Moderate	4	
	1485–1667	Less moderate	3	
Slope (degree)	0–3.95	Very good	7	5.64
	3.96–8.83	Good	6	
	8.84–17.07	Less moderate	3	
	17.08–28.36	Poor	2	
	28.37–77.79	Very poor	1	
Elevation (m)	36–144	Very good	7	4.68
	145–215	Good	6	
	216–335	Moderate	4	
	336–944	Very poor	1	
Soil	Ferric Luvisols	Good	6	3.54
	Chromic Vertisols	Highly moderate	5	
	Chromic Luvisols	Moderate	4	

(continued on next page)

Table 12.4 Input thematic layers and computed weights—cont'd

Factors	Sub-classes	Category	Rank	Weight (%)
Lithology	Sandstone and conglomerate	Very good	7	2.63
	Shale	Good	6	
	Quartzite	Highly moderate	5	
	Limestone	Moderate	4	
	Others	Less moderate	3	
	Dolerite	Poor	2	
	Granite and granite gneiss	Very poor	1	
Land use land cover (LULC)	Waterbody	Very good	7	1.74
	Sand	Good	6	
	Agricultural land	Highly moderate	5	
	Open forest	Moderate	4	
	Dense forest	Poor	2	
	Others	Very poor	1	

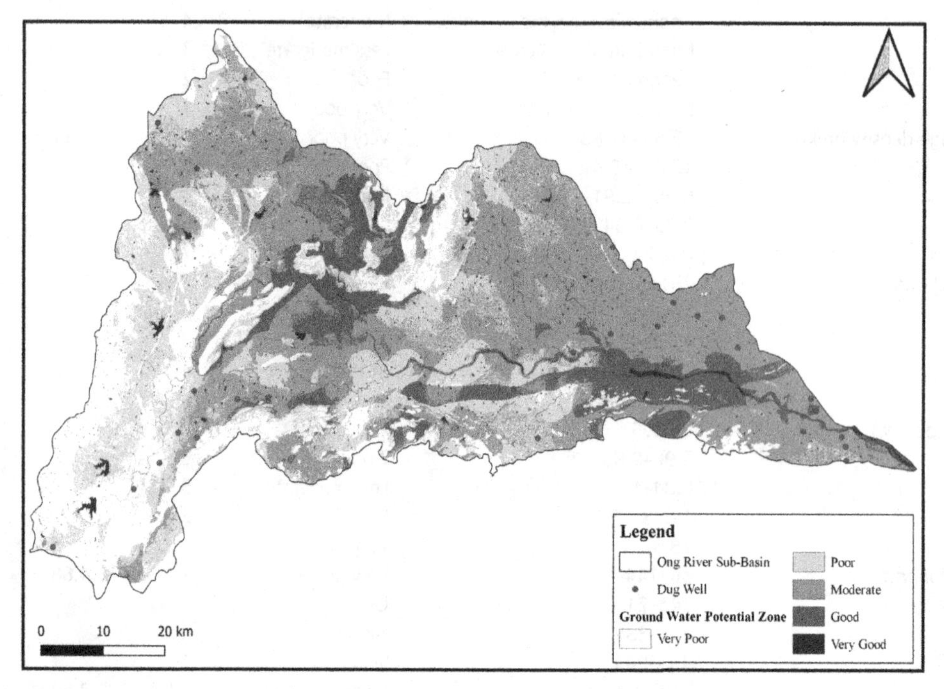

FIGURE 12.14 Groundwater potential zone (GWPZ) map of the Ong sub-basin.

The generated GWPZ map was grouped into five classes ranging from very poor to very good potential zones (Fig. 12.14; Table 12.5). These areas are characterized by rugged terrain with hills and valleys, geomorphology situated at relatively higher altitudes, lower rainfall, and lineament density (Shekhar and Pandey, 2015; Magesh et al., 2012). The groundwater level data was accessed for this region to validate the GWPZ map. The mean annual water level data recorded

Table 12.5 Estimated area under various groundwater potential zone (GWPZ) classes.

	Area			
GWPZ class	**Area (km²)**	**Area (%)**	**Number of dug wells**	**Mean depth (mbgl)**
Very poor	1002.56	19.44	2	6.22
Poor	1319.27	25.59	8	6.45
Moderate	2115.77	41.04	13	4.41
Good	657.58	12.75	8	2.74
Very good	59.82	1.16	–	–

below the ground surface was used. The locations of 31 dug well data were used in this study (Fig. 12.14). The corresponding number of dug wells and the mean water level data for each GWPZ category are shown in Table 12.5. The study revealed that the majority of the dug wells (13) were overlapping with the moderate GWPZ category having a mean water level of 4.41 m. The number of dug wells under the good and poor category were equal, indicating a mean value of 2.74 m and 6.45 m respectively. The mean water level observed in the very poor GWPZ category was 6.22 m. However, no dug well was observed in the very good category.

The AHP-derived weights indicated a well-accepted value with a CR value of 0.1. Similar weighting criteria have also been used in various studies in other river basins (Kumar et al., 2014). AHP-based weights indicated maximum value for the presence of surface water followed by geomorphology and lineament and least weights for LULC.

The Ong sub-basin is dominated by cropland followed by open and dense forest. The study revealed that the majority of the study area, ~45% falls under the poor and very poor groundwater potential zone category, whereas only ~14% of the area has a good and very good groundwater potential zone. The dug well data for this region was used for validation, well corroborated with the GWPZ map. The dug wells in the moderate and good GWP zones indicated a lower depth (below the ground surface) than the poor and very poor GWP zones which was consistent with the outcome of the study reported in Nasik district, Maharashtra (Singh et al., 2013). The study revealed that the middle-north and east part of the Ong sub-basin is very high and highly suitable because of the distribution of cultivated land which has a high infiltration rate. This is in agreement with a similar outcome reported for the Kalu river basin, Theni and Nalgonda districts, Andhra Pradesh (Magesh et al., 2012; Prasad et al., 2008). Our study demonstrated that the areas close to surface water bodies and with lower elevation range were carrying highly suitable groundwater potential, which is in sync with the outcome reported for study conducted in Delhi (Mallick et al., 2015). Further, the study revealed that ~40% area was under moderately suitable GWPZ due to suitable slope and pediplain terrain.

12.5 Conclusion

The groundwater potential zones of the Ong sub-basin were assessed using multi-criteria decision analysis. The study revealed that only ~1.1.6% of the study area consists of the very good groundwater potential zone and the area ~12.75% consists of good groundwater potential. In contrast, about ~25.59% and ~19.44% of the total study area consists of poor and very poor groundwater potential. These areas of poor and very poor groundwater potential fall in western, southern and northern parts of the basin. The majority of the study area ~41% lies under the moderate groundwater potential zone. The areas under the very good groundwater potential zone are mostly characterized by surface water bodies. The areas under a good groundwater potential zone are mostly agriculture-dominated and characterized by flat terrain, lower elevation, drainage density, high lineament density, good annual rainfall, and coarse-textured soil. These areas mostly fall in the eastern and some of the northern part of the basin. Although several water harvesting structures as reservoirs are constructed in this sub-basin, most of them are small and insufficient to facilitate the irrigation demand during the dry period. The outcome of the study provides baseline information on groundwater potential in the Ong sub-basin and shall enable better planning on sustained utilization of resources.

References

Abd Manap, Nampak, M., Pradhan, H., Lee, B., Sulaiman, S., ., W.N.A, Ramli, M.F, 2014. Application of probabilistic-based frequency ratio model in groundwater potential mapping using remote sensing data and GIS. Arab. J. Geosci. 7 (2), 711–724.

Agarwal, E., Agarwal, R., Garg, R.D., Garg, P.K., 2013. Delineation of groundwater potential zone: an AHP/ANP approach. J. Earth Syst. Sci. 122 (3), 887–898.

Akanbi, O.A., 2018. Hydrogeological characterization and prospect of basement Aquifers of Ibarapa region, southwestern Nigeria. Appl. Water Sci. 8 (3), 89.

Arulbalaji, P., Padmalal, D., Sreelash, K., 2019. GIS and AHP techniques based delineation of groundwater potential zones: a case study from southern Western Ghats, India. Sci. Rep. 9 (1), 1–17.

Barik, K.K., Jeet, R., Annaduari, R., Tripathy, J.K, 2016. Hydrogeological mapping and identification of groundwater recharge potential zone of Reamal Block Deogarh District, Odisha: a geospatial technology approach. Int. J. Adv. Remote Sens. GIS 5 (6), 1829–1843.

Bates, R.L., Jackson, J.A., et al., 1984. In: Dictionary of Geological Terms, 584. Anchor Books, American Geological Institute.

Behera, M.D., Biradar, C., Das, P., Chowdary, V.M., 2019. Developing quantifiable approaches for delineating suitable options for irrigating fallow areas during dry season: a case study from Eastern India. Environ. Monit. Assess. 191 (3), 1–18.

Bhanja, S.N., Mukherjee, A., Rodell, M., Wada, Y., Chattopadhyay, S., Velicogna, I., … Famiglietti, J.S., 2017. Groundwater rejuvenation in parts of India influenced by water-policy change implementation. Sci. Rep. 7 (1), 1–7.

Central Ground Water Board (CGWB), Ground Water Yearbook 2013–14, 2014. http://cgwb.gov.in/Documents/Ground%20Water%20Year%20Book%202013-14.pdf.

Central Ground Water Board (CGWB) (2020). Ground Water Year Book 2019–2020. http://www.indiaenvironmentportal.org.in/files/file/GROUND%20WATER%20YEAR%20BOOK%202019-2020%20Odisha.pdf.

Chowdhury, A., Jha, M.K., Chowdary, V.M., Mal, B.C., 2009. Integrated remote sensing and GIS-based approach for assessing groundwater potential in West Medinipur district, West Bengal, India. Int. J. Remote Sens. 30 (1), 231–250.

Das, P., Behera, M.D., Patidar, N., Sahoo, B., Tripathi, P., Behera, P.R., ... Krishnamurthy, Y.V.N., 2018. Impact of LULC change on the runoff, base flow and evapotranspiration dynamics in eastern Indian river basins during 1985–2005 using variable infiltration capacity approach. J. Earth Syst. Sci. 127 (2), 1–19.

Garai, S., Das, P., 2021. Performance of frequency ratio approach for mapping of groundwater prospect areas in an area of mixed topography. In: Shit, P.K., Bhunia, G.S., Adhikary, P.P., Dash, C.J. (Eds.). Groundwater and Society: Applications of Geospatial Technology. Springer, Cham, Switzerland AG, pp. 221–246.

Gleick, P.H., 1989. Climate change, hydrology, and water resources. Rev. Geophys. 27 (3), 329–344.

Golkarian, A., Naghibi, S.A., Kalantar, B., Pradhan, B., 2018. Groundwater potential mapping using C5. 0, random forest, and multivariate adaptive regression spline models in GIS. Environ. Monit. Assess. 190 (3), 149.

Gupta, M., Srivastava, P.K., 2010. Integrating GIS and remote sensing for identification of groundwater potential zones in the hilly terrain of Pavagarh, Gujarat, India. Water Int. 35 (2), 233–245.

Halder, S., Roy, M.B., Roy, P.K., 2020. Fuzzy logic algorithm based analytic hierarchy process for delineation of groundwater potential zones in complex topography. Arab. J. Geosci. 13 (13), 1–22.

Jha, M.K., Chowdary, V.M., Chowdhury, A., 2010. Groundwater assessment in Salboni Block, West Bengal (India) using remote sensing, geographical information system and multi-criteria decision analysis techniques. Hydrogeol. J. 18 (7), 1713–1728.

Kulkarni, H., Aslekar, U., Joshi, D., 2018. Specific yield of unconfined aquifers in revisiting efficiency of groundwater usage in agricultural systems. In: Saha, D., Marwaha, S., Mukherjee, A. (Eds.). Clean and Sustainable Groundwater in India. Springer, Singapore, pp. 125–137.

Kumar, T., Gautam, A.K., Kumar, T., 2014. Appraising the accuracy of GIS-based multi-criteria decision making technique for delineation of groundwater potential zones. Water Resour. Manag. 28 (13), 4449–4466.

Lee, S., Hong, S.-M., Jung, H.-S., 2018. GIS-based groundwater potential mapping using artificial neural network and support vector machine models: the case of Boryeong city in Korea. Geocarto. Int. 33 (8), 847–861.

Magesh, N.S., Chandrasekar, N., Soundranayagam, J.P., 2012. Delineation of groundwater potential zones in Theni district, Tamil Nadu, using remote sensing, GIS and MIF techniques. Geosci. Front. 3 (2), 189–196.

Mallick, J., Singh, C.K., Al-Wadi, H., Ahmed, M., Rahman, A., Shashtri, S., &Mukherjee, S., 2015. Geospatial and geostatistical approach for groundwater potential zone delineation. Hydrol. Process. 29 (3), 395–418.

NGC (National Geographic Society), 2019. Surface water. National Geographic. https://www.nationalgeographic.org/encyclopedia/surface-water/.

ODriscoll, M., Clinton, S., Jefferson, A., Manda, A., McMillan, S., 2010. Urbanization effects on watershed hydrology and in-stream processes in the southern United States. Water 2 (3), 605–648.

Oh, H.-J., Kim, Y.-S., Choi, J.-K., Park, E., Lee, S., 2011. GIS mapping of regional probabilistic groundwater potential in the area of Pohang City, Korea. J. Hydrol. 399 (3–4), 158–172.

Prasad, R.K., Mondal, N.C., Banerjee, P., Nandakumar, M.V, Singh, V.S., 2008. Deciphering potential groundwater zone in hard rock through the application of GIS. Environ. Geol. 55 (3), 467–475.

Rahmati, O., Pourghasemi, H.R., Melesse, A.M., 2016. Application of GIS-based data driven random forest and maximum entropy models for groundwater potential mapping: a case study at Mehran Region, Iran. Catena 137, 360–372.

Rao, C.H., 2002. Sustainable use of water for irrigation in Indian agriculture. Econ. Pol. Wkly. 1742–1745.

Saaty, T.L., 1980. The Analytic Hierarchy Process. McGraw Hill, New York.

Saaty, T.L., 2008. Decision making with the analytic hierarchy process. Int. J. Serv. Sci. 1 (1), 83–98.

Shekhar, S., Pandey, A.C., 2015. Delineation of groundwater potential zone in hard rock terrain of India using remote sensing, geographical information system (GIS) and analytic hierarchy process (AHP) techniques. Geocarto. Int. 30 (4), 402–421.

Singh, D.K., Singh, A.K., 2002. Groundwater situation in India: problems and perspective. Int. J. Water Resour. Dev. 18 (4), 563–580.

Singh, P., Thakur, J.K., Kumar, S., 2013. Delineating groundwater potential zones in a hard-rock terrain using geospatial tool. Hydrol. Sci. J. 58 (1), 213–223.

Thabile, G., Das, D.M., Raul, S.K., Subudhi, C.R., Panigrahi, B., 2020. Assessment of groundwater potential in the Kalahandi District of Odisha (India) using remote sensing, geographic information system and analytical hierarchy process. J. Indian Soc. Remote Sens. 48, 1–15. https://doi.org/10.1007/s12524-020-01188-3.

Waikar, M.L., Nilawar, A.P., 2014. Identification of groundwater potential zone using remote sensing and GIS technique. Int. J. Innov. Res. Sci. Eng. Technol. 3 (5), 12163–12174.

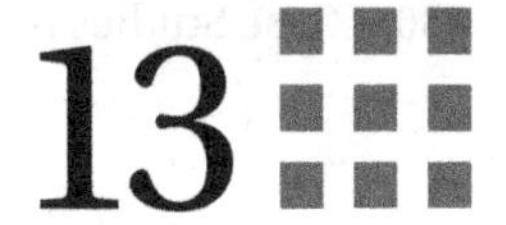

Innovative trend analysis of groundwater resources under changing climate in Malda district, India

Tapash Mandal, Kunal Chakraborty and Snehasish Saha

DEPARTMENT OF GEOGRAPHY AND APPLIED GEOGRAPHY, UNIVERSITY OF NORTH BENGAL, DARJEELING, INDIA

13.1 Introduction

Groundwater is the most valuable natural resource under the surface layer of the earth, which is regarded as a natural endowment of the environment, and it is the primary source of fresh water, accounting for roughly 30% of all freshwater available on the earth (Elbeih, 2015; Mallick et al., 2015). Groundwater is essential for the survival of not only human and animal life, but also the vegetative cover that covers the earth's surface, and it is used to provide water for the household, agricultural, and industrial uses and also it ensures agricultural productivity, food security, and sustainable livelihood which leading towards the economic development for a country (Qadir et al., 2007; Sasakova et al., 2018; Pande et al., 2021). Groundwater storage is found primarily in pore spaces in aquifers, aquitards, aquicludes, and aquifuges. Based on permeability and porosity, earth formations are classified as an aquifer, aquitard, aquiclude, and aquifuge. These are the relative classes based on the availability of water in an area, and basically, there are two types of aquifers, confined and unconfined aquifers (Khadri and Moharir, 2016; Moharir et al., 2020). The open water surface of an unconfined aquifer and the static level of a well drilling into a confined aquifer are referred to as the water table or water level. Water levels fluctuate in response to the pace of water recharge and outflow (Sahoo and Jha, 2017; Marques et al., 2020). Significantly, if the groundwater recharge rate is higher than the withdrawal rate, the groundwater level will rise; alternatively, if the rate of groundwater withdrawal is higher, the water level would decline.

The level of groundwater changes owing to both the natural and human-induced factors (Döll et al., 2012; Bhanja et al., 2018; Alipour et al., 2018; Zakwan, 2019; Arshad and Umar, 2020). Large-scale contamination, fast depletion owing to maximum abstraction, and saltwater intrusion have posed severe threats to this replenishable natural resource (Mukherjee et al., 2018; Chinchmalatpure et al., 2019). Natural influences include variations in rainfall and

Case Studies in Geospatial Applications to Groundwater Resources. DOI: https://doi.org/10.1016/B978-0-323-99963-2.00004-3

temperature patterns, whereas human causes include changes in land-use patterns, irrigation types, discharge of industrial effluents, groundwater extraction, and recharge, and so on (Mukate et al., 2020). The usual recharge rate of subsurface water fluctuates from months (shallow aquifers) to million years (desert aquifers) with very deep roots. Consequently, in many regions, the groundwater extraction rate exceeds the natural restoration rate (Han et al., 2017; Hu et al., 2019). In recent years, it has been shown that both the quality and magnitude of groundwater in developing nations are declining (Ghosh et al., 2015; MacDonald et al., 2016; Hamed et al., 2018; Patra et al., 2018). Natural or artificial replenishment of groundwater is possible. Water infiltration after rainfall occurrences, fluvial seepages, infiltration from reservoirs, irrigated lands, and other water bodies are all examples of natural recharge. Rainfall harvesting, check dams, reducing saltwater intrusion are all examples of artificial recharge. Continuous population growth and their rising requirements, and industrial growth, have necessitated a significant increase in groundwater pumping rates (Zeidan, 2017; Pande et al., 2021). For a few decades, the declining amount of rainfall affects the groundwater table and the level in declining stage worldwide (Mandal et al., 2021).

With a decreasing trend in monsoon rainfalls, the agricultural industry in most semiarid areas of India confronts significant water scarcity and crop production risks. India is preferably more dependent on groundwater than any other country on the planet, with an annual withdrawal scenario of 200,000 to 300,000 million cubic meters. According to the World Bank, India consumes 25% of the worldwide demand for groundwater which is about 600,000 million cubic meters per year; nevertheless, more than 90% of this water is used for agriculture (Choudhury et al., 2021). Around 80% of India's population relies on groundwater for drinking and agriculture, and demand is steadily increasing as the country's population grows (Gupta and Onta, 1997). Groundwater use for irrigation has increased dramatically across the country due to sophisticated pumping technology and their low cost. Overutilization is causing a gap between rising demand and water availability, impeding the process of sustainable development. If current trends continue, over 60% of India's aquifers will be in a state of water stress in the next 20 years (Briscoe and Malik, 2008). However, during the last several decades, rapid growth of population and groundwater use in most of the country's regions has resulted in groundwater extraction at a pace considerably faster than natural recharge, resulting in a drop in groundwater level. This example also shows that changes in groundwater levels are driven by both natural and anthropogenic factors (Zakwan, 2021).

Hydrological modifications can create significant disruption to predict the trend and nature of groundwater (Naughton et al., 2017). Without proper monitoring of climatic, geological, and hydrological substances, it is challenging to identify the techniques to analyze the long- or short-term trends (Cosgrove and Loucks, 2015). Precipitation, temperature, river flow, and groundwater level are these climatic and hydrogeological elements relevant to identifying the trend as the global climatic changes have occurred (Citakoglu and Minarecioglu, 2021). Many researchers have introduced commonly used Spearman's Rho test and nonparametric Mann Kendall test to detect the nature and trends among hydrologic variables (Ashraf et al., 2021; Mallick et al., 2021; Zakwan, 2021). Fluvial discharges and their declining trends of the Yangtze River in China show the changes in climatic conditions in hydrological parameters resulting in scarcity of water and flood situation in the different areas of the Yangtze River basin (Lu et al., 2003).

Zakwan and Ara (2019) have described the monthly rainfall of Bihar in India, and it seems to be in a declining trend over the whole region. Artificial neural network models have been introduced to monitor the groundwater fluctuations based on monthly precipitation data by Coulibaly et al. (2001). Zakwan (2021) analyzed the groundwater level trend in Rajasthan, India, using the innovative trend analysis (ITA).

Malda district is confronting major agricultural transformations under the changing climatic conditions. Here the ITA (Şen, 2012, 2017) method has been applied for assessing the trend of climatic and hydrological variables to find out the fluctuations and depletion of GWD. This method at present has gained popularity as it is used in many scientific pieces of research. The most commonly used statistical techniques for the trend analysis Mann (Mann, 1945)-Kendall test and modified Mann-Kendall (Hamed et al., 2018) test have been introduced to clear the reliability of the ITA method. Sens's slope estimator (Sen, 1968) is also used to detect the state of the magnitude of changes in GWD in the study area depending on the climate changes and their impact.

13.2 Study area

Malda district experiences a particularly hot and humid monsoon climate lies in Gangetic West Bengal in India and also known as the gateway of North Bengal. The physiography of the district is divided into three major categories, that is, Tal, Diara, and Barind. Tal is located west of the Mahananda river and north of the Kalindri river. It is a low-lying region prone to flooding due to rising water levels in the Ganges, Mahananda, and Kalindri rivers. Diara region consists of newer alluvial deposits by the fluvial action of the Ganges. Palaeochannels still exist in this region as a trace of plenty of groundwater resources. The undulations of consecutive rises and depressions, seamed with small rivers in the valleys, are a distinctive feature of the Barind tract and consist of older alluviums. The climate of the Malda area is characterized by hot summers, abundant rain, and a humid atmosphere throughout the year. The district may be split into four distinct seasons based on temperature variations, rainfall, humidity, and winds; as hot-summer season comprising March to May month, monsoon season comprising month June to September, retreating monsoon with the month of October to November, and winter Season with the month of December to February. The average temperature of the district is approximately 30°C, with an average diurnal variation of around 6°C and an average rainfall of 308.7 mm. The district is located amid 25°32′08″ N to 24°40′20″ N latitude, and 88°28′10″ E to 87°45′50″E longitude. Murshidabad district lies in the south, Uttar Dinajpur district and Dakshin Dinajpur district are situated in the north, the international border with Bangladesh is to the east, and Jharkhand and Bihar state is to the west. The district has a total area of 3733 km^2. The depth of the water table in any place is determined by various elements such as physiography, climate, and the porosity of the sub-strata. The water table in the district's uplands and lowlands is likely to differ significantly. The distribution of moisture in the various soil strata also influences the depth of the groundwater table. Geological factors influence groundwater level, much as the height of the water table is influenced by subsurface relief. The relief of the water table is continually changing in connection to the state of the soil and groundwater balance.

13.3 Database and methodology

Seasonal groundwater data for 32 dug wells in Malda district was obtained from the India Water Resource Information System (IWRIS) site (https://www.indiawris.gov.in/wris/#/) for the period 1996–2017. After extracting the data, it was observed that certain stations had a significant amount of missing value; therefore, we did not include them. The dataset was then submitted to a detect error/missing check, and it was discovered that 1% of the data in some stations was missing, as approximated using multiple imputation methods. As a result, those stations with missing results were eliminated from the study; consequently, the dataset of 32 dug wells was employed. The India-water site was used to collect seasonal GWD data from 32 stations between 1996 and 2017. The various methodologies used in this study has been discussed in the following section.

13.3.1 Autocorrelation function (ACF)

Autocorrelation is typically regarded as the most challenging problems when evaluating and identifying time series data patterns. The variance of the Mann-Kendall (MK) test statistic increases the degree of serial dependency (autocorrelation), and positive serial correlation in a time series data increases the Type I error (false positive) and detects a significant trend when there is no trend. As a result, the presence of serial correlation for all data series was investigated initially in this study using the lag-1 autocorrelation coefficient (r_k) at a 0.05 significant level for the two-tailed (upward and downward) test.

$$r_k = \frac{\sum_{k=1}^{N-K} (x_p - \overline{x_p})(x_{p+k} - \overline{x_{p+k}})}{\left[\sum_{k=1}^{n-k} (x_p - \overline{x_p})^2 (x_{t+k} - \overline{x_{p+k}})^2\right]^{0.5}} \tag{13.1}$$

where r_k is the Autocorrelation function of time series x_t at lag k, x_p is observed data flow series, $\bar{x}$ is the mean of time series (x_p), N denote the total length of x_p time-series, k is the maximum lag.

The alternative hypothesis is that the true r_k must be other than zero, but it might be positive or negative because the test is two-tailed. If r_k lies between the larger and smaller confidence interval margins, the time series data are regarded as sequentially associated; otherwise, the data are assumed to be serially independent.

13.3.2 Innovative trend analysis

Z. Sen introduced the ITA approach in 2012 to find trends in hydrometeorological data. This graphical trend identification approach is more reliable in detecting time series sub-trends. Also, unlike the MK./mMK and Spearman row (SR) tests, this technique does not need the assumption of data nonlinearities, serially independent data, and the optimal number of data. The ITA has gained the most popularity in time series analysis because of its comparative advantage. A time series is divided into two subsets with an equal amount of data points in ITA. These two subsets are rearranged and represented in a Cartesian coordinate system, with the first subset plotted on the horizontal axis (X-axis) and the second subset plotted on the vertical axis (Y-axis). The graph is divided into two identical triangles by a straight line of 1:1 (45°), which represents "no trend" in the data series. When the data points are scattered above the 45° line, the trend is upward; similarly, the trend is downward if the data points are scattered below the 1:1 line. The slope of ITA (Şen, 2012), on the other hand, is calculated using the following equation

$$\beta = \frac{1}{n}\sum_{i=1}^{n}\frac{10(x_j - x_k)}{\bar{x}} \tag{13.2}$$

where β represents the ITA slope, n represents the range of individual subset of the time series, $x_j$and x_k denotes the values of the successive sub-set, $\bar{x}$ is the mean of the 1st sub-set (X_j).

The positive value of *B* in the time series detects a rising trend, whereas the negative value detects a falling trend. Assume the quality of the observed original time series is odd. In such a situation, the initial observation is discarded before being divided into two parts so that the more recent time-series data may be fully used.

13.3.3 Mann-Kendall test

Mann-Kendall (Mann 1945) test statistic (S) of the series x_1,x_2,x_3..., and$x_n$are as computed by the following formula

$$S = \sum_{k=1}^{n-1}\sum_{j=k+1}^{n}\text{sign}(x_j - x_k) \tag{13.3}$$

where n is the number of data points, sign signifies the signum work,x_j and x_k denotes the data points of time j and k,

$$\begin{aligned} &= +1 \text{ if } x_j - x_k > 0 \\ \text{Sign}(x_j - x_k) &= 0 \text{ if } x_j - x_k = 0 \\ &= -1 \text{ if } x_j - x_k < 0 \end{aligned} \tag{13.4}$$

Variance of S, VAR(S) is calculated as:

$$\text{VAR}(S) = \frac{1}{18}\left\{n(n-1)(2n+5) - \sum_{i=1}^{g} t_i(t_i - 1)(2t_i + 5)\right\} \tag{13.5}$$

where g specifies the number of connected groups, and t_i defines the degree of the connections. The linked groups are a grouping of sample data having comparable values.

Time-series data Statistic (S) is identified by Kendall's τ (tau), using the following formula

$$\tau = \frac{S}{D} \tag{13.6}$$

where

$$D = \sqrt{\frac{1}{2}n(n-1) - \frac{1}{2}\sum_{i=1}^{g} t_i(t_i - 1)}\sqrt{\frac{1}{2}n(n-1)}$$

when n> 10, the standardized test measurement Z is developed using the estimation of (S) and VAR(S).

$$Z = \begin{cases} \frac{S-1}{\sqrt{VAR(S)}}, \text{ if } S > 0 \\ 0, \text{ if } S = 0 \\ \frac{S+1}{\sqrt{VAR(S)}}, \text{ if } S < 0 \end{cases} \tag{13.7}$$

The positive and negative values of Z denote rising and falling trends in the time series, respectively. (H_0) is rejected if the estimation of Z is larger than 1.96, which is generated from the normal ordinary distribution table, to test the null hypothesis (H_a) of no trend against the alternative hypothesis (H_a) of an upward or downward pattern at the 0.05 level of significance.

13.3.4 Modified Mann-Kendall test

Modified VAR (S) statistics can be calculated as

$$VAR(S) = \left(\frac{n\,(n-1)(2n+5)}{18}\right) \cdot \left(\frac{n}{n_e^*}\right) \tag{13.8}$$

here, the correction factor $\left(\frac{n}{n_e^*}\right)$ is adjusted to the auto-correlated data as:

$$\left(\frac{n}{n_e^*}\right) = 1 + \left(\frac{2}{n^3 - 3n^2 + 2n}\right) \sum_{f=1}^{n-1} (n-f)(n-f-1)(n-f-2)\,\rho_e(f) \tag{13.9}$$

$\rho_e(f)$ is the autocorrelation function between observation ranks and can be calculated as

$$\rho(f) = 2\sin\left(\frac{\pi}{6}\,\rho_e(f)\right) 12$$

13.3.5 Sen's slope estimator

The degree of change was estimated using Sen's slope (Sen, 1968) estimator (slope$Q$). The slope Q may be calculated with N pairs of data as follows

$$Q_i = \frac{x_k - x_j}{k - j}, \quad i = 1, 2, \cdots\cdots N, \; k > j \tag{13.10}$$

where x_k and x_j represent values of data at k, j times and; Q_i is the median of a slope.

13.4 Results and discussion

13.4.1 Descriptive statistics

Table 13.1 shows various statistical measures of annual GWD for 32 locational sites, including mean minimum, mean maximum, mean, standard deviation (SD), coefficient of variation (CV), skewness (CS), and kurtosis (CK). The average annual GWD ranged from 10.535 mbgl ± 5.325 (Bulbulchandi) to 4.6125 mbgl ± 1.22 (Ranikamat), with a CV of 21.32% and 39.32%, respectively,

Table 13.1 Descriptive statistics of the annual ground water depth (mbgl) of different site of Malda district (1996–2017).

Name of the site	Minimum	Maximum	Mean	SD	CV	Skewness	Kurtosis
Pakuahat pz	3.27	16.41	5.50	3.64	66.04	2.31	4.17
Chanchal pz	3.66	8.60	5.65	1.19	20.99	0.49	0.31
Malatipur	4.49	8.13	6.05	1.15	18.98	0.44	−1.14
Battaly 1	2.74	6.24	4.76	0.94	19.75	−0.49	0.19
Bholanathpur	1.40	3.70	2.50	0.54	21.41	−0.13	0.24
Milky	3.24	6.03	4.74	0.74	15.56	−0.20	−0.64
Nimaisarai	4.64	8.12	6.99	0.88	12.56	−0.88	0.69
Agampur	3.53	9.12	5.45	1.44	26.48	1.06	0.62
Deotala	1.56	2.87	2.23	0.40	17.91	0.01	−1.03
Gazole	2.79	16.20	5.21	3.54	67.91	2.34	4.79
Mashaldighi	3.94	8.97	5.43	1.24	22.79	1.43	2.01
Magura	1.82	8.47	4.97	1.92	38.64	0.04	−0.51
Bulbulchandi	5.33	10.53	7.65	1.63	21.32	0.12	−1.16
Habibpur	2.09	4.62	3.47	0.64	18.53	−0.14	−0.26
Makuli	3.17	5.41	4.19	0.76	18.21	0.12	−1.41
Ranikamat	1.22	4.61	2.73	1.07	39.32	0.06	−1.33
Tulshihata	2.22	5.41	3.97	0.90	22.69	−0.06	−0.75
Baroduari	2.47	6.79	4.85	1.14	23.58	−0.22	−0.10
Daulatpur	3.15	6.02	5.00	0.84	16.82	−1.05	0.19
Sujapur	2.81	4.95	3.80	0.48	12.64	0.17	0.42
Mothabari	2.17	6.90	4.18	1.42	34.02	0.52	−0.92
Dariapur	1.76	4.53	3.43	0.67	19.44	−0.72	0.52
New 16 mile	1.56	5.74	3.49	1.08	31.02	0.34	−0.27
Pirganj	2.02	6.94	4.51	1.36	30.19	0.01	−0.81
Manikchak	3.46	5.60	4.58	0.47	10.23	0.10	1.02
Nurpur pz	2.93	6.09	4.24	0.80	18.99	0.66	0.41
Malda Pz	3.47	7.95	6.60	1.15	17.46	−0.97	1.10
Malda Town	2.81	6.88	4.89	0.81	16.53	0.04	2.38
Mochia	3.51	8.42	5.62	1.28	22.83	0.45	0.23
Debipur	3.13	5.18	4.32	0.42	9.81	−0.79	2.12
Ratua	3.95	5.73	4.80	0.57	11.95	0.21	−1.27
Samsi	3.27	6.80	4.93	0.77	15.69	0.14	0.64

while the district mean annual GWD is 7.6490 mbgl ± 2.2340. The SD of annual GWD ranged from 3.64% (CV 66.04%) (Pakuahatpz) to 0.40% (CV 17.91%) (Deotala). The average minimum and maximum GWDs ranged from 1.22 mbgl (Ranikamat) to 5.33 mbgl (Bulbulchandi) and 2.87 mbgl (Deotala) to 16.41 mbgl (Pakuahatpz), respectively. The CV of the datasets reveals a medium to a high level of variability in GWD in the research area, ranging from 9.81% (Debipur) to 66.04% (Pakuahatpz). Skewness is a metric for determining the degree of symmetry or asymmetry in a dataset. According to the findings, the skewness of Malda's annual GWD ranged from −1.05 (Daulatpur) to 2.31 (Malda) (Pakuahat pz). The time series dataset's Kurtosis, on the other hand, ranged from −1.41 (Makuli) to 4.17. (Pakuahat pz). Table 13.1 furthermore depicts that the data are skewed since the skewness coefficient (Pearson's) value is not less than zero. Most of

Table 13.2 Details of groundwater depth trend of premonsoon season of Malda district.

Sl. No.	Name of the station	Slope IT	Z statistics of MK/mMK	Sen's slope	ITD
1	Agampur	−0.32	−1.58	−0.15	▼
2	Baroduari	0.07	1.97*	0.09	▲
3	Battaly 1	0.07	0.99	0.04	▲
4	Bholanathpur	−0.15	−0.12	−0.03	▼
5	Bulbulchandi	−0.42	−2.54*	−0.37	▼
6	Chanchal pz	0.42	0.84	0.10	▲
7	Dariapur	0.02	0.71	0.04	▲
8	Daulatpur	0.04	1.86	0.06	▲
9	Debipur	−0.08	−2.68*	−0.04	▼
10	Deotala	−0.09	−2.03*	−0.09	▼
11	Gazole	−0.01	−2.20*	−0.17	▼
12	Habibpur	0.03	1.14	0.02	▲
13	Magura	0.05	0.65	0.09	▲
14	Makuli	−0.24	−3.05*	−0.16	▼
15	Malatipur	0.13	1.84	0.16	▲
16	Malda Pz	0.13	3.71*	0.17	▲
17	Malda Town	0.13	0.17	0.02	▲
18	Manikchak	−0.04	−0.59	−0.03	▼
19	Mashaldighi	−0.06	−1.95	−0.08	▼
20	Milky	0.02	2.09*	0.03	▲
21	Mochia	0.10	1.64	0.10	▲
22	Mothabari	0.18	1.33	0.13	▲
23	New 16 mile	0.18	−0.11	0.02	▲
24	Nimaisarai	−0.25	−2.60*	−0.23	▼
25	Nurpur pz	0.25	2.57*	0.06	▲
26	Pakuahat pz	0.25	1.47	0.10	▲
27	Pirganj	0.08	0.99	0.09	▲
28	Ranikamat	−0.27	−2.82*	−0.25	▼
29	Ratua	0.01	0.23	0.00	▲
30	Samsi	0.05	1.07	0.03	▲
31	Sujapur	0.01	1.83	0.03	▼
32	Tulshihata	0.07	3.02*	0.08	▲

* The trend is significant at 95% confidence level. *ITD*, innovative trend detection. [shaded] The trend was detected by modified Mann-Kendall test based on lag-1 autocorrelation.

the datasets are positively skewed; the dataset's mean, median, and mode are also different. The kurtosis is not mesokurtic (normal distribution) since the values are less and higher than zero; rather, the kurtosis defines platykurtic (thin tails) and leptokurtic (fat tails), that is, non-normal distribution.

13.4.2 The trend in groundwater depth

13.4.2.1 The trend in the premonsoon season

Table 13.2 shows the ITA, MK/mMK, and Sen's slope analysis of GWD fluctuation of 32 dug well stations for the premonsoon season of the Malda district. The ITA represents that about 62% of the dug well stations have fallen under the increasing and about 39% of the stations have fallen under the decreasing trend of the fluctuation of GWD. Seasonal variability of GWD depends especially on the climatic variability of any region, and in the case of the present study area, the rainfall intensity and the recharge rate of groundwater in the premonsoonal months are not very high. But, the increasing tendency of the GWD is higher than the declining rates for this particular season. The integration of all the methods shows that 39.60% (12 no's) stations indicated the statistically significant value at the 95% confidence level. Under the significant values, a positive trend is found in Baroduari, Milky, Nurpur, and Tulshitala stations, and in Bholanathpur, Bulbulchandi, Debipur, Deotala, Gazole, Makuli, Nimaisarai, and Ranikamat stations, a negative trend is found. From the Sen's slope analysis, significant decreasing trends are seen in Blubulchandi (-0.37 m/year), Debipur (-0.04 m/year), Deotala (-0.09 m/year), Gazole (-0.17 m/year), Makuli (-0.16 m/year), Nimaisarai (-0.23 m/year), Ranikamat (-0.25 m/year), and Sujapur (0.03 m/year) and significant increasing trends in the stations of Baroduari (0.09 m/year), Malda pz (0.17 m/year), Milky (0.03 m/year), Nurpur pz (0.06 m/year), and Tulshitala (0.08 m/year). The highest increasing trend is at Malda pz and lowest decreasing trend found at Bulbulchandi station. The number of meander scars is scattered in this region as paleochannel, and the perennial nature of these cut-offs balances the recharge rate of the groundwater geographically. Fig. 13.1(A–C) shows the premonsoon groundwater fluctuation using the innovative trend, which demarcates the highest increasing rate (0.05–0.18 m/year), especially in the Diara region of the district comprising over the very low slopes. The Sen's slope map indicates that the highest rate (0.05–0.17 m/year) of increase of GWD for the particular season over the low-lying surface of the land in the north, that is, Tal region.

13.4.2.2 The trend in the monsoon season

From the analysis of ITA, it can be easily assessed that the increasing trend of the GWD in the monsoon season is about 94%, and the decreasing trend is only about 6% (Table 13.3). Plenty of monsoonal rainfall over the region, low lying landform configuration, a very gentle slope increases the runoff time, and the groundwater recharged through the infiltration. However, many wetlands in this region possess a good amount of rainwater and balance the groundwater's recharge rates. About 60% of the stations indicate the statistically significant increase and decreasing trend rate at a 95% confidence level. Baroduari, Battaly 1, Bholanthpur, Chanchal pz, Dariapur, Daulatpur, Habibpur, Malatipur, Malda pz, Malda Town, Milky, Mochia, Mothabari, Nimaisari, Nurpur pz, Ratua, and Tulshitala stations have fallen under the significant positive value, and only the Mashaldighi station has fallen under the significant negative value which indicates that the positivity rate or the increasing rate is overpowering on the decreasing on in this particular season. The ITA findings also revealed that Chanchal pz (0.16 m/year), Dariapur (0.15 m/year), Malatipur (0.19 m/year), Malda (0.20 m/year), Milky (0.15 m/year), Mochia (0.15 m/year), and Ratua (0.21 m/year) had notable positive trends. Tulshitala (0.15 m/year), with noticeable declining trends in Mashaldighi -0.09 m/year and Ranikamat -0.08 m/year, respectively. Sen's slope indicates a maximum increasing trend at Ratua and a decreasing trend at Mashaldighi, as determined by the Z statistics at the 0.05 level of significance. From Fig. 13.2 (D–F), it can be said that the Barind tract comprises the decreasing trend (-0.11–0.08 m/year) of GWD than Tal and Diara region (0.09–0.27 m/year) significantly. Sen's slope estimator map also

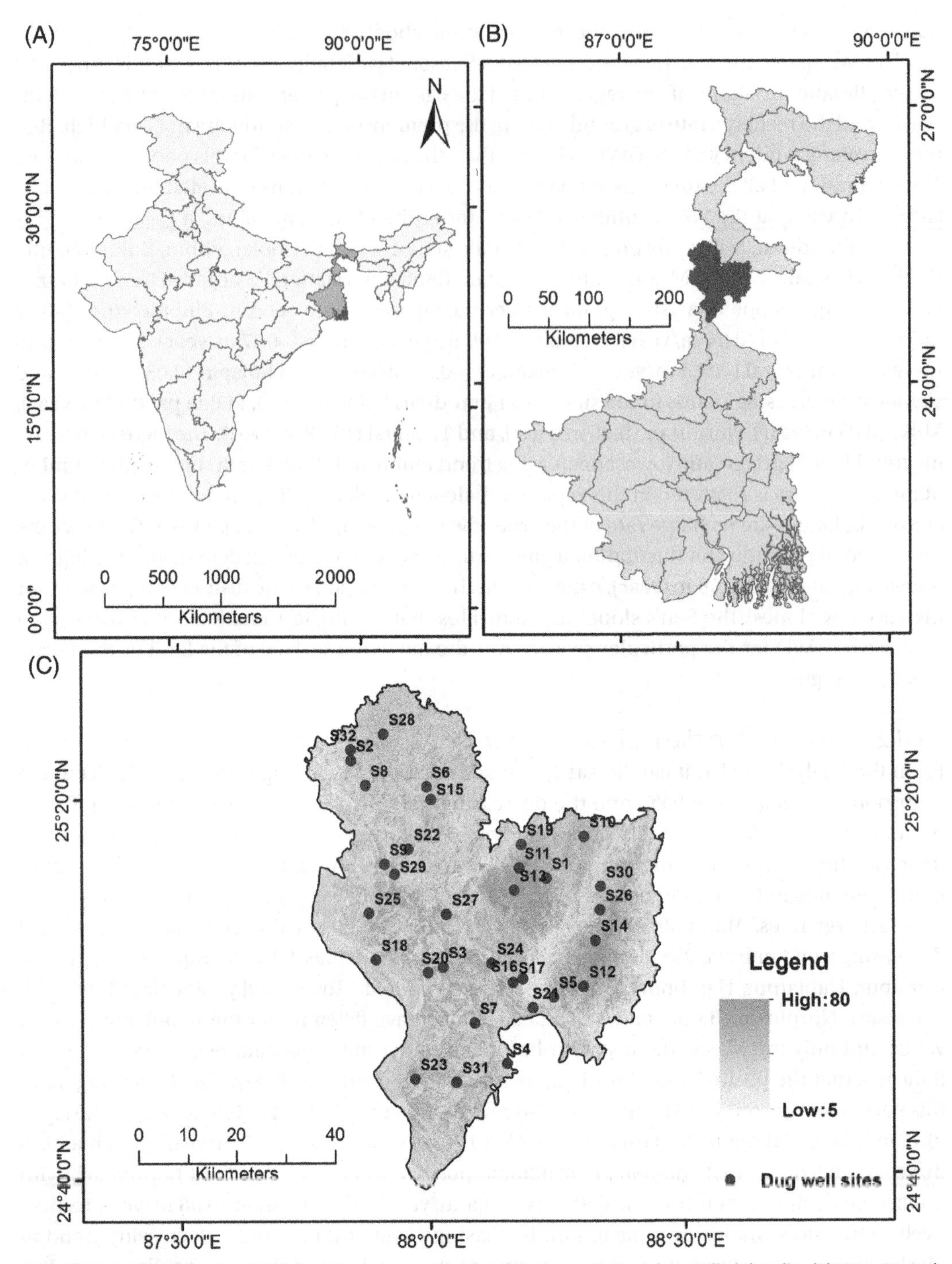

FIGURE 13.1 Study area map with location of groundwater depth monitoring dug wells (A) India, (B) West Bengal, and (C) Malda.

Table 13.3 Details of groundwater depth trend of monsoon season of Malda district.

Sl. No.	Name of the station	Slope IT	Z statistics of MK/mMK	Sen's slope	ITD
1	Agampur	0.07	0.03	0.01	▲
2	Baroduari	0.12	2.40*	0.11	▲
3	Battaly 1	0.12	2.06*	0.09	▲
4	Bholanathpur	0.10	2.79*	0.09	▲
5	Bulbulchandi	0.02	0.21	0.02	▲
6	Chanchal pz	0.02	4.25*	0.16	▲
7	Dariapur	0.19	3.22*	0.15	▲
8	Daulatpur	0.12	3.24*	0.14	▲
9	Debipur	0.03	1.35	0.06	▲
10	Deotala	0.02	0.37	0.01	▲
11	Gazole	0.27	1.83	0.12	▲
12	Habibpur	0.16	2.82*	0.08	▲
13	Magura	0.10	1.16	0.11	▲
14	Makuli	0.02	0.96	0.04	▲
15	Malatipur	0.16	2.54*	0.19	▲
16	Malda Pz	0.16	4.34*	0.20	▲
17	Malda Town	0.16	2.03*	0.07	▲
18	Manikchak	0.04	0.48	0.00	▲
19	Mashaldighi	−0.11	−2.23*	−0.09	▼
20	Milky	0.15	5.80*	0.15	▲
21	Mochia	0.18	2.59*	0.16	▲
22	Mothabari	0.11	4.69*	0.14	▲
23	New 16 mile	0.11	2.40*	0.10	▲
24	Nimaisarai	0.09	2.85*	0.13	▲
25	Nurpur pz	0.09	3.30*	0.12	▲
26	Pakuahat pz	0.09	1.75	0.08	▲
27	Pirganj	0.09	1.30	0.03	▲
28	Ranikamat	−0.07	−1.83	−0.08	▼
29	Ratua	0.22	4.17*	0.21	▲
30	Samsi	0.06	1.58	0.09	▲
31	Sujapur	0.03	0.56	0.02	▲
32	Tulshihata	0.11	5.53*	0.15	▲

* The trend is significant at 95% confidence level. *ITD*, innovative trend detection. [shaded] The trend was detected by modified Mann-Kendall test based on lag-1 autocorrelation.

demarcates the same scenario as the ITA slope of increasing and decreasing trend. Geographical location and the climatic condition of the study area are predominantly controlling the nature of the groundwater fluctuations in the monsoonal months.

13.4.2.3 The trend in the postmonsoon season

Table 13.4 and Figs. 13.3(A–C) and 13.4 show the results of ITA slope, MK/mMK, and Sen's slope of postmonsoon GWD fluctuations of 32 selected dug well stations. Most stations (84.40%) indicated increasing trends (ITA), although about 56.30% of increasing and decreasing trends are statistically significant at the 0.05 significance level. On the other hand, approximately 15.60% of

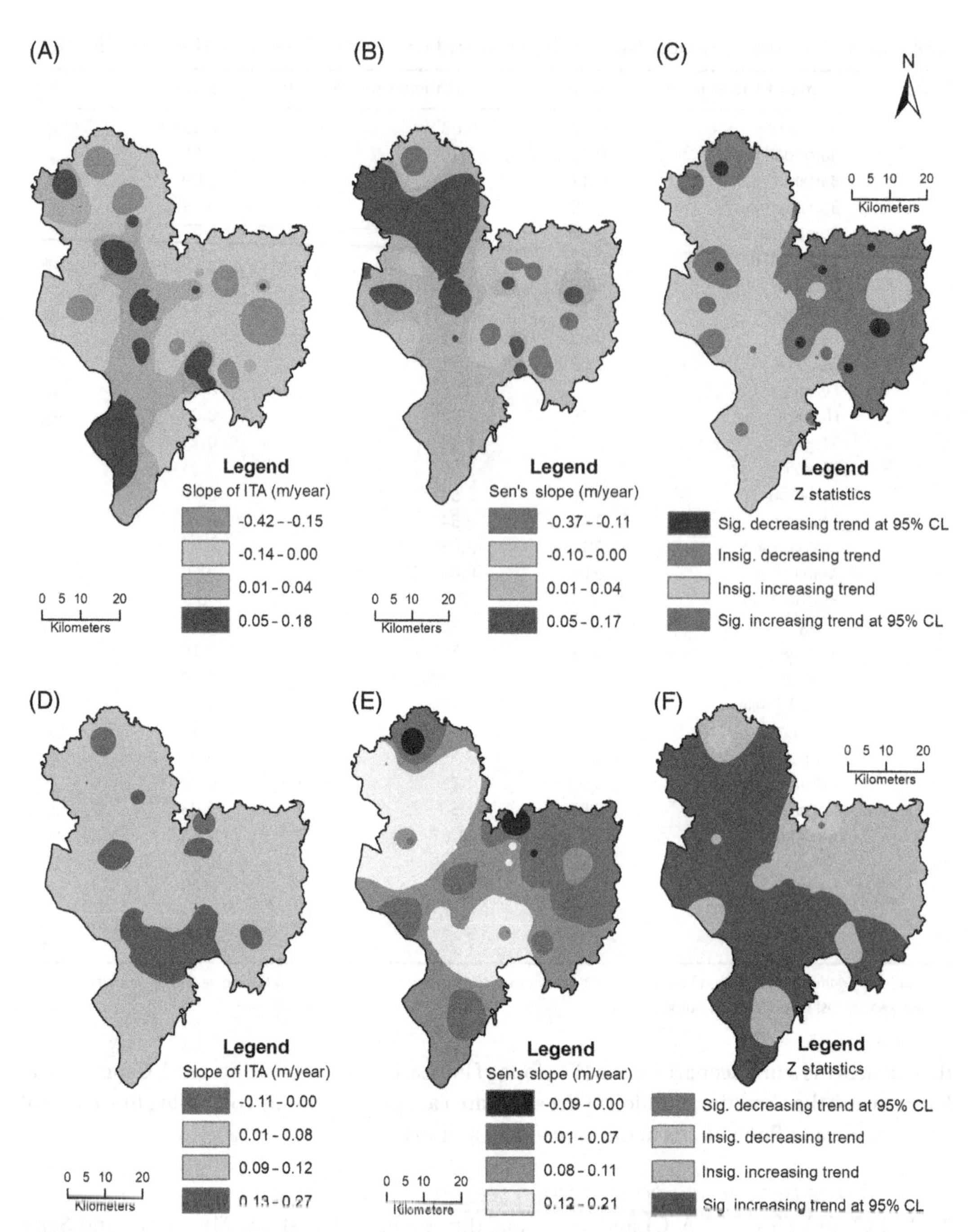

FIGURE 13.2 Distribution of ITA slope, Sen's slope and Z statistics of (A–C) premonsoon and (D–F) monsoon.

Table 13.4 Details of groundwater depth trend of postmonsoon season of Malda district.

Sl. No.	Name of the station	Slope IT	Z statistics of MK/mMK	Sen's slope	ITD
1	Agampur	0.03	1.07	0.08	▲
2	Baroduari	0.22	3.19*	0.19	▲
3	Battaly 1	0.22	3.61*	0.17	▲
4	Bholanathpur	0.10	2.00*	0.07	▲
5	Bulbulchandi	−0.01	−0.68	−0.02	▼
6	Chanchal pz	0.01	3.50*	0.24	▲
7	Dariapur	0.00	0.37	0.03	▲
8	Daulatpur	0.06	1.69	0.05	▲
9	Debipur	0.02	2.17*	0.03	▲
10	Deotala	−0.02	−0.87	−0.01	▼
11	Gazole	0.28	1.39	0.13	▲
12	Habibpur	0.06	1.44	0.05	▲
13	Magura	0.04	1.07	0.06	▲
14	Makuli	0.07	1.38	0.03	▲
15	Malatipur	0.15	3.64*	0.13	▲
16	Malda Pz	0.15	3.95*	0.12	▲
17	Malda Town	0.15	3.98*	0.12	▲
18	Manikchak	0.05	2.62*	0.04	▲
19	Mashaldighi	−0.03	−0.68	−0.02	▼
20	Milky	0.16	3.72*	0.014	▲
21	Mochia	0.23	3.72*	0.14	▲
22	Mothabari	0.17	3.38*	0.18	▲
23	New 16 mile	0.17	0.79	0.06	▲
24	Nimaisarai	−0.01	−1.38	−0.05	▼
25	Nurpur pz	0.00	2.80*	0.13	▲
26	Pakuahat pz	0.00	2.88*	0.10	▲
27	Pirganj	0.18	3.30*	0.22	▲
28	Ranikamat	−0.01	−0.34	−0.01	▼
29	Ratua	0.08	3.30*	0.07	▲
30	Samsi	0.15	3.59*	0.13	▲
31	Sujapur	0.03	1.80	0.04	▲
32	Tulshihata	0.19	4.51*	0.15	▲

* The trend is significant at 95% confidence level. *ITD*, innovative trend detection. The trend was detected by modified Mann-Kendall test based on lag-1 autocorrelation.

the stations indicated substantial negative trends over the study period. A significant positive trend is found in the Baroduari, Battaly 1, Bholanthpur, Debipur, Malatipur, Malda pz, Malda Town, Manikchak, Milky, Mochia, Mothabari, Nurpur pz, Pakuahat pz, Pirganj, Ratua, Samsi, and Tulshitala stations. Highest increasing trend through the analysis of Sen's slope is found in Chanchal pz (0.24 m/year), and the lowest increasing trend is found in Debipur (0.03 m/year). Overall Tal region and in the northern part of the district, and a portion of the Diara region covering the western portion of the district possess the significant trend of increasing and decreasing at a 95% confidence level. Sen's slope distribution map indicates that the increasing trend (0.14–

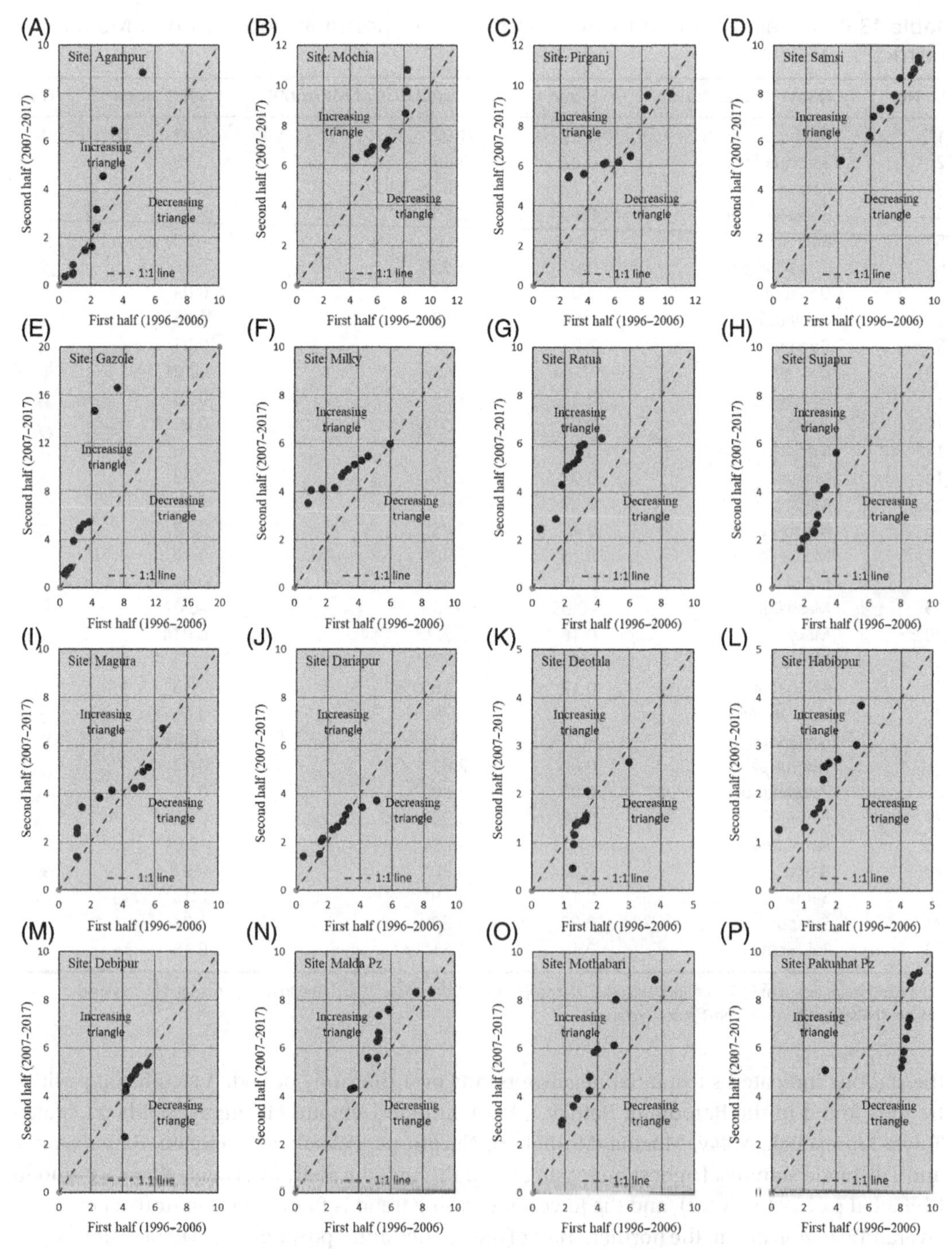

FIGURE 13.3 ITA graph of some selected dug well site from different season of Malda district (A–D) premonsoon, (E–H) monsoon, (I–L) postmonsoon, and (M–P) winter.

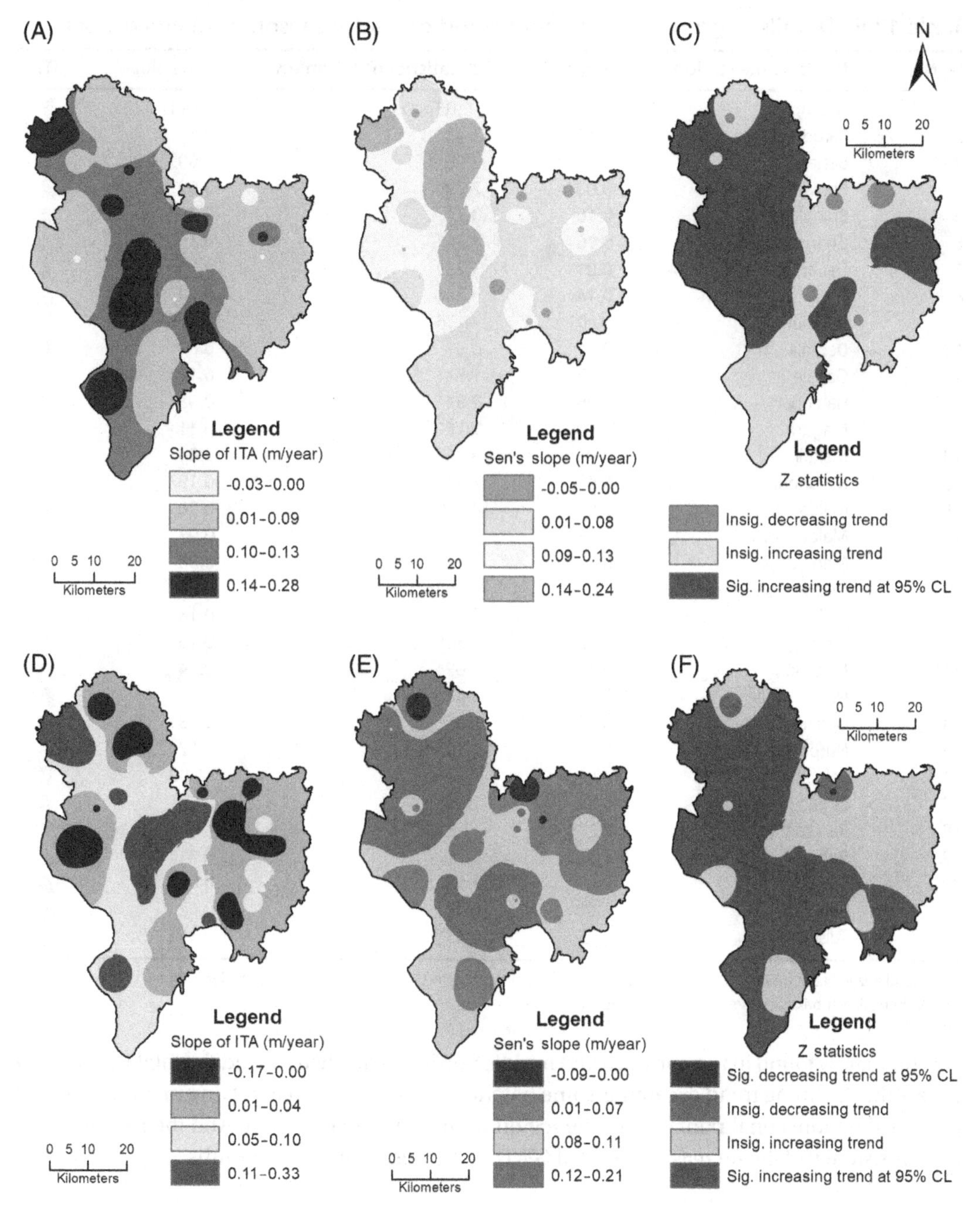

FIGURE 13.4 Distribution of ITA slope, Sen's slope and Z statistics of (A–C) postmonsoon and (D–F) winter.

Table 13.5 Details of groundwater depth trend of winter season of Malda district

Sl. No.	Name of the station	Slope IT	Z statistics of MK/mMK	Sen's slope	ITD
1	Agampur	−0.17	−0.03	−0.01	▼
2	Baroduari	0.17	2.40*	0.11	▲
3	Battaly 1	0.17	2.06*	0.09	▲
4	Bholanathpur	0.03	2.79*	0.09	▲
5	Bulbulchandi	0.16	0.21	0.02	▲
6	Chanchal pz	0.16	4.25*	0.16	▲
7	Dariapur	0.01	3.22*	0.15	▲
8	Daulatpur	0.14	3.24*	0.14	▲
9	Debipur	−0.01	−1.35	−0.06	▼
10	Deotala	0.01	0.37	0.01	▲
11	Gazole	0.33	1.83	0.12	▲
12	Habibpur	0.06	2.82*	0.08	▲
13	Magura	0.10	1.16	0.11	▲
14	Makuli	0.08	0.96	0.04	▲
15	Malatipur	0.09	2.54*	0.19	▲
16	Malda Pz	0.09	4.34*	0.20	▲
17	Malda Town	0.09	2.03*	0.07	▲
18	Manikchak	0.02	0.48	0.00	▲
19	Mashaldighi	−0.03	−2.23*	−0.09	▼
20	Milky	0.07	5.80*	0.15	▲
21	Mochia	0.15	2.59*	0.16	▲
22	Mothabari	0.13	4.69*	0.14	▲
23	New 16 mile	0.13	2.40*	0.10	▲
24	Nimaisarai	0.10	2.85*	0.13	▲
25	Nurpur pz	0.10	3.30*	0.12	▲
26	Pakuahat pz	−0.10	−1.75	−0.08	▼
27	Pirganj	0.21	1.30	0.03	▲
28	Ranikamat	−0.06	−1.83	−0.08	▼
29	Ratua	0.02	4.17*	0.21	▲
30	Samsi	0.07	1.58	0.09	▲
31	Sujapur	0.03	0.56	0.02	▲
32	Tulshihata	0.15	5.53*	0.15	▲

* The trend is significant at 95% confidence level. *ITD*, innovative trend detection. [shaded] The trend was detected by modified Mann-Kendall test based on lag-1 autocorrelation.

0.24 m/year) is found in the northern and middle portion of the district, and the total Barind tract covers the declining trend except for some patches. Tal is typically a low-lying marshy land and persists the monsoonal rainfall over the region in this particular season. And the results reveal the consistency of the increasing trend of GWD in the postmonsoonal months.

13.4.2.4 The trend in the winter season

The majority of the stations (84.40%) observed increasing trends over the study period according to ITA's winter GWD trends (1996–2017) (Table 13.5). At 95% confidence level, 59.4% of them are statistically significant. Except for a few isolated regions like Agampur, Debipur, Mashaldighi,

Pakuahat pz, and Ranikamat stations, these increasing tendencies can be detected across the study area. Agampur (slope = -0.17 m/year) and Pirganj (slope = +0.21 m/year), had the highest declining and increasing trend (ITA) respectively. Sen's slope analysis found that the average increase rate over Malda during the winter season is 0.014 m/year. The spatial-temporal variation derived from Sen's slope is depicted in Fig. 13.3(D–F), showing uniformity in increasing tendencies throughout Malda's northwest and southern parts, that is, Tal and Diara region. However, Mashaldighi (-0.09 m/year) has had the largest declining trend, and Ratua (+0.21 m/year) has the highest increasing trend. Except for the Barind tract, the overall region has fallen under the increasing trend of GWD at 95% confidence level.

13.5 Conclusion

This study investigated the regional variability and trends of the seasonal depth of groundwater resources across the Malda district. From the observation of the trend of GWD, it is possible to conclude that the ITA analysis indicates an inclining and a declining tendency for all of the stations, and the Z statistic strengthens its reliability. However, an increasing tendency has been seen at the regional level in recent decades in the Tal region and the Barind tract of the district. The postmonsoon and winter seasons show overall inclining tendencies, whereas the premonsoon season shows both positive and negative trends. Groundwater seasonal depth patterns that are consistent are incredibly unusual. However, the findings showed that certain dug well site stations had shown a notable negative trend in long-term seasonal GWD. The monsoon (June–September), postmonsoon (October–November), and winter (December–February) seasons were dominated by an inclining tendency, while the premonsoon (March–May) season had both decreasing and increasing tendencies. Some stations show no trend using the MK/mMK test, but innovative trend analysis detects a significant trend in those observations. However, such rigorous research improves water resource management and risk management and gives clear insight into complicated dynamic processes. For the micro-level analysis, other parameters related to groundwater must also be included.

Acknowledgments

The authors are highly indebted to DSTBT, WB for funding to conduct the field works, which really helped ground the truthing of facts.

Conflict of interest

The authors express that there is no conflict of interest directly or indirectly.

References

Alipour, A., Hashemi, S., Shokri, S.B.S., Moravej, M., 2018. Spatio-temporal analysis of groundwater level in an arid area. Int. J. Water 12 (1), 66–81. doi:10.1504/IJW.2018.090185.

Arshad, I., Umar, R., 2020. Status of urban hydrogeology research with emphasis on India. Hydrogeol. J. 28 (2), 477–490. doi:10.1007/s10040-019-02091-z.

Ashraf, M.S., Ahmad, I., Khan, N.M., Zhang, F., Bilal, A., Guo, J., 2021. Streamflow variations in monthly, seasonal, annual and extreme values using Mann-Kendall, Spearmen's Rho and innovative trend analysis. Water Res. Manage. 35 (1), 243–261. doi:10.1007/s11269-020-02723-0.

Bhanja, S., Mukherjee, A., Ramaswamy, R., Scanlon, B., Malakar, P., Verma, S., 2018. Long-term groundwater recharge rates across India by in situ measurements. Hydrol. Earth Syst. Sci. Discuss. 12, 1–19. doi:10.5194/hess-2018-313.

Briscoe, J., Malik, R.P., 2008. India's water economy: bracing for a turbulent future. Water 11 (Issue 10).

Chinchmalatpure, A.R., Gorain, B., Kumar, S., Camus, D.D., Vibhute, S.D., 2019. Research developments in saline agriculture. J.C. Dagar, Indian Council of Agricultural Research; Rajender Kumar Yadav, Central Soil Salinity Research Institute; Parbodh C. Sharma, Central Soil Salinity Research Institute, Springer Nature Singapore Pte Ltd. 2019. Research Developments in Saline Agriculture doi:10.1007/978-981-13-5832-6.

Choudhury, M., Jyethi, D.S., Dutta, J., Purkayastha, S.P., Deb, D., Das, R., Roy, G., Sen, T., Bhattacharyya, K.G., 2021. Investigation of groundwater and soil quality near to a municipal waste disposal site in Silchar, Assam, India. Int. J. Energy Water Resour. 23, 0123456789. doi:10.1007/s42108-021-00117-5.

Citakoglu, H., Minarecioglu, N., 2021. Trend analysis and change point determination for hydro-meteorological and groundwater data of Kizilirmak basin. Theoretic. Appl. Climatol. 22, 0123456789. doi:10.1007/s00704-021-03696-9.

Cosgrove, W.J., & Loucks, D.P. (2015). Water management: Current and future challenges and research directions. Water Resour. Res. 51, 4823–4839. doi:10.1002/2014WR016869

Coulibaly, P., Anctil, F., Aravena, R., Bobée, B., 2001. Artificial neural network modeling of water table depth fluctuations. Water Resour. Res. 37 (4), 885–896. doi:10.1029/2000WR900368.

Döll, P., Hoffmann-Dobrev, H., Portmann, F.T., Siebert, S., Eicker, A., Rodell, M., Strassberg, G., Scanlon, B.R., 2012. Impact of water withdrawals from groundwater and surface water on continental water storage variations. J. Geodyn. 59–60, 143–156. doi:10.1016/j.jog.2011.05.001.

Elbeih, S.F., 2015. An overview of integrated remote sensing and GIS for groundwater mapping in Egypt. Ain Shams Eng. J. 6 (1), 1–15. doi:10.1016/j.asej.2014.08.008.

Ghosh, A., Tiwari, A.K., Das, S., 2015. A GIS based DRASTIC model for assessing groundwater vulnerability of Katri Watershed, Dhanbad, India. Model. Earth Syst. Environ. 1 (3), 1–14. doi:10.1007/s40808-015-0009-2.

Gupta, A.D, Onta, P.R, 1997. Aménagement durable des ressources en eaux souterraines. Hydrol. Sci. J. 42 (4), 565–582. doi:10.1080/02626669709492054.

Hamed, Y., Hadji, R., Redhaounia, B., Zighmi, K., Bâali, F., El Gayar, A., 2018. Climate impact on surface and groundwater in North Africa: a global synthesis of findings and recommendations. Euro-Mediterr. J. Environ. Integr. 3 (1), 1–15. doi:10.1007/s41207-018-0067-8.

Han, D., Currell, M.J., Cao, G., Hall, B., 2017. Alterations to groundwater recharge due to anthropogenic landscape change. J. Hydrol. 554, 545–557. doi:10.1016/j.jhydrol.2017.09.018.

Hu, Z., Zhou, Q., Chen, X., Chen, D., Li, J., Guo, M., Yin, G., Duan, Z., 2019. Groundwater depletion estimated from GRACE: a challenge of sustainable development in an arid region of Central Asia. Remote Sens. 11 (16). doi:10.3390/rs11161908.

Khadri, S.F.R., Moharir, K., 2016. Characterization of aquifer parameter in basaltic hard rock region through pumping test methods: a case study of Man River basin in Akola and Buldhana districts Maharashtra India. Model. Earth Syst. Environ. 2 (1), 1–18. doi:10.1007/s40808-015-0047-9.

Lu, X.X., Ashmore, P., Wang, J.F., 2003. Seasonal water discharge and sediment load changes in the Upper Yangtze, China. Mountain Res. Dev. 23 (1), 56–64. doi:10.1659/0276-4741(2003)023[0056:SWDASL]2.0.CO;2.

MacDonald, A.M., Bonsor, H.C., Ahmed, K.M., Burgess, W.G., Basharat, M., Calow, R.C., Dixit, A., Foster, S.S.D., Gopal, K., Lapworth, D.J., Lark, R.M., Moench, M., Mukherjee, A., Rao, M.S., Shamsudduha, M., Smith, L., Taylor, R.G., Tucker, J., Van Steenbergen, F., Yadav, S.K., 2016. Groundwater quality and depletion in the Indo-Gangetic Basin mapped from in situ observations. Nat. Geosci. 9 (10), 762–766. doi:10.1038/ngeo2791.

Mallick, J., Singh, C.K., Al-Wadi, H., Ahmed, M., Rahman, A., Shashtri, S., Mukherjee, S., 2015. Geospatial and geostatistical approach for groundwater potential zone delineation. Hydrol. Process. 29 (3), 395–418. doi:10.1002/hyp.10153.

Mallick, J., Talukdar, S., Alsubih, M., Salam, R., Ahmed, M., Kahla, N.B, Shamimuzzaman, M., 2021. Analysing the trend of rainfall in Asir region of Saudi Arabia using the family of Mann-Kendall tests, innovative trend analysis, and detrended fluctuation analysis. Theor. Appl. Climatol. 143 (1–2), 823–841. doi:10.1007/s00704-020-03448-1.

Mandal, T., Saha, S., Das, J., Sarkar, A., 2021. Groundwater depletion susceptibility zonation using TOPSIS model in Bhagirathi river basin, India. Model. Earth Syst. Environ. 12, 0123456789. doi:10.1007/s40808-021-01176-7.

Mann, H.B., 1945. Non-parametric test against trend. Econometrica 13 (3), 245–259. http://www.economist.com/node/18330371?story%7B_%7Did=18330371.

Marques, E.A.G., Silva Junior, G.C., Eger, G.Z.S., Ilambwetsi, A.M., Raphael, P., Generoso, T.N., Oliveira, J., Júnior, J.N., 2020. Analysis of groundwater and river stage fluctuations and their relationship with water use and climate variation effects on Alto Grande watershed, Northeastern Brazil. J. South Am. Earth Sci. 103, 102723. doi:10.1016/j.jsames.2020.102723.

Mukate, S.V., Panaskar, D.B., Wagh, V.M., Baker, S.J., 2020. Understanding the influence of industrial and agricultural land uses on groundwater quality in semiarid region of Solapur, India. Environment, Development and Sustainability, 22. Springer Netherlands doi:10.1007/s10668-019-00342-3.

Mukherjee, A., Bhanja, S.N., Wada, Y., 2018. Groundwater depletion causing reduction of baseflow triggering Ganges river summer drying. Sci. Rep. 8 (1), 1–9. doi:10.1038/s41598-018-30246-7.

N. Moharir, K., B. Pande, C., Kumar Singh, S., Abarca Del Rio, R., 2020. Evaluation of analytical methods to study aquifer properties with pumping test in Deccan Basalt Region of the Morna River Basin in Akola District of Maharashtra in India. In: Groundwater Hydrology, pp. 1–12. doi:10.5772/intechopen.84632.

Naughton, O., Johnston, P.M., McCormack, T., Gill, L.W., 2017. Groundwater flood risk mapping and management: examples from a lowland karst catchment in Ireland. J. Flood Risk Manage. 10 (1), 53–64. doi:10.1111/jfr3.12145.

Pande, C., Moharir, K., Pande, R., 2021. Assessment of morphometric and hypsometric study for watershed development using spatial technology: a case study of Wardha river basin in Maharashtra, India. Int. J. River Basin Manage. 19 (1), 43–53. doi:10.1080/15715124.2018.1505737.

Patra, S., Sahoo, S., Mishra, P., Mahapatra, S.C., 2018. Impacts of urbanization on land use /cover changes and its probable implications on local climate and groundwater level. J. Urban Manage. 7 (2), 70–84. doi:10.1016/j.jum.2018.04.006.

Qadir, M., Sharma, B.R., Bruggeman, A., Choukr-Allah, R., Karajeh, F., 2007. Non-conventional water resources and opportunities for water augmentation to achieve food security in water scarce countries. Agric. Water Manage. 87 (1), 2–22. doi:10.1016/j.agwat.2006.03.018.

Sahoo, S., Jha, M.K., 2017. Numerical groundwater-flow modeling to evaluate potential effects of pumping and recharge: implications for sustainable groundwater management in the Mahanadi delta region, India. Hydrogeol. J. 25 (8), 2489–2511. doi:10.1007/s10040-017-1610-4.

Sasakova, N., Gregova, G., Takacova, D., Mojzisova, J., Papajova, I., Venglovsky, J., Szaboova, T., Kovacova, S., 2018. Pollution of surface and ground water by sources related to agricultural activities. Front. Sustain. Food Syst. 2 (July), 12–21. doi:10.3389/fsufs.2018.00042.

Sen, P.K., 1968. Estimates of the regression coefficient based on Kendall's Tau. J. Am. Stat. Assoc. 63 (324), 1379–1389.

Şen, Z., 2012. Innovative trend analysis methodology. J. Hydrol. Eng. 17 (9), 1042–1046. doi:10.1061/(asce)he.1943-5584.0000556.

Şen, Z., 2017. Innovative trend significance test and applications. Theor. Appl. Climatol. 127 (3–4), 939–947. doi:10.1007/s00704-015-1681-x.

Zakwan, M., 2019. Comparative analysis of the novel infiltration model with other infiltration models. In: Mohammad Zakwan (Ed.), Water Environ. J. 33 (4), 620–632. doi:10.1111/wej.12435.

Zakwan, M., 2021. Groundwater Resources Development and Planning in the Semi-Arid Region doi:10.1007/978-3-030-68124-1.

Zakwan, M., Ara, Z., 2019. Statistical analysis of rainfall in Bihar. Sustain. Water Resour. Manage. 5 (4), 1781–1789. doi:10.1007/s40899-019-00340-3.

Zeidan, B.A., 2017. Groundwater degradation and remediation in the Nile Delta aquifer. In: Şen, Z. (Ed.), Handbook Environ. Chem., 55, pp. 159–232. doi:10.1007/698_2016_128.

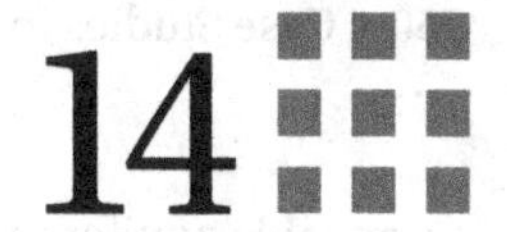

Assessing vulnerability of groundwater resource in urban and sub-urban areas of Siliguri, North Bengal (India): A special reference to LULC alteration

Mantu Das, Baidurya Biswas and Snehasish Saha

DEPARTMENT OF GEOGRAPHY AND APPLIED GEOGRAPHY, UNIVERSITY OF NORTH BENGAL, RAJA RAMMOHUNPUR, DARJEELING, WEST BENGAL, INDIA

14.1 Introduction

Groundwater is an important natural resource in our natural physico-environmental settings. It is a valuable resource not only in terms of economic and agricultural point of view, but also it has immense power to control environmental balance. Groundwater is a key supply of drinking water, and it also helps to sustain the ecological value of many locations (Mishra et al., 2014; Kumar et al., 2005). Groundwater management has placed a greater emphasis on water quality (Manap et al., 2012; Neshat et al., 2014). Generally, percolation of pollutants destroys groundwater system and provokes groundwater vulnerability. Vulnerability is often thought of as an "intrinsic" quality of a groundwater system, based on its susceptibility to human and/or environmental influences (Kumar et al., 2013). In the mid-19th century, the term groundwater vulnerability was pronounced highly in order to raise awareness of the dangers of groundwater pollution (Kumar et al., 2013). Pollutant attenuation is not included in groundwater vulnerability since it solely deals with the hydrogeological context. Natural hydrogeological variables have varying effects on various contaminants, depending on their interactions and chemical qualities. However, groundwater is a vulnerable natural resource in order to pollution due to overexploitation and mismanagement (Borevsky et al., 2004; Pedreira et al., 2015; Job, 2010; Moghaddam et al., 2018).

The importance of the protection of groundwater and management has been acknowledged by almost all the researchers. The primary drivers of groundwater quality degradation in the research region are agricultural pesticides and wastewater. Insecure municipal waste, landfills on

Case Studies in Geospatial Applications to Groundwater Resources. DOI: https://doi.org/10.1016/B978-0-323-99963-2.00006-7

permeable aquifer units, as well as unregulated sewage discharge cause detrimental impact on quality of groundwater. The range of pollutants plus their mixing in surface water and groundwater are a danger to groundwater quality as a result of surface and groundwater movement. Therefore, the area's groundwater susceptibility should be identified in order to safeguard the groundwater.

Siliguri sub-division is a complex spatial extension, it means rural agriculture dominated land and urbanized paved surface take place within small spatial extension. Here, anthropogenic activities take place and have played very crucial role towards groundwater vulnerability through dumping practice in the periphery of Siliguri town, using chemical fertilizers and land use and land over (LULC) alteration has immense power to change the groundwater quality.

The prime objective of the present study is the identification of the groundwater vulnerability zones following the semi-automated DRASTIC model under geographic information system (GIS) environment. Spatio-temporal land use changes are the dominant controlling factor for their vulnerability. So, to identify the change scenario of landscape (both natural and anthropogenic features) over time and simultaneously future trend also that have a role to manage the vulnerability status of the groundwater. The study's ultimate goal would be to advise management strategies for maintaining and conserving the area's significant groundwater resources.

14.2 Brief description of the study area

Siliguri sub-division is located in the foothills of Darjeeling district in West Bengal, India. It has four blocks, namely Matigara, Naxalbari, Kharibari, and Phansidewa. Total geographical area of the Siliguri sub-division is approximately 752 km^2 (Fig. 14.1) with a population of 971,120 (Census of India, 2011). According to census 2011, most of the population lives in rural areas (55.11%), but it is noteworthy that 44.89% population mainly resides in Siliguri town. Geologically, versatile piece of land is covered with older alluvium, newer alluvium, Daling, Siwalik, and Lower Gondwana rock structure. Dramatic land use transformations have been observed in Siliguri sub-division with special reference to urbanization and deforestation practices. That is why groundwater trends to be vulnerable to contamination.

14.3 Study materials and methodology

14.3.1 Data used

For the study of groundwater vulnerability assessment, following assessment procedures were considered, that is, process-based simulation models, the statistical calculation-based methods (Harbaugh et al., 2000), and the overlay and index method (Dixon, 2004; Kumar et al., 2013). In GIS, overlay and index approaches are simple to use mostly for regional scale. As a corollary, such methodologies are very often used. DRASTIC is the most extensively utilized of these approaches (Aller et al., 1987; Plymale and Angle, 2002; Fritch et al., 2000; Shukla et al., 2000; Huan et al., 2012; Yin et al., 2012; El-Naqa et al., 2006; Pacheco and Fernandes, 2013; Saidi et al., 2010, 2011; Mimi et al., 2012; Javadi et al., 2011a, b; Ettazarini, 2006). For the present study, both primary

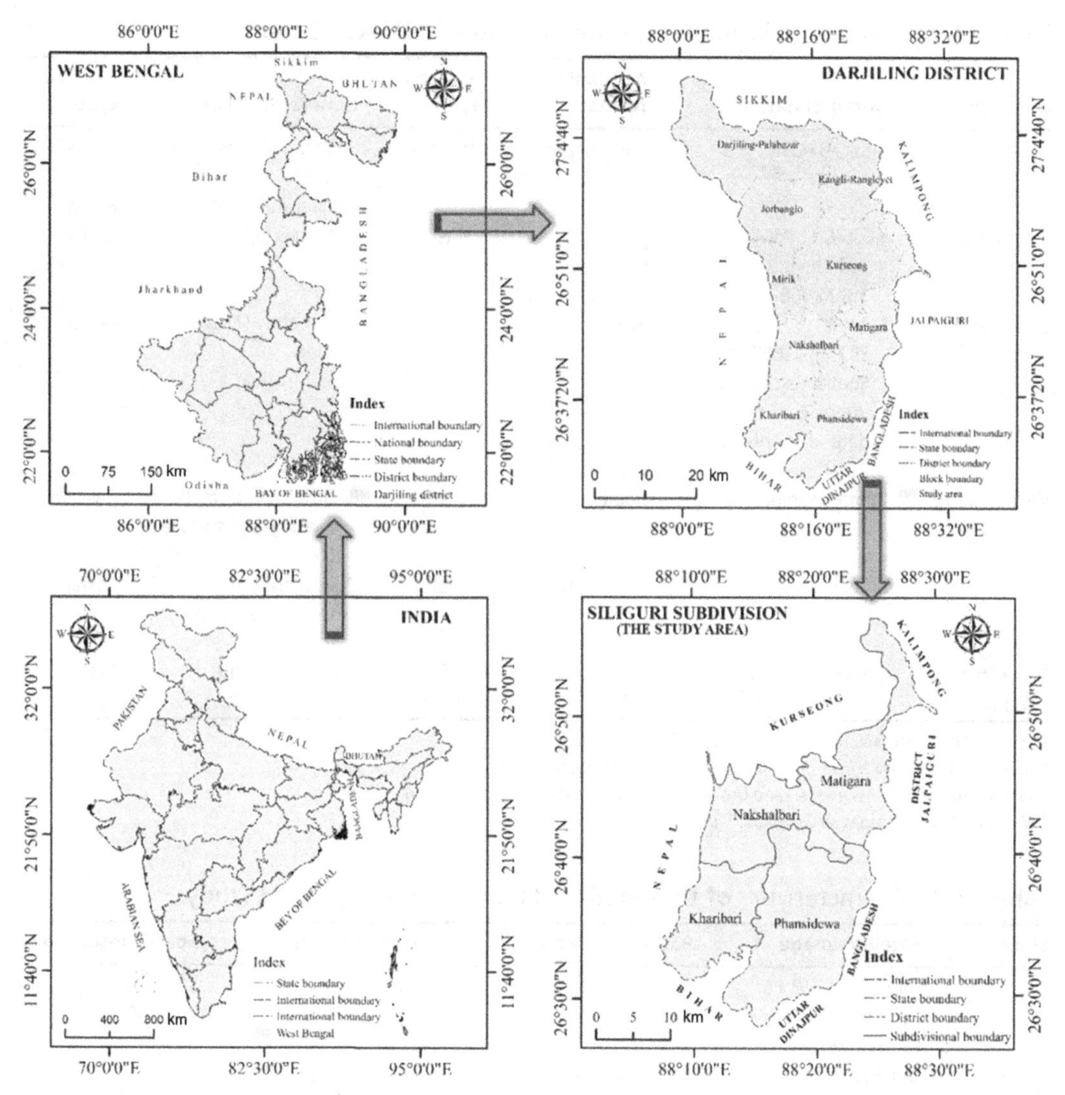

FIGURE 14.1 Locational identity of the study area.

and secondary data sources were used for constructing the groundwater vulnerability zones in Siliguri sub-division (Table 14.1). However, Table 14.2 provide the detailed description of data used in the present study.

14.3.2 DRASTIC data layers

In raster GIS, the allocated layers for the seven DRASTIC parameters were calculated based on standard weight and rating (Table 14.2).

Table 14.1 Detail description about the data used in DRASTIC model.

Data type	Detail of data	Available format	Extracted layer	Generated layer	GIS data type
Lithology map[b]	Bhukosh-GSI	ESRI shapefile	Lithology	Impact of vadose zone (I)	Raster
				Aquifer media (A)	Raster
Soil map[c]	NBSS & LUP WB Soil-Sheet 3 Scale: 1:500,000	JPG	Soil types	Soil types (S)	Raster
SRTM-DEM[a]	Entity ID: SRTM1N26E088V3 Spatial resolution-30 metre. Publication date: 23 September 2014	TIFF	Elevation	Topography (T) average slope	Raster
Well data (m)	Primary survey	Point data	Isopleth	Depth to water table from the soil surface (D)	Raster
HC (m/s)	Primary survey	Point data	Isopleth	Conductivity (hydraulics) of the aquifer (C)	Raster
Net recharge (mm)[d]	WHYMAP GWR	ESRI shapefile	Net recharge	Net recharge (R)	Raster

Source: [a]USGS Earth Explorer (https://earthexplorer.usgs.gov/).
[b] Bhukosh – Geological Survey of India (https://bhukosh.gsi.gov.in/).
[c] National Bureau of Soil Survey & Land Use Planning (https://esdac.jrc.ec.europa.eu/).
[d] WHYMAP GWR (https://www.whymap.org/).

Table 14.2 Characteristics of the satellite data used in the present study.

Sl. no.	Satellite image	Path	Raw	Date of acquisition	Spatial resolution
1	Landsat 8 OLI/TIRS	139	41, 42	23 November 2020	30 × 30
2	Landsat 7ETM+	139	41, 42	4 November 2010	30 × 30
3	Landsat 7ETM+	139	41, 42	12 December 2000	30 × 30
4	Landsat 5 TM	139	41, 42	23 December 1990	30 × 30

Source: USGS Earth explorer.

Depth of water (D)

Depth of the water table starting from earth surface is an important parameter for this vulnerability assessment study based on model analysis. In the present study, total 35 sample wells were selected. Interpolation technique for generating the raster map with 30 m pixel size under the ArcGIS Geostatistical Analyst extension was used following the Krigging algorithm (Kumar, 2007; Gundogdu and Guney, 2007). Depth of water table (D) measures the distance a pollutant must travel from the surface to reach the groundwater. In DRASTIC model, "D" factor has been considered the most powerful parameter for groundwater vulnerability assessment. Therefore, it has given higher rank based on standard Aller's weightage.

Depth of groundwater level is indicative of longer time for contamination and vice versa (Aller et al., 1987; Neshat et al., 2014). Based on field survey data, the Siliguri sub-division has been categorized into three groundwater table zone (for details see Table 14.4).

Net recharge (R)

Net recharge is an amount of water that percolates into the earth surface and reaches to the underground water table (Rahman, 2008; Crosbie et al., 2015; Moghaddam et al., 2018). This parameter has important role for the groundwater vulnerability. Moreover, the net recharge is directly proportional to the groundwater vulnerability. Groundwater recharge depends on several factors, that is, surface characteristics, land use pattern, lithological properties, etc. Here, two net groundwater recharge group were developed, that is, 7–10 in/year and greater than 10 in/year. In Siliguri sub-division, most of the geographical areas are experiencing good quality of net recharge (Table 14.5).

Aquifer media (A)

The permeability of each stratum of media is utilized to create an aquifer media rating. More the water, and hence more toxins enter the aquifers when permeability is maximum. As a result, high permeability results into high vulnerability scores (Mondal et al., 2018). For better clarification please see Table 14.6.

Soil media (S)

The transfer of contaminants with water through the surface soil towards the groundwater is influenced by the soil medium. Some interactions with soil being one media can influence the sorts of reactions that can occur when it comes to altering groundwater quality. The structure of the soil surface might influence sorption events. Different types of soils are better habitat to microorganisms engaged in biodegrading the pollutant in the rating scale (Aller et al., 1987). This ranking method focuses on the contaminant's transit to be mixed up with the aquifer rather than other qualities. The studied area is covered with five types of soil category viz. W009, W007, W006, W004, and W002 (Table 14.7). In the studied area, most dominant soil type was W009 which covered 66.33 % of total geographical area.

Topography (T)

The slope of the ground surface is reciprocated by topography in the DRASTIC model created from a digital elevation model using an SRTM DEM (30 m spatial resolution). The terrain of the surface impacts groundwater vulnerability as the gradient of the land is a significant component for deciding that whether the pollutant discharged flowing as run-off will permeate the aquifer or not. The contamination is less likely to become run-off on a low slope, and hence more likely to permeate the aquifer. Based on slope (%), the Siliguri sub-division were grouped into five topographic class based on their influencing capacity. It is observed that the second topographic group (2–6 slope in %) was dominant topographic class followed by foremost class, that is, 0–2 slope in % and has immense power to reduce the groundwater vulnerability (Table 14.8).

Table 14.3 Standard weight and rating of individual parameter and sub-categories using in DRASTIC model.

Depth to water (m)		Recharge (mm)		Topography (slope %)		Conductivity (m/day)		Aquifer media		Vadose zone material		Soil media	
Range	Rating	Range	Rating	Range	Rating	Range	Rating	Range	Rating	Range	Rating	Range	Rating
0–1.5	10	0–50.8	1	0–2	10	0.04–4.1	1	Massive shale	2	Confining layer	1	Thin or absent	10
1.5–4.6	9	50.8–101.6	3	2–6	9	4.1–12.3	2	Metamorphic/ ign eous	3	Silt/clay	3	Gravel	10
4.6–9.1	7	101.6–177.8	6	6–12	5	12.3–28.7	4	Weathered metamorphic igneous	4	Shale	3	Sand	9
9.1–15.2	5	177.8–254	8	12–18	3	28.7–41	6			Limestone	3	Peat	8
15.2–22.8	3	>254	9	>18	1	41–82	8	Glacial till	5	Sandstone	6	Shrinking clay	7
22.8–30.4	2					>82	10	Bedded sandstone, limestone	6	Bedded limestone, sandstone	6	Sandy loam	6
>30.4	1											Loam	5
								Massive sandstone	6	Sand and gravel	6	Silty loam	4
								Massive limestone	8	W. silt		Clay loam	3
								Sand and gravel	8	Sand and gravel	8	Muck	2
								Basalt	9	Basalt	9	No shrinking clay	1
								Karsts limestone	10	Karsts limestone	10		
DRASTIC weight: 5		DRASTIC weight: 4		DRASTIC weight: 1		DRASTIC weight: 3		DRASTIC weight: 3		DRASTIC weight: 5		DRASTIC weight: 2	

Source: Aller et al. (1987).

Table 14.4 Spatial variation of groundwater table at postmonsoon season.

Sl. no.	Depth to water table (ft.)	Area (km^2)	Area (%)
1	0–5	108.30	14.40
2	5–15	584.40	77.71
3	15–30	59.28	7.88

Source: Primary survey, September– October 2021.

Table 14.5 Spatial variation of the Net recharge.

Sl. no.	Net recharge (In/year)	Area (km^2)	Area (%)
1	7–10	16.09	2.14
2	>10	735.80	97.86

Source: WHYMAP GWR.

Impact of the vadose zone (I)

Data for unsaturated zone lithology were extracted from the lithology map, provided by the Geological survey of India (GSI), India and were used in construction of this layer. Nine classes were identified based on distinct lithological setup (Table 14.9) as shown in Fig. 14.3. Dominant vadose class is Baikunthapur formation (48.19% of spatial coverage) consist of sand, silt and clay with calcareous concretions.

Hydraulic conductivity (H)

Groundwater is constantly in motion, and hydraulic conductivity measures an aquifer's capacity to transport water (Saha and Alam, 2014). As a result, the pace at which contaminants flow to the aquifer is determined by this component (Aller et al., 1987; Mondal et al., 2018). Once a pollutant enters the aquifer, these components become channels for fluid flow, as well as pathways for contaminant migration. Hydraulic conductivity has a positive relationship with vulnerability. Total geographical area has been categorized into three distinct hydraulic conductivity zones, that is, 1-100 m/s, 100-300 m/s and >300 m/s with spatial coverage of 83.14 %, 15.18 %, and 1.68 % respectively (Table 14.10). It is noteworthy that major part of the total geographical area has lower rate of hydraulic conductivity (<100 m/s).

Each layer was categorized for various scales of rating when all the relevant levels were constructed. Thus, the DRASTIC index was calculated by multiplying the obtained numerals by the weight-factor.

14.3.3 Methodology of the DRASTIC vulnerability model

The DRASTIC is such a model which follows layer overlaying technique for knowing the vulnerability of aquifers using prescheduled seven hydrogeological parameters (Fig. 14.2). Fig. 14.3A-G shows the parameters of DRASTIC method for determining the groundwater vulnerability. According to the conventional model, seven factors are the leading factors (Aller et al., 1987), that is, D, R, A, S, T, I, and C. At first, seven layers of raster type were prepared using 30-metre spatial resolution. Adhering the typical range of ten-graded relative scale (Table 14.2), these parameters were summed up in the equation (r). The more the value, the more is the chances of

Table 14.6 Types of aquifer media in Siliguri sub-division.

Age	Group	Formation	Aquifer media	Area (km^2)	Area (%)
Pleistocene	Older alluvium	Duars	Red and orange colour highly oxidized soil (AM1)	6.50	0.86
Meghalayan	Newer alluvium	Present day deposits	Sand, silt, and clay (AM2)	40.81	5.43
Pleistocene	Older alluvium	Duars	Brown and yellowish colour highly oxidized soil (AM3)	175.40	23.33
Holocene	Newer alluvium	Shaugaon	Sand, silt, and clay (AM4)	38.89	5.17
Proterozoic	Daling	Reyang	Quartz arenite, black slate, cherty phyllite (AM5)	15.30	2.03
Proterozoic	Daling	Gorubathan	Chlorite sericite schist and quartzite (AM6)	0.34	0.04
Holocene	Newer alluvium	Jalpaiguri	Feebly oxidized sand, silt, and clay (AM7)	99.82	13.27
Pleistocene - Holocene	Older alluvium	Baikunthapur	Sand, silt, clay with calcareous concretions (AM8)	362.40	48.19
Pliocene - Pleistocene	Siwalik	Kamarchinwa	Sandstone, clay, shale, conglomerate (AM9)	11.44	1.52
Permian	Lower Gondwana	Damuda	Sandstone, shale with minor coal (AM10)	1.05	0.14

Source: Geological Survey of India (GSI).

Table 14.7 Types of soil media in the Siliguri Sub-division.

Sl. no.	Soil types	Descriptions	Area (km^2)	Area (%)
1	W009	Coarse loamy, Aquic Udifluvents associated with fine loamy, Fluventic Eutrochepts	498.72	66.33
2	W007	Fine loamy, Fluventic Eutrochepts associated with coarse loamy, Aquic Udifluvents	119.08	15.84
3	W006	Coarse loamy, Umbric Dystrochrepts associated with fine loamy, Fluventic Dystrochrepts	108.72	14.46
4	W004	Loamy-skeletal, Typic Haplumbrepts associated with loamy-skeletal, Typic Udorthents	9.61	1.28
5	W002	Coarse loamy, Typic Udorthents associated with loamy-skeletal, Typic Dystrochrepts	15.78	2.10

Source: National Bureau of Soil Survey & Land Use Planning (NBSS & LUP).

Table 14.8 Identifying the topographic features in Siliguri sub-division.

Sl. no.	Topography (Slope %)	Area (km^2)	Area (%)
1	0–2	305.30	40.61
2	2–6	350.70	46.45
3	6–12	60.73	8.08
4	12–18	6.49	0.86
5	>18	28.58	3.80

Source: SRTM DEM, 2014.

contamination and vice versa (Table 14.3). Each parameter was assigned to a weighting factor (w) between 1 and 5 and which is indicating the relative significance of each component (Table 14.11). The following is the formula for the linear equation scheduled for the total factors (Kihumba et al., 2016; Mondal et al., 2018):

$$V\ (\text{intrinsic}) = D_rD_w + R_rR_w + A_rA_w + S_rS_w + T_rT_w + I_rI_w + C_rC_w \tag{14.1}$$

where "V(intrinsic)" denotes intrinsic nature of vulnerability, superscript "r" is a 10-graded relative-factor, and superscript "w" denotes a weight of 5-graded type.

Groundwater contamination vulnerability increases as the V(intrinsic) index rises. The groundwater vulnerability index mapping is divided into multiple classes, each showing a similar level of groundwater contamination vulnerability (Hamza et al., 2015).

14.3.4 LULC classification

The changing nature of LULC was identified by analyzing the multiband raster images. Spectral information, such as color band, is one of the main types of information used for the purpose of image classification, as well as interpretation. In this context, image conversion from FCC (false color composite) to TCC (true color composite also called natural color composite) map has been developed through the systematic image classification technique. Image classification

Table 14.9 Classification of the vadose layers under Siliguri sub-division.

Age	Group	Formation	Type of vadose layer	Area	Area (%)
Proterozoic	Daling	Reyang	Quartz-arenite, black slate, phyllite (IVZ1) (cherty)	15.30	2.03
Proterozoic	Daling	Gorubathan	Chlorite based sericite schist and quartzites (IVZ2)	0.34	0.04
Pliocene - Pleistocene	Siwalik	Kamarchinwa	Sandstone, clay, shale, conglomerate (IVZ3)	11.44	1.52
Permian	Lower Gondwana	Damuda	Sandstone, shale with minor coal (IVZ4)	1.05	0.14
Pleistocene	Alluvium (Older)	Duars	Brown and yellowish colour highly oxidized soil (IVZ5)	175.38	23.32
Holocene	Alluvium (Newer)	Jalpaiguri	Feebly oxidized sand, silt, and clay (IVZ6)	99.82	13.28
Pleistocene	Older alluvium	Duars	Red and orange colour highly oxidized soil (IVZ7)	6.50	0.86
Meghalayan	Newer alluvium	Present day deposits	Sand, silt, and clay (IVZ8)	79.72	10.60
Pleistocene - Holocene	Older alluvium	Baikunthapur	Sand, silt, clay with calcareous concretions (IVZ9)	362.38	48.19

Source: Geological Survey of India (GSI).

Table 14.10 Spatial variation of the hydraulic conductivity (HC).

Sl. no.	HC(m/s)	Area (km^2)	Area (%)
1	1–100	625.12	83.14
2	100–300	114.14	15.18
3	>300	12.63	1.68

Source: Data collected through primary survey and processed by present authors.

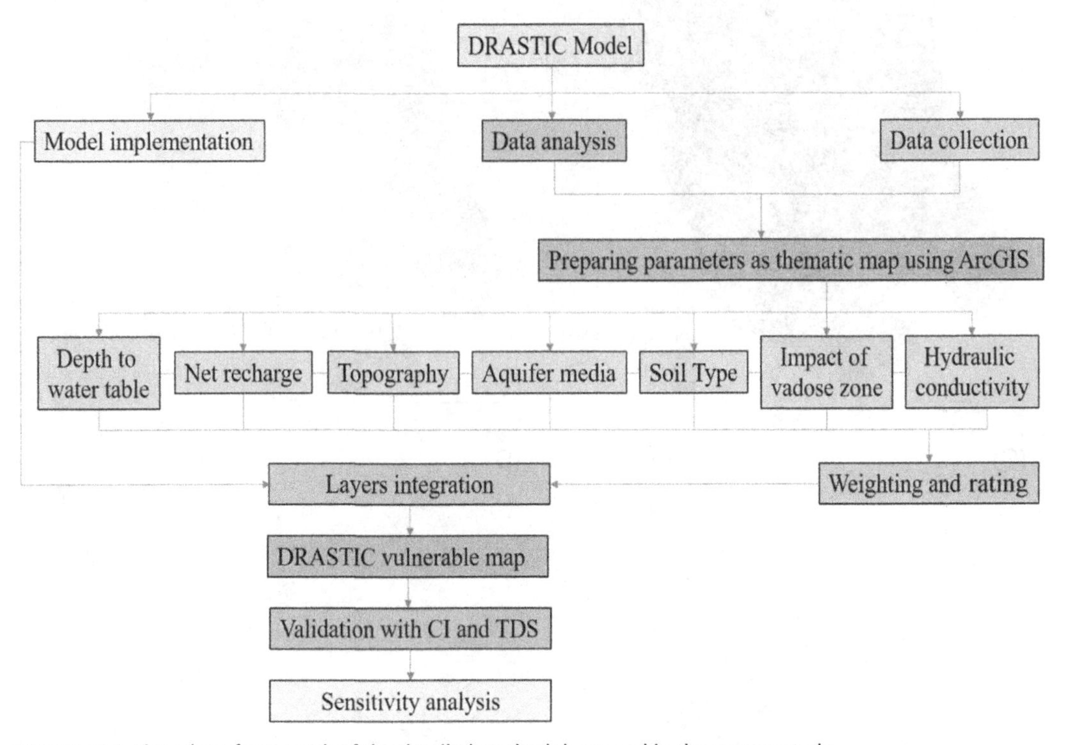

FIGURE 14.2 Flow chart framework of the detailed methodology used in the present study.

was performed for the automatic categorization of pixels having same reflectance range for the specific LULC class (Lillesand et al., 2015). It is noticed that image classification method has been used as a tool in mapping from remotely sensed data (Foody, 1999). Supervised classification following selection of training sites towards categorization (Lillesand et al., 2015) was considered. There are many methods available that are being used to implement the supervised image classification, that is, parallelepiped classification, K-nearest neighbor, minimum distance to mean classification, Gaussian maximum likelihood, linear discriminant, features space, etc. (Lillesand et al., 2015; Bhatta, 2011; Reddy, 2008). Maximum likelihood classifier (Foody et al., 1992) for LULC classification using ArcMap (10.5) software was chosen. Maximum likelihood/Bayesian method was taken as classifier algorithm to evaluate both the variance and covariance of the spectral response patterns and each pixel is assigned to the class with highest possibility of

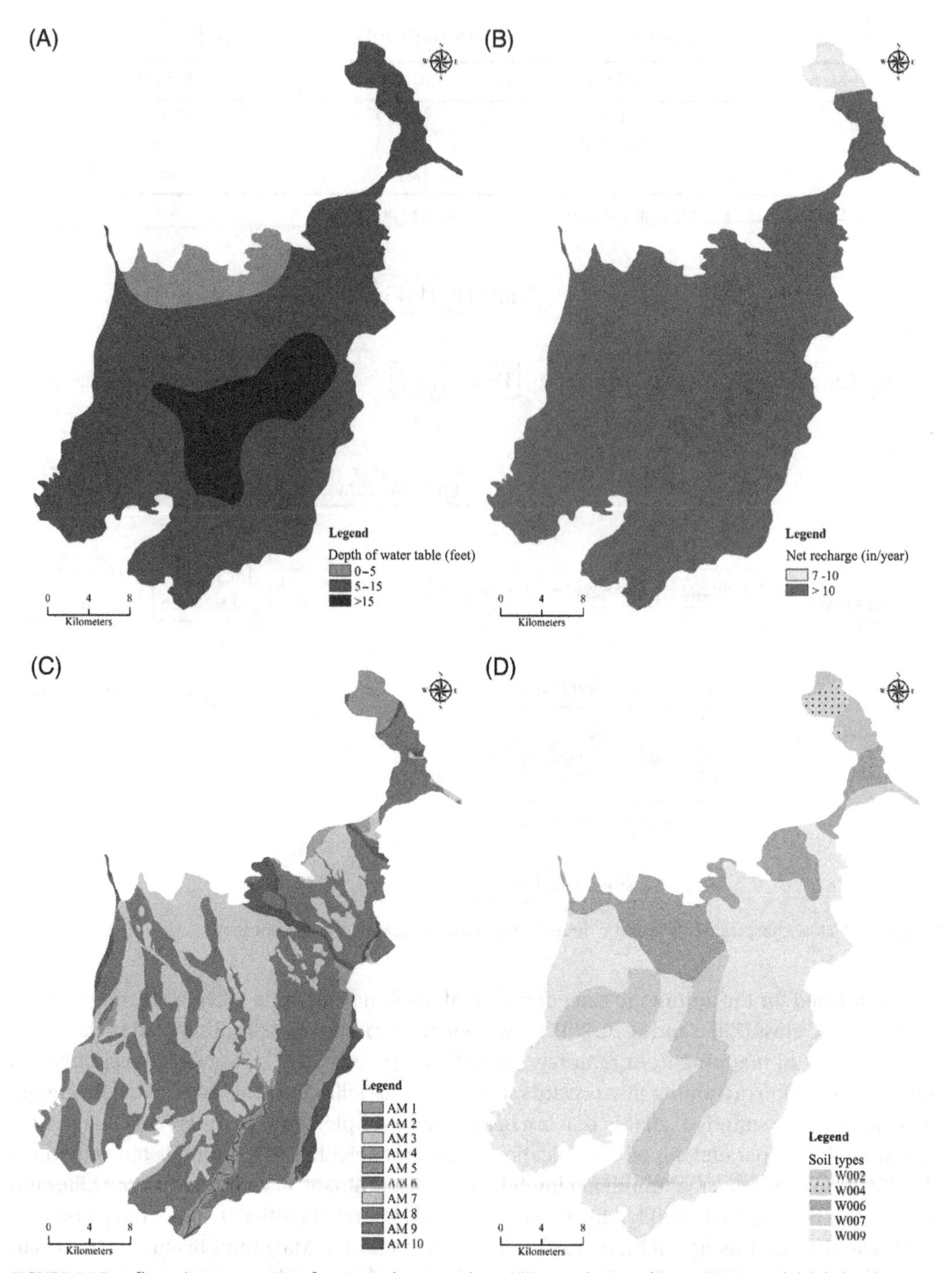

FIGURE 14.3 Influencing parameters for groundwater-vulnerability analysis under DRASTIC model (A) depth to the water table, (B) net recharge, (C) aquifer as media, (D) soil as media, (E) topography, slope %, (F) impact of vadose zone, and (G) hydraulic conductivity.

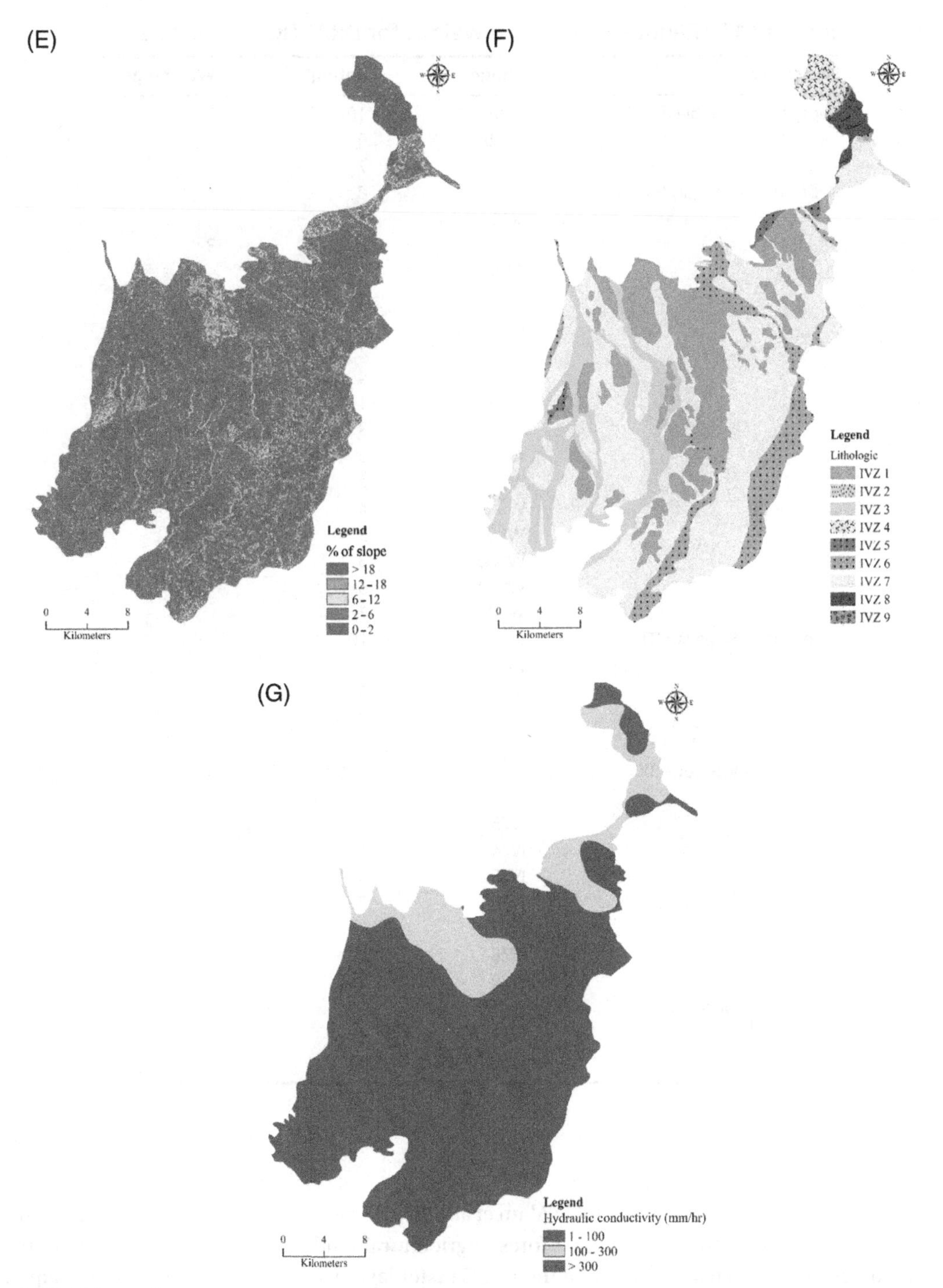

FIGURE 14.3, cont'd.

Table 14.11 Ranges, rating, and weight for DRASTIC parameters.

Parameters	Ranges	Rating*	Weightage*
Depth to water table (ft.) (D)	0–5	10	5
	5–15	9	
	15–30	7	
Net recharge (in/year) (R)	7–10	8	4
	>10	9	
Aquifer media (A)	AM1	2	3
	AM2	1	
	AM3	2	
	AM4	1	
	AM5	2	
	AM6	4	
	AM7	1	
	AM8	3	
	AM9	5	
	AM10	2	
Soil type (S)	W009	4	2
	W007	3	
	W006	5	
	W004	6	
	W002	7	
Topography (Slope %) (T)	0–2	10	1
	2–6	9	
	6–12	5	
	12–18	3	
	>18	1	
Impact of vadose zone (I)	IVZ1	3	5
	IVZ2	3	
	IVZ3	4	
	IVZ4	4	
	IVZ5	5	
	IVZ6	2	
	IVZ7	2	
	IVZ8	2	
	IVZ9	6	
Hydraulic conductivity (mm/h) (C)	1–100	1	3
	100–300	2	
	>300	4	

* Assigning rank based on Table 3.

association (Shalaby and Tateishi, 2007; Alam et al., 2020). Seven-fold LULC classes, that is, water body, sand deposit, dense forest, open forest, agricultural land, bare soil, and settlement area were determined. Postclassification of the LULC raster layer was cross verified with the help of Google Earth pro software in a systematic statistical way.

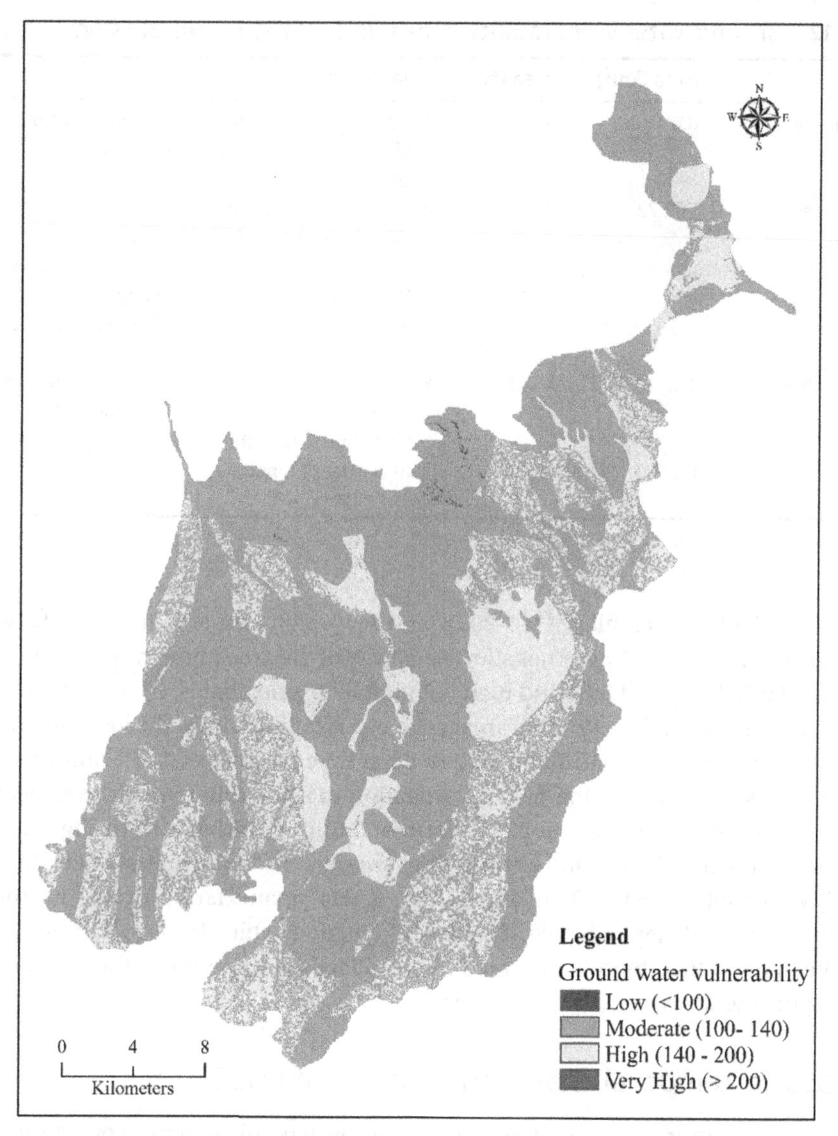

FIGURE 14.4 Groundwater vulnerability zone of Siliguri sub-division using DRASTIC model.

14.4 Results and discussion

14.4.1 Vulnerability of groundwater status (DRASTIC model based)

The calculated DRASTIC index values considering seven parameters provide status of existing vulnerability of Siliguri sub-division. The range of DRASTIC index computed for the study area are ranging from 98 to 329.1 which helped to classify the study area into vulnerability zones (Fig. 14.4). The vulnerability zones are the low vulnerability (<100) type, moderate vulnerability

Table 14.12 Groundwater vulnerability zones under Siliguri sub-division.

GWVZ*	Area (km^2)	Area (%)	Village names
Low vulnerability (1–100)	0.47	0.06	Tarabarir Chhat, Dumriguri Chhat, Bara Adalpur Dwitiya Khanda, Panchakulguri, Dhemal, Khaprail, Khoklong, and Lalsara Chhat
Moderate vulnerability (101–140)	559.72	74.89	Atal, Sivok Hill Forest, Barabhita, Sittong Forest, Nipania, Uttar Bagdogra (CT), Hoda Bhitar Chhat, Fulbar, Chhota Chenga, Dalkajhar Forest, Churaman, Damdama, Nembutari, Ghusuru, Subal, Rangia, Lohasing, Baunibhita, Tarabarir Chhat, Purbba Madati, and Belgachi
High vulnerability (141–200)	187.18	25.05	Ranidanga, Lalman (CT), Nirmmal, Mahideb, Rangapani, Haribhita, Dalkajhar Forest, Dhambhita, Rajajhar, Shaibhita, Bara Pathuram, Bhushibhita, and Kuchia
Very high vulnerability (>200)	0.0009	0.0001	Spatial extension not a single village

* *GWVZ*, Groundwater vulnerability zone.

(100–140) type, high vulnerability (140–200), and very high vulnerability (>200) type. Thus, classification of the vulnerability zones shows majority of the area under moderate vulnerability type, that is, 559.72 km^2 or 74.89% and high vulnerability zone, that is, 25.05% of the study area (Table 14.12). Comparatively very small areal coverage is falling within the low vulnerability and very high vulnerability zones and share of their areal cover as follows 0.06% and 0.0001% of the study area. It has observed during field investigation that the different anthropogenic factors were affecting groundwater vulnerability in form of dumping garbage, throwing various pollutant materials in the river Mahananda, tea gardens area, etc. There are many villages falling under very highly vulnerable zone like Ranidanga, Lalman, Rangapani Haribhita, etc. due their higher population density and higher degree of pollutants creation, which leads to high vulnerability of groundwater. Low vulnerability is observed in the villages of Khaprail, Lalsara Chhat, Dhemal, Panchakulguri, and Khoklong Tarabarir Chhat.

14.4.2 Spatio-temporal changing scenario of LULC

Land use and land cover pattern on the earth surface is dynamic in nature over time. Transformation rate of LULC types depend on natural as well as anthropogenic factors. Natural factors generally alter the land use types in a rhythmic way but anthropogenic factors have immense power to drastic change within short period of time. In Anthropocene, human induced functions such as urbanization, deforestation, population pressure, ill management of natural resources can directly control the earth surface structure or functions. In the present study, 40 years temporal study was considered for change analysis of LULC pattern (Fig. 14.5). Six types of LULC category have been executed using supervised image classification with nearest neighbor algorithm, that is, settlement area, sand deposit, tea garden area, forest land, water bodies, and agricultural field. From data analysis (Fig. 14.6), four land use type, that is, settlement, sand

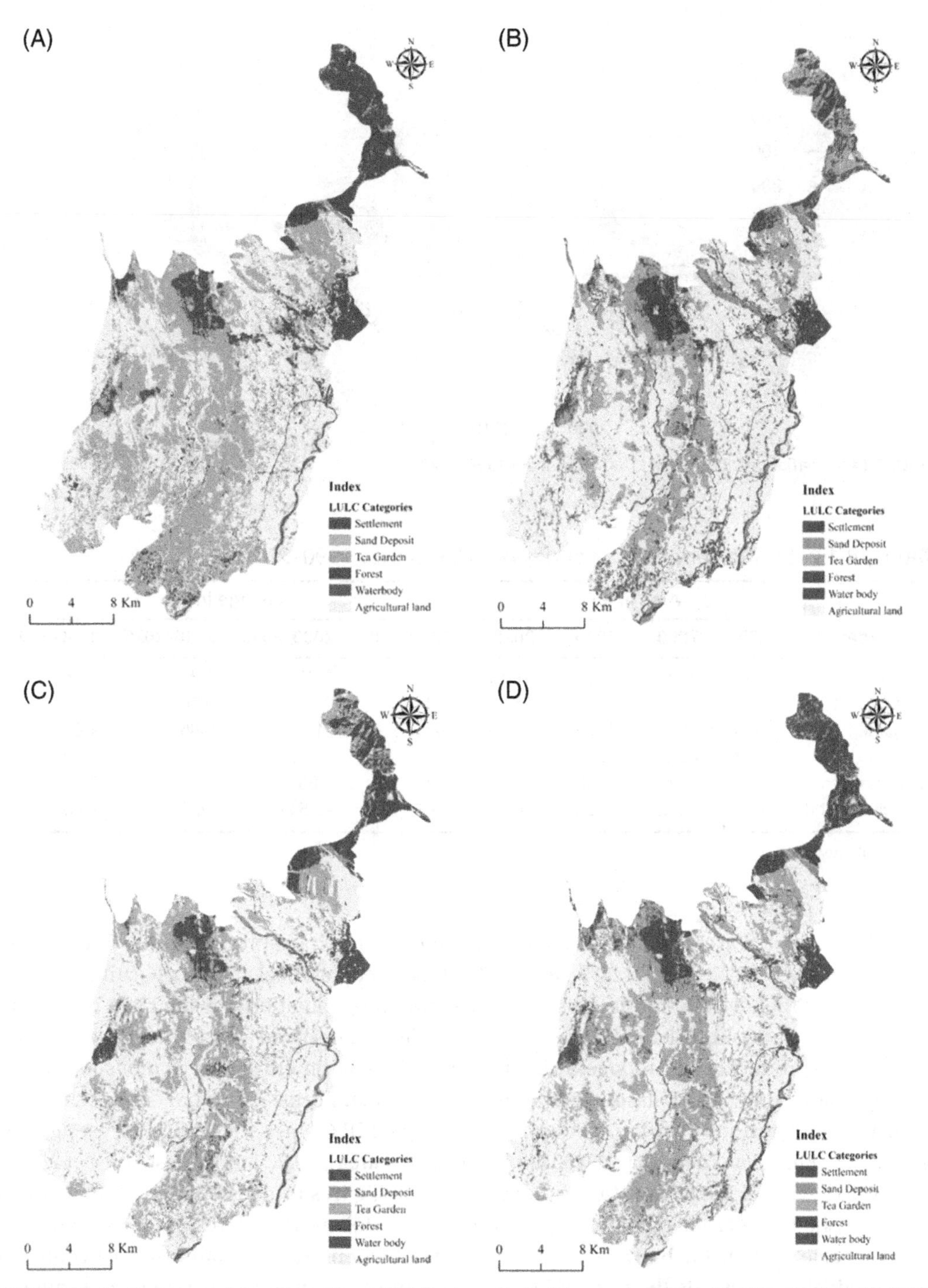

FIGURE 14.5 Spatio-temporal transformation of LULC categories in Siliguri sub-division (A) 2020, (B) 2010, (C) 2000, and (D) 1990.

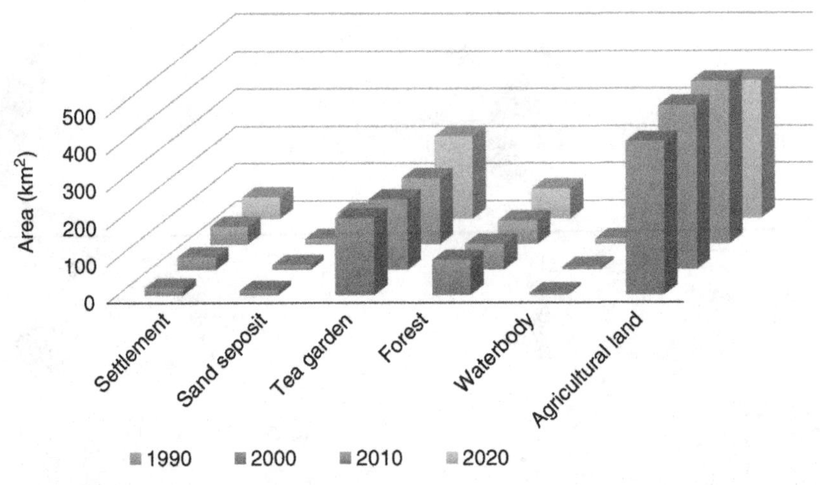

FIGURE 14.6 Spatial distribution of LULC types from 1990 to 2020.

Table 14.13 Spatio-temporal changes of LULC types, 1990–2020.

	Area in km²				Change in %			
LULC types	**1990**	**2000**	**2010**	**2020**	**1990–2000**	**2000–2010**	**2010–2020**	**1990–2020**
Settlement	19.65	34.53	48.14	57.69	1.98	1.81	1.27	5.06
Sand deposit	12.92	13.39	14.74	15.79	0.06	0.18	0.14	0.38
Tea garden	206.97	188.16	176.46	220.52	−2.50	−1.56	5.86	1.80
Forest	95.04	69.26	63.10	79.72	−3.43	−0.82	2.21	−2.04
Water body	5.69	7.26	14.24	9.25	0.21	0.93	−0.66	0.47
Agricultural land	411.67	439.35	435.25	368.98	3.68	−0.54	−8.81	−5.68

Source: Satellite images, 1990–2020.

deposit, tea garden area, and water bodies have changed in a positive direction. Their spatial distribution was 5.06%, 0.38%, 1.80%, and 0.47% of total geographical area in a positive direction under study (Table 14.13). Continuous pressure of human population was the main controlling force in Siliguri and sub-urban periphery region. Similarly, forest and agricultural land losses are the negative product of anthropogenic interference. Huge sediment supply in monsoon season through Mahananda and Balason river system was the main cause of spatial enhancement. In 2020, under Siliguri sub-division, spatial coverage of the LULC types were 57.69 km² (settlement area), 15.79 km² (sand deposit), 220.52 km² (tea garden area), 79.72 km² (forest land), 9.25 km² (water bodies), and 368.98 km² (agricultural field). In Siliguri sub division, the settlement area has flourished in a faster rate and same path also has followed by forest land but there has direct inverse relationship between forest land and settlement. Such a relationship can enhance the sustainability towards the unhealthy environment of our mother earth and consequently groundwater vulnerability also will go towards less vulnerable in near future.

14.4.3 Relationship between LULC change and groundwater vulnerability

Anthropogenic activities on the land surface are extremely hazardous to groundwater (Lerner and Harris, 2009). Groundwater vulnerability depends on several factors viz. depth to water table, net recharge, aquifer media, soil types, topography, impact of vadose zone, and hydraulic conductivity based on DRASTIC model, but there are lots of other influencing factors according to their magnitude. Among these, LULC is considered one of them for deteriorating the groundwater vulnerability. Generally, groundwater is the main source of water to fulfil human needs like drinking, irrigation, and household needs. But continuously changing its (mother earth) originality towards artificial structure which directly or indirectly influences the groundwater table and quality also. Both diffuse and point sources of pollution can affect groundwater, but microbiological ones that can cause sickness and occasionally even death which is the most dangerous to people's health (Lerner and Harris, 2009). To increase agricultural production, farmers use chemical fertilizers without sustainable practices. Fertilizers used to boost short-term soil fertility can worsen the quality of groundwater (Singh et al., 2010). Deforestation is an important factor for groundwater vulnerability. Other agrochemicals, in addition to nitrates, are frequent groundwater pollutants, following their percolability and reaction power with the aquifer matrix and soil (Holman et al., 2008). Hence, the output model shows Patiram Jote, Khalpara, Panchanan colony, and adjacent areas experiencing high vulnerability to groundwater due to its quality (Fig. 14.4). Untreated natural environment is comparatively less vulnerable because less human interferences, that is, less waste dumping practice. Such a situation can be observed in the DRASTIC output map. In the Anthropocene, increased urbanization reduces infiltration, which has an impact on groundwater recharge and storage, and also affects the groundwater quality (Mishra et al., 2014, 2015).

14.5 Ground truth validation of the groundwater vulnerability map

As the accuracy of the model output for such multi-criteria decision making (MCDM) technique is of prime priority. The area under the receiver operating characteristic (ROC), that is, area under curve (AUC) was done taking the inventory data (groundwater quality data, that is, TDS and chloride ions (Cl^-) at several sample sites) is a well-established technique for testing such MCDM models because of its relevance to forecast (Pourghasemi et al., 2012; Tehrany et al., 2013; Das, 2020, 2021).Generally chloride ions (Cl^-) and TDS are often excessive in polluted water but comparatively deficient in groundwater (Ozler, 2002; Mohammadi et al., 2012; Moghaddam et al., 2018) expressed as fresh in character. Therefore, these two parameters have taken into consideration to validate the DRASTIC output. Total 30 sample stations were taken to validate the ground reality (Photo plate 1) and to promote for management strategy of the DRASTIC output. Tasters graded the palatability of drinkable water likewise: excellent, <300 mg/L; good, between 300 and 600 mg/L; fair, between 600 and 900 mg/L; bad,

PHOTO PLATE 1 Field photographs (A) well water using for drinking, (B) measurement of water table with reference to well at Matigara (reference site 1), (C) LULC type—paddy field besides Gaya ganga tea estate during postmonsoon, and (D) LULC type—tea plantation.

between 900 and 1200 mg/L; and worst being designated as unacceptable with higher than 1200 mg/L (Bruvold and Ongerth, 1969; WHO, 1996). Because of its tasteless character, water with exceptionally less TDS concentrations may also be unpleasant. Concentration of chloride with more than 250 mg/L, on the other hand, can cause a perceptible taste in water (WHO, 2003). However, the total 106 samples of well data (TDS and Cl^-) were collected to validate the DRASTIC output map. ROC results showed that in case of Cl^-, DRASTIC output was validated with 91.50 % confidence level (AUC - 0.915) but in case of TDS, DRASTIC output was validated with 89.20 % of confidence level (AUC - 0.892) (Photo plate 2). Therefore, based on Fig. 14.7, the model output can be accepted in view of scientific point and will be used for planning and management purpose.

PHOTO PLATE 2 Field photographs (A) biomedical waste at NBMC (near emergency ward). This source area of untreated biomedical waste, (B) dumping practice (Siliguri dumping ground) leads to greater contamination of groundwater, (C) polluted river bed at Mahananda river near air view more, and (D) cleaning process is going on towards sustainability at Mahananda river near air view more, Siliguri.

14.6 Conclusion

The alteration of LULC with increased population of the Siliguri urban area and sub-urban areas has great impact on groundwater vulnerability. The vulnerability has assessed through the employment of DRASTIC model and the model divided the study area into four zones of vulnerability under GIS environment considering raster layers of various seven parameters. Majority of the study areas (74.89%) falling with medium vulnerability zone which helps to decision makers about present condition and steps to be taken for sustainable development of that area. From the study of LULC changes over decades have shown that the total settlement area has increased from 19.65 km^2 in 1990 to 57.69 km^2 in 2020 which is nearly threefold. The huge increase in settlement leads to increase of population and consequent pollution. In the similar way tea garden areas have increased significantly. The decrease of forest covers areas over a period of 20 years represent potential of increase pollution. Combining all the alterations of LULC

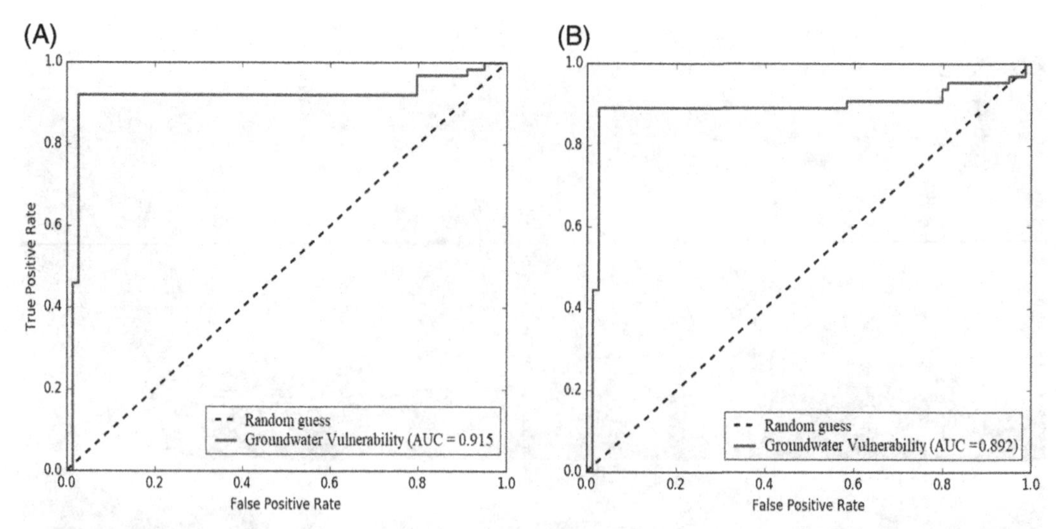

FIGURE 14.7 Validation of ground truth using ROCs (A) through chloride ions (Cl^-) and (B) through total dissolved solid (TDS). N = 106.

and the assessment using DRASTIC model help to understand the medium to high vulnerability has observed in those areas where the maximum alteration of LULC have been taken place. This analysis of groundwater vulnerability will help planners of Siliguri urban and sub-urban areas to select the areas or sites to priorities the controlling measures and waste dumping sites.

Acknowledgement

The authors are in debt to NASA for DEMs, USGS for acquiring satellite Landsat imageries, and GSI on receipt of geomorphology map. Similarly, they are in debt to FAO on receiving soil records. By heart they are thankful to their teachers of the Department of Geography and Applied Geography, University of North Bengal for moral supports to carry out innovative works. The first author is very much in debt to UGC for the fellowship being the JRF for funding research works progressively.

Conflict of the interest

The authors declare that they have no conflict of interest directly or indirectly or any third-party interest towards their work to the best of their belief and knowledge.

References

Alam, A., Bhat, M.S., Maheen, M., 2020. Using Landsat satellite data for assessing the land use and land cover change in Kashmir valley. GeoJournal 85 (6), 1529–1543. https://doi.org/10.1007/s10708-019-10037-x.

Aller, L., Bennett, T., Lehr, J., Petty, R.J., Hackett, G., 1987. DRASTIC: A Standardized System for Evaluating Ground Water Pollution Potential Using Hydrogeologic Settings. US Environmental Protection Agency, Washington, DC, p. 455.

Bhatta, B., 2011. Remote Sensing and GIS, second ed Oxford University Press, Oxford.

Borevsky, B., Yazvin, L., Margat, L., 2004. Importance of groundwater for water supply. In: Zektser, IS, Everett, LG (Eds.). In: Groundwater Resources of the World and their US IHP-VI, Series on Groundwater, 6. UNESCO, Paris, pp. 20–24.

Bruvold, W.H., Ongerth, H.J., 1969. Taste quality of mineralized water. J. Am. Water Works Assoc. 61 (4), 170–174.

Census of India, 2011. 2011: Instruction Manual for Houselisting and Housing Census. Registrar General of India, New Delhi. http://censusindia.gov.in/2011-manuals/Index_hl.html.

Crosbie, R.S., Davies, P., Harrington, N., Lamontagne, S., 2015. Ground truthing groundwater-recharge estimates derived from remotely sensed evapotranspiration: a case in South Australia. Hydrogeol. J. 23 (2), 335–350. doi:10.1007/s10040-014-1200-7.

Das, S., 2020. Flood susceptibility mapping of the Western Ghat coastal belt using multi-source geospatial data and analytical hierarchy process (AHP). Remote Sens. Appl. 20, 100379. doi:10.1016/j.rsase.2020.100379.

Das, M., Parveen, T., Ghosh, D., Alam, J., 2021. Assessing groundwater status and human perception in drought-prone areas: a case of Bankura-I and Bankura-II blocks, West Bengal (India). Environ. Earth Sci. 80 (18), 1–23. doi:10.1007/s12665-021-09909-8.

Dixon, B., 2004. Prediction of ground water vulnerability using an integrated GIS-based neuro-fuzzy techniques. J. Spatial Hydrol. 4 (2), 1–38.

El-Naqa, A., Hammouri, N., Kuisi, M., 2006. GIS-based evaluation of groundwater vulnerability in the Russeifa area, Jordan. Revistamexicana cienciasgeol. 23 (3), 277–287.

Ettazarini, S., 2006. Groundwater pollution risk mapping for the Eocene aquifer of the OumEr-Rabia basin, Morocco. Environ. Geol. 51 (3), 341–347. doi:10.1007/s00254-006-0330-1.

Foody, G.M., 1999. The continuum of classification fuzziness in thematic mapping. Photogramm. Eng. Remote Sensing 65, 443–452.

Foody, G.M., Campbell, N.A., Trodd, N.M., Wood, T.F., 1992. Derivation and applications of probabilistic measures of class membership from the maximum-likelihood classification. Photogramm. Eng. Remote Sensing 58 (9), 1335–1341.

Fritch, T.G., McKnight, C.L., Jr, Y., J., C., Arnold, J.G, 2000. An aquifer vulnerability assessment of the Paluxy aquifer, central Texas, USA, using GIS and a modified DRASTIC approach. Environ. Manag. 25 (3), 337–345. doi:10.1007/s002679910026.

Gundogdu, K.S., Guney, I., 2007. Spatial analyses of groundwater levels using universal kriging. J. Earth Syst. Sci. 116 (1), 49–55.

Hamza, S.M., Ahsan, A., Imteaz, M.A., Rahman, A., Mohammad, T.A., Ghazali, A.H, 2015. Accomplishment and subjectivity of GIS-based DRASTIC groundwater vulnerability assessment method: a review. Environ. Earth Sci. 73 (7), 3063–3076. doi:10.1007/s12665-014-3601-2.

Harbaugh, A.W., Banta, E.R., Hill, M.C., & McDonald, M.G. (2000). Modflow-2000, the U.S. Geological Survey Modular Ground-Water Model-User Guide to Modularization Concepts and the Ground-Water Flow Process. Open-File Report. U. S. Geological Survey, 00-92, 134.

Holman, I.P., Whelan, M.J., Howden, N.J., Bellamy, P.H., Willby, N.J., Rivas-Casado, M., McConvey, P., 2008. Phosphorus in groundwater: an overlooked contributor to eutrophication? Hydrol. Process. Int. J. 22 (26), 5121–5127. doi:10.1002/hyp.7198.

Huan, H., Wang, J., Teng, Y., 2012. Assessment and validation of groundwater vulnerability to nitrate based on a modified DRASTIC model: a case study in Jilin City of northeast China. Sci. Total Environ. 440, 14–23. doi:10.1016/j.scitotenv.2012.08.037.

Javadi, S., Kavehkar, N., Mousavizadeh, M.H., Mohammadi, K., 2011a. Modification of DRASTIC model to map groundwater vulnerability to pollution using nitrate measurements in agricultural areas. J. Agric. Sci. Technol. 13 (2), 239–249.

Javadi, S., Kavehkar, N., Mohammadi, K., Khodadadi, A., Kahawita, R., 2011b. Calibrating DRASTIC using field measurements, sensitivity analysis and statistical methods to assess groundwater vulnerability. Water Int. 36 (6), 719–732. doi:10.1080/02508060.2011.610921.

Job, A.C., 2010. Groundwater Economics. CRC Press, Boca Raton, p. 650.

Kihumba, A.M., Vanclooster, M., Longo, J.N., 2016. Assessing groundwater vulnerability in the Kinshasa region, DR Congo, using a calibrated DRASTIC model. J. Afr. Earth Sci. 126, 13–22. doi:10.1016/j.jafrearsci.2016.11.025.

Kumar, R., Singh, R.D., Sharma, K.D., 2005. Water resources of India. Curr. Sci. 794–811. https://www.jstor.org/stable/24111024.

Kumar, S., Thirumalaivasan, D., Radhakrishnan, N., 2013. GIS based assessment of groundwater vulnerability using drastic model. Arab. J. Sci. Eng. 39 (1), 207–216. doi:10.1007/s13369-013-0843-3.

Kumar, V., 2007. Optimal contour mapping of groundwater levels using universal kriging: a case study. Hydrol. Sci. J. 52 (5), 1038–1050. doi:10.1623/hysj.52.5.1038.

Lerner, D.N., Harris, B., 2009. The relationship between land use and groundwater resources and quality. Land Use Policy 26, S265–S273. doi:10.1016/j.landusepol.2009.09.005.

Lillesand, T.M., Kiefer, R.W., Chipman, J.W., 2015. Remote Sensing and Image Interpretation, seventh ed. Wiley, USA.

Manap, M.A., Nampak, H., Pradhan, B., Lee, S., Sulaiman, W.N.A., Ramli, M.F., 2014. Application of probabilistic-based frequency ratio model in groundwater potential mapping using remote sensing data and GIS. Arabian J. Geosci. 7 (2), 711–724.

Mimi, Z.A., Mahmoud, N., Madi, M.A, 2012. Modified DRASTIC assessment for intrinsic vulnerability mapping of karst aquifers: a case study. Environ. Earth Sci. 66 (2), 447–456. doi:10.1007/s12665-011-1252-0.

Mishra, N., Khare, D., Gupta, K.K., Shukla, R., 2014. Impact of land use change on groundwater: a review. Adv. Water Resour. Prot. 2 (28), 28–41.

Mishra, N., Khare, D., Kumar, S., 2015. Impact of land use change on groundwater recharge in upper Ganga canal command. Int. J. Appl. Eng. Res. 10 (33), 24284–24288.

Moghaddam, M.H.R., Rouhi, M.N., Sarkar, S., Rahimpour, T., 2018. Groundwater vulnerability assessment using the DRASTIC model under GIS platform in the Ajabshir Plain, southeast coast of Urmia Lake, Iran. Arab. J. Geosci. 11 (19), 1–15. doi:10.1007/s12517-018-3928-1.

Mohammadi, Z., Zare, M., Sharifzade, B., 2012. Delineation of groundwater salinization in a coastal aquifer, Bousheher, South of Iran. Environ. Earth Sci. 67 (5), 1473–1484. doi:10.1007/s12665-012-1591-5.

Mondal, I., Bandyopadhyay, J., Chowdhury, P., 2018. A GIS based DRASTIC model for assessing groundwater vulnerability in Jangalmahal area, West Bengal, India. Sustain. Water Resour. Manag. 5 (2), 557–573. doi:10.1007/s40899-018-0224-x.

Neshat, A., Pradhan, B., Pirasteh, S., Shafri, H.Z.M., 2014. Estimating groundwater vulnerability to pollution using a modified DRASTIC model in the Kerman agricultural area, Iran. Environ. Earth Sci. 71 (7), 3119–3131.

Ozler, M.H., 2002. Hydrochemistry and salt-water intrusion in the Van aquifer, East Turkey. Environ. Geol. 43, 759–775.

Pacheco, F.A., Fernandes, L.F.S., 2013. The multivariate statistical structure of DRASTIC model. J. Hydrol. 476, 442–459.

Pedreira, R., Kallioras, A., Pliakas, F., Gkiougkis, I., Schuth, C., 2015. Groundwater vulnerability assessment of a coastal aquifer system at river Nestos eastern Delta, Greece. Environ. Earth Sci. 73, 6387–6415. doi:10.1007/s12665-014-3864-7.

Plymale, C.L., & Angle, M.P., 2002. Groundwater Pollution Potential of Fulton County, Ohio. Ohio Department of Natural Resources Division of Water, Water Resources Section, Groundwater Pollution Potential Report No. 45.

Pourghasemi, H.R., Mohammady, M., Pradhan, B., 2012. Landslide susceptibility mapping using index of entropy and conditional probability models in GIS: Safarood Basin, Iran. Catena 97, 71–84. doi:10.1016/j.catena.2012.05.005.

Rahman, A., 2008. A GIS based DRASTIC model for assessing groundwater vulnerability in shallow aquifer in Aligarh, India. Appl. Geography 28 (1), 32–53. doi:10.1016/j.apgeog.2007.07.008.

Reddy, M.A., 2008. Remote Sensing and Geographical Information System. BSP, Hyderabad, India, pp. 1–453.

Saha, D., Alam, F., 2014. Groundwater vulnerability assessment using DRASTIC and Pesticide DRASTIC models in intense agriculture area of the Gangetic plains, India. Environ. Monit. Assess. 186 (12), 8741–8763.

Saidi, S., Bouri, S., Dhia, H.B, 2010. Groundwater vulnerability and risk mapping of the Hajeb-jelma aquifer (Central Tunisia) using a GIS-based DRASTIC model. Environ. Earth Sci. 59 (7), 1579–1588. doi:10.1007/s12665-009-0143-0.

Saidi, S., Bouri, S., Dhia, H.B., Anselme, B., 2011. Assessment of groundwater risk using intrinsic vulnerability and hazard mapping: application to Souassi aquifer, Tunisian Sahel. Agric. Water Manag. 98 (10), 1671–1682. doi:10.1016/j.agwat.2011.06.005.

Shalaby, A., Tateishi, R., 2007. Remote sensing and GIS for mapping and monitoring land cover and land-use changes in the North-western coastal zone of Egypt. Appl. Geography 27 (1), 28–41.

Shukla, S., Mostaghimi, S., Shanholt, V.O., Collins, M.C., Ross, B.B., 2000. A county-level assessment of ground water contamination by pesticides. Groundwater Monit. Remediat. 20 (1), 104–119. doi:10.1111/j.1745-6592.2000.tb00257.x.

Singh, S., Singh, C., Mukherjee, S., 2010. Impact of land-use and land-cover change on groundwater quality in the Lower Shiwalik hills: a remote sensing and GIS based approach. Open Geosci. 2 (2), 124–131. doi:10.2478/v10085-010-0003-x.

Tehrany, M.S., Pradhan, B., Jebur, M.N., 2013. Spatial prediction of food susceptible areas using rule based decision tree (DT) and a novel ensemble bivariate and multivariate statistical models in GIS. J. Hydrol. 504, 69–79. doi:10.1016/j.jhydrol.2013.09.034.

WHO, 1996. Guidelines for Drinking-Water Quality. Guidelines for Drinking-Water Quality, Vol. 2. Health Criteria and Other Supporting Information. World Health Organization, Geneva WHO/SDE/WSH/03.04/16.

WHO, 2003. Chloride in Drinking-Water. Background Document for Preparation of WHO Guidelines for Drinking-Water Quality. World Health Organization, Geneva (WHO/SDE/WSH/03.04/3).

Yin, L., Zhang, E., Wang, X., Wenninger, J., Dong, J., Guo, L., Huang, J., 2012. A GIS-based DRASTIC model for assessing groundwater vulnerability in the Ordos Plateau, China. Environ. Earth Sci. 69 (1), 171–185.

Groundwater fluctuation and agricultural insecurity: A geospatial analysis of West Bengal in India

Santu Guchhait[a], Gour Dolui[a], Subhrangsu Das[b] and Nirmalya Das[a]

[a]DEPARTMENT OF GEOGRAPHY, PANSKURA BANAMALI COLLEGE (AUTONOMOUS), PANSKURA, WEST BENGAL, INDIA [b]DEPARTMENT OF GEOGRAPHY, UTKAL UNIVERSITY, ODISHA, INDIA

15.1 Introduction

Groundwater is a very important natural resource, predominantly in areas of arid to semi-arid environmental conditions, for example, Iran (Chezgi et al., 2016). Indian plateau track and it fulfils a large proportion of water demands for different purposes such as irrigation, natural growth of vegetation, domestic and drinking uses, etc. (Morris et al., 2003; Sharma et al., 2017; Shah et al., 2019). Anywhere in the world, the groundwater level is not fixed at all; it is fluctuating all over the year in different seasons (Walker and Salt, 2012). The seasonal variability of the available precipitation and resultant infiltration are the major responsible factors for such water level fluctuation (Okkonen and Kløve, 2010; Cai and Ofterdinger, 2016). This seasonal fluctuation of groundwater can be responsible for changing natural growth or agriculture production insecurity in any region (Zaveri et al., 2016; Gezie, 2019), especially in the areas that are mostly dependent on rainfall and underground water for agriculture (Misra, 2014; Kløve et al., 2014; Gao et al., 2022).

Groundwater is one of the major sources of irrigation but rapid depletion of groundwater and seasonal fluctuation is becoming a matter of concern across the world (Siebert et al., 2010; Sarkar et al., 2020; Bhattarai et al., 2021). In India that is very predominant as groundwater provides more than 60% of available water for irrigation (Gandhi and Bhamoriya, 2011; Saha et al., 2018). The increasing demand for groundwater in the Indian agriculture system helps to drop water levels by more than 8 m on average for the last 30 decades (Rodell et al., 2009; Aeschbach-Hertig and Gleeson, 2012; Sekhri, 2013). Both deep and very shallow groundwater levels will affect agricultural production (Mukherji et al., 2013). Reducing groundwater levels can be depriving the water availability to crops; also, this would be reducing the soil moisture condition and thus the cost of production will defiantly increase due to pumping expenses

Case Studies in Geospatial Applications to Groundwater Resources. DOI: https://doi.org/10.1016/B978-0-323-99963-2.00002-X

(Balasubramani et al., 2019). On the other hand, a very shallow water level almost near to the surface can create an over-saturation condition in the agriculture field and leads to a water logging situation (Chowdary et al., 2008). So, in both conditions crop production can be hampered. Therefore, it is always significant to know the groundwater fluctuation of any particular region to understand the variability of crop production, cost of production, the demand for irrigation, etc. (Jain et al., 2021).

This study aims to understand the seasonal fluctuation of the groundwater level of West Bengal and the impacts of these changes in water level on overall crop production in different seasons. It will clarify the agriculture insecurity and uncertainty associated with groundwater depletion and availability in the diverse landscape of West Bengal.

15.2 Study Area

The state of West Bengal is situated in Eastern India. Its total geographical area is 88,752 km^2 which extends between 21°31′ 00″ to 27°33′15″ N and 85°40′20″ to 89°33′00″ E (Fig. 15.1). West Bengal is the second-highest densely populated (1029 persons per km^2) state in India. West Bengal shares its borders with Nepal, Bhutan, and Sikkim in the north, Assam and Bangladesh in the east, Bihar and Jharkhand in the west, and Orissa in the southwest. The state of West Bengal is divided into 19 administrative districts in 2011. But at present, four new districts, for example, Alipurduar, Kalimpong, Jhargram, and Paschim Bardhaman are added from splitting the Jalpaiguri, Darjeeling, Paschim Medinipur, and Bardhaman respectively. From the Himalayas in the north to the Bay of Bengal in the south, the state consists mostly of riverine plains. Hooghly river, the distributaries of the Ganga are an important river of West Bengal. Several tributary rivers like Mayurakshi, Ajoy, Damodar, Dwarakeswar, Shilabati, Rupnarayan, Kangsabati, Teesta, Torsha, Jaldhaka, and Mahananda, etc. also flow through the state. Although this state has a water scarcity problem. The climate in West Bengal varies greatly. The southern part of West Bengal has a tropical wet-dry (Savannah) climate, while the northern part has a humid subtropical climate. West Bengal's average annual rainfall and temperature are 175 cm and 30°C, respectively although it varies widely. The districts Darjeeling, Jalpaiguri, and Cooch Behar experienced heavy rainfall (above 250 cm), while the districts Bankura and Purulia received very minimum annual rainfall (below 125 cm). Rainfall occurs primarily during the monsoon season. The districts Bankura and Purulia suffered from frequent droughts condition due to low rainfall and high temperatures.

There are two types of hydro-geological formations in the state of West Bengal, for example, fissured hard rocks and porous alluvial formations. West Bengal's total annual groundwater recharge and net groundwater availability are 29.33 BCM and 26.56 BCM respectively. West Bengal's total 61.41% area is under cultivation of which 42.83% land is irrigated. The cropping intensity of this state is 177%. It is the largest producer of rice and jute; and the second-largest producer of potatoes in India. Approximately, 21% of West Bengal's GDP is generated by the agricultural sector, which is declining over time. The level of groundwater depleting consistently in the West Bengal due to continuous extraction of groundwater for agricultural irrigation and

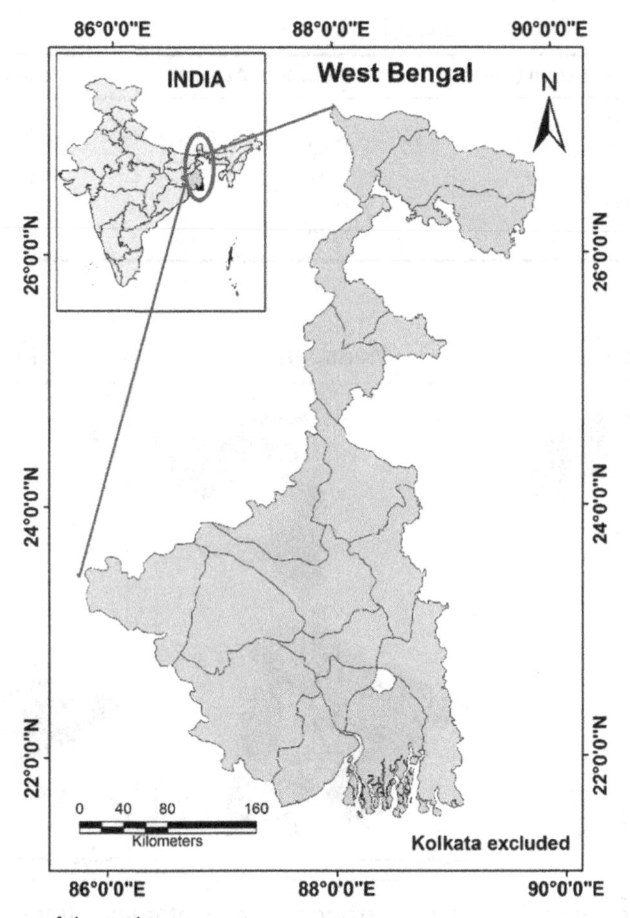

FIGURE 15.1 Location map of the study area.

other different purposes. So, sustainable groundwater planning and management are very essential for the future agricultural production of West Bengal.

15.3 Data and methods

The present study is mainly based on secondary data collected for the period 2015–2020 against 18 districts (excluding Kolkata) of West Bengal. Groundwater levels data are composed of Central Ground Water Board, Eastern region, Kolkata, and crop production data from Crop Production Statistics Information Systems (CPSIS) for the same period. The depth of the groundwater table below ground level (mbgl) is taken for April (premonsoon), August (monsoon), and November (postmonsoon) (Table 15.1). All crops are considered for district production (premonsoon and postmonsoon) excluding coconut, as its seasonal data are not available. For analysis, the five-year mean (2015–2020) depth of the groundwater table (mbgl) is calculated for

Table 15.1 Number of wells analyzed.

Year	Premonsoon (April)	Monsoon (August)	Postmonsoon (November)
2015–16	1338	1478	1583
2016–17	1375	1542	1490
2017–18	1389	1439	1457
2018–19	1392	1420	1379
2019–20	1312	1320	1359

Source: CPSIS.

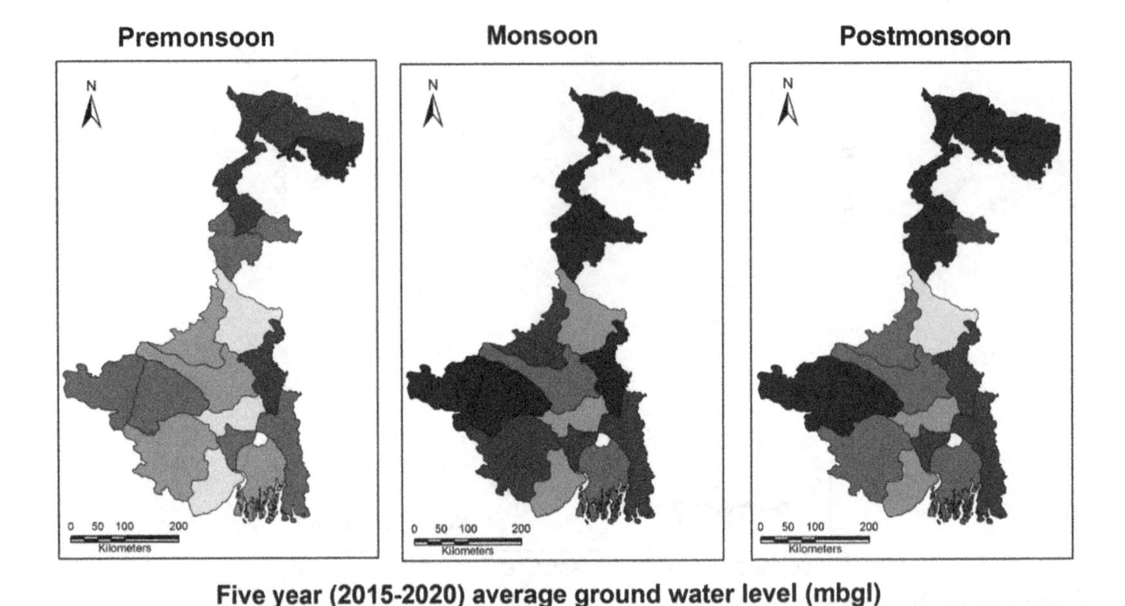

FIGURE 15.2 Spatial variation of mean depth of ground water level for premonsoon, monsoon, and postmonsoon.

each season and seasonal change (pre and postmonsoon) of the groundwater depth for each district has been estimated using simple subtraction methods. The same process is applied for crop production analysis. All choropleth maps are visualized using ArcMAP (v10.1) software; statistical calculations and cartograms are prepared in MS-Excel 2016.

15.4 Results and discussion

15.4.1 Spatial variation of mean depth of ground water table

The mean depth of groundwater level (mbgl) of each district was calculated for the premonsoon, monsoon, and postmonsoon seasons (2015–2020). The map of premonsoon (Fig. 15.2) shows that the mean depth of water level remained below 10 m from the surface for Purba Medinipur, Hugli,

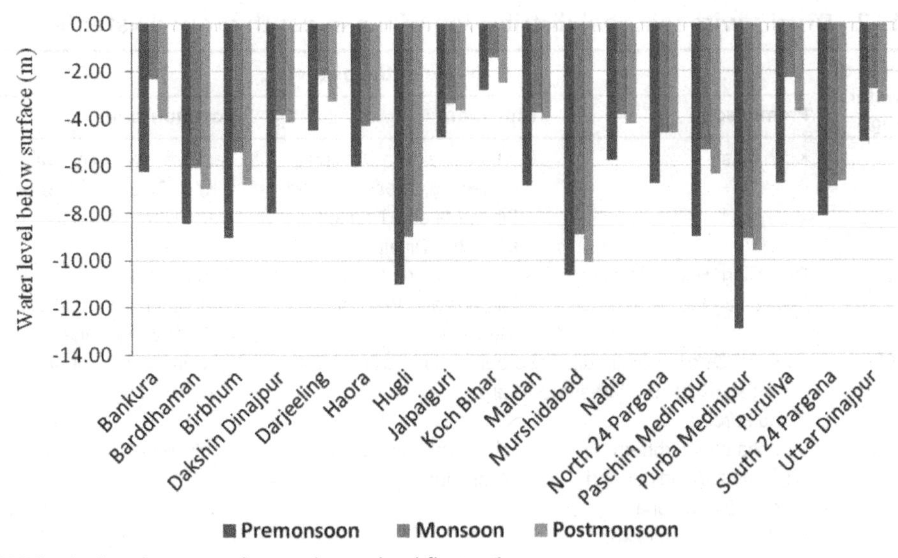

FIGURE 15.3 District-wise seasonal groundwater level fluctuation.

and Murshidabad districts. The majority of the districts have groundwater levels below 6.00–10.00 m depth. Only Koch Bihar has a water level near the surface between 4.00 m depth (Fig. 15.3). During the monsoon period, the number of districts with near-surface water levels (less than 4 m) increased in comparison to the premonsoon period. All the six districts of North Bengal and Bankura, Puruliya, and Nadia districts of South Bengal experienced water levels between 4 m depth. No district has a mean water level beyond 10 m depth during the monsoon period. On the other hand, the water level of Paschim Medinipur and Birbhum district falls during the postmonsoon period to the next range of mean depth from the range of 4.00–6.00 mbgl. Murshidabad experienced a water level below 10 m depth (Table 15.2).

15.4.2 Spatial variation of average crop production

Average crop production of each district was calculated for the premonsoon and postmonsoon period (2015–2020). The map (Fig. 15.4) of average crop production during premonsoon shows that the most of the districts (Bankura, Birbhum, Dakshin Dinajpur, Darjeeling, Haora, Jalpaiguri, Purulia, South 24 Pargana, Barddhaman, Hugli, Koch Bihar, Maldah, North 24 Pargana, Purba Medinipur, and Uttar Dinajpur) of West Bengal experienced relatively low production below 6 ('00000 metric ton) (Fig. 15.5). Only Nadia, Paschim Medinipur, and Murshidabad districts have crop production between 6 and 12 ('00000 metric ton). On the other hand, the image of crop production become reverses during postmonsoon season. Most of the districts (Bankura, Barddhaman, Birbhum, Hugli, Jalpaiguri, Koch Bihar, Murshidabad, Paschim Medinipur, and Uttar Dinajpur) experienced crop production more than 12 ('00000 metric ton). But, Darjeeling have low production (below 300,000 metric ton) for both the period of pre- and postmonsoon (Table 15.3).

Table 15.2 District wise seasonal distribution of mean depth groundwater level.

Mean depth of groundwater table (mbgl)	Name of the districts		
	Premonsoon	Monsoon	Postmonsoon
<4.00	Koch Bihar	Bankura, Dakshin Dinajpur, Darjeeling, Jalpaiguri, Koch Bihar, Maldha, Nadia, Puruliya and Uttar Dinajpur	Bankura, Darjeeling, Jalpaiguri, Koch Bihar, Maldha, Puruliya and Uttar Dinajpur
4.01–6.00	Darjeeling, Haora, Jalpaiguri, Nadia and Uttar Dinajpur	Birbhum, Haora, North 24 Pargana and Paschim Medinipur	Barddhaman, Birbhum, Dakshin Dinajpur, Haora, Nadia and North 24 Pargana
6.01–8.00	Bankura, Dakshin Dinajpur, Maldah, North 24 Pargana and Puruliya	Barddhaman and South 24 Pargana	Paschim Medinipur and South 24 Pargana
8.01–10.00	Barddhaman, Birbhum, Paschim Medinipur and South 24 Pargana	Hugli, Murshidabad and Purba Medinipur	Hugli and Purba Medinipur
>10.01	Hugli, Murshidabad and Purba Medinipur		Murshidabad

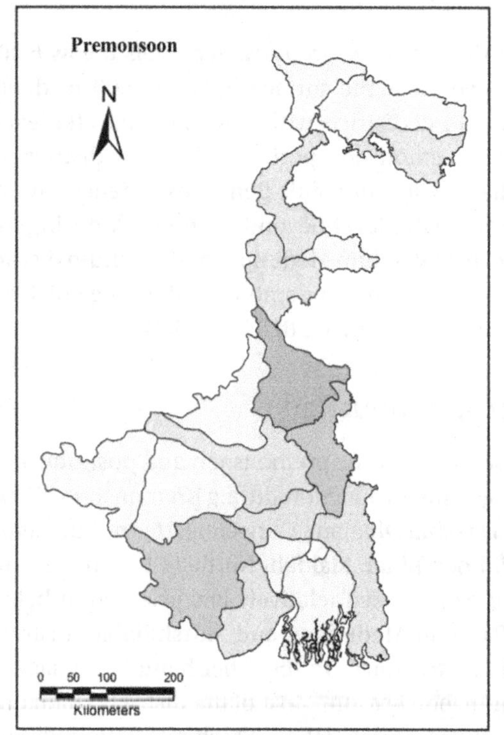

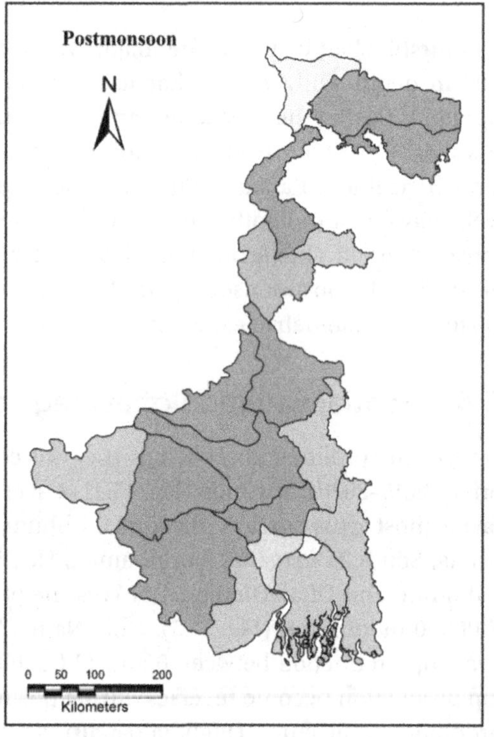

FIGURE 15.4 Spatial variation of average crop production for premonsoon and postmonsoon.

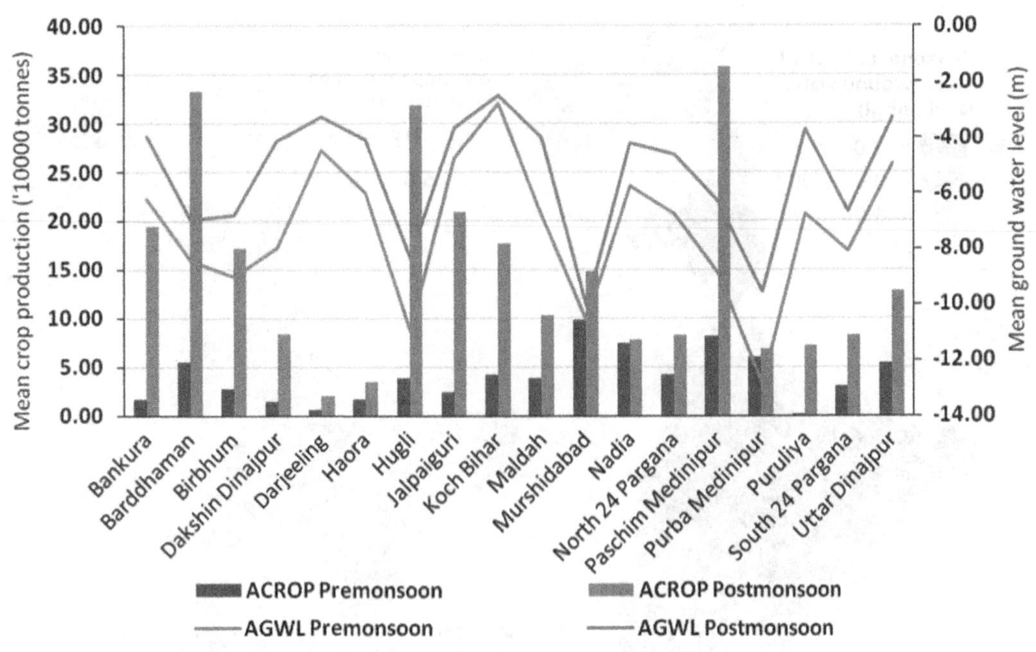

FIGURE 15.5 District-wise mean ground water level and mean crop production.

Table 15.3 District wise seasonal variation of average crop production.

Average crop production ('100,000 tones)	Name of the districts	
	Premonsoon	**Postmonsoon**
<3.00	Bankura, Birbhum, Dakshin Dinajpur, Darjeeling, Haora, Jalpaiguri, Purulia and South 24 Pargana	Darjeeling
3.01–6.00	Barddhaman, Hugli, Koch Bihar, Maldah, North 24 Pargana, Purba Medinipur and Uttar Dinajpur	Haora
6.01–9.00	Nadia and Paschim Medinipur	Dakshin Dinajpur, Nadia, North 24 Pargana, Purba Medinipur, Purulia and South 24 Pargana
9.01–12.00	Murshidabad	Maldah
>12.01		Bankura, Barddhaman, Birbhum, Hugli, Jalpaiguri, Koch Bihar, Murshidabad, Paschim Medinipur and Uttar Dinajpur

15.4.3 Seasonal fluctuation of mean ground water depth and crop production

Seasonal fluctuation of groundwater is natural but the degree of fluctuation is a matter of concern. Due to high rainfall and infiltration, groundwater level rise upward during the postmonsoon period and drops at the time of postmonsoon. Besides groundwater, crop production also

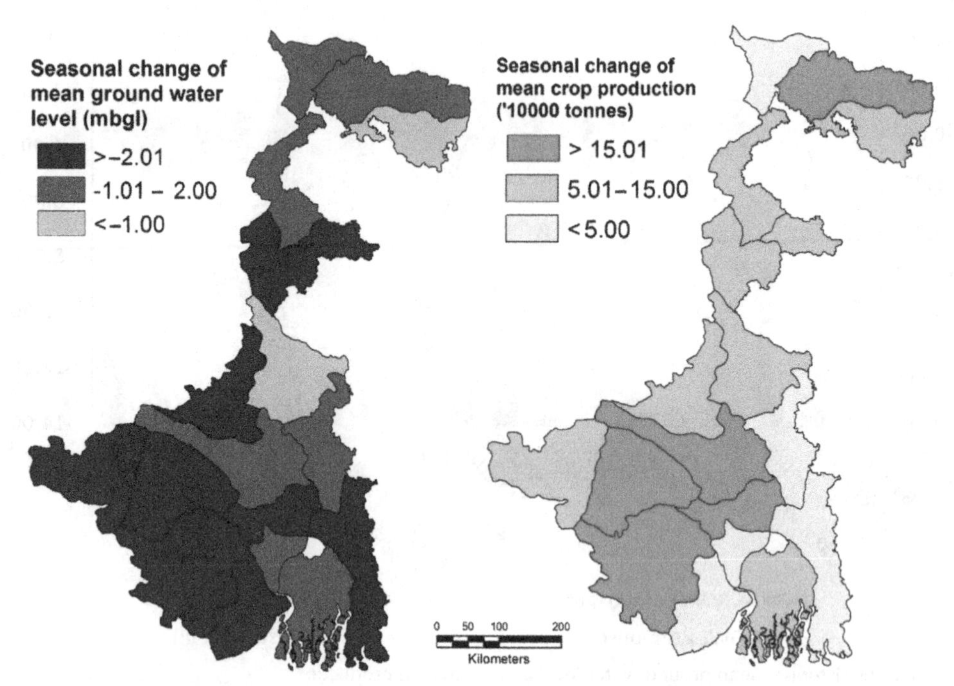

FIGURE 15.6 Seasonal changes of mean groundwater level and mean crop production.

varies with seasonal change based on the water availability for irrigation. The spatial variation of groundwater availability is also responsible for the spatial variation of crop production.

Fig. 15.6 depicts that the district of Koch Bihar and Murshidabad have a seasonal change in mean groundwater level below 1 m. Most of the districts of South Bengal including Bankura, Puruliya, Purba Medinipur, Hugli, Birbhum, and North 24 Parganas experienced more than 2 m of mean groundwater fluctuation. The rest of the districts have 1–2 m of fluctuation of groundwater. Two districts of North Bengal, Dakshin Dinajpur and Maldah have the highest fluctuation of five-year mean groundwater level (Fig. 15.6). Murshidabad and Koch Bihar have the minimum (less than 0.5 m) fluctuation.

Seasonal change in crop production (5-year mean) was also found in different districts of West Bengal. Five districts namely Jalpaiguri, Bardhhaman, Hugli, Bankura, and Paschim Medinipur have more than 1,500,000 metric ton of crop production during the postmonsoon period compared to premonsoon production. Darjeeling, Haora, Nadia, Purba Medinipur, and North 24 Parganas have a minimum change of crop production of fewer than 500,000 metric ton. The rest of the districts have a seasonal change of 5–1,500,000 metric ton of mean crop production (Fig. 15.7).

The boxplot (Fig. 15.8) represents seasonal as well as five years (2015–2020) variation of groundwater levels. Each bar represents a seasonal period (premonsoon, monsoon, postmonsoon) and the height of the bar represents the range of groundwater level from 2015 to 2020. During premonsoon, the level of groundwater falls and rises in monsoon months. All the districts

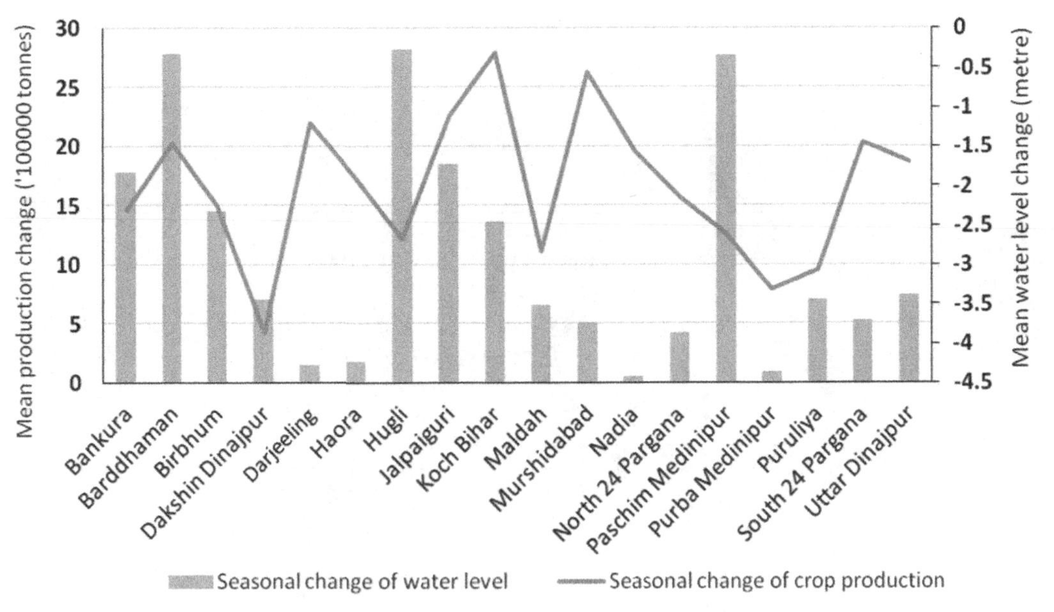

FIGURE 15.7 Seasonal change of water level and crop production.

show the same pattern except Haora and Hugli. Both the districts have the highest groundwater level in postmonsoon and monsoon months. From the boxplot (Fig. 15.9), it is found that the fluctuation of the yearly mean groundwater level of Jalpaiguri was high during monsoon months. The maximum fluctuation was found in Paschim and Purba Medinipur during the postmonsoon period. Fig. 15.10 indicates the positive relationship between total crop production and groundwater levels of all the districts in both premonsoon and postmonsoon seasons. That means, the availability of groundwater significantly influences the crop production in the study area either premonsoon or postmonsoon seasons and thus, districts with higher groundwater fluctuation have a higher difference in crop production. So, climate change and consequent rainfall instability may hamper the groundwater recharge and the depth of the water table. As a result, crop production and agriculture insecurity can be seen even with the presence of available pumping stations. So, we need to conserve and sustainable the use of groundwater resources.

15.5 Conclusion

Through this analysis, we attempted to identify the relationship between groundwater fluctuation and crop production. In this study, district-level data on agriculture production and groundwater levels of West Bengal from 2015 to 2020 have been used for analysis. To understand seasonal changes and relationships, both the variable database has been prepared based on five-year mean values. By analyzing seasonal fluctuation of groundwater availability and variation in crop production, this study reveals the positive correlation between these two aspects. Either premonsoon or postmonsoon, available groundwater impacts crop production positively in

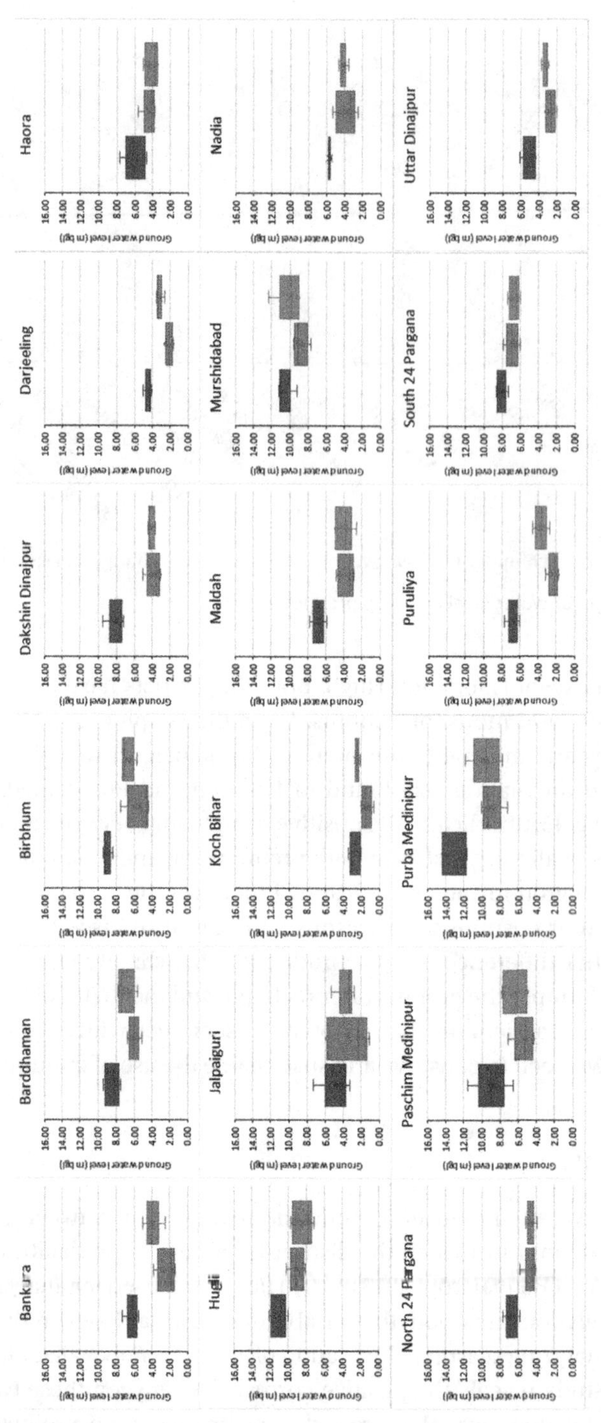

FIGURE 15.8 Boxplot showing district wise seasonal groundwater fluctuation.

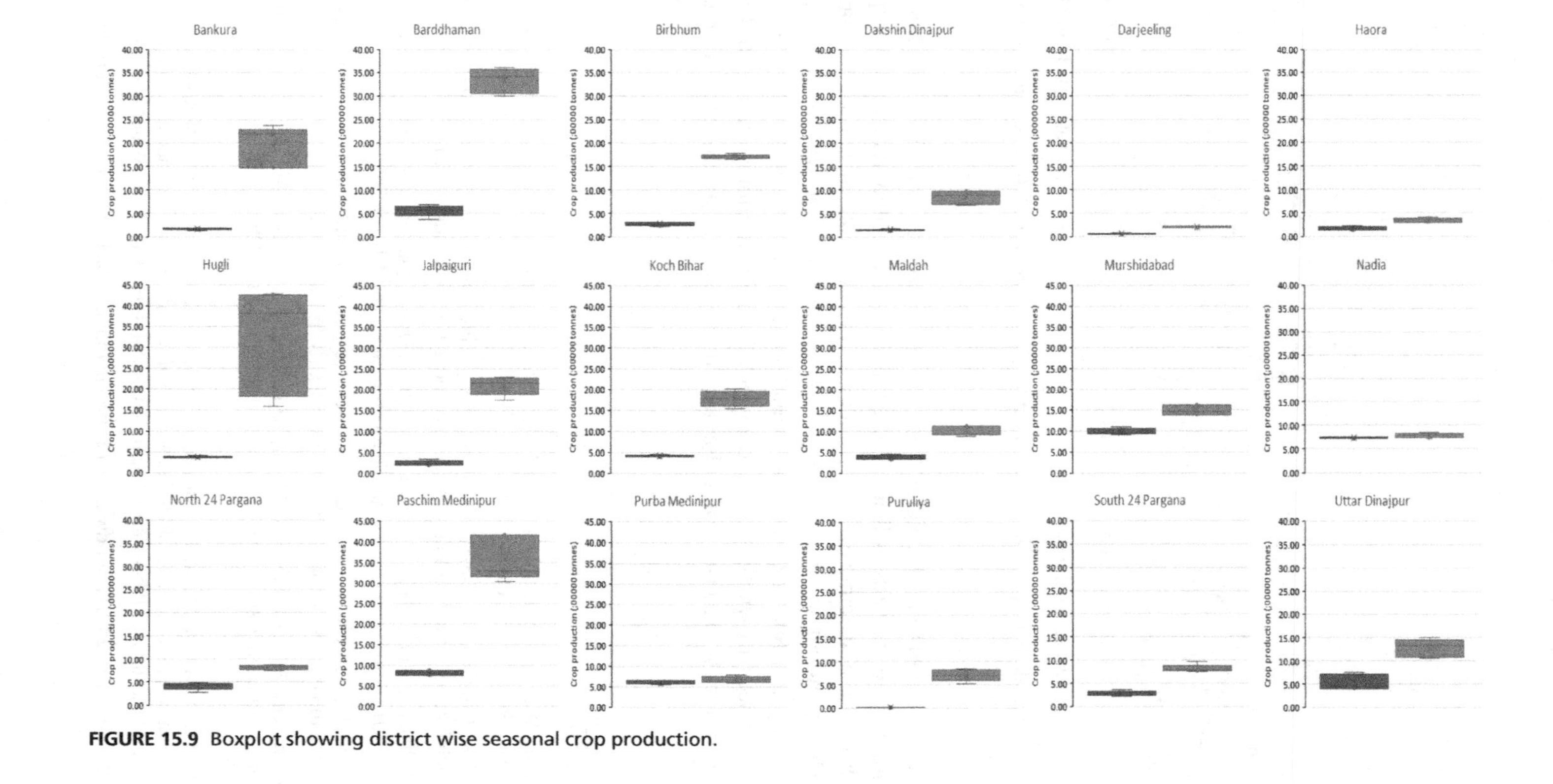

FIGURE 15.9 Boxplot showing district wise seasonal crop production.

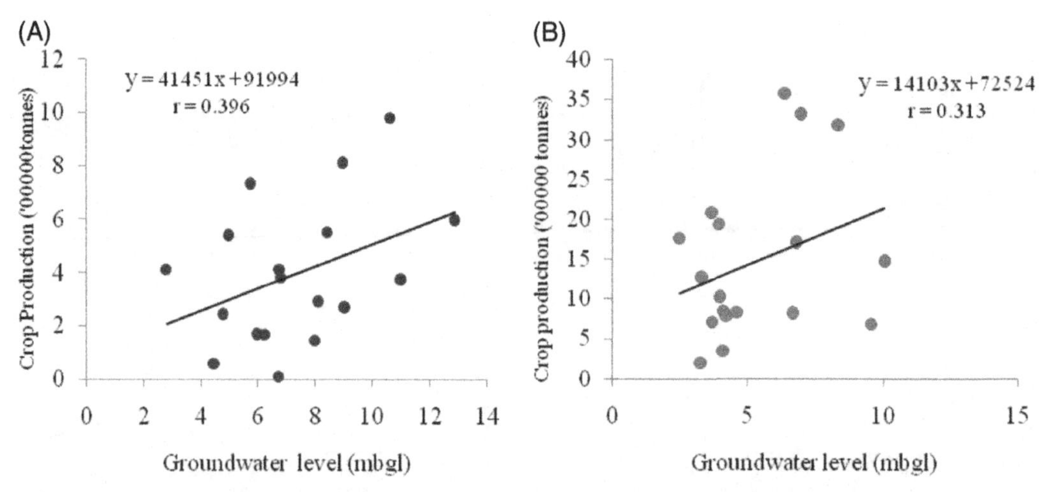

FIGURE 15.10 Showing relationship between groundwater and crop production. (A) premonsoon and (B) postmonsoon.

most of the districts and thus, the higher fluctuation in water levels indicates higher differences in crop production. So, it clearly indicates if the groundwater depletion increases crop insecurity defiantly will increase, whether irrigation facilities are available or not. As the agriculture in West Bengal is dependent on monsoon only for one crop, the other two crops are dependent on groundwater irrigation. So, if groundwater availability fluctuates more severely crop production and agriculture insecurity will be increased, especially in the premonsoon season. Therefore, we need to build awareness to conserve and the sustainable use of groundwater resources for future humanity.

References

Aeschbach-Hertig, W., Gleeson, T., 2012. Regional strategies for the accelerating global problem of groundwater depletion. Nat. Geosci. 5 (12), 853–861.

Balasubramani, K., Gomathi, M., Kumaraswamy, K., 2019. Evaluation of groundwater resources in Aiyar Basin: A GIS approach for agricultural planning and development. Geosfera Indonesia 4 (3), 302–310.

Bhattarai, N., Pollack, A., Lobell, D.B., Fishman, R., Singh, B., Dar, A., Jain, M., 2021. The impact of groundwater depletion on agricultural production in India. Environ. Res. Lett. 16 (8), 085003.

Cai, Z., Ofterdinger, U., 2016. Analysis of groundwater-level response to rainfall and estimation of annual recharge in fractured hard rock aquifers, NW Ireland. J. Hydrol. 535, 71–84.

Chezgi, J., Pourghasemi, H.R., Naghibi, S.A., Moradi, H.R., Kheirkhah Zarkesh, M., 2016. Assessment of a spatial multi-criteria evaluation to site selection underground dams in the Alborz Province, Iran. Geocarto Int. 31 (6), 628–646.

Chowdary, V.M., Chandran, R.V., Neeti, N., Bothale, R.V., Srivastava, Y.K., Ingle, P., Ramakrishnan, D., Dutta, D., Jeyaram, A., Sharma, J.R., Singh, R., 2008. Assessment of surface and sub-surface waterlogged areas in irrigation command areas of Bihar state using remote sensing and GIS. Agric. Water Manage. 95 (7), 754–766.

Gandhi, V.P., Bhamoriya, V., 2011. Groundwater Irrigation in India. India Infrastructure Report, 90.

Gao, F., Sun, S., Yao, N., Yang, H., Cheng, B., Luan, X., Wang, K., 2022. Identifying the impact of crop distribution on groundwater resources carrying capacity in groundwater-depended agricultural regions. Agric. Water Manage. 264, 107504.

Gezie, M., 2019. Farmer's response to climate change and variability in Ethiopia: a review. Cogent Food Agric. 5 (1), 1613770.

Jain, P., Raina, G., Sinha, S., Malik, P., Mathur, S., 2021. Agrovoltaics: step towards sustainable energy-food combination. Bioresour. Technol. Rep. 15, 100766.

Kløve, B., Ala-Aho, P., Bertrand, G., Gurdak, J.J., Kupfersberger, H., Kværner, J., Muotka, T., Mykrä, H., Preda, E., Rossi, P., Uvo, C.B., 2014. Climate change impacts on groundwater and dependent ecosystems. J. Hydrol. 518, 250–266.

Misra, A.K., 2014. Climate change and challenges of water and food security. Int. J. Sustain. Built Environ. 3 (1), 153–165.

Morris, B.L., Lawrence, A.R., Chilton, P.J.C., Adams, B., Calow, R.C., & Klinck, B.A. (2003). Groundwater and its susceptibility to degradation: a global assessment of the problem and options for management. UNEP; British Geological Survey; United Kingdom. Department for International Development; Belgium.

Mukherji, A., Rawat, S., Shah, T., 2013. Major insights from India's minor irrigation censuses: 1986–87 to 2006–07. Econ. Pol. Weekly 7, 115–124.

Okkonen, J., Kløve, B., 2010. A conceptual and statistical approach for the analysis of climate impact on ground water table fluctuation patterns in cold conditions. J. Hydrol. 388 (1-2), 1–12.

Rodell, M., Velicogna, I., Famiglietti, J.S., 2009. Satellite-based estimates of groundwater depletion in India. Nature 460 (7258), 999–1002.

Saha, D., Marwaha, S., Mukherjee, A., 2018. Groundwater resources and sustainable management issues in India. In: Clean and Sustainable Groundwater in India. Springer, Singapore, pp. 1–11.

Sarkar, T., Kannaujiya, S., Taloor, A.K., Ray, P.K.C., Chauhan, P., 2020. Integrated study of GRACE data derived interannual groundwater storage variability over water stressed Indian regions. Groundw. Sustain. Dev. 10, 100376.

Sekhri, S., 2013. Missing Water: Agricultural Stress and Adaptation Strategies in Response to Groundwater Depletion in India. Department of Economics, University of Virginia, Charlottesville, VA.

Shah, B., Kansara, B., Shankar, J., Soni, M., Bhimjiyani, P., Bhanushali, T., Sircar, A., 2019. Reckoning of water quality for irrigation and drinking purposes in the Konkan geothermal provinces, Maharashtra, India. Groundw. Sustain. Dev. 9, 100247.

Sharma, R.K., Yadav, M., Gupta, R., 2017. Water quality and sustainability in India: challenges and opportunities. In: Chemistry and Water. Elsevier, Amsterdam, pp. 183–205.

Siebert, S., Burke, J., Faures, J.M., Frenken, K., Hoogeveen, J., Döll, P., Portmann, F.T., 2010. Groundwater use for irrigation: a global inventory. Hydrol. Earth Syst. Sci. 14 (10), 1863–1880.

Walker, B., Salt, D., 2012. Resilience Thinking: Sustaining Ecosystems and People in a Changing World. Island Press, Washington.

Zaveri, E., Grogan, D.S., Fisher-Vanden, K., Frolking, S., Lammers, R.B., Wrenn, D.H., Prusevich, A., Nicholas, R.E., 2016. Invisible water, visible impact: groundwater use and Indian agriculture under climate change. Environ. Res. Lett. 11 (8), 084005.

16

Assessment of groundwater quality for irrigation purposes: A case study of Hooghly District, West Bengal, India

Sadik Mahammad[a], Md. Mofizul Hoque[a], Aznarul Islam[a] and Arijit Majumder[b]

[a]DEPARTMENT OF GEOGRAPHY, ALIAH UNIVERSITY, KOLKATA, INDIA [b]DEPARTMENT OF GEOGRAPHY, JADAVPUR UNIVERSITY, KOLKATA, INDIA

16.1 Introduction

At present, the dependency on natural resources is continuously increasing because of the rapid industrialization and the population growth that treats the conservation and sustainable uses of the natural resources to humanity (Kaur et al., 2017; Abbasnia et al., 2019). Groundwater meets the requirements of water for drinking, domestic, irrigation, and industrial purposes, especially in arid and semi-arid regions (Alwan et al., 2019). The overexploitation and pollution with various chemical and biological sources is shrinking the uses of surface water worldwide that accelerate the dependency for human consumption on groundwater (Singh et al., 2006; Bhat et al., 2016). The assessment of groundwater quality is important as well as its quantity. In India, groundwater is a major source of irrigation therefore, a special focus is needed in terms of its sustainable uses (Bhat et al., 2016). The amount of rainfall, characteristics of the geological structure, and aquifer mineralogy play a significant role to measure the groundwater quality of the region (Abbasnia et al., 2019). Apart from that, anthropological activities such as agricultural practices, industrial, municipal, as well as residential activities can alter the quality of groundwater (Subramani et al., 2005). As a result, groundwater quality can change the physical and chemical characteristics of the soil, which affects the crop yield (Simsek and Gunduz, 2007; Ghazaryan et al., 2020). Therefore, the assessment of groundwater quality for irrigation purposes demonstrates the sources of groundwater pollutants and allows to use the of suitable water in agricultural activities, which may help the planners for sustainable management of water resources in a region.

Several studies concentrated to evaluate the quality of groundwater for drinking, domestic, and irrigation purposes (Ibrahim, 2019; Honarbakhsh et al., 2019; Abdullah et al., 2019; Ramesh and Elango, 2012; Kumari et al., 2019; Rufino et al., 2019; Kurdi and Eslamkish, 2017; Luo et al., 2021). Hydrogeochemical information is a significant requirement for the evaluation of the

Case Studies in Geospatial Applications to Groundwater Resources. DOI: https://doi.org/10.1016/B978-0-323-99963-2.00003-1

groundwater quality of an area for drinking and irrigation uses (Narasaiah and Rao, 2021). Consequently, many of the researchers offered irrigation water quality indices such as sodium absorption ratio (SAR), sodium percentage (%Na), residual sodium carbonate (RSC), magnesium hazard (MH), permeability index (PI), potential salinity (PS), residual sodium bicarbonate (RSBC), and Kelley's ratio (KR) to assess the suitability of groundwater for irrigation purposes (Kumar et al., 2009; Li et al., 2013; Adimalla et al., 2018; Adimalla et al., 2020; Aravinthasamy et al., 2020; Balasubramani et al., 2020; Ghazaryan et al., 2020; Mthembu et al., 2021; Naidu et al., 2021; Zemunac et al., 2021). Moreover, to assess the irrigation water quality based on these individual parameters many difficulties and inaccuracies occur. Hence, comprehensive parameters-based approaches have been developed to overcome the difficulties and it may help to minimize a large number of analyses into a single unit (Mahammad et al., 2022). Therefore, the irrigation water quality index (IWQI) has been used to quantify the irrigation water quality (Yıldız and Karakuş, 2020).

The Hooghly district is located in the lower Gangetic plain and it is well known for its agricultural activities. The district is facing serious constraints due to the paucity of groundwater and its quality since past few decades. According to Central Groundwater Board (CGWB, 2006, 2017), the district comprises 12 semi-critical community development (C.D.) blocks (Arambagh, Chanditala-I, Chanditala-II, Dhaniakhali, Goghat-I, Jangipara, Khanakul-I, Pandua, Polba-Dadpur, Purshura, Singur, Tarakeswar) in 2013 whereas, in 2004 it was only two (Goghat-I and Pandua). Anxiously, in 2013 only one critical C.D. block (Goghat-II) of West Bengal was located in Hooghly district, which placed the district in the water stress category (CGWB, 2017). Apart from that various C.D. blocks (Balagarh, Chanditala-II, Dhaniakhali, Goghat-I, Haripal, Khanakul-I, Khanakul-II, Pandua, Polba-Dadpur, Serampur-Uttarpara, and Singur) of the study area were affected by arsenic contaminations (CGWB, 2017). Several studies concerning groundwater quality such as arsenic contamination have been executed including the present study area (Acharyya and Shah, 2007; Pal and Mukherjee, 2009, 2010). In addition, the stages of groundwater development and its consequences on agricultural practices were analyzed by Das et al. (2021) and Majumder and Sivaramakrishnan (2014). Besides, Mahammad and Islam (2021a) assessed the suitability of groundwater for drinking purposes based on the concentration of physical-chemical parameters. However, the comprehensive study on the suitability of groundwater for irrigation in the present study area has not been attempted so far. Therefore, the main objectives of the present study are (1) to evaluate the hydrochemical parameters of groundwater and (2) to assess the groundwater quality for irrigation purposes using the IWQI method.

16.2 Study area

The study area (Hooghly district) is situated in the lower Gangetic plain of West Bengal covering an area of ~3150 km^2. It extends from 22°35′36″N to 23°13′30″N latitude and from 87°30′16″E to 88°30′15″E longitude (Fig. 16.1). The study area is situated in the stable shelf zone in the western flank of Bengal basin categorized by homoclinal sedimentary sequence dipping towards ESE (Sengupta, 1972). The quaternary alluvium comprises the study area in two forms, older alluvium

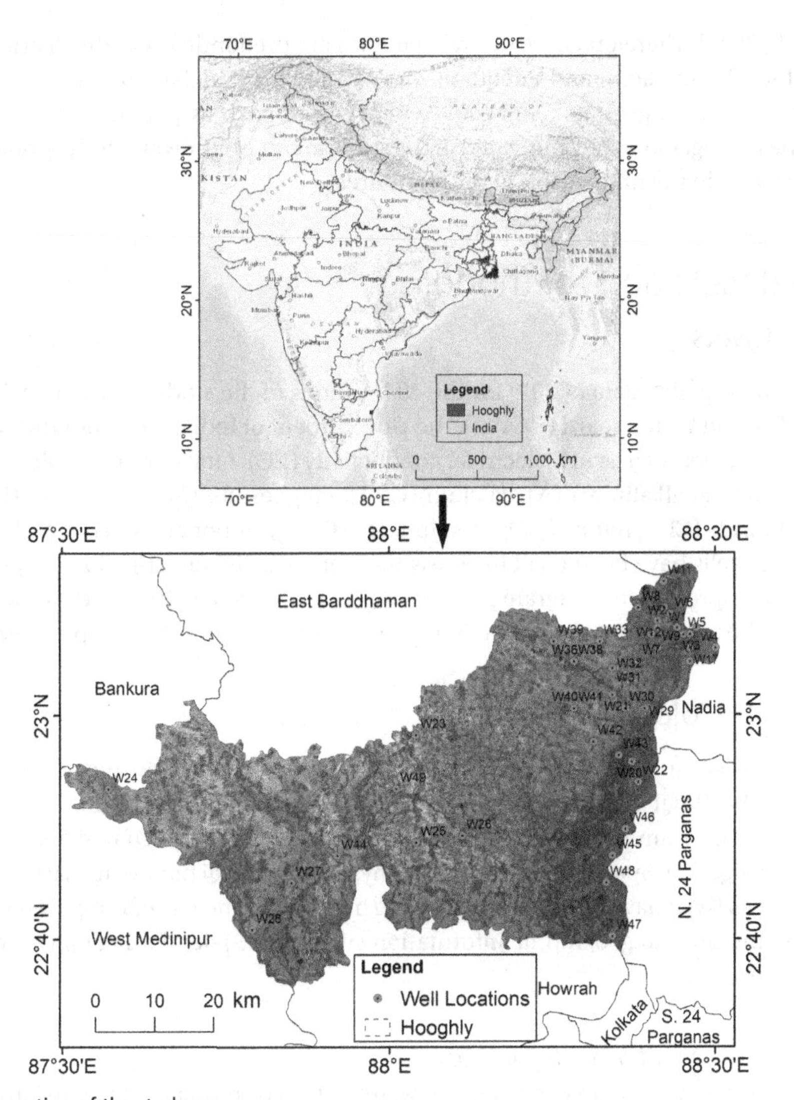

FIGURE 16.1 Location of the study area.

and newer alluvium (Patra et al., 2018). The older alluvium is highly oxidized and calcareous and is confined in the northwest part of the district in the Goghat-I C.D. block (Das and Pal, 2020). The rest of the district is covered by newer alluvium, which consists of sand with occasional gravel, kankar, and silt (Mahammad and Islam, 2021b). The Hooghly River, the Damodar River, the Dwarakeswar River, and the Mundeswari River are the main rivers draining the study area. The study area consists of two geomorphological divisions such as plains and uplands which are separated by the Dwarakeswar River (Patra et al., 2016). The study area experiences a hot moist sub-humid climate and received a mean annual rainfall of 1350 mm (Patra et al., 2016). The ustic soil moisture and hyperthermic soil temperature are occupied in the study area

(Singha et al., 2020). Therefore, agricultural activities are predominant in the district due to the presence of fertile soil, land, and adequate water resources (Singha et al., 2020). Kharif crops including rice, sesame, jute, etc., and *rabi* (winter) crops such as potato, mustard, pulse, and different types of vegetables are cultivated in the study area. Groundwater is the principal source of irrigation water in the *rabi* season in the study area.

16.3 Datasets and methodology

16.3.1 Datasets

The groundwater quality data of 2019 for the 49 locations of the study area were collected from the Central Ground Water Board (CGWB), the Government of India. The groundwater physical-chemical parameters such as pH, electrical conductivity (EC), total dissolved solids (TDS), total hardness (TH), total alkalinity (TA), calcium (Ca^{2+}), magnesium (Mg^{2+}), sodium (Na^{+}), potassium (K^{+}), chloride (Cl^{-}), nitrate (NO_3^{-}), sulphate (SO_4^{2-}), carbonate (CO_3^{2-}), and bicarbonate (HCO_3^{-}) of 49 wells have been used to assess the water quality for irrigation purposes. All the parameters are expressed in milligram per liter (mg/L) unit except EC (μs/cm). The dug well, and tube wells are the two types of sample wells from where the groundwater samples were collected.

16.3.2 Methodology

In order to assess the suitability of groundwater quality for irrigation purposes, the present chapter used the IWQI based on physical-chemical parameters and SAR. Apart from that, physical-chemical parameters were also employed to evaluate the IWQI to detect groundwater suitability for irrigation. Spatial distributions of physical-chemical parameters, irrigation quality indices as well as IWQI have been portrayed using inverse distance weighting (IDW) and kriging interpolation methods in geographic information system (GIS) depending on outliers limits of the datasets (Fig. 16.2).

16.3.2.1 Irrigation water quality indices

Groundwater is the major source of irrigation water in the study area that necessitates irrigation water quality assessment (Ghosh et al., 2017). In the assessment of groundwater suitability for irrigation, total salinity, and sodium play a crucial role (Adimalla et al., 2018; Sarkar and Islam, 2019). Hence, the following irrigation indices are analyzed to determine the suitability of groundwater for irrigation purposes.

16.3.2.1.1 Sodium absorption ratio

Sodium absorption ratio is used to determine sodium concentration with respect to calcium and magnesium. It is calculated using Eq. (16.1) following Todd and Mays (2004).

$$SAR = \frac{Na^{+}}{\sqrt{(Ca^{2+} + Mg^{2+})/2}} \tag{16.1}$$

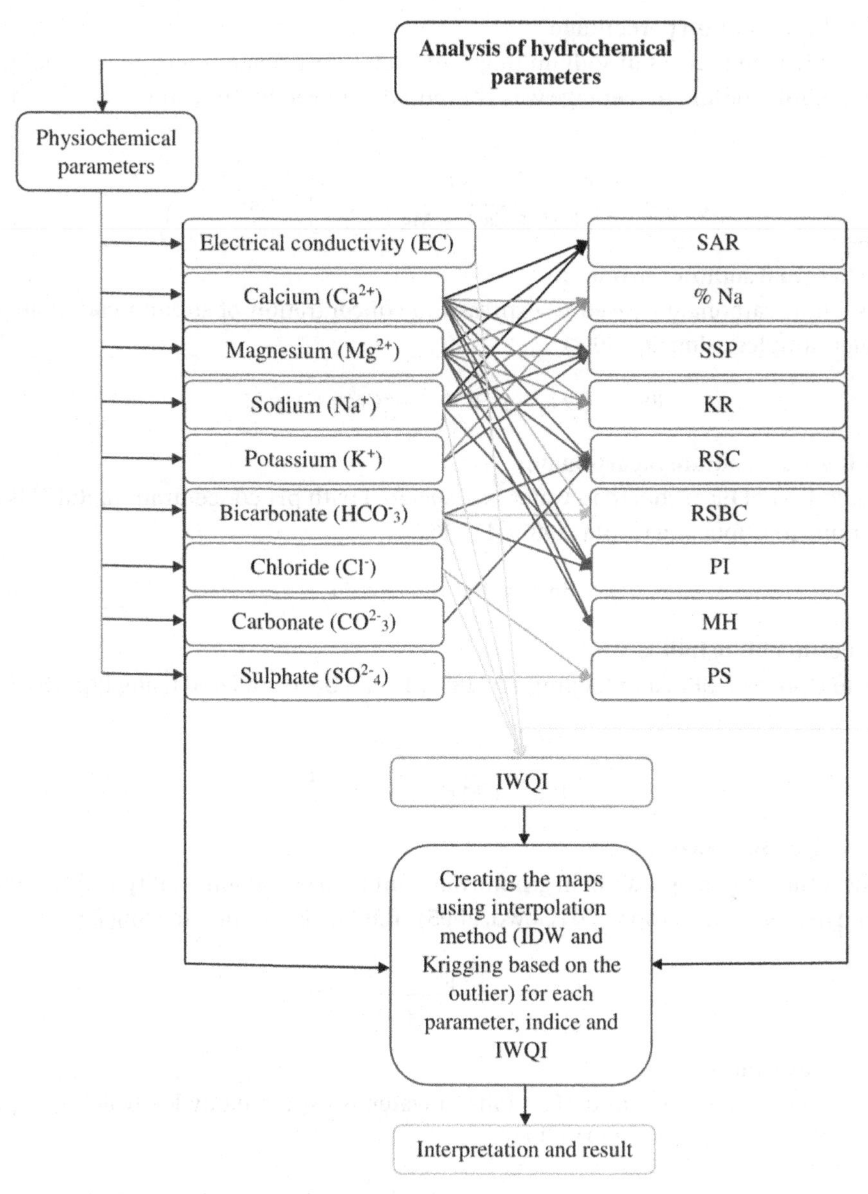

FIGURE 16.2 Methodological flowchart of the study.

16.3.2.1.2 Sodium percentage

Assessment of %Na is calculated according to Richards (1954) and Wilcox (1955) as mentioned in Eq. (16.2). A high concentration of %Na can reduce the permeability of the soil and affect plant growth (Zhou et al., 2021).

$$\% \text{Na} = \frac{(\text{Na}^+ + \text{K}^+)}{(\text{Ca}^{2+} + \text{Mg}^{2+} + \text{Na}^+ + \text{K}^+)} \times 100 \tag{16.2}$$

16.3.2.1.3 Soluble sodium percentage

A comparatively high percent of sodium in groundwater for irrigation may reduce soil permeability. The soluble sodium percentage was computed followed by Todd and Mays (2004) using Eq. (16.3).

$$SSP = \left[\frac{Na^{+}}{Na^{+} + Ca^{2+} + Mg^{2+} + K^{+}}\right] \times 100 \tag{16.3}$$

16.3.2.1.4 Residual sodium carbonate

Residual sodium carbonate is used to express the concentration of sodium carbonate in the groundwater samples using Eq. (16.4).

$$RSC = (CO_3^{2-} + HCO_3^{-}) - (Ca^{2+} + Mg^{2+}) \tag{16.4}$$

16.3.2.1.5 Residual sodium bicarbonate

The concentration of bicarbonate in water is associated with pH concentration. RSBC is calculated according to Gupta and Gupta (1987) (Eq. 16.5).

$$RBSC = (HCO_3^{-} - Ca^{2+}) \tag{16.5}$$

16.3.2.1.6 Permeability index

According to Doneen (1964) and Raghunath (1987) PI can be formulated using Eq. (16.6).

$$PI = \frac{\left(Na^{+} + \sqrt{HCO_3^{-}}\right)}{(Ca^{2+} + Mg^{2+} + Na^{+})} \times 100 \tag{16.6}$$

16.3.2.1.7 Magnesium hazard

A high concentration of magnesium in groundwater influences the soil quality and converts it to alkaline and decreases the crop yield (Gowd, 2005). The MH is computed using Eq. (16.7).

$$MH = \frac{(Mg^{2+})}{(Ca^{2+} + Mg^{2+})} \times 100 \tag{16.7}$$

16.3.2.1.8 Kelley's ratio

The balance among Na^{+}, Ca^{2+}, and Mg^{2+} ions in water, is expressed by KR which is expressed using Eq. (16.8) following Kelley (1940, 1951).

$$KR = \frac{Na^{+}}{(Ca^{2+} + Mg^{2+})} \tag{16.8}$$

16.3.2.2 Irrigation water quality index

Irrigation water quality index was developed for measuring the status of irrigation water quality (Meireles et al., 2010). Here, it is measured based on the selected water quality parameters (HCO_3^{-}, Cl^{-}, Na^{+}, SAR, EC) compared with their specific standard values or limiting values of each class. The irrigation water quality has been calculated using Eq. (16.7) through their limiting

Table 16.1 Limiting values of parameters for q_i calculation.

HCO_3^- (meq /L)	Cl^- (meq /L)	Na^+ (meq /L)	SAR (meq /L)1/2	EC (μS/cm)	Q_I
$1 \leq HCO_3^- < 1.5$	$1 \leq Cl^- < 4$	$2 \leq Na^+ < 3$	$2 \leq SAR < 3$	$200 \leq EC < 750$	85–100
$1.5 \leq HCO_3^- < 4.5$	$4 \leq Cl^- < 7$	$3 \leq Na^+ < 6$	$3 \leq SAR < 6$	$750 \leq EC < 1500$	60–85
$4.5 \leq HCO_3^- < 8.5$	$7 \leq Cl^- < 10$	$6 \leq Na^+ < 9$	$6 \leq SAR < 12$	$1500 \leq EC < 3000$	35–60
$HCO_3^- < 1$ or $HCO_3^- \geq 8.5$	$Cl^- < 1$ or $Cl^- \geq 10$	$Na^+ < 2$ or $Na^+ \geq 9$	$SAR < 2$ or $SAR \leq 12$	$EC < 200$ or $EC \geq 3000$	0–35

Source: Meireles et al. (2010).

Table 16.2 Weights of each parameter for IWQI.

Parameters	w_i
EC	0.211
Na^+	0.204
HCO_3^-	0.202
Cl^-	0.194
SAR	0.189
Total	**1**

Source: Meireles et al. (2010).

values (q_i) (Table 16.1) and weights (w_i) (Table 16.2).

$$IWQI = \sum_{i=1}^{n} q_i w_i \tag{16.9}$$

where $qi = q_{max} - \left[\frac{\{(x_{ij} - x_{inf}) \times q_{imap}\}}{x_{amp}}\right]$ where q_{max} represents the maximum value of qi for each class

x_{ij} represents the observed value of each parameter

x_{inf} represents the minimum limit of the class

q_{iamp} represents the class amplitude

x_{amp} is class amplitude in which the parameter lies

The upper limit was considered to be the highest value determined in the physical-chemical and chemical analysis of the water samples to evaluate x_{amp} of the last class of each parameter.

w_i is calculated using the equation (Eq. 16.10).

$$w_{i=} \frac{\sum_{j=1}^{k} F_j A_{ij}}{\sum_{j=1}^{k} \sum_{i=1}^{n} F_i A_{ij}} \tag{16.10}$$

16.3.2.3 Spatial interpolation

The spatial interpolation of the physical-chemical parameters, irrigation water indices, and IWQI expresses the nature and spatial variations of the concentration of groundwater suitability. In this regard, the two widely accepted interpolation methods (IDW, and kriging) are used depending on the presence of outliers or variability in the datasets. The datasets having outliers or variability is suitable for the IDW interpolation method while the datasets without outliers or variability are preferable for the kriging method (Krause, 2012). The quartile (Q) method was used for the derivation of outliers in the datasets following Sarkar et al. (2021) using Eqs. (16.11) and (16.12).

Table 16.3 Permissible limits of physical-chemical parameters for irrigation.

Parameters (unit)	Permissible limits
pH	6.5–8.5
EC (μS/cm)	3000
TDS (mg/L)	2000
Cl^- (mg/L)	355
NO_3^- (mg/L)	30
HCO_3^- (mg/L)	255

Source: FAO (1985).

Moreover, when there were no such outliers found, the coefficient of variation (CV) was also computed to find out the nature of variability in the data structure. The higher CV ($\geq$ 50%) indicates higher variability in the datasets.

$$\text{Upper limit} = Q3 + 1.5\,\text{IQR} \tag{16.11}$$

$$\text{Lower limit} = Q1 - 1.5\,\text{IQR} \tag{16.12}$$

where IQR stands for interquartile range.

The analysis reveals that the majority of the parameters (EC, Ca^{2+}, Na^+, K^+, Cl^-, SO_4^{2}, CO_3^{2-}, HCO_3^-, SAR, SSP, KR, RSC, RSBC, MH, and IWQI) show outliers in the data structures. Only a few parameters (Mg^{2+}, %Na, and PI) do not contain outliers, however, they depict higher CV ($\geq$50%). Thus, for the portrayal of the spatial variations of all the physical-chemical parameters and indices, the IDW method was applied.

16.4 Results and discussion

16.4.1 Physical-chemical parameters

Several physical-chemical parameters such as EC, Na^+, K^+, HCO_3^-, CO_3^{2-}, Cl^-, and SO_4^{2-} have been used to determine the IWQI of the sampled groundwater. The permissible limits of physical-chemical parameters such as pH, EC, TDS, Cl^-, NO_3^-, and HCO_3^- of the groundwater samples for irrigation uses have been defined according to FAO, 1985 (Table 16.3). It reveals that concentrations of NO_3^- of all the wells lie below the permissible limit, while 44.9% of wells exceed the permissible limit of HCO_3^-. Besides, fewer wells regarding pH (10.20%), EC (2.04%), TDS (2.04%), and Cl^- (2.04%) exceed the permissible limits. Hence, based on the concentration of physical-chemical parameters, it is also found that the water quality of W31 was worst among the wells.

Electrical conductivity is a significant measure for the salinity hazard of the groundwater as it is related to the number of dissolved salts in the water (Houatmia et al., 2016). In the present study, EC values vary from 317 μS/cm to 4449 μS/cm with an average of 883.22 μS/cm and standard deviation of 781.32 μS/cm (Table 16.4). According to FAO (1985), the standard of EC for irrigation is 3000 (Table 16.3). In the study area, only 2.04% of the sample water exceeds the limit of the suitability of irrigation water. Moreover, based on the classification scheme of

Table 16.4 Statistical summary of physical-chemical parameters and irrigation water quality indices.

Parameters (unit)	Minimum	Maximum	Average	SD	CV
EC (μS/cm)	317.00	4449.00	883.22	789.42	89.38
Ca^{2+} (mg/L)	6.00	104.00	26.24	19.80	75.43
Mg^{2+} (mg/L)	4.00	59.00	26.67	13.53	50.72
Na^{+} (mg/L)	11.00	395.00	74.59	72.27	96.89
HCO_3^- (mg/L)	67.00	659.00	275.12	137.70	50.05
Cl^- (mg/L)	14.00	425.00	85.12	78.02	91.65
K^+ (mg/L)	1.00	246.00	19.98	46.05	230.51
CO_3^{2-} (mg/L)	0.00	96.00	6.37	19.56	307.25
SO_4^{2-} (mg/L)	0.00	67.00	12.65	15.88	125.47
SAR	0.41	14.63	2.58	2.71	105.07
%Na	10.46	90.02	45.24	18.74	41.42
SSP (%)	10.14	89.46	42.48	18.02	42.43
KR	11.29	848.58	106.99	134.68	125.88
MH (%)	7.35	92.40	62.87	19.88	31.62
RSC (meq/L)	−3.3	8.53	1.22	2.51	206.01
RSBC (meq/L)	−1.29	9.21	3.20	2.24	69.95
PI (%)	41.70	111.55	76.98	16.53	21.47

EC, 6.01% (3), 32.65% (16), and 61.22% (30) of well's water are unsuitable, permissible, and good quality, respectively for irrigation suitability (Table 16.5 and Fig. 16.3).

The concentration of Ca^{2+} ranges from 6 to 104 mg/L with a 26.25 mg/L mean value and its CV (75.43) is higher than SD (19.8) (Table 16.4). Based on the Ca^{2+} concentration value five potentiality zone have been created such as very low (<25), low (25–50), medium (50–75), high (75–100), and very high (>100) (Fig. 16.4a). According to this classification, 67.35% (33), 22.45% (11), 6.12% (3), 2.04% (1), and 2.04% (1) of wells come under the very low, low, medium, high, and very high zone respectively (Table 16.5). And the concentration of Mg^{2+} varies from 4 to 59 mg/L with a 26.67 mg/L mean value and its CV (50.72) is higher than SD (13.53) (Table 16.4). According to the Mg^{2+} concentration value, five potentiality zone have been created such as very low (<20), low (20–24), medium (24–28), high (28–32), and very high (>32) (Fig. 16.4b). Based on this classification, 36.73% (18), 14.29% (7), 10.20% (5), 4.08% (2), and 34.69% (17) of wells come under the very low, low, medium, high, and very high zone respectively (Table 16.5).

The concentration of Na^+ ranges from 11 to 395 mg/L with a 74.6 mg/L mean value and its CV (96.89) is higher than SD (72.27) (Table 16.4). Based on the Na^+ concentration value five potentiality zone have been created such as very low (<80), low (80–160), medium (160–240), high (240–320), and very high (>320) (Fig. 16.5a). According to this classification, 69.39% (34), 18.37% (9), 8.16% (4), 2.04% (1), and 2.04% (1) of wells come under the very low, low, medium, high, and very high zone respectively (Table 16.5). Moreover, the concentration of K^+ ranges from 1 to 246 mg/L with a 19.98 mg/L mean value and its CV (230.54) is very high than SD (46.05) (Table 16.4). According to the K^+ concentration value, five potentiality zone have been created such as very low (<50), low (50–100), medium (100–150), high (150–200), and very high (>200)

Table 16.5 Classification scheme of physical-chemical parameters of groundwater.

Classification scheme	Categories	Range	Number of wells	Percentage
EC (µS/cm)	Excellent	<250	0	0.00
	Good	250–750	30	61.22
	Permissible	750–2250	16	32.65
	Unsuitable	>2250	3	6.12
Ca^{2+} (mg/L)	very low	<25	33	67.35
	low	25–50	11	22.45
	medium	50–75	3	6.12
	high	75–100	1	2.04
	very high	>100	1	2.04
Mg^{2+} (mg/L)	very low	<20	18	36.73
	low	20–24	7	14.29
	medium	24–28	5	10.20
	high	28–32	2	4.08
	very high	>32	17	34.69
Na^{+} (mg/L)	very low	<80	34	69.39
	low	80–160	9	18.37
	medium	160–240	4	8.16
	high	240–320	1	2.04
	very high	>320	1	2.04
K^{+} (mg/L)	very low	<50	46	93.88
	low	50–100	1	2.04
	medium	100–150	0	0.00
	high	150–200	0	0.00
	very high	>200	2	4.08
HCO_3^{-} (mg/L)	very low	<150	7	14.29
	low	150–300	26	53.06
	medium	300–450	11	22.45
	high	450–600	4	8.16
	very high	>600	1	2.04
CO_3^{2-} (mg/L)	very low	<15	43	87.76
	low	15–30	3	6.12
	medium	30–45	0	0.00
	high	45–60	1	2.04
	very high	>60	2	4.08
Cl^{-} (mg/L)	very low	<100	35	71.43
	low	100–200	11	22.45
	medium	200–300	16	32.65
	high	300-400	2	4.08
	very high	>400	1	2.04
SO_4^{2-} (mg/L)	very low	<15	31	63.27
	low	15–30	13	26.53
	medium	30–45	2	4.08
	high	45–60	1	2.04
	very high	>60	2	4.08

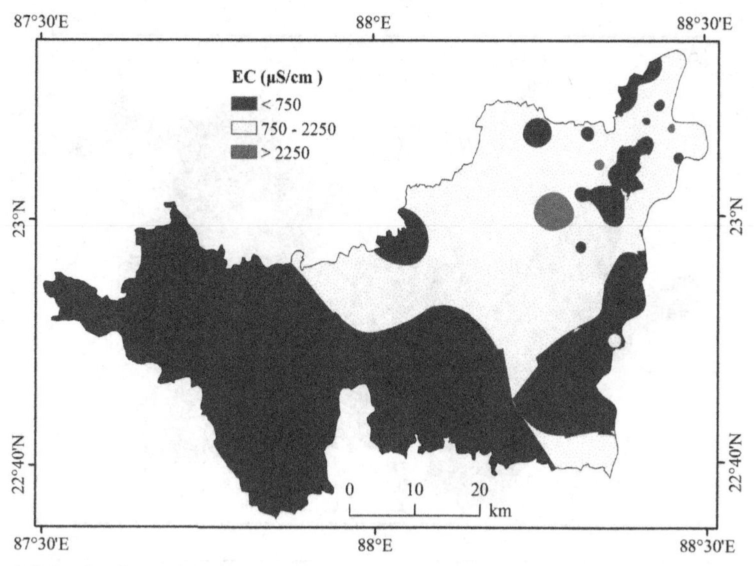

FIGURE 16.3 Spatial distribution of EC.

(Fig. 16.5b). Based on this classification, 93.88% (46), 2.04% (1), and 4.08% (2) of wells come under the very low, low, and very high zone respectively (Table 16.5).

The concentration of HCO_3^- varies from 67 to 659 mg/L with a 275.12 mg/L mean value and its CV (50.05) is lower than SD (137.7) (Table 16.4). According to the HCO_3^- concentration value, five potentiality zone have been created such as very low (<150), low (150-300), medium (300-450), high (450-600), and very high (>600) (Fig. 16.6a). Based on this classification, 14.29% (7), 53.06% (26), 22.45% (11), 8.16% (4), and 2.04% (1) of wells come under the very low, low, medium, high, and very high zone respectively (Table 16.5). And the concentration of CO_3^{2-} ranges from 0 to 96 mg/L with a 6.38 mg/L mean value and its CV (307.25) is very high than SD (19.56). According to the CO_3^{2-} concentration value, five potentiality zone have been created such as very low (<15), low (15-30), medium (30-45), high (45-60), and very high (>60) (Fig. 16.6b). Based on this classification, 87.76% (43), 6.12% (3), 0 (0), 2.04% (1), and 4.08% (2) of wells come under the very low, low, high, and very high zone respectively (Table 16.5).

The concentration of Cl^- ranges from 14 to 425 mg/L with an 85.12 mg/L mean value and its CV (91.65) is higher than SD (78.02) (Table 16.4). Based on the Cl^- concentration value five potentiality zone have been created such as very low (<100), low (100-200), medium (200-300), high (300-400), and very high (>400) (Fig. 16.7a). According to this classification, 71.43% (35), 22.45% (11), 32.65% (16), 4.08% (2), and 2.04% (1) of wells come under the very low, low, medium, high, and very high zone respectively (Table 16.5). And the concentration of SO_4^{2-} varies from 0 to 67 mg/L with a mean value of 12.65 mg/L and CV (125.47) (Table 16.4). According to the SO_4^{2-}concentration value, five potentiality zone have been created such as very low (<15), low (15-30), medium (30-45), high (45-60) and very high (>60) (Fig. 16.7b). Based on this classification, 63.27% (31), 26.53% (13), 4.08% (2), 2.04% (1), and 4.08% (2) of wells come under the very low, low, high, and very high zone respectively (Table 16.5).

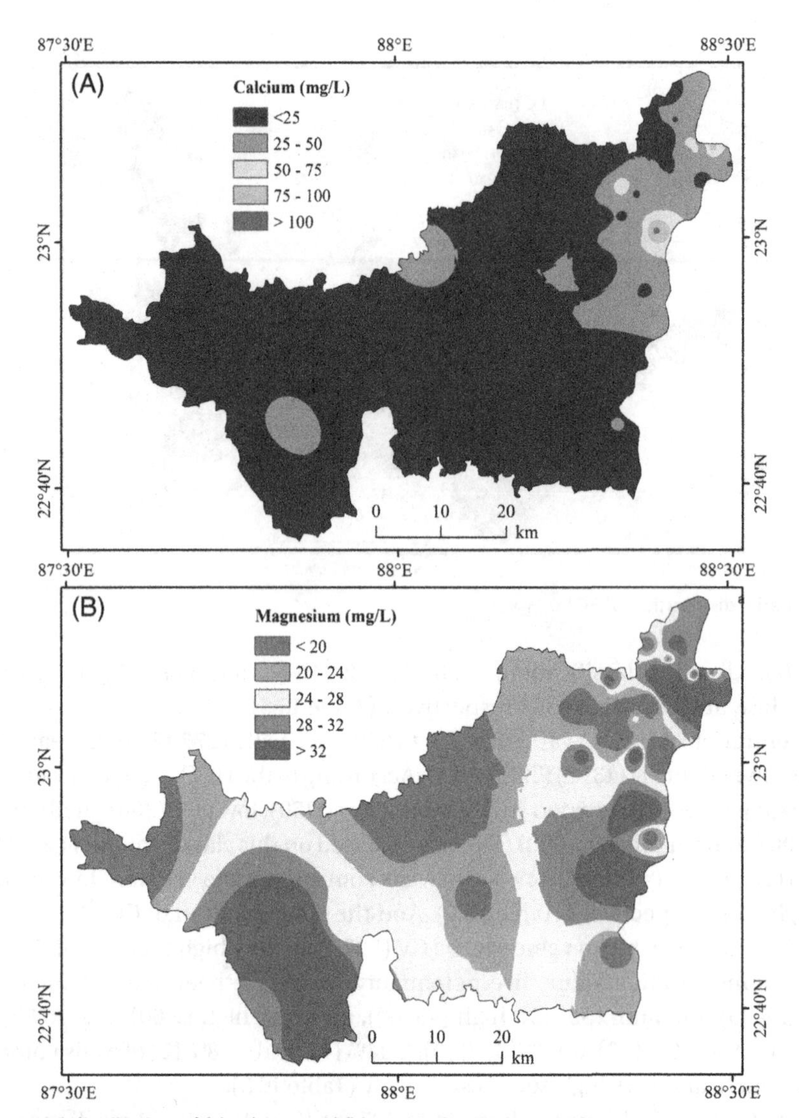

FIGURE 16.4 Spatial distribution of (a) calcium and (b) magnesium concentration.

16.4.2 Groundwater suitability for irrigation

Sodium absorption ratio is a widely used indicator to measure the suitability of water for irrigation uses. It can express the degree of cations exchange in the soil in terms of replacing the absorption magnesium with sodium and calcium occurs (Chen et al., 2019). In the present study, the concentration of SAR was found to vary from 0.41 to 14 with a standard deviation of 2.68 (Table 16.4). The concentration of SAR has been categorized into three classes such as good (<6), doubtful (6–9), and unsuitable (>9) in which, 91.84% of the groundwater sample is recorded in good class, 4.08% of the sample well-considered as doubtful and 4.08% of the wells falls in

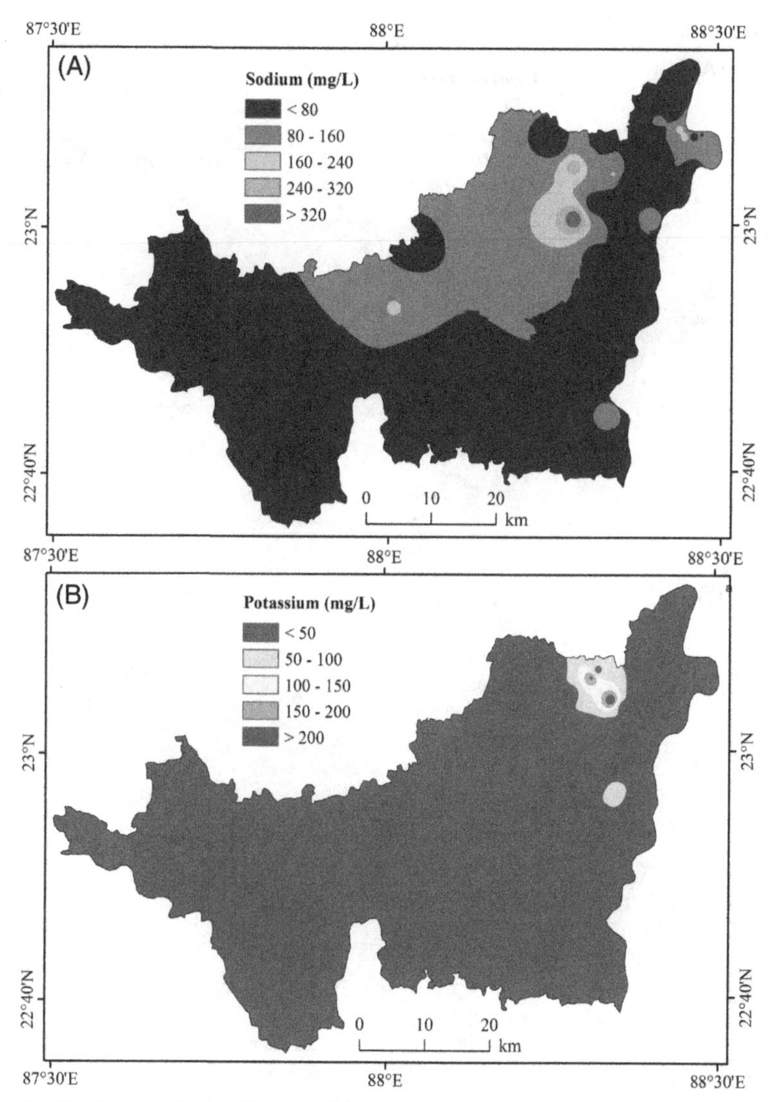

FIGURE 16.5 Spatial distribution of (a) sodium and (b) potassium concentration.

the unsuitable class (Table 16.6) (Fig. 16.8a). According to the classification of SAR proposed by Richards (1954), 47 sample wells of the present study area belong to excellent quality while only two sample wells have good suitability for irrigation uses. It is also observed that none of the sample well is located in the moderate and poor suitability of groundwater for irrigations.

The percentage of sodium content is an important indicator to determine the suitability of groundwater for agricultural purposes (Wilcox, 1948). The groundwater with a high concentration of %Na deteriorates the water quality and affects the crop yields. It is observed that the value of %Na ranges from 10.45 to 90.02 with a mean of 45.24 and a standard deviation of 18.75 (Table 16.4). The spatial distribution of %Na has been categorized into four classes including

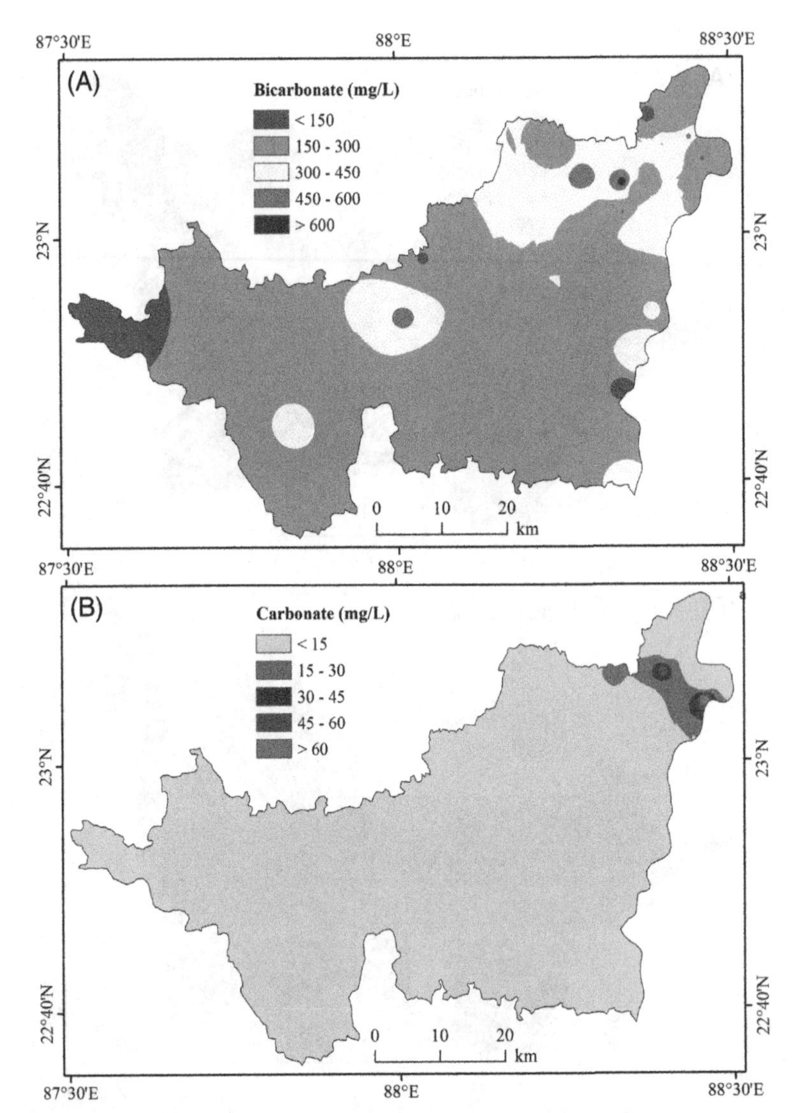

FIGURE 16.6 Spatial distribution of (a) bicarbonate and (b) carbonate.

excellent (0–20), good (20–40), permissible (40–60), doubtful (60–80), and unsuitable (>80) (Fig. 16.8b). From the spatial variation of the %Na, it is observed that 6.12% of the samples are considered as excellent, 38.78% of sample wells accounts for good, 32.65% considered as permissible, 18.37% of the sample well is observed as doubtful, and 4.08% of the sample well is considered as unsuitable for irrigation purposes (Table 16.6).

The present study reveals that the value of SSP ranges from 10.14 to 89.46 with a mean of 42.48 and a standard deviation of 18.02 (Table 16.4). In the present study, the concentration of SSP has been classified into four groups such as excellent (0–20), good (20–40), permissible

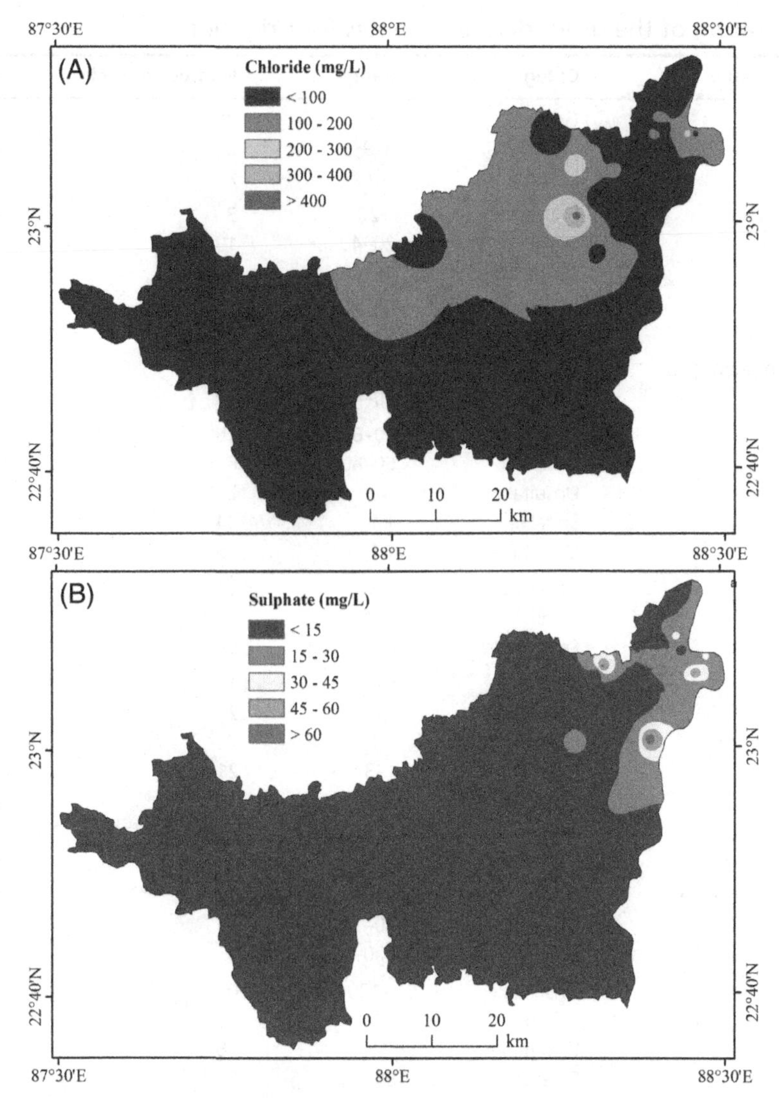

FIGURE 16.7 Spatial distribution of (a) chloride and (b) sulphate.

(40–60), doubtful (60–80), and unsuitable (>80) (Fig. 16.9a). The spatial variation of the SSP reveals that 8.16% of the samples account for excellent quality, 44.9% of sample wells fall in the good quality, 30.61% are considered as permissible quality, 12.24% of the sample well is observed as doubtful quality, and 4.08% of the sample well is considered as unsuitable quality for irrigation (Table 16.6).

The Kelley's ratio is a significant indicator to determine the suitability of irrigation water by analyzing the concentration of sodium in irrigation water. The high concentration of sodium delays the rate of plant growth (Kelley et al., 1940). The groundwater with a KR value greater than 50 is considered unsuitable for irrigation. In the present study, the concentration of KR ranges

Table 16.6 Rating of the groundwater samples for irrigation.

Classification scheme	Categories	Range	Number of wells	Percentage
SAR	Good	<6	45	91.84
	Suitable	6–9	2	4.08
	Unsuitable	>9	2	4.08
Salinity potential (SP)	Excellent	<20	3	6.12
	Good	20–40	19	38.78
	Permissible	40–60	16	32.65
	Doubtful	60–80	9	18.37
	Unsuitable	>80	2	4.08
Magnesium absorption ratio (MAR)	Excellent	<20	4	8.16
	Good	20–40	22	44.9
	Permissible	40–60	15	30.61
	Doubtful	60–80	6	12.24
	Unsuitable	>80	2	4.08
KR	Suitable	<1	34	69.39
	Unsuitable	>1	15	30.61
RSC (meq/L)	Excellent	<2	34	69.39
	Good	2–4	9	18.37
	Permissible	4–6	9	6.12
	Doubtful	6–8	1	2.04
	Unsuitable	>8	2	4.08
RSBC (meq/L)	Excellent	<1	7	14.29
	Good	1–3	21	42.86
	Permissible	3–5	10	20.41
	Doubtful	5–7	7	14.29
	Unsuitable	>7	4	8.16
Residual sodium carbonate (RSC) in meq/L	Good	<80	26	53.06
	Moderate	80–100	19	38.78
	Poor	100–120	4	8.16
	Unsuitable	>120	0	0
Soluble sodium percentage (SSP)	Suitable	<50	11	22.45
	Unsuitable	>50	38	77.55
IWQI	Excellent	85–100	0	0
	Good	70–85	13	26.53
	Poor	55–70	30	61.22
	Very poor	40–55	4	8.16
	Unsuitable	0–40	2	4.08

from 0.11 to 8.49 with a mean of 1.07 and a standard deviation value of 1.33 (Table 16.4). In the present study, all the samples fall in a suitable class (Fig. 16.9b) (Table 16.6).

Residual sodium concentration is a useful indicator to determine the presence of sodium cations in the groundwater. The concentration of RSC ranges from -3.3 to 8.53 in the study region. The average and standard deviation of the concentration of RSC is observed as 1.22 and 2.51 respectively (Table 16.4). The classification of the RSC concentration has been made

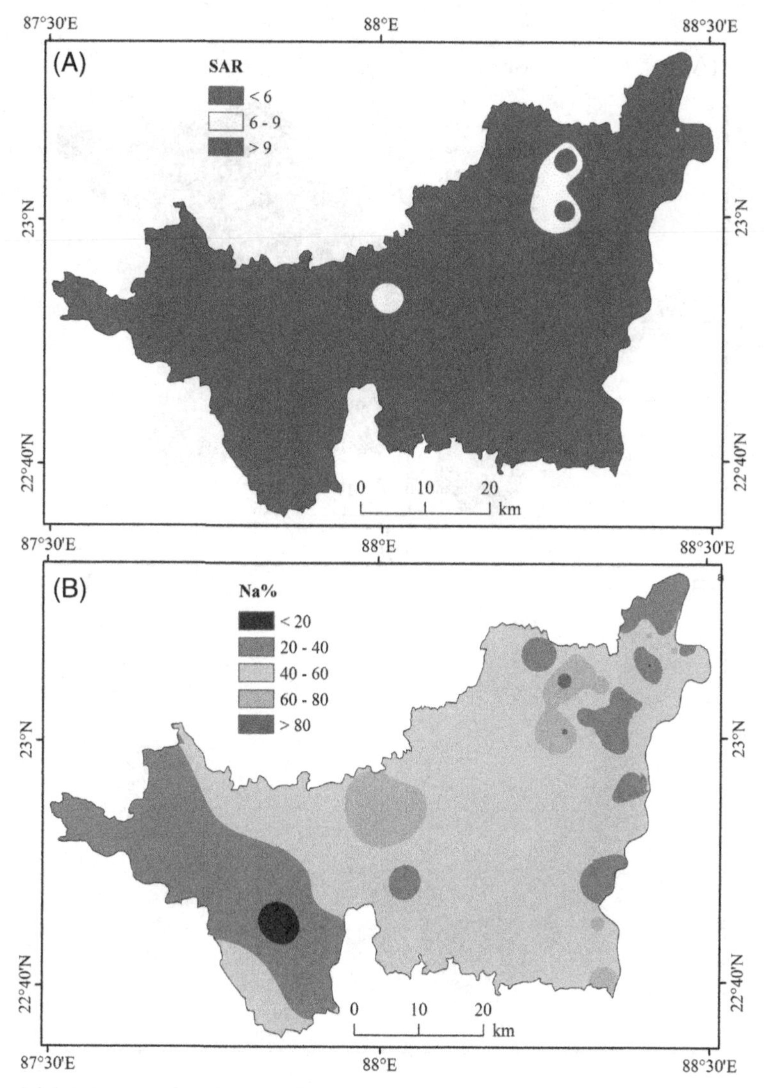

FIGURE 16.8 Spatial distribution of (a) SAR and (b) %Na.

into three groups such as excellent (<2), good (2–40), permissible (4–6), doubtful (6–8), and unsuitable (>8). The results depict that 14.29% of the sample well are considered as excellent, 18.37% samples are shown as good, 6.12% samples are charaterised as permissible and 2.04% of samples designated as doubtful category. Additionally, 8.16% of the groundwater samples are considered in unsuitable category in terms of irrigation suitability of the study area (Table 16.6 and Fig. 16.10a).

Furthermore, RSBC concentration was found to vary from −1.29 to 9.21 with an average value of 3.2 and a standard deviation of 2.24 (Table 16.4). In the present study area, the concentration

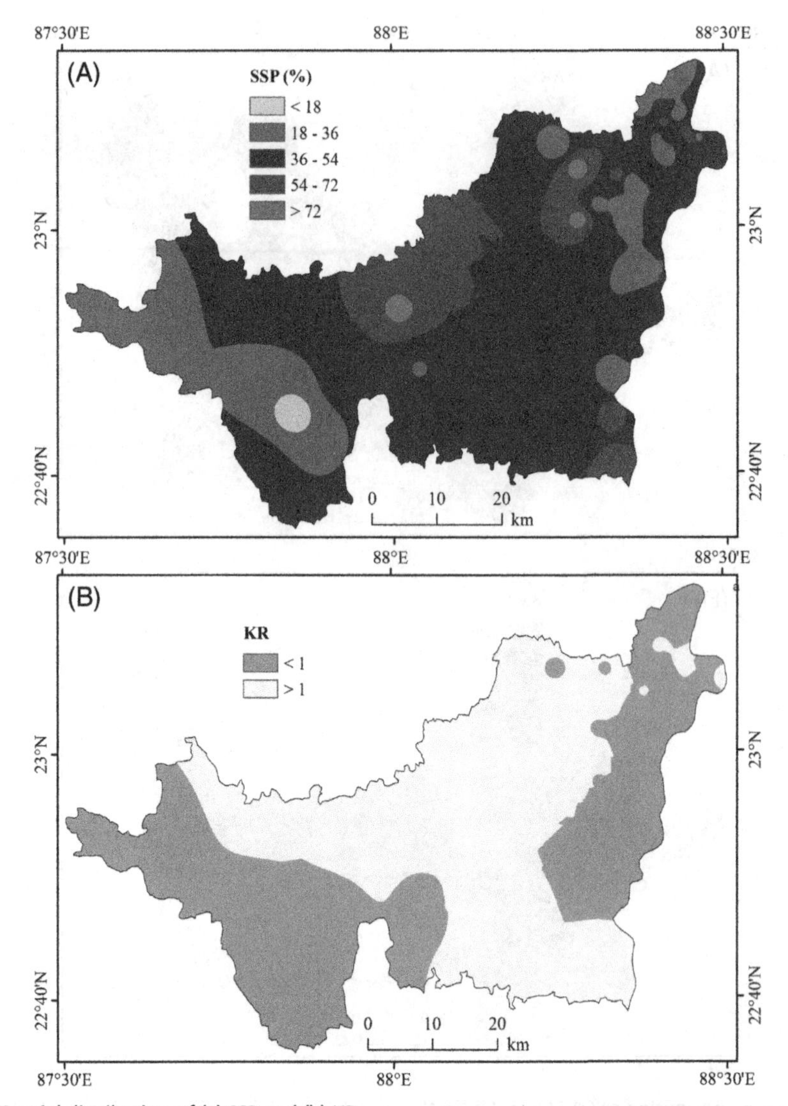

FIGURE 16.9 Spatial distribution of (a) SSP and (b) KR.

of the RSBC is grouped into two classes—suitable and unsuitable. The suitable class (<5) comprises 79.6% of the sample wells while the unsuitable class includes 20.4 % of the sample wells (Table 16.6 and Fig. 16.10b).

In the present chapter, the PI was analyzed which ranges from 41.70 to 111.55 with a mean of 76.98 and a standard deviation of 16.53 (Table 16.4). Doneen (1964) classified PI into four classes as good (<80), moderate (80–100), poor (100–120), and unsuitable (>120) for the suitability of irrigation water. In the study, the PI value was observed in three classes such as good, moderate, and poor. The PI concentration reveals that 53.08% of the sample wells are considered as good

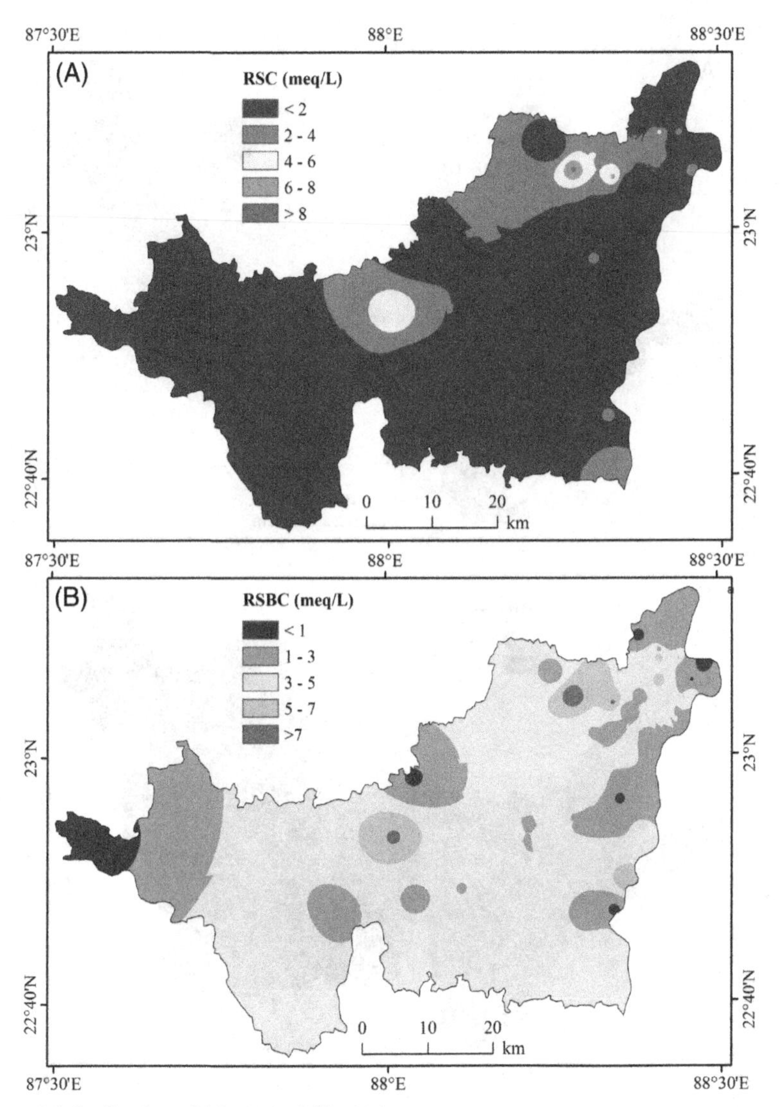

FIGURE 16.10 Spatial distribution of (a) RSC and (b) RSBC.

classes and 38.78% of the sample water belongs to the moderate class while 8.16% of the sample water designates as the poor class for irrigation uses (Table 16.6 and Fig. 16.11a).

The MH value varies from 7.35 to 92.4 with a mean of 62.87 and a standard deviation of 19.68 in the study area (Table 16.4). According to Narasaiah and Rao (2021), the threshold value of the MH for irrigation purposes has been considered as 50. Therefore, the MH concentration, beyond 50 is observed as unsuitable for irrigation. The spatial distribution of MH has been grouped into two classes including good (<50) and unsuitable (>50). The concentration of the MH depicts that 22.45% of the sample wells are designated as good, and 77.55% of the sample wells are considered unsuitable for irrigation purposes (Table 16.6 and Fig. 16.11b).

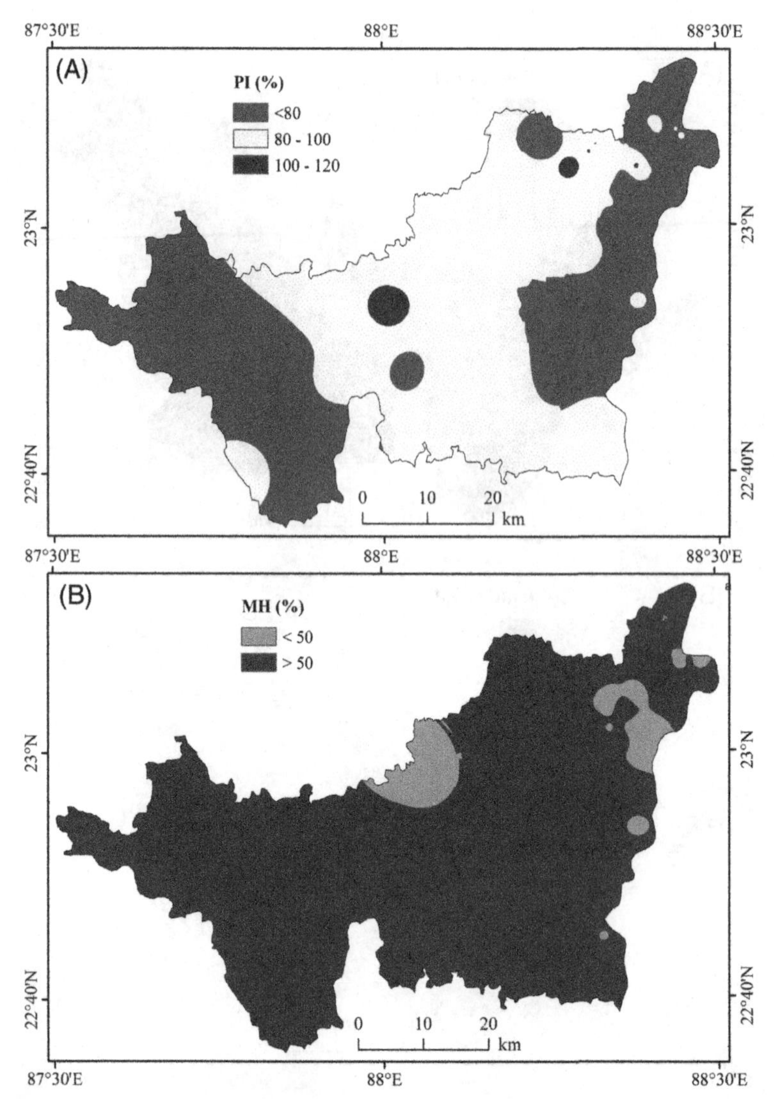

FIGURE 16.11 Spatial distribution of (a) PI and (b) MH.

The IWQI was used to quantify the degree of water quality in a single value for irrigation suitability. In the present study, the calculated value of IWQI was found to range between 23.73 and 79.50 with a mean of 64.15 (Table 16.3). The higher IWQI represents the higher water quality for irrigation uses. In the study area, the categorization of the IWQI has been made into four classes such as good (70–85), poor (55–70), very poor (40–55), and unsuitable (0–40). The results depict that 13 wells (26.53%) are considered as a good category while only 2 wells (4.08%) are designated as unsuitable for irrigation. Additionally, the poor and very poor water quality class comprises 30 wells (61.22%) and 4 wells (8.16%) respectively (Table 16.4). Considering the

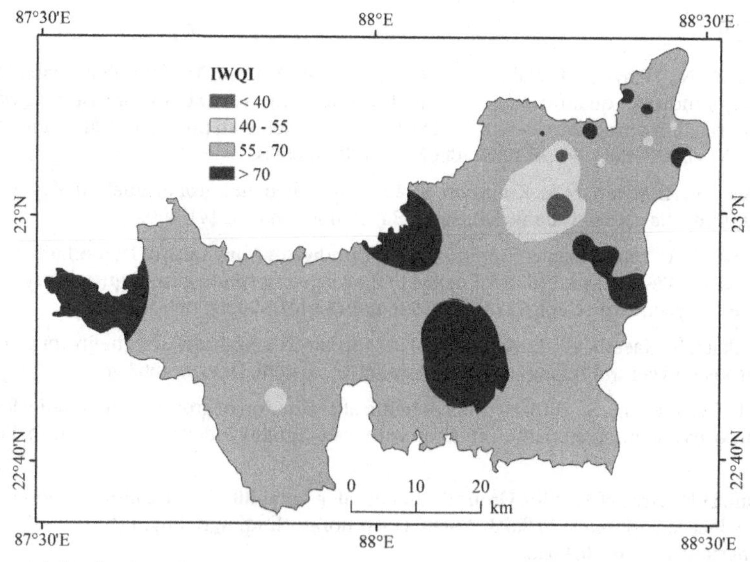

FIGURE 16.12 Spatial distribution of IQWI.

IWQ standard, none of the water samples came under the excellent water quality (85–700) for irrigation in terms of the IWQI category. Therefore, most of the wells in the study area comprise the water with poor to unsuitable category for irrigation. Moreover, the spatial distribution of IWQI shows that unsuitable water quality for irrigation is observed in the eastern part of the study area while a good category of IWQI has been found in the middle potation, western and eastern fringe of the study area. Besides, a very poor category of IWQI is concentrated in the northeast and southwest patches and the rest of the study area comprises poor quality sample groundwater (Fig. 16.12).

16.5 Conclusion

Groundwater is the major source of irrigation water especially in *rabi* season in the study area. The study aims to evaluate the physical-chemical parameters of groundwater and determine the groundwater quality for irrigation purposes using IWQI. The calculated IWQI value ranges from 23.73 to 79.50 with a mean of 64.15. The weightage of EC, Na^+, HCO_3^-, Cl^-, and SAR has been assigned as 0.211, 0.204, 0.202, 0.194, and 0.189 respectively to compute IWQI. The result reveals that 13 wells (26.53%) are considered as good category. Only 2 wells (4.08%) are observed in the unsuitable class for irrigation. However, 30 wells (61.22%) account for the poor class for irrigation suitability. The class of very poor water quality includes 4 wells (8.16%) of the study area. Therefore, it is also observed that most of the wells in the study area comprise the water with poor to unsuitable category for irrigation purposes. Thus, more attention should be paid to groundwater quality assessment which may help the planners for sustainable management of water resources.

References

Abbasnia, A., Yousefi, N., Mahvi, A.H., Nabizadeh, R., Radfard, M., Yousefi, M., Alimohammadi, M., 2019. Evaluation of groundwater quality using water quality index and its suitability for assessing water for drinking and irrigation purposes: case study of Sistan and Baluchistan province (Iran). Hum. Ecol. Risk Assess. Int. J. 25 (4), 988–1005. doi:10.1080/10807039.2018.1458596.

Abdullah, T.O., Ali, S.S., Al-Ansari, N.A., Knutsson, S., 2019. Hydrogeochemical evaluation of groundwater and its suitability for domestic uses in Halabja Saidsadiq Basin. Iraq Water 11 (4), 690.

Acharyya, S.K., Shah, B.A., 2007. Arsenic-contaminated groundwater from parts of Damodar fan-delta and west of Bhagirathi River, West Bengal, India: influence of fluvial geomorphology and quaternary morphostratigraphy. Environ. Geol. 52 (3), 489–501. doi:10.1007/s00254-006-0482-z.

Adimalla, N., Dhakate, R., Kasarla, A., Taloor, A.K., 2020. Appraisal of groundwater quality for drinking and irrigation purposes in central Telangana, India. Groundw. Sustain. Dev. 10, 100334.

Adimalla, N., Li, P., Venkatayogi, S., 2018. Hydrogeochemical evaluation of groundwater quality for drinking and irrigation purposes and integrated interpretation with water quality index studies. Environ. Proc. 5 (2), 363–383.

Alwan, I.A., Karim, H.H., Aziz, N.A., 2019. Groundwater aquifer suitability for irrigation purposes using multi-criteria decision approach in Salah Al-Din Governorate/Iraq. Agri. Eng. 1 (2), 303–323. doi:10.3390/agriengineering1020023.

Aravinthasamy, P., Karunanidhi, D., Rao, N.S., Subramani, T., Srinivasamoorthy, K., 2020. Irrigation risk assessment of groundwater in a non-perennial river basin of South India: implication from irrigation water quality index (IWQI) and geographical information system (GIS) approaches. Arab. J. Geosci. 13 (21), 1–14.

Balasubramani, K., Murthy, K.R., Gomathi, M., Kumaraswamy, K., 2020. Integrated assessment of groundwater resources in a semi-arid watershed of South India: implications for irrigated agriculture. GeoJournal 85 (6), 1701–1723.

Bhat, M.A., Grewal, M.S., Rajpaul, R., Wani, S.A., Dar, E.A., 2016. Assessment of groundwater quality for irrigation purposes using chemical indices. Ind. J. Ecol. 43 (2), 574–579.

CGWB, 2006. Dynamic Ground Water Resources of India. Ministry of Water Resources, River Development & Ganga Rejuvenation, Government of India, Faridabad.

CGWB, 2017. Dynamic Ground Water Resources of India. Ministry of Water Resources, River Development & Ganga Rejuvenation, Government of India, Faridabad.

Chen, J., Huang, Q., Lin, Y., Fang, Y., Qian, H., Liu, R., Ma, H., 2019. Hydrogeochemical characteristics and quality assessment of groundwater in an irrigated region, Northwest China. Water 11 (1), 96.

Das, B., Pal, S.C., 2020. Assessment of groundwater recharge and its potential zone identification in groundwater-stressed Goghat-I block of Hugli District, West Bengal, India. Environ. Dev. Sustain. 22 (6), 5905–5923.

Das, J., Rahman, A.S., Mandal, T., Saha, P., 2021. Exploring driving forces of large-scale unsustainable groundwater development for irrigation in lower Ganga River basin in India. Environ. Dev. Sustain. 23 (5), 7289–7309. doi:10.1007/s10668-020-00917-5.

Doneen, L.D., 1962, June. The influence of crop and soil on percolating water. In: Proceedings of 1961 Biennial Conference on Groundwater Recharge, pp. 156–163.

FAO, 1985. In:Water Quality for Agriculture, 29. Food and Agriculture Organization of the United Nations, Rome, p. 174.

Ghazaryan, K., Movsesyan, H., Gevorgyan, A., Minkina, T., Sushkova, S., Rajput, V., Mandzhieva, S., 2020. Comparative hydrochemical assessment of groundwater quality from different aquifers for irrigation purposes using IWQI: a case-study from Masis province in Armenia. Groundw. Sustain. Dev. 11, 100459. doi:10.1016/j.gsd.2020.100459.

Ghosh, S., Gorain, S., Mondal, B., 2017. Spatio-temporal variations and links between irrigation and agricultural development in an eastern Indian State. Irrig. Drain. 66 (5), 784–796.

Gowd, S.S., 2005. Assessment of groundwater quality for drinking and irrigation purposes: a case study of Peddavanka watershed, Anantapur District, Andhra Pradesh, India. Environ. Geol. 48 (6), 702–712.

Gupta, S.K., Gupta, I.C., 1987. Management of Saline Soils and Water. Oxford and IBH Publishing Company Pvt. Ltd., New Delhi.

Honarbakhsh, A., Tahmoures, M., Tashayo, B., Mousazadeh, M., Ingram, B., Ostovari, Y., 2019. GIS-based assessment of groundwater quality for drinking purpose in northern part of Fars province, Marvdasht. J. Water Supply Res. Technol.-Aqua 68 (3), 187–196.

Houatmia, F., Azouzi, R., Charef, A., Bédir, M., 2016. Assessment of groundwater quality for irrigation and drinking purposes and identification of hydrogeochemical mechanisms evolution in Northeastern, Tunisia. Environ. Earth Sci. 75 (9), 746.

Ibrahim, M.N., 2019. Assessing groundwater quality for drinking purpose in Jordan: application of water quality index. J. Ecol. Eng. 20 (3).

Kaur, T., Bhardwaj, R., Arora, S., 2017. Assessment of groundwater quality for drinking and irrigation purposes using hydrochemical studies in Malwa region, southwestern part of Punjab, India. Appl. Water Sci. 7 (6), 3301–3316. doi:10.1007/s13201-016-0476-2.

Kelley, W.P., 1951. Alkali Soils: Their Formation, Properties and Reclamation. Reinhold Publishing Corporation, New York, p. 176.

Kelley, W.P., Brown, S.M., Leibig, G.I., 1940. Chemical effects of saline irrigation water on soils. Soil Sci. 49, 95–107.

Krause, E., 2012. Dealing with extreme values in kriging. ESRI ArcGIS. https://www.esri.com/arcgis-blog/products/arcgis-desktop/analytics/dealingwith-extreme-values-in-kriging/.

Kumar, S.K., Rammohan, V., Sahayam, J.D., Jeevanandam, M., 2009. Assessment of groundwater quality and hydrogeochemistry of Manimuktha River basin, Tamil Nadu, India. Environ. Monit. Assess. 159 (1), 341–351.

Kumari, M.K.N., Sakai, K., Kimura, S., Yuge, K., Gunarathna, M.H.J.P., 2019. Classification of groundwater suitability for irrigation in the Ulagalla tank Cascade landscape by GIS and the analytic hierarchy process. Agronomy 9 (7), 351. doi:10.3390/agronomy9070351.

Kurdi, M., Eslamkish, T., 2017. Hydro-geochemical classification and spatial distribution of groundwater to examine the suitability for irrigation purposes (Golestan Province, north of Iran). Paddy Water Environ. 15 (4), 731–744.

Li, P., Wu, J., Qian, H., 2013. Assessment of groundwater quality for irrigation purposes and identification of hydrogeochemical evolution mechanisms in Pengyang County, China. Environ. Earth Sci. 69 (7), 2211–2225. doi:10.1007/s12665-012-2049-5.

Luo, Y., Xiao, Y., Hao, Q., Zhang, Y., Zhao, Z., Wang, S., Dong, G., 2021. Groundwater geochemical signatures and implication for sustainable development in a typical endorheic watershed on Tibetan plateau. Environ. Sci. and Pollut. Res. 28 (35), 48312–48329. doi:10.1007/s11356-021-14018-x.

Mahammad, S., Islam, A., 2021a. Evaluating the groundwater quality of Damodar Fan Delta (India) using fuzzy-AHP MCDM technique. Appl. Water Sci. 11 (7), 1–17.

Mahammad, S., Islam, A., 2021b. Identification of palaeochannels using optical images and radar data: a study of the Damodar Fan Delta, India. Arab. J. Geosci. 14 (17), 1–22.

Mahammad, S., Islam, A., Shit, P. K., 2022. Geospatial assessment of groundwater quality using entropy-based irrigation water quality index and heavy metal pollution indices. Environ. Sci. and Pollut. Res. 1-24. doi:10.1007/s11356-022-20665-5.

Majumder, A., Sivaramakrishnan, L., 2014. Ground water budgeting in alluvial Damodar fan delta: a study in semi-critical Pandua block of West Bengal, India. Int. J. Geol. Earth Environ. Sci. 4 (3), 23–37.

Meireles, A.C.M., Andrade, E.M.D., Chaves, L.C.G., Frischkorn, H., Crisostomo, L.A., 2010. A new proposal of the classification of irrigation water. Rev. Ciênc. Agronôm. 41, 349–357.

Mthembu, P.P., Elumalai, V., Senthilkumar, M. Wu, J., 2021. Investigation of Geochemical Characterization and Groundwater Quality with Special Emphasis on Health Risk Assessment in Alluvial Aquifers, South Africa. Int. J. Environ. Sci. Technol. 18, 3711–3730. doi:10.1007/s13762-021-03129-0.

Naidu, S., Gupta, G., Singh, R., Tahama, K., Erram, V.C., 2021. Hydrogeochemical processes regulating the groundwater quality and its suitability for drinking and irrigation purpose in parts of coastal Sindhudurg District, Maharashtra. J. Geol. Soc. Ind. 97 (2), 173–185.

Narasaiah, V., Rao, B.V., 2021. Groundwater quality of a hard rock aquifer in the Subledu Basin of Khammam district, India. Appl. Water Sci. 11 (6), 1–21. doi:10.1007/s13201-021-01424-2.

Obiefuna, G.I., Sheriff, A., 2011. Assessment of shallow ground water quality of Pindiga Gombe Area, Yola Area, NE, Nigeria for irrigation and domestic purposes. J. Environ. Earth Sci. 3 (2), 131–141.

Pal, T., Mukherjee, P.K., 2009. Study of subsurface geology in locating arsenic-free groundwater in Bengal delta, West Bengal, India. Environ. Geol. 56 (6), 1211–1225.

Pal, T., Mukherjee, P.K., 2010. Search for groundwater arsenic in Pleistocene sequence of the Damodar River flood plain. Model. Earth Syst. Environ. 2 (4), 1–11.

Paliwal, K.V., 1972. Irrigation with Saline Water. In: Monogram No. 2 (New Series). IARI, New Delhi, p. 198.

Patra, S., Mishra, P., Mahapatra, S.C., 2018. Delineation of groundwater potential zone for sustainable development: a case study from Ganga Alluvial Plain covering Hooghly district of India using remote sensing, geographic information system and analytic hierarchy process. J. Cleaner Prod. 172, 2485–2502.

Patra, S., Mishra, P., Mahapatra, S.C., Mithun, S.K., 2016. Modelling impacts of chemical fertilizer on agricultural production: a case study on Hooghly district, West Bengal, India. J. Geosci. 64 (1–4), 109–112. doi:10.1007/s40808-016-0223-6.

Pophare, A.M., Balpande, U.S., Nawale, V.P., 2018. Hydrochemistry of groundwater in Suketi River Basin, Himachal Himalaya, India. J. Geosci. Res. 3, 67–83.

Raghunath, H.M., 1987. Groundwater. Wiley Eastern Ltd., New Delhi.

Ramesh, K., Elango, L., 2012. Groundwater quality and its suitability for domestic and agricultural use in Tondiar river basin, Tamil Nadu, India. Environ. Monit. Assess. 184 (6), 3887–3899.

Richards, L.A., 1954. Diagnosis and Improvement of Saline and Alkaline Soils. U.S. Salinity Laboratory, U.S. Department of Agriculture Handbook No. 60 U.S. Government Printing Office, Washington, DC.

Rufino, F., Busico, G., Cuoco, E., Darrah, T.H., Tedesco, D., 2019. Evaluating the suitability of urban groundwater resources for drinking water and irrigation purposes: an integrated approach in the Agro-Aversano area of Southern Italy. Environ. Monit. Assess. 191 (12), 1–17. doi:10.1007/s10661-019-7978-y.

Sarkar, B., Islam, A., 2019. Assessing the suitability of water for irrigation using major physical parameters and ion chemistry: a study of the Churni River, India. Arab. J. Geosci. 12 (20), 1–16.

Sarkar, B., Islam, A., Majumder, A., 2021. Seawater intrusion into groundwater and its impact on irrigation and agriculture: evidence from the coastal region of West Bengal, India. Reg. Stud. Mar. Sci. 44, 101751.

Sengupta, S., 1972. Geological framework of the Bhagirathi-Hooghly basin. The Bhagirathi-Hooghly Basin. RD Press, Kolkata, pp. 3–8.

Simsek, C., Gunduz, O., 2007. IWQ index: a GIS-integrated technique to assess irrigation water quality. Environ. Monit. Assess. 128 (1), 277–300. doi:10.1007/s10661-006-9312-8.

Singh, K.P., Malik, A., Mohan, D., Singh, V.K., Sinha, S., 2006. Evaluation of groundwater quality in northern Indo-Gangetic alluvium region. Environ. Monit. Assess. 112 (1), 211–230. doi:10.1007/s10661-006-0357-5.

Singha, C., Swain, K.C., Swain, S.K., 2020. Best crop rotation selection with GIS-AHP technique using soil nutrient variability. Agriculture 10 (6), 213.

Subramani, T., Elango, L., Damodarasamy, S.R., 2005. Groundwater quality and its suitability for drinking and agricultural use in Chithar River Basin, Tamil Nadu, India. Environ. Geol. 47 (8), 1099–1110. doi:10.1007/s00254-005-1243-0.

Todd, D.K., Mays, L.W., 2004. Groundwater Hydrology. John Wiley & Sons, New York.

Wilcox, L., 1948. The quality of water for irrigation use (Technical Bulletin 1962). United State Department of Agriculture, Washington DC, USA.

Wilcox, L., 1955. Classification and Use of Irrigation Waters (No. 969). US Department of Agriculture, Washington DC, USA.

Yıldız, S., Karakuş, C.B., 2020. Estimation of irrigation water quality index with development of an optimum model: a case study. Environ. Dev. Sustain. 22 (5), 4771–4786.

Zemunac, R., Savic, R., Blagojevic, B., Benka, P., Bezdan, A., Salvai, A., 2021. Assessment of surface and groundwater quality for irrigation purposes in the Danube-Tisa-Danube hydrosystem area (Serbia). Environ. Monit. Assess. 193 (8), 1–19.

Zhao, X., Guo, H., Wang, Y., Wang, G., Wang, H., Zang, X., Zhu, J., 2021. Groundwater hydrogeochemical characteristics and quality suitability assessment for irrigation and drinking purposes in an agricultural region of the North China plain. Environ. Earth Sci. 80 (4), 1–22. doi:10.1007/s12665-021-09432-w.

Zhou, Y., Li, P., Chen, M., Dong, Z., Lu, C., 2021. Groundwater quality for potable and irrigation uses and associated health risk in southern part of Gu'an County, North China Plain. Environ. Geochem. Health 43 (2), 813–835. doi:10.1007/s10653-020-00553-y.

17

Geo-spatial assessment of groundwater drought risk zone due to drought propagation in the Upper Dwarakeshwar River Basin (UDRB), West Bengal

Ujjal Senapati[a], Debasish Talukdar[a], Dipankar Saha[a] and Tapan Kumar Das[b]

[a]DEPARTMENT OF GEOGRAPHY, COOCHBEHAR PANCHANAN BARMA UNIVERSITY, COOCH BEHAR, WEST BENGAL, INDIA [b]DEPARTMENT OF GEOGRAPHY, COOCH BEHAR COLLEGE, COOCH BEHAR, WEST BENGAL, INDIA

17.1 Introduction

A constant and serious lack of precipitation in an area for a time span is usually known as drought (Zargar et al., 2011). Droughts are typically divided into four types, as seen in Fig. 17.1, namely meteorological drought, agricultural drought (soil moisture drought), hydrological drought (stream flow, groundwater drought), and socioeconomic drought (Belal et al., 2012; Van Loon, 2015). Drought might be caused by both natural and anthropogenic elements. The term "groundwater drought" (GWD) was first used by Rutulis in 1987 (Rutulis, 1989) and later expounded by Calow et al. (1997). The actual understanding of GWD speaks about the effects of a meteorological drought upon the groundwater framework. Groundwater recharge as well as storage of groundwater and discharge may get disturbed due to the absence of rainwater (Calow et al., 1997; Rutulis, 1989). GWD is a kind of hydrological drought that can be expressed when the groundwater level remains below normal or lower during the spring season (Van Loon, 2015; Marchant and Bloomfield, 2018). A GWD is defined as a period of declining groundwater levels that causes water-related issues. Due to variations in groundwater conditions and groundwater demand for humans and the environment, the amount of groundwater decrease that would be termed as drought varies regionally and locally. Drought-related groundwater shortages or increased extraction can result in

Case Studies in Geospatial Applications to Groundwater Resources. DOI: https://doi.org/10.1016/B978-0-323-99963-2.00019-5

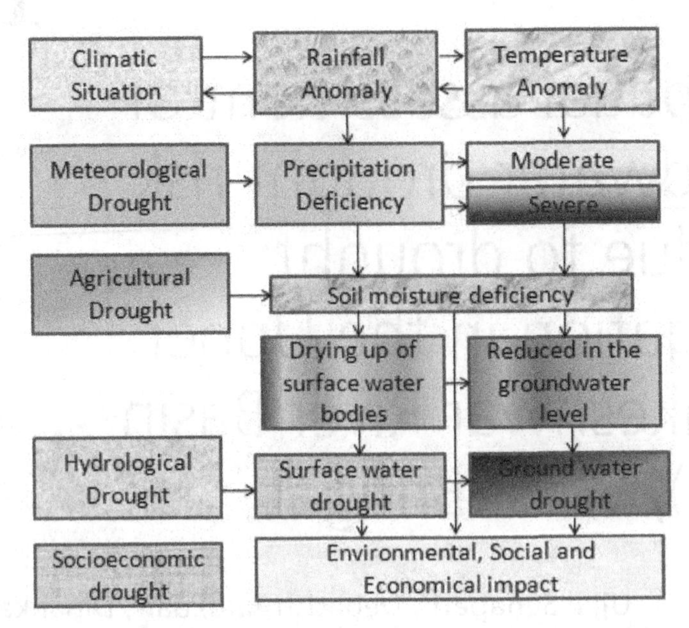

FIGURE 17.1 Different types of droughts and their progression are represented in this diagram (Prepared by the author).

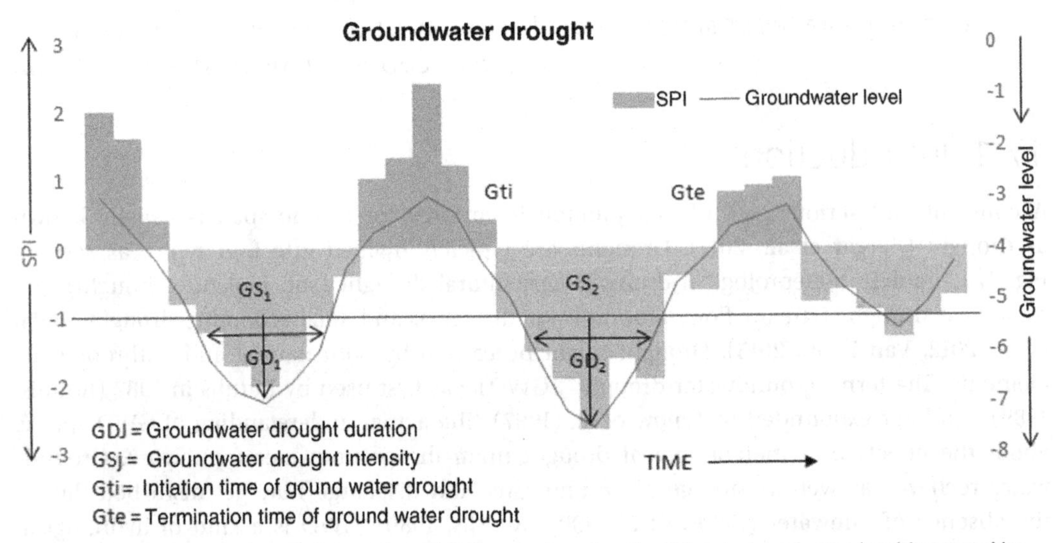

FIGURE 17.2 A diagram of groundwater drought character by using the SPI and groundwater level (prepared by the author).

lower water levels and flows in lakes, streams, and other water bodies. Groundwater is an essential natural resource during a drought. Characteristics of GWD have been shown in the Fig. 17.2.

Whenever such a situation is prevailing in any area, the long-term stress can be seen upon the social, economic, and environmental scenario (Shahid and Hazarika, 2010). Consequently, the water scarcity in the boreholes, directly and indirectly, disturbs the sufficient water supply to the common people; restrictes use water for agricultural and industrial sector; decline in groundwater discharge to the groundwater-fed (Effluent) rivers and wetlands, creating pressure on ecological balance and lead loss of other valuable amenities (Lange et al., 2016). This propagation of effects through the hydrological framework (Peters et al., 2003), which dynamically influences further and bigger aquifer system, infers that groundwater might be utilized as a drought mitigation measure during beginning phases of a drought. It additionally, the usual mitigation measures become failed when the GWD occurs at the later phases of meteorological drought (Calow et al., 1997). GWD may reduce the phreatic groundwater levels (Mishra and Singh, 2010), hampering the access of water shallow wells, tube well and springs.

Groundwater plays a significant role all around the world, more specifically in semi-arid and arid regions where most of the domestic as well as productive water demand is fulfilled by the groundwater supply. If lack of water resources is observed during a drought, groundwater may be an alternate or additional supply of water during the periods of surface-water drought. In recent decades, climate change and global warming have resulted in a precipitation shortfall and an increase in evapotranspiration as a result of a temperature rise directly altering a region's groundwater recharge. Therefore, researches on GWD, are analyzing the variability of groundwater in different countries and regions such as, United Kingdom (Bloomfield and Marchant, 2013), England (Marchant and Bloomfield, 2018), African region (Villholth et al., 2013), China (Han et al., 2019), South Germany and the Central Netherlands (Kumar et al., 2016), Bangladesh (Shahid and Hazarika, 2010), GRACE region (Thomas et al., 2017). Similarly, GWD research is being done regionally in India. Such as Karnataka (The Ghataprabha river basin) (Pathak et al., 2018), Orissa (Panda et al., 2007), West Bengal (The Shilabati river basin) (Halder et al., 2020). Groundwater provides a critical support to the nation like India. As shown in a research paper, extensive pumping of groundwater for irrigation has resulted in a long-term GWD (Asoka and Mishra, 2019). Groundwater tables are declining rapidly in some parts of western and southern India, leading to GWD (Bhanja et al., 2017). Two surveys of groundwater changes were conducted across India, using the Gravity Recovery and Climate Experiment (GRACE) satellite-based model. Ground reality verification studies of the models have been conducted using 5000 and 19,000 observation wells, respectively. In recent times, the Indian state of West Bengal, especially Bankura, Purulia, Jhargram, and West Midnapore districts have been experiencing continuous drought (Bhunia et al., 2020; Senapati et al., 2021; Senapati and Das, 2021) and the water level in the wells has decreased due to drying up of water bodies and increasing temperature graph levels (Halder et al., 2020; Raha and Gayen, 2020).

Sometimes due to the lack of proper groundwater management system and over dependency upon groundwater can cause vulnerable condition for groundwater. To get rid of this problem little consideration needs to be paid for appropriate groundwater management system in this region. Continuous attention to groundwater management and raising consciousness among people may improve the groundwater problem (Calow et al., 1997). But it goes without saying that GWDR assessment has not been done, especially in Indian regions. If the GWDR zone map of the region is prepared before the onset of GWD, then the disaster management team can make a significant contribution. Presently, the integration of remote sensing (RS) and geographic information system (GIS) strategies play an important role in identifying hazard, vulnerability, and risk in the study of disaster management. GIS-based multi-criteria decision analysis (MCDA) technique is a frequently used (Han et al., 2019; Shahid and Hazarika, 2010) standard tool for allocating weight to the factors based on their relative importance using subjective knowledge, previous information, and site-specific settings of a GWDR study. Such as, Weighted Overlay Analysis (WOA) (Villholth et al., 2013).

Thus, the objective of this study is the identification of GWDR zone in the upper Dwarakeshwar River basin (UDRB), West Bengal, using GIS-based weighted overlay analysis technique for the management and planning of groundwater resources. In areas with low rainfall, people are more dependent on groundwater, in those areas unplanned groundwater withdrawal and climate change reduce groundwater recharge, which increases environmental stress on the area. So it is essential to take the necessary steps like rainwater harvesting and artificial groundwater structure development to integrate groundwater and drought management using the methodology of this study over other arid and semi-arid regions.

17.2 Study area

The Dwarakeswar river is one of the most important eastern flowing rivers in Bengal's western region. The Upper Dwarakeshwar River Basin (UDRB) is located in the middle portion of the Purulia and Bankura districts and the basin has an elliptical form, which actually originated in the Purulia district's Tilboni Hill and it is a subsystem of the Ganga-Bhagirathi River system. This river basin is part of the "Chotanagpur Plateau" geomorphological zone. This rain-fed river is located between the latitudes of 23° 08′ 58.80″ and 23° 31′ 55.88″ north, and the longitudes of 86° 30′ 52.43″ and 87° 09′ 13.34″ east. It covers a total area of 1934 km^2. This area is entirely made up of pre-Cambrian crystalline and alluvium deposits. The basin region straddles the administrative boundaries of Bankura's Chhatna, Bankura-I, Bankura-II, Indpur, Onda, Gangajalghati, and Saltora Blocks and it covers the blocks of Chhatna, Bankura-I, Bankura-II, Indpur, Onda, Gangajalghati, and Saltora in the Purulia district (Fig. 17.3). Every year, the study area experiences drought, which generates numerous problems in terms of economic activity. The climate in the region is subtropical, with scorching hot summers in March and June. The average annual rainfall is from 1300 mm to 1550 mm, with the highest amount (about 80%) falling during the monsoon season from June to September, and the average temperature is 25–28°C. In May, the high temperature can reach 46°C, while the low temperature in the winter can fall below 7°C. The agricultural activity employs roughly the majority of the population. The principal crop in this region is Aman paddy, which is harvested during the Kharif season. On the other hand, Rabi crops are wheat, mustard, till, and potato.

17.3 Methodology

Groundwater drought risk (GWDR) mapping has been created utilizing a composite mapping analysis approach in a conventional GIS in an ArcGIS 10.3.1 environment, based on their relative weights. Several thematic layers depicting various factors controlling GWDR are layered and

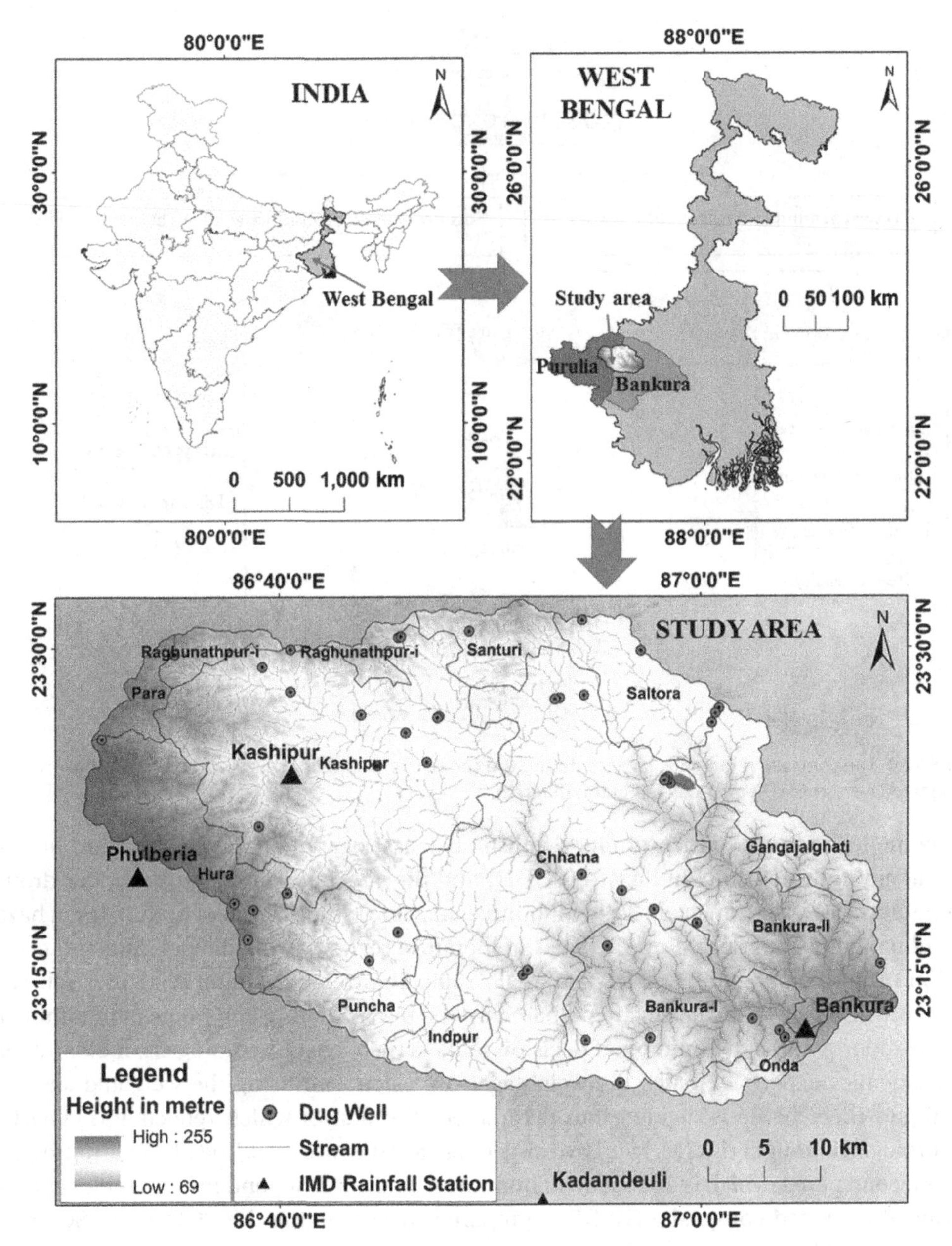

FIGURE 17.3 Location map of study area.

mathematically integrated using a simple linear algorithm and an accompanying weighting system to generate a geographically distributed measure of GWDR across the Upper Dwarakeswar River Basin (UDRB) region. Fig. 17.4 shows a schematic diagram of the components and thematic layers contained in the model.

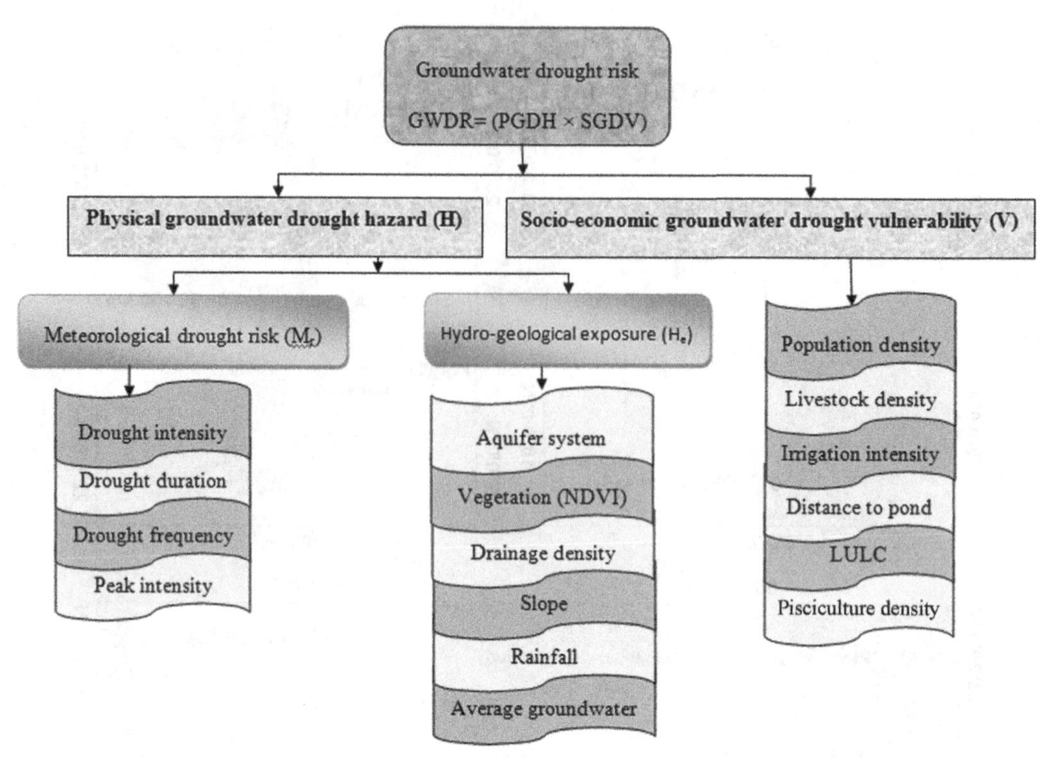

FIGURE 17.4 The controlling thematic layers entering the composite mapping analysis and the resulting aggregated layers in GWDR are depicted in this diagram.

The model attempts to incorporate a holistic risk assessment taking into account both the physical groundwater drought hazard (PGDH) and the socio-economic groundwater drought vulnerability (SGDV) (Han et al., 2019; Villholth et al., 2013). The criteria of both types of hazard and vulnerability are selected on the basis of literature review, availability of data and experts opinion (Abbasi et al., 2019; MacDonald et al., 2009; Sharma and Mujumdar, 2017; Yeh et al., 2009). Individual thematic layers are prepared on the basis of rating and range maintaining the relative importance as per opinion of decision makers (geologists, hydrologists, meteorologists and academics experts). Finally, the weightages are calculated using the weighted aggregate technique. The scheme is divided into two panels, the first of which reflects the PGDH via meteorological drought risk (M_r) and hydro-geological exposure (H_e). The SGDV is represented by the second panel, which is a function of human groundwater dependency for their economic activity. The needed composite GWDR is prepared by combining the PGDH and SGDV. The following algorithm is using for calculating this GWDR.

$$\mathrm{GWDR} = (\mathrm{PGDH} \times \mathrm{SGDV})$$

where GWDR = groundwater drought risk, PGDH = physical groundwater drought hazard and SGDV = socio-economic groundwater drought vulnerability.

$$\mathrm{PGDH} = (Mr + He)$$

Mr = meteorological drought risk and He = hydro-geological exposure.

Before create algorithm, each variable classes range has been scored on an ordinal scale from 1 to 100, representing "minimum" and "maximum" on the other hand, subclasses rating has been graded from 1 to 5, indicating "very low," "low," "moderate," "high," and very high" risk, respectively. Finally, the weight has been calculated by multiplying the range and rating. M*r*, H*e*, and SGDV has been calculated by using weighted sum strategy is expressed in the following way:

$$\mathrm{AA} = \sum_{i=1}^{n} (x^k \times y^l)$$

where AA expressed the M*r*, H*e*, and SGDV. x^k is represent the range of each factor class, y^l is denotes the rating of each factor subclass, and *n* signifies the total number of thematic layers. GWDR mapping databases are taken from freely accessible extracts. Table 17.1 lists all the data and its details. The data was collected at a variety of spatial resolutions; however, it was all transposed onto a common pixel with a resolution of 30 m.

17.4 Physical Groundwater Drought Hazard (PGDH)

The PGDH risk consolidates the danger of meteorological drought season and the intrinsic hydro-geological inclination to drought spell. The hydro-geological drought spell inclination is related with the potential for groundwater recharge and the inborn aquifer productivity.

17.5 Meteorological drought risk (M*r*)

Meteorological drought is a recurrence phenomenon. A general meteorological drought risk index (M), as a marker of the intensity, peak intensity, frequency, and duration of the drought has been determined for the UDRB region. Utilizing the precipitation information related with 9 meteorological stations is extricated from the soil and water assessment tool (SWAT) created from climate forecast system reanalysis (CFSR) gridded dataset. From the precipitation information, standardized precipitation index (SPI) is created. SPI was created by Mckee et al. (1993) to monitor dry and wet spells. To consider a correlation with more conventional perception-based examinations, the investigation has been limited to precipitation as the most significant parameter for drought. It has been noted that SPI-based climatic factors such as drought intensity, peak intensity, drought frequency, and drought duration are important parameters to assess meteorological drought indicators. The aggregated SWAT dataset consists of 9 meteorological stations' rainfall amounts on a daily basis, covering the 35 years period from 1979 to 2014. SPI has been shown to be the most suitable and reliable method for use calculating multi-time scale metrological as well as agricultural drought index in various places of the world. The departure of total rainfall (x) from the long-term rainfall mean (x) is calculated using the input data. After the process is completed, the total deviation is divided by the rainfall standard deviation (δ), which is calculated as follows: (Salehnia et al., 2017).

$$\mathrm{SPI} = \frac{\mathrm{X_i} - \bar{\mathrm{X}}}{\delta}$$

Table 17.1 Meta-data and its details in GWDR.

Data set	Data description	Source	Data type/method	Thematic layer
Meteorological drought risk	36-year daily SWAT rainfall data associated with 9 meteorological stations from 1979 to 2014, developed from Climate Forecast System Reanalysis (CFSR) gridded data set.	Global Weather Data for SWAT (https://globalweather.tamu.edu/)	Point data, interpolation (IDW) method	Drought intensity Drought duration Drought frequency Peak intensity
Hydro-geological exposure	Aquifer systems of India (scale—1:50,000)	Central Ground water Board (CGWB), Ministry of Water Resources, Government of India.	Soft copy, scanned, georeferencing, and digitization	Aquifer system
	LANDSAT 8 (spatial resolution 30 m) (LC08_L1TP_139044_20181220_20181227_01_T1)	USGS EarthExplorer (https://earthexplorer.usgs.gov/)	TIFF, (Nir - red)/(Nir + red)	NDVI
	SRTM DEM (spatial resolution 30 m) (n23_e086_1arc_v3, n23_e087_1arc_v3)		DEM, GIS processing	Drainage density Slope
	21-year groundwater data associated with 50 dug well point from 1996 to 2017	Central Ground Water Board (CGWB), India. (http://cgwb.gov.in/)	Point data, interpolation (IDW) method	Groundwater level
	Average rainfall data of Bankura (CWC), Kadamdeli, Phulberia, and Kashipur meteorological stations.	India Meteorological Department (IMD), Pune (https://dsp.imdpune.gov.in/)	Point data, interpolation (IDW) method	Rainfall
Socio-economic groundwater drought vulnerability	LANDSAT 8 (Resolution 30 m) (LC08_L1TP_139044_20181220_20181227_01_T1)	USGS EarthExplorer (https://earthexplorer.usgs.gov/)	TIFF, maximum likelihood classification	LULC
	Total population/total area	Census of India, 2011 (https://censusindia.gov.in)	Spatial data, quantities technique	Population density
	Irrigated area/total area	District Statistical Handbook, Bankura and Puruliya, in 2014		Irrigation density
	Total agricultural area/total area			Agricultural area density
	No of livestock/total area			Livestock density
	Piscicultural area/total area			Pisciculture density

The long-term rainfall is then fitted to the probability distribution and then transformed into the normal distribution to the mean SPI for the location and the desired period is zero,

$$SPI = \frac{a - M}{\delta}$$

where "a" is the individual gamma distribution, "M" is mean and δ is the standard deviation of rainfall. According to Mckee et al. (1993) negative SPI values denote more drought risk.

17.6 Parameters used in meteorological drought risk assessment

This study considers drought intensity (D_I), drought duration (D_D), drought peak intensity (P_I), and drought frequency (D_F) for evaluation of meteorological drought risk assessment. The description of parameters is as follows:

17.6.1 Drought intensity (D_I)

According to Dupigny-Giroux (2001) departure (down) of a SPEI from its' normal value can be stated as drought intensity. According to Abbasi et al. (2019), a drought event is defined as a period when the SPEI is continuously negative and SPI reaches a value of -1.0 or less. So, here I_D denotes the value of SPEI which is less than the value of -1.0. Lesser the value of SPI more will be the intensity of drought (Ghosh, 2019).

17.6.2 Drought duration (D_D)

Spinoni et al. (2014) defines the duration of drought in a fine tune with the help of run theory. Drought event starts when the SPEI is continuously negative and reaches to the intensity of –1.0 or less while the event ends when the SPEI becomes positive. Thus, duration of drought is the continuous negative dimension of SPEI (Abbasi et al., 2019; Ghosh, 2019).

17.6.3 Drought frequency (D_F) or occurrence rate (%)

The number of droughts per 35 years calculated using following formula:

$$D_{Fj,35} = \frac{M_j}{j.m} \times 100\,(\%)$$

where $D_{Fj},_{35}$ is the frequency of droughts for timescale j in 35 years; Nj is the number of months with droughts for time scale j in the n-year set; j is time scale (3 months); n is the number of years in the data set (Wang et al., 2014; Ghosh, 2019).

Table 17.2 Rating, range scheme and weights for meteorological drought risk elements in the base scenario.

Meteorological drought risk parameters	Sub-classes	Rating	Range	Weight
Drought intensity	−1.372 to −1.349	5	25	125
	−1.348 to −1.335	4		100
	−1.334 to −1.321	3		75
	−1.320 to −1.303	2		50
	−1.302 to −1.278	1		25
Drought duration	15.00 to 15.42	1	35	35
	15.43 to 15.75	2		70
	15.76 to 16.06	3		105
	16.07 to 16.41	4		140
	16.42 to 16.90	5		175
Drought frequency	−2.722 to −2.478	5	20	100
	−2.477 to −2.293	4		80
	−2.292 to −2.137	3		60
	−2.136 to −2.013	2		40
	−2.012 to −1.893	1		20
Peak intensity	12.5 to 13.12	1	20	20
	13.13 to 13.61	2		40
	13.62 to 14.08	3		60
	14.09 to 14.64	4		80
	14.65 to 15.42	5		100

17.6.4 Peak intensity (P_I),

The SPI can easily be used to determine peak intensity. The maximum drought intensity refers to the highest level of drought, also known as peak intensity.

All thematic maps have been prepared based on this SPI calculated information by using the world wide acceptable inverse distance weighted (IDW) interpolation method of the ArcGis software. The categorization and weighting scheme used to determine the relative contribution of the factors considered in the meteorological drought risk assessment are listed in Table 17.2. Fig. 17.5 shows the different thematic layers of meteorological drought risk.

17.7 Hydro-geological exposure (He)

The capability of hydro-geological elements is a major determining factor in GWDR. Hydrogeology is the branch of geology concerned with the distribution and movement of groundwater in the Earth's crusts. Here we have made the GWDR zone using the factors aquifer system, normalized difference vegetation index (NDVI), drainage density, rainfall, groundwater level, and slope. The aquifer determines the recharge capacity as well as the groundwater storage capacity.

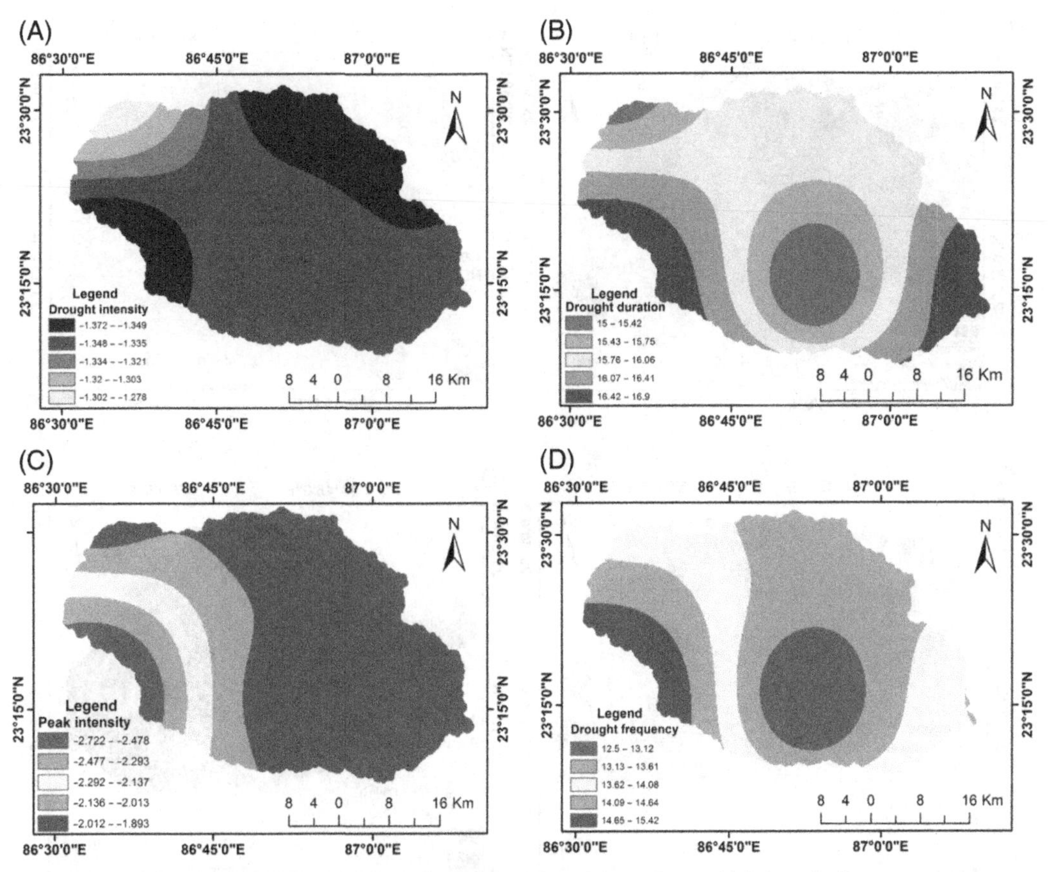

FIGURE 17.5 (A) drought intensity, (B) drought duration, (C) peak intensity, and (D) drought frequency.

The greater the aquifer thickness, the larger the improved infiltration and thus groundwater recharge storage capacity, and vice versa. It is considered that good plant cover, as measured by a high long-term average NDVI improves infiltration and thus recharge, whereas poor vegetation cover obstructs recharge and increases surface runoff. High drainage density areas will have limited penetration, resulting in insufficient runoff, and vice versa. Lower density values are therefore more beneficial for increased groundwater recharge, and greater weights are applied. Direct infiltration from net rainfall causes recharge in a distributed manner. In semi-arid and dry locations, recharging may be controlled by targeted water entry from ephemeral water bodies. The area has a significant capacity for groundwater it is not particularly exposed to drought and vice versa. The differential distribution of net rainfall between overland flow and soil infiltration is influenced by topography, or rather terrain slope. A steep slope indicates that there will be less recharge. The categorization and weighting scheme used to determine the relative contribution of the factors considered in the hydro-geological exposure assessment are listed in Table 17.3 and Fig. 17.6 shows the different thematic layers of hydro-geological drought exposure.

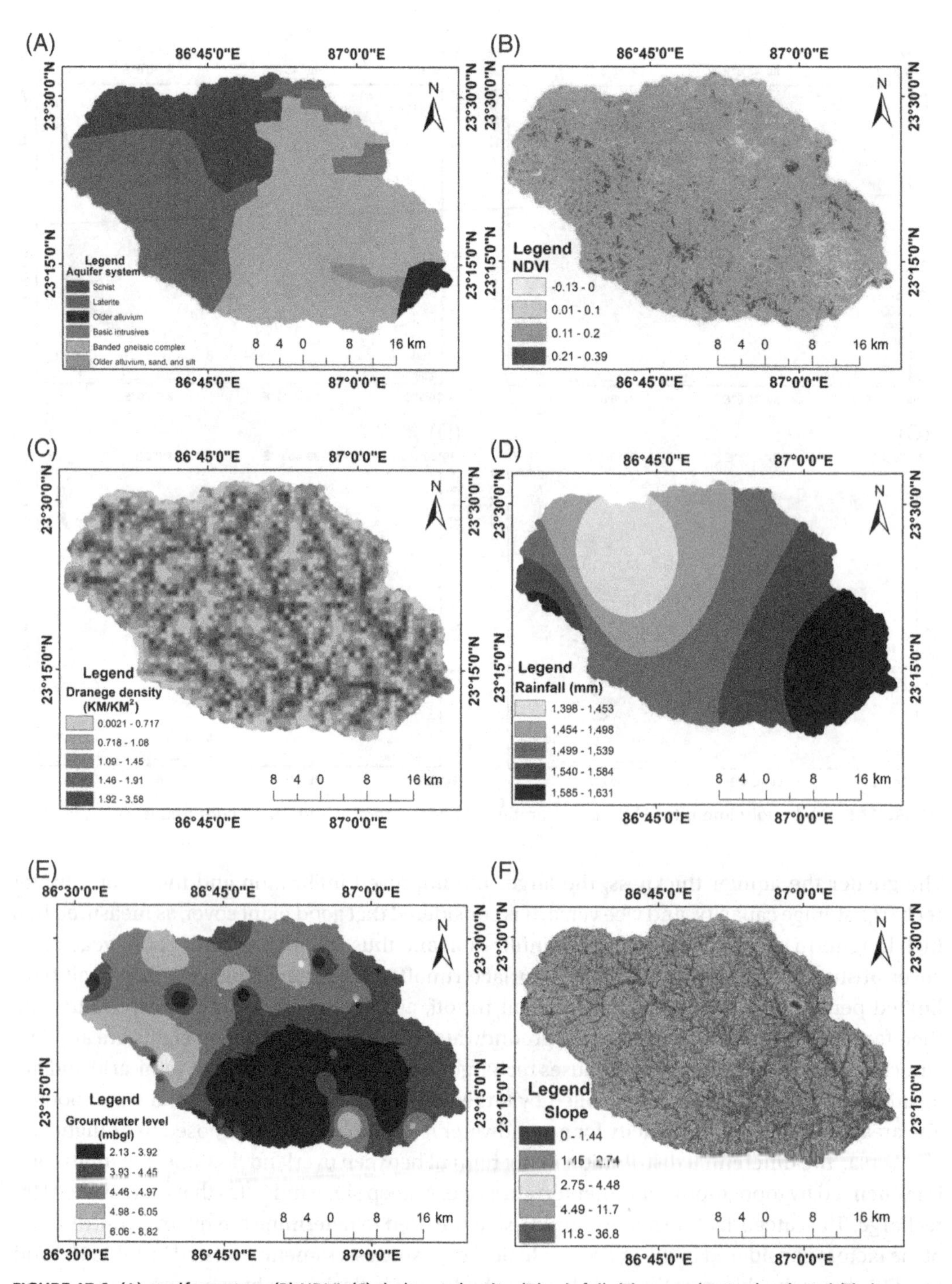

FIGURE 17.6 (A) aquifer system, (B) NDVI, (C) drainage density, (D) rainfall, (E) groundwater level, and (F) slope.

Table 17.3 Rating, range scheme and weights for hydro-geological exposure elements in the base scenario.

Hydro-geological drought exposure	Sub-classes	Rating	Range	Weight
Aquifer system	Schist	4	25	100
	Laterite	4		100
	Older alluvium	1		25
	Basic intrusive	5		125
	Banded gneissic complex	3		75
	Older alluvium, sand, and silt	2		50
NDVI	−0.13 to 0	5	15	75
	0.01 to 0.1	4		60
	0.11 to 0.2	3		45
	0.21 to 0.39	2		30
Drainage density	0.0021–0.717	5	15	75
	0.718–1.08	4		60
	1.09–1.45	3		45
	1.46–1.91	2		30
	1.92–3.58	1		15
Rainfall	1398–1453	5	20	100
	1454–1498	4		80
	1499–1539	3		60
	1540–1584	2		40
	1585–1631	1		20
Groundwater level	2.13–3.92	1	20	20
	3.93–4.45	2		40
	4.46–4.97	3		60
	4.98–6.05	4		80
	6.06–8.82	5		100
Slope	0–1.44	1	5	5
	1.45–2.74	2		10
	2.75–4.48	3		15
	4.49–11.7	4		20
	11.8–36.8	5		25

17.8 Socio-economic Groundwater Drought Vulnerability (SGDV)

Drought vulnerability assessment connected with groundwater reliance is highly dependent on human socioeconomic conditions. The majority of groundwater demand in UDRB is driven by distributed use for residential, livestock, and irrigation. As a result, basic indices for population density, irrigation density, agricultural area density, livestock density, pisciculture density, as well as land use and landcover, represent the socio-economic groundwater dependence. Increased SGDV is associated with higher population, as more people must essentially share

Table 17.4 Rating, range scheme and weights for socio-economic groundwater drought vulnerability elements in the base scenario.

Socio-economic groundwater drought vulnerability	Sub-classes	Rating	Range	Weight
Land use and landcover	Natural vegetation	1	25	25
	Water bodies	2		50
	Settlement	3		150
	Fallow land	4		200
	Agricultural land	5		250
Population density	375–428	1	20	20
	429–482	2		40
	483–535	3		60
	536–589	4		80
	590–642	5		100
Irrigation density	0.0587–0.182	1	20	20
	0.183–0.306	2		40
	0.307 -0.429	3		60
	0.43–0.553	4		80
	0.554–0.676	5		100
Agricultural area density	0.0462–0.33	1	15	15
	0.331–0.614	2		30
	0.615–0.898	3		45
	0.899–1.18	4		60
	1.19–1.47	5		75
Livestock density	250.1–299.6	1	15	15
	299.7–349.1	2		30
	349.2–398.5	3		45
	398.6–448	4		60
	448.1–497.5	5		75
Pisciculture density	0.023–0.024	1	5	5
	0.025–0.031	2		10
	0.032–0.044	3		15
	0.045–0.054	4		20
	0.055–0.064	5		25

the same resource. Population density is an important component of human susceptibility, during drought is also reflected in groundwater dependence. Similarly, more dependence on groundwater irrigation means a higher GWDV. It might be claimed that having groundwater irrigation infrastructure in place can help with drought resistance. Larger GWDV is associated with higher livestock density. In the same way, if the amount of agricultural area density in a region is more, more groundwater depends on it. Pisciculture density also creates pressure on groundwater. on the other hand, land use and landcover is also a significant parameter to assess socio-economic vulnerability. The nature of land characteristics is influencing the use of groundwater. For example, if the agricultural land of a region is more, the use of groundwater increases. Similarly, if there is forest land, groundwater storage is higher. Again, the rate of groundwater uptake in populated areas is higher than infiltration. Wetlands play a significant role in groundwater storage. In this way, the vulnerability of the social-economic GWD has been assessed with the factors used here. Table 17.4 shows the classification and weighting scheme that has been used to calculate the relative contribution of the factors evaluated in the SGDV and Fig. 17.7 shows the different thematic layers of SGDV.

17.9 Result

The distribution pattern of meteorological drought risk in the UDRB region is depicted in Fig. 17.8. The outcome is comparable to UDRB's earlier drought risk mapping. The present method has the benefit of using a regional climate replica that incorporates experiential climate information to determine reliability, physical dependability, and complete exposure of the region. The hydro-geological exposure map and the SGDV map are shown in Figs. 17.9 and 17.10, respectively. The meteorological drought risk map illustrates the climatic sensitivity of the study area. It is seen that during meteorological drought groundwater has facing tremendous pressure. In the study area, the drought intensity ranges from (-1.372 to -1.278), peak intensity ranges from (12.5 to 15.42), frequency ranges from (-2.722 to -1.893), and duration ranges from (15.00 to 16.90) Figure 5 shows the different thematic layers of meteorological drought risk Fig. 17.5. Drought duration and intensity has been the dominant factor of meteorological drought. These findings support a general trend in between drought and groundwater levels. Where dry seasons and drought periods might last for several seasons, it has been observed that there the groundwater level is constantly declining. The hydro-geological exposure map represents the potential groundwater discharge zone. By which the overall condition of groundwater in a region can be known and the budget of groundwater can be calculated. The components of hydro-geological exposure have been influencing the hydrological drought. In this study area, the aquifer system is divided into six aquifer zone. They are respectively, Schist, Laterite, Older alluvium, Basic intrusive, Banded gneissic complex, and Older alluvium, sand and silt. NDVI ranges from -0.13 to 0.39, Drainage density ranges from 0.0021 to 3.58, Rainfall ranges from 1398 to 1631, Groundwater level ranges from 2.13 to 8.82, Slope ranges from 0 to 36.8. The aquifer system, NDVI, drainage density, and rainfall are all dominant factors to assess the hydro-geological exposure map Fig. 17.6. The SGDV map represents the individual and societal awareness and ability to drought preparedness related to groundwater reliance. It is possible to observe in this map how the various economic activities of the people increase the rate of groundwater vulnerability during drought. This map has been prepared by combining the combined effects of landuse and landcover which is divided into five classes, such as Natural vegetation, Water bodies, Settlement, Fallow land and Agricultural land. Population density ranges from 375 to 642, irrigation density ranges from 0.0587 to 0.676, agricultural area density ranges from 0.0462 to 1.47, livestock density ranges from 250.1 to 497.5 and pisciculture density ranges from 0.023 to 0.064 Fig. 17.7. It is seen that map is dominated by the land use and landcover, population density, irrigation density, agricultural area density. Here all map is divided into 5 parts viz, very low, low, moderate, high, and very high. SGDV represent 307 km^2 stand for 16%, 98 km^2 5%, 391 km^2 20%, 895 km^2 46%,

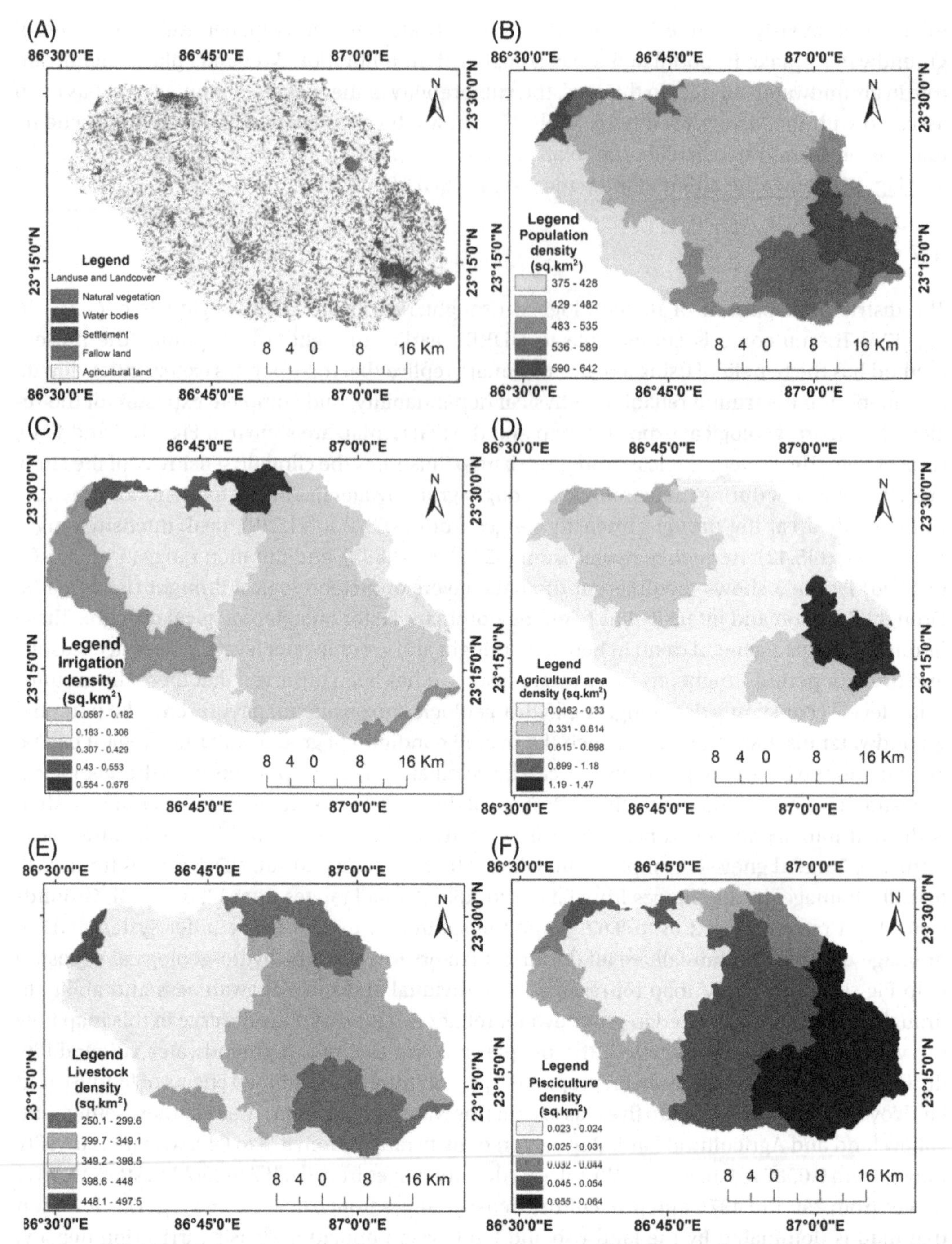

FIGURE 17.7 (A) land use and landcover, (B) population density, (C) irrigation density, (D) agricultural area density, (E) livestock density, and (F) pisciculture density. Impact of climate change on groundwater (SPI vs. groundwater data).

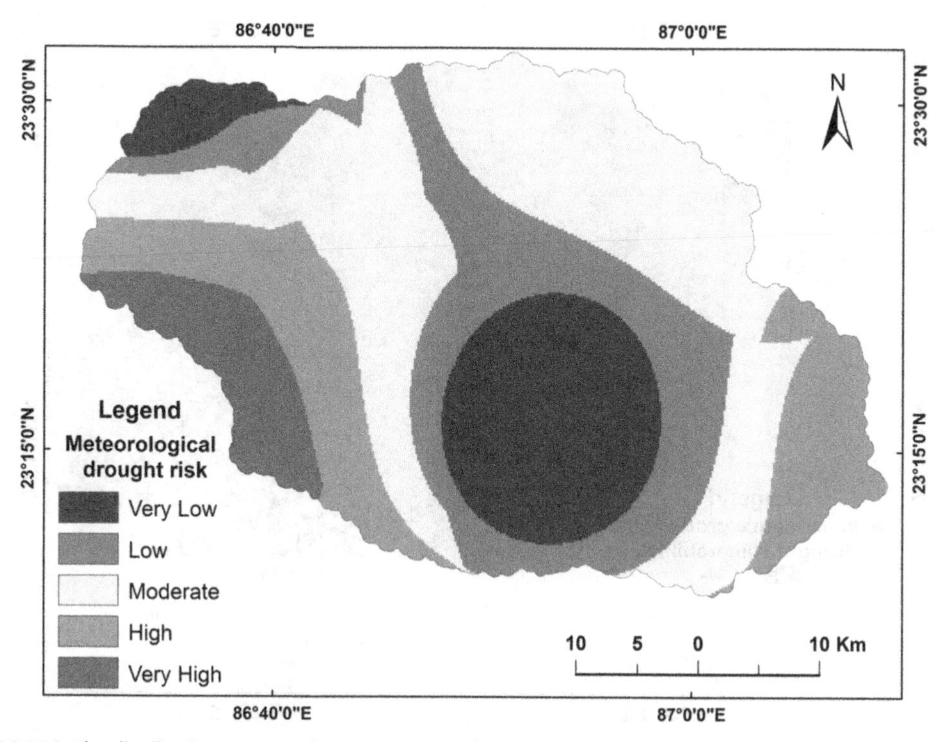

FIGURE 17.8 The distribution pattern of meteorological drought risk in the UDRB region.

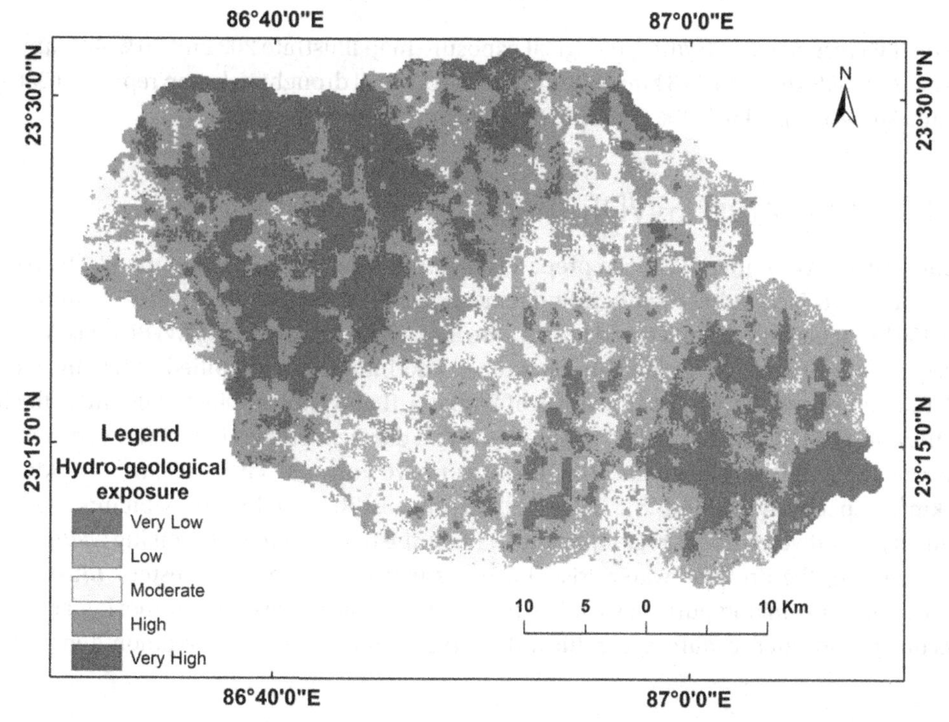

FIGURE 17.9 The hydro-geological exposure map.

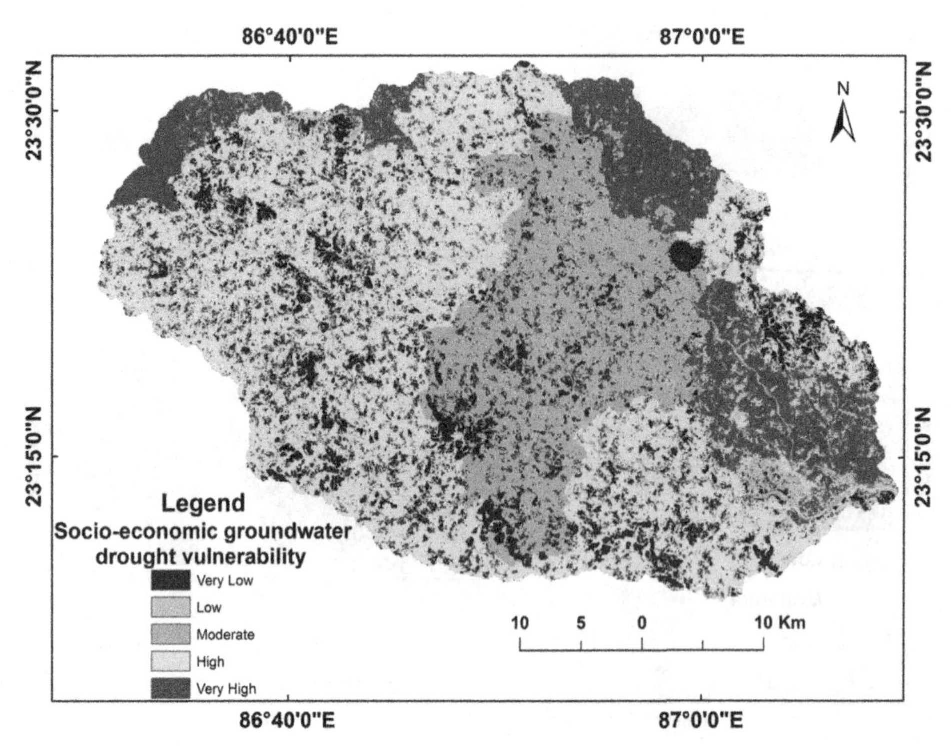

FIGURE 17.10 The socio-economic groundwater drought vulnerability map.

243 km^2 13% respectively. Hydro-geological exposure map illustrate 202 km^2 10%, 471 km^2 24%, 380 km^2 20%, 508 km^2 26%, 373 km^2 19% and meteorological drought risk map represent 330 km^2 17%, 389 km^2 20%, 742 km^2 38%, 340 km^2 18%, 133 km^2 7% area, respectively.

17.10 Integrated GWDR map

The integrated GWDR map has been prepared by combining the SGDV and the PGDH parameters. These two factors have same influencing capacity to predict the GWDR. Saltora, Para, Hura, Kashipur, Raghunathpur-I, and Puncha are located in the very high GWDR areas. Whereas the regions of Chhatna, Gangajalghati, Bankura-I, and Indpur are identified as having a low to moderate risk of a GWD. Most of the areas of Bankura-II and Santuri blocks fall under the high GWDR zone. GWDR map has been classified into five different classes (Fig. 17.11) very low, low, moderate, high, and very high, covering the area of 270 km^2, 437 km^2, 495 km^2, 477 km^2 and 255 km^2, respectively (Table 17.5). These findings suggest that the local inhabitants are highly reliant on groundwater. In this region, groundwater is the major source of drinking water. During the dry season, the drinking water crisis becomes worse. People and livestock must rely on groundwater for drinking purposes while surface water typically dries up, imposing additional stress on groundwater resources. Significantly population increase and irrigation density leads

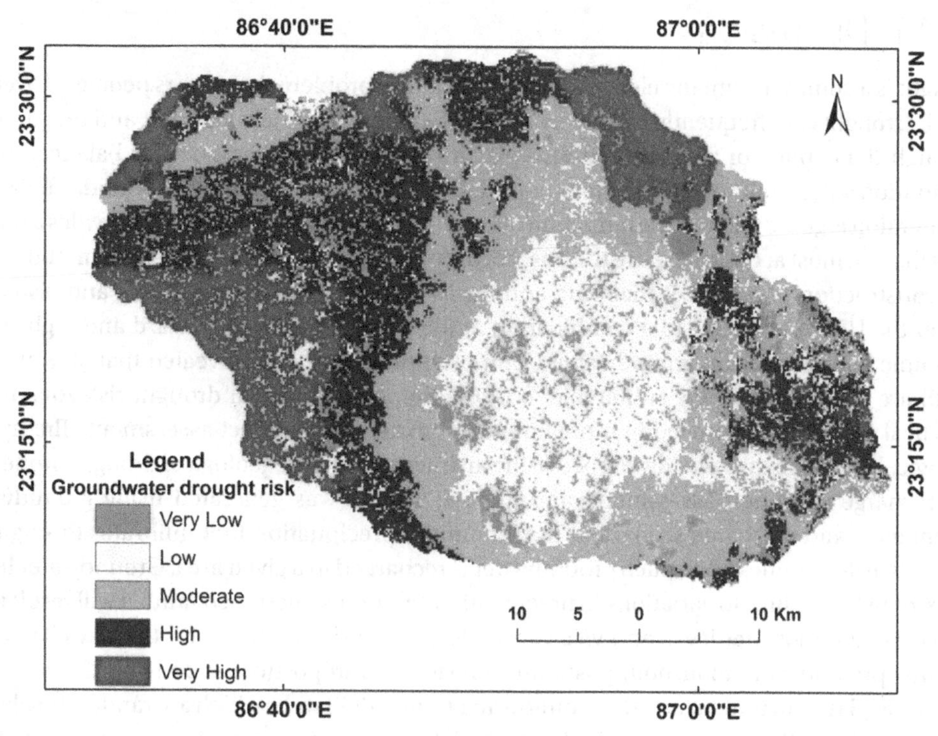

FIGURE 17.11 Spatial distribution of integrated Groundwater Drought Risk (GWDR) zone map of UDRB region.

Table 17.5 Area and characteristics of different groundwater drought risk zone of UDRB.

Zone	Area	%	Characteristic
Very low	270	13.9607	Areas are very weakly affected by groundwater drought risk
Low	437	22.59566	Areas are slightly and weakly affected by groundwater drought risk
Moderate	495	25.59462	Areas are intermediately affected by groundwater drought risk
High	477	24.66391	Areas are highly affected by groundwater drought risk
Very high	255	13.18511	Areas are strongly affected by groundwater drought risk

to high SGDV. In the Upper Dwarakeshwar River Basin, a predominantly high proportion of people reside in areas with a high risk of GWD; this has to be taken into account and accentuated in vulnerability reduction and drought management. Drought propensity and livestock density are correlated. When rural agriculture is hindered by droughts, local communities experience financial hardship. At that time, the income generated by domesticated livestock helps the local people to deal with a few of their financial difficulties. Animal husbandry has emerged as an alternative source of livelihood that depends on the native vegetation and natural landscape.

17.11 Discussion

Drought is a common climatic element as well as a major problem that affects people all over the world. Droughts are frequently caused by a lack of proper water development and preparation, although the climate of a region can also play a key role. Whatever basis for balanced water distribution is probably to be lapsed for a while following the start of a drought period, after which the attention is going to be on the reduction of harsh impacts. The current study provides a map of areas that are most at GWDR before the onset of drought. This chapter describes the methodology and construction of a GWDR mapping and decision support tool for analysis and visualization in the UDRB area, due to a combination of high climatic drought hazard and high socio-economic dependence on groundwater. Composite mapping study revealed that 38% areas of UDRB are relatively strongly vulnerable to GDR. The meteorological drought risk zone map's analytical capacity was proved in a preliminary climate change effect assessment. The hydrogeological exposure map was developed using an integrated hydrogeological drought proneness and recharge condition assessment, and the SGDV map was generated using groundwater dependency and discharge scenarios. The amount of precipitation that infiltrates through the soil surface determines how much groundwater is recharged in a given area. Groundwater levels are extremely sensitive to variations in precipitation situations. Increased rainfall will result in an increase in groundwater level, and vice versa. Mean groundwater levels for the 13 wells in four seasons (premonsoon, monsoon, postmonsoon (Rabi), and postmonsoon (Kharif) have been visually displayed to understand the groundwater trend. All 13 wells which are randomly selected from each block. Three types of groundwater variations such as seasonal variations, short-term variations and secular variations are seen here and the groundwater table of all wells is gradually decreasing shown in Fig. 17.12. Fig. 17.13 describes the rise and fall of groundwater with the rainfall anomaly index. The prolonged drought resulted in lower groundwater levels and in turn invites the GWD. So, this study shows that the region is slowly moving towards GWD conditions. Newly developed GWDR map will greatly help prevent GWD. Therefore, drought management and water-security initiatives, particularly for basic rural water supply, must pay special attention to these high-risk areas, which encompass considerable groundwater-dependent populations Drought management and water-security initiatives, particularly for basic rural water supply, demand special attention.

17.12 Validation

The GWD was declared based on the assessment of the long-term time series assessment of groundwater table measurement results. It has been validated based on rainfall and groundwater data collected from the India Meteorological Department (IMD) in Pune and the Central Ground Water Board (CGWB) in India. Data has been taken here from 1996 to 2017. The RAI has been utilized to determine the meteorological drought years in this region. RAI is a widely used standardized index for the measurement of recent weather anomalies and spatial and temporal representations of drought in a region. On the other hand, randomly selecting one well in each block and analyzing its time series shows that the linear trend of each well is negative.

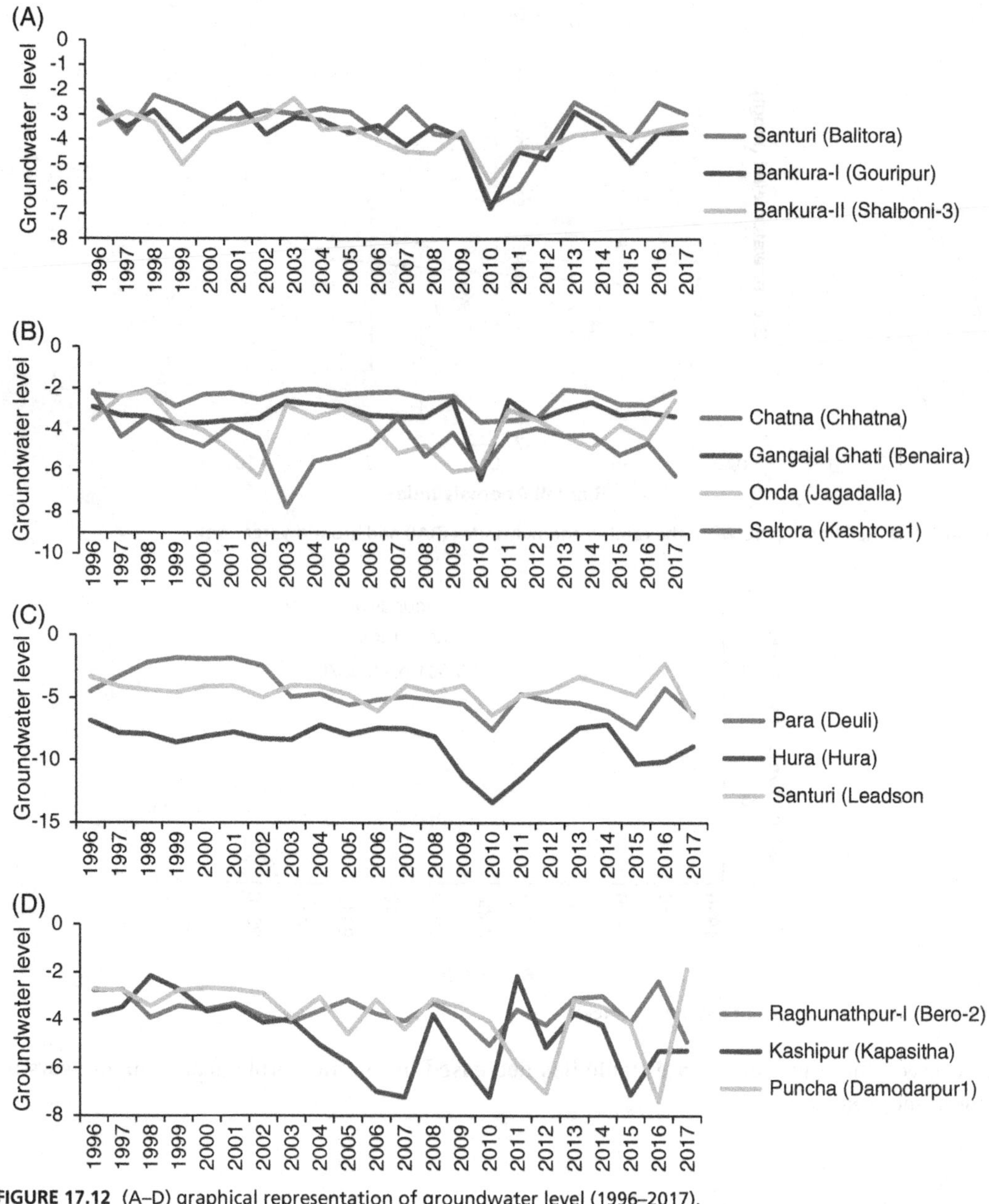

FIGURE 17.12 (A–D) graphical representation of groundwater level (1996–2017).

Groundwater's missing data is considered to be the average value of the three surrounding wells. Finally, the wells in each block have been averaged and drawn a trend line (Fig. 17.14), which is representing the whole region. This shows that the groundwater level of the whole region is slowly declining. Fig. 17.15 shows a linear analysis of the groundwater table with RAI. Where it

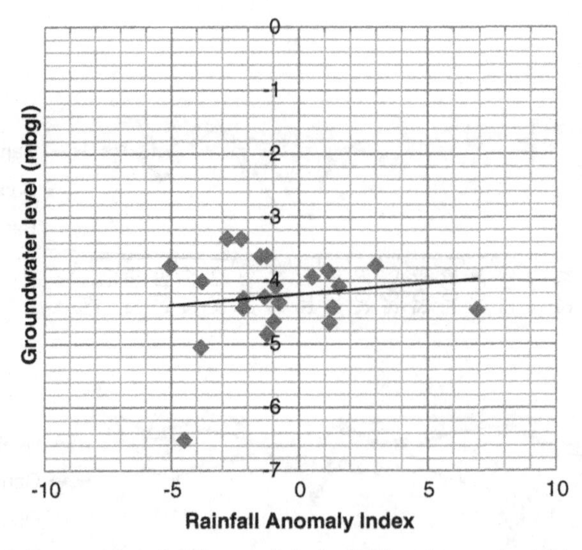

FIGURE 17.13 Relationship between the rainfall anomaly index (RAI) and groundwater level.

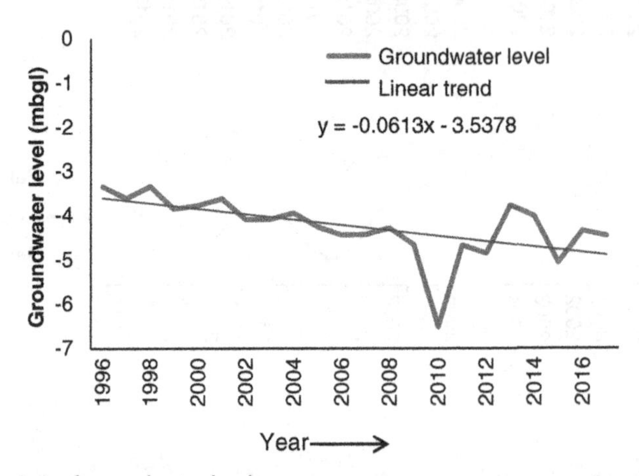

FIGURE 17.14 Trend analysis of groundwater levels.

is observed that the groundwater table has decreased in the meteorological drought years or subsequent years.

17.13 Conclusion

A GWDR determination and mapped by using different type of thematic data and finally all composite parameters have been integrated by the help of GIS modeling. The UDRB has been facing GWDs that last for several months or even seasons, because of massive groundwater withdrawal for water supply. Groundwater will provide a key shield in the early stages of a drought, but during and after a lengthy drought, certain vulnerable groundwater systems may

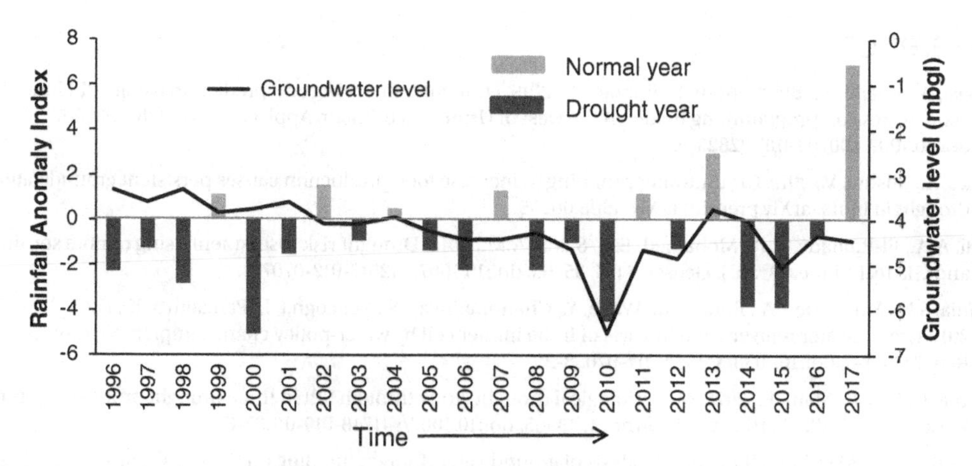

FIGURE 17.15 Graphical representation of the rainfall anomaly index (RAI) and groundwater level (1996–2017), relationship between normal years, drought years and changes of groundwater level.

become desiccated and lose access. In such a situation, the best method is to comprehend the hydro-geological processes at function and their expected consequences, identify vulnerable populations and places. Ensure long-term resource usage through competent and proactive management at regional and local scales. It's worth noting that groundwater resources are only depleted over time due to a lack of infiltration recharge, and that their regeneration necessitates long and abundant rainfall. Even if there is a prolonged drought, there is no major impact on groundwater disposable resources or detrimental effects on intakes. But if the drought persists for a long time or the frequency is too high, then there is a lot of pressure on the groundwater which slowly decreases the groundwater table. As a result, in areas that experience severe groundwater crisis during droughts, those areas need to artificially increase groundwater storage at normal times. Some artificial groundwater recharge structures are constructed such as structural dams, recharge pits and shafts, check dams, percolation pond with wells, gravity head recharge wells, water absorption trenches, horizontal dams, rock-filled earthworks, and farm ponds. During monsoons it must be used to collect and stored excess surface runoff and increase groundwater level by storing rainwater. In this way groundwater can be used properly during drought and GWD can be monitored and assessed. This will play an important role in the development of human society.

Acknowledgement

The authors are sincerely grateful to the Indian Meteorological Department (IMD), the United States Geological Survey (USGS), the Central Groundwater Board (CGWB). The authors also extend their thanks to anonymous reviewers for the valuable constructive comments and suggestions.

Conflict of interest

The authors declare no competing interests.

Reference

Abbasi, A., Khalili, K., Behmanesh, J., Shirzad, A., 2019. Drought monitoring and prediction using SPEI index and gene expression programming model in the west of Urmia Lake. Theor. Appl. Climatol. 138, 553–567. doi:10.1007/s00704-019-02825-9.

Asoka, A., Mishra, V., 2019. Groundwater pumping to increase food production causes persistent groundwater drought in India. arXiv preprint arXiv:1908.00255.

Belal, A.A., El-Ramady, H.R., Mohamed, E.S., Saleh, A.M., 2012. Drought risk assessment using remote sensing and GIS techniques. Arab. J. Geosci. 7 (1), 35–53. doi:10.1007/s12517-012-0707-2.

Bhanja, S.N., Mukherjee, A., Rodell, M., Wada, Y., Chattopadhyay, S., Velicogna, I., Pangaluru, K., Famiglietti, J.S., 2017. Groundwater rejuvenation in parts of India influenced by water-policy change implementation. Sci. Rep. 7 (1), 1–7. doi:10.1038/s41598-017-07058-2.

Bhunia, P., Das, P., Maiti, R., 2020. Meteorological drought study through SPI in three drought prone districts of West Bengal, India. Earth. Syst. Environ. 4, 43–55. doi:10.1007/s41748-019-00137-6.

Bloomfield, J.P., Marchant, B.P., 2013. Analysis of groundwater drought building on the standardised precipitation index approach. Hydrol. Earth Syst. Sci. 17 (12), 4769–4787. doi:10.5194/hess-17-4769-2013.

Calow, R.C., Robins, N.S., Macdonald, A.M., Macdonald, D.M.J., Gibbs, B.R., Orpen, W.R.G., Mtembezeka, P., Andrews, A.J., Appiah, S.O., 1997. Groundwater management in drought-prone areas of Africa. Int. J. Water Resour. Dev. 13 (2), 241–262. https://doi:10.1080/07900629749863.

Dupigny-Giroux, L.A., 2001. Towards characterizing and planning for drought in Vermont: part-I: a climatological perspective. J. Am. Water Resour. Assoc. 37 (3), 505–525. doi:10.1111/j.1752-1688.2001.tb05489.x.

Ghosh, K.G., 2019. Spatial and temporal appraisal of drought jeopardy over the Gangetic West Bengal, eastern India. Geoenviron. Disasters 6, 1. doi:10.1186/s40677-018-0117-1.

Halder, S., Roy, M.B., Roy, P.K., 2020. Analysis of groundwater level trend and groundwater drought using standard groundwater level Index: a case study of an eastern river basin of West Bengal, India. SN Appl. Sci. 2 (3), 1–24. doi:10.1007/s42452-020-2302-6.

Han, Z., Huang, S., Huang, Q., Leng, G., Wang, H., Bai, Q., Zhao, J., Ma, L., Wang, L., Du, M., 2019. Propagation dynamics from meteorological to groundwater drought and their possible influence factors. J. Hydrol. 578, 124102. doi:10.1016/j.jhydrol.2019.124102.

Kumar, R., Musuuza, J.L., Loon, A.F.V., Teuling, A.J., Barthel, R., Ten Broek, J., Mai, J., Samaniego, L., Attinger, S., 2016. Multiscale evaluation of the standardized precipitation index as a groundwater drought indicator. Hydrol. Earth Syst. Sci. 20 (3), 1117–1131. doi:10.5194/hess-20-1117-2016.

Lange, B., Holman, I., Bloomfield, J.P., 2016. A framework for a joint hydro-meteorological-social analysis of drought. Sci. Total Environ. 578, 297–306. doi:10.1016/j.scitotenv.2016.10.145.

MacDonald, A.M., Calow, R.C., MacDonald, D.M., Darling, W.G., Dochartaigh, B.E., 2009. What impact will climate change have on rural groundwater supplies in Africa? Hydrol. Sci. J. 54 (4), 690–703. doi:10.1623/hysj.54.4.690.

Marchant, B.P., Bloomfield, J.P., 2018. Spatio-temporal modelling of the status of groundwater droughts. J. Hydrol. 564, 397–413. doi:10.1016/j.jhydrol.2018.07.009.

McKee, T.B., Doesken, N.J., Kleist, J., 1993. The relationship of drought frequency and duration to time scales. In: Proceedings of the 8th Conference on Applied Climatology, 17. American Meteorological Society, Boston, MA, pp. 179–183.

Mishra, A.K., Singh, V.P., 2010. A review of drought concepts. J. Hydrol. 39 (1-2), 202–216. doi:10.1016/j.jhydrol.2010.07.012.

Panda, D.K., Mishra, A., Jena, S.K., James, B.K., Kumar, A., 2007. The influence of drought and anthropogenic effects on groundwater levels in Orissa, India. J. Hydrol. 343 (3-4), 140–153. doi:10.1016/j.jhydrol.2007.06.007.

Pathak, A.A., Dodamani, B.M., 2018. Trend analysis of groundwater levels and assessment of regional groundwater drought: Ghataprabha River Basin, India. Nat. Resour. Res. 28 (3), 631–643. doi:10.1007/s11053-018-9417-0.

Peters, E., Torfs, P.J.J.F., Van Lanen, H.A., Bier, G., 2003. Propagation of drought through groundwater—a new approach using linear reservoir theory. Hydrol. Process. 17 (15), 3023–3040. doi:10.1002/hyp.1274.

Raha, S., Gayen, S.K., 2020. Simulation of meteorological drought using exponential smoothing models: a study on Bankura District, West Bengal, India. SN Appl. Sci. 2, 909. doi:10.1007/s42452-020-2730-3.

Rutulis, M., 1989. Groundwater drought sensitivity of southern Manitoba. Can. Water Resour. J. 14 (1), 18–33. doi:10.4296/cwrj1401018.

Salehnia, N., Alizadeh, A., Sanaeinejad, H., Bannayan, M., Zarrin, A., Hoogenboom, G., 2017. Estimation of meteorological drought indices based on AgMERRA precipitation data and station-observed precipitation data. J. Arid Land 9 (6), 797–809. doi:10.1007/s40333-017-0070-y.

Senapati, U., Das, T.K., 2021. Assessment of basin-scale groundwater potentiality mapping in drought-prone upper Dwarakeshwar River basin, West Bengal, India, using GIS-based AHP techniques. Arab. J. Geosci. 14 (11), 1–22. doi:10.1007/s12517-021-07316-8.

Senapati, U., Raha, S., Das, T.K., Gayen, S.K., 2021. A Composite assessment of agricultural drought susceptibility using analytic hierarchy process: case study of western region of West Bengal. In: Agriculture, Food and Nutrition Security. Springer, Cham, pp. 15–40. doi:10.1007/978-3-030-69333-6_2.

Shahid, S., Hazarika, M.K., 2010. Groundwater drought in the northwestern districts of Bangladesh. Water Resour. Manag. 24 (10), 1989–2006. doi:10.1007/s11269-009-9534-y.

Sharma, S., Mujumdar, P., 2017. Increasing frequency and spatial extent of concurrent meteorological droughts and heatwaves in India. Sci. Rep. 7 (1), 1–9. doi:10.1038/s41598-017-15896-3.

Spinoni, J., Naumann, G., Carrao, H., Barbosa, P., Vogt, J., 2014. World drought frequency, duration, and severity for 1951-2010. Int. J. Climatol. 34 (8), 2792–2804. doi:10.1002/joc.3875.

Thomas, B.F., Famiglietti, J.S., Landerer, F.W., Wiese, D.N., Molotch, N.P., Argus, D.F., 2017. GRACE groundwater drought index: Evaluation of California Central Valley groundwater drought. Remote Sens. Environ. 198, 384–392. doi:10.1016/j.rse.2017.06.026.

Van Loon, A.F., 2015. Hydrological drought explained. Wiley Interdisciplinary Rev. Water 2 (4), 359–392. doi:10.1002/wat2.1085.

Villholth, K.G., Tøttrup, C., Stendel, M., Maherry, A., 2013. Integrated mapping of groundwater drought risk in the Southern African Development Community (SADC) region. Hydrogeol. J. 21 (4), 863–885. doi:10.1007/s10040-013-0968-1.

Wang, Q., Wu, J., Lei, T., He, B., Wu, Z., Liu, M., Mo, X., Geng, G., Li, X., Zhou, H., Liu, D., 2014. Temporal-spatial characteristics of severe drought events and their impact on agriculture on a global scale. Quat. Int. 349, 10–21. doi:10.1016/j.quaint.2014.06.021.

Yeh, H.-F., Lee, C.-H., Hsu, K.-C., Chang, P.-H., 2009. GIS for the assessment of the groundwater recharge potential zone. Environ. Geol. 58, 185–195. doi:10.1007/s00254-008-1504-9.

Zargar, A., Sadiq, R., Naser, B., Khan, F.I., 2011. A review of drought indices. Environ. Rev. 19 (NA), 333–349. doi:10.1139/a11-013.

Assessment of groundwater level fluctuations in and around Ranchi district, Jharkhand using geospatial datasets and methods

Pranav Pratik and Priyank Pravin Patel

DEPARTMENT OF GEOGRAPHY, PRESIDENCY UNIVERSITY, KOLKATA

18.1 Introduction

Groundwater forms the reserve of sub-surface water lying beneath the surface at depth in permeable rocks called aquifers. Groundwater is the most important water resource on earth (Villeneuve et al., 1990). It is a dynamic and replenishable natural resource. But in terrains composed of harder rocks, the availability of groundwater is of limited extent due to limited permeability (Bhunia et al., 2012). Groundwater resource development is key in India, as 56.6% of its population depends on agriculture and allied activities (as per the Census of India of 2011). In sustaining agriculture, groundwater plays an important role, particularly in the dry seasons. It is estimated that about 45% of the irrigation water requirement is met from groundwater sources, while across the country more than 90% of the rural populace and nearly 30% of the urban population depends on groundwater for drinking and domestic requirements (Reddy et al., 1996). This dependence on groundwater has continuously increased due to the adoption of high yielding varieties (HYV) seeds for crops, especially for cereal crops like, rice and wheat, and multi-cropping patterns. As a result, just surface water resources were not able to fulfil this increased and continuous demand and the need of a timely, assured water supply has led to the over-exploitation of groundwater resources (Tiwari et al., 2009; Naik and Awasthi, 2003). Increasing population and agricultural activities have not only created more demand for groundwater resources due to the inadequate availability of surface water resources, but have also opened pathways for polluting groundwater resources (Das et al., 2022). Proper groundwater utilization assessment therefore plays a pivot role in determining locations of water supply and monitoring wells, and in controlling groundwater pollution (Foster, 1987; Fetter, 2001). The environmental concerns related to the groundwater quality generally focuses on the impact of pollution and quality degradation on human health. Nearly two-thirds of all ailments in India, such as

Case Studies in Geospatial Applications to Groundwater Resources. DOI: https://doi.org/10.1016/B978-0-323-99963-2.00015-8

jaundice, cholera, diarrhea and dysentery, typhoid, etc. are caused by the consumption of polluted water and these water-borne diseases claim nearly 1.5 million lives annually in the country (Ghazali, 1992). Hence, groundwater level fluctuation assessment is essential for management of groundwater resources and subsequent land use planning (Rupert, 2001; Babiker et al., 2005).

The most reliable and standard methods for determining locations and thickness of aquifers and other subsurface information is through test drilling and stratigraphy analysis but these require high capital investment, skilled labor and are time-taking (Todd, 1980; Fetter, 2001). However, integrated studies using conventional surveys along with satellite image data and geographical information system (GIS) tools, are useful in this respect, providing quite high accuracy results and also reducing any bias in the ascertained outputs (Bhunia et al., 2012). Previously, many researchers have used remote sensing and GIS technique to define the spatial distribution of groundwater potential zones on the basis of geomorphology and other associated parameters (Krishnamurty and Srinivas, 1996; Ravindran and Jeyaram, 1997; Sree Devi et al., 2001; Gupta and Patel, 2021). Lineaments are natural, linear surface elements, interpreted directly from satellite imagery (Garza and Slade, 1986; Parizek et al., 1967) and are often used in water resource investigations (Boyer and McQueen, 1964; Brown, 1994; Dhakate et al., 2008). The sub-surface hydrologic environment, has a primary influence on groundwater movement and can be used to potentially delineate the future groundwater extraction areas (Ravindran, 1997), while the quality of groundwater is as important as that of quantity (Rahman, 2003). Previously, groundwater vulnerability maps have been prepared in and around Ranchi using the DRASTIC method (Jha and Sebastian, 2005).

18.2 The study area

Ranchi district lies in the southeast-central part of the state of Jharkhand. Earlier it was a much larger district, but has recently been divided into two—the present district of Ranchi and the southern portion that has been made into a new district of Khunti. The district borders the West Bengal district of Puruliya on its east, while on its north it is bordered by Ramgarh, Hazaribag, and Chhatra, to the west by Latehar, Lohardaga, and Gumla and to the southeast by Seraikela-Kharsawan. The district also contains the capital city of Jharkhand, also called Ranchi, situated almost in the geographic center of the district.

The district is primarily a tabular landmass, in geologic and geomorphic terms. It has a quite even flat surface with isolated hillocks known as Tongri. Hills lying on the west have elevation above 800 meters and those lying in east have elevation less than 75 meters. The average elevation of the district is 650 meters but western portion is relatively higher than eastern part. The entire area is full of *tanrs* and *dons* on account of the rolling topography. Tanrs are the comparatively highlands and Dons are lower lands. The entire region is an integral part of the Chotanagpur Plateau. Geologically the area is comprised of Archean granites, gneisses, and schists. The district has varied hydrogeological characteristics due to which ground water potential differs from one region to another. Two types of aquifers are found, that is, weathered aquifer and fractured aquifers. Thickness of weathered aquifers varies from 10 m to 25 m in granite terrain and 30 m to 60 m in lateritic terrain. In weathered aquifer ground water occurs in unconfined condition while in fractured aquifer ground water occurs in semi confined to confined condition. The district experiences pleasant climatic condition due to its relatively higher elevation. December temperatures are lowest around 10°C and May the highest at 37°C, on average. The average annual rainfall is 1375 mm, with more than 80% of this falling in the summer monsoon months. The soils of the district are mostly of the residual type. High temperature and high rainfall have led to the formation of lateritic type of soils from weathering and leaching of the Archean metamorphic complex. Some alluvial soils are present along the river valleys.

18.3 Objectives

This brief paper seeks to investigate the following facets:

- Assessment of groundwater recharge, present utilization status and balance available in Ranchi district.
- Establishing relations between the volume changes of groundwater with related changes in the water table and thereby estimating the volume of water extracted and recharged during specific periods throughout the year, across the last few decades.

18.4 Datasets and methods

The following study is mostly done based on secondary data collected from various Central and State Government offices. The various data used and their sources are as follows:

a. Data used in the work is from 1996 to 2014. This data pertains to the position of the groundwater table below the surface as recorded four times a year. The assessment periods are as follows:
   - ➢ After completion of Rabi crop, postwinter level (during January/February).
   - ➢ During the premonsoon stage (in May)—where the water table drops to its greatest depth from the surface.
   - ➢ During the monsoon (August)—where the water table is nearest to the surface.
   - ➢ In the postmonsoon stage (November).

Using the above four primary variables, the following secondary parameters have been tabulated for each year for each of the sampled locations:

- ➢ Change in depth from premonsoon depth to monsoon level of same year which implies groundwater rise due to recharge.
- ➢ Change in depth from monsoon to postmonsoon level of same year which implies groundwater drop mostly due to utilization during Kharif crop.
- ➢ Change in depth from postmonsoon to postwinter level of next year which implies groundwater drop mainly due to utilization during Rabi crop.

- Change in depth from monsoon level to postwinter level of next year which implies groundwater drop due to total utilization from kharif and rabi crops mainly.
- Change in depth from postwinter level of next year to premonsoon level of next year denoting the further drop during dry period.
- Change in depth from monsoon level of this year to premonsoon level of next year denoting the cumulative decrease in groundwater depth due to total abstraction or utilization.

While, as stated above, the level of the groundwater (or its depth from the ground surface, measured as "mbgl"—meters below ground level) rises and falls normally following the general trend, there are the odd months and instances when a reverse may be noted, if unseasonal rain may have occurred or utilization/abstraction amounts may have varied.

a. District level groundwater information and reports from Ground Water Directorate, Water Resources Department, Government of Jharkhand and Central Groundwater Board.
b. Collating different reports together and preparing thematic maps to show the groundwater utilization status of the different blocks in Ranchi district.

The whole work was divided into three phases:

In the first phase the crude data received from the Central Groundwater Board has been processed to get the change in groundwater level during various period in a year. The isopleth maps representing the depth to groundwater surface is prepared using GIS software like Mapinfo Professional GIS and Arc-GIS. A larger area is sampled than just the locations situated within the confines of the present Ranchi District boundary, in order to avoid the edge-effect that this would have produced while generating the isopleth surfaces. Thus, parts of the surrounding districts of Ranchi district have also been included in the analysis and the groundwater level data for a total of 61 sites have been collected from the relevant sources, then tabulated and finally enumerated and mapped. Using these, the change in water table depth from the surface is analyzed for the respective periods within each year and across the years. This gave the trend of utilization for the area, which has imprint of the socioeconomic attributes of the region. For instance, a more than normal water level drop in between November and March indicates the cultivation of Rabi crop in the area and hence more groundwater utilization. Lastly, the percentage of area under various depth zones is calculated to get a comprehensive picture of depth of groundwater in the whole district.

In the second phase, cartograms are used to quantitatively represent the above spatial distribution of the groundwater fluctuations. Groundwater trends and situation at a large number of sites across the district are represented and interpreted briefly. The situation within the largest urban settlement of Ranchi is also analyzed as here the demand for water is very high.

In the last phase, wireframe surface for annual monsoonal recharge (AMR) and annual withdrawal (AW) is created by using SURFER-11 software. These surfaces are then overlain over each other to get the mismatch (difference surfaces) between the AMR and AW. A higher AMR surface than the AW surface shows a net gain in the groundwater level and subsequent volume (if computed over the areal coverage of this depth gain) in the region while conversely a higher AW surface than the AMR surface shows the net loss and reveals areas where the abstraction might be outstripping the replenishment and thus help to identify possible locales for concern

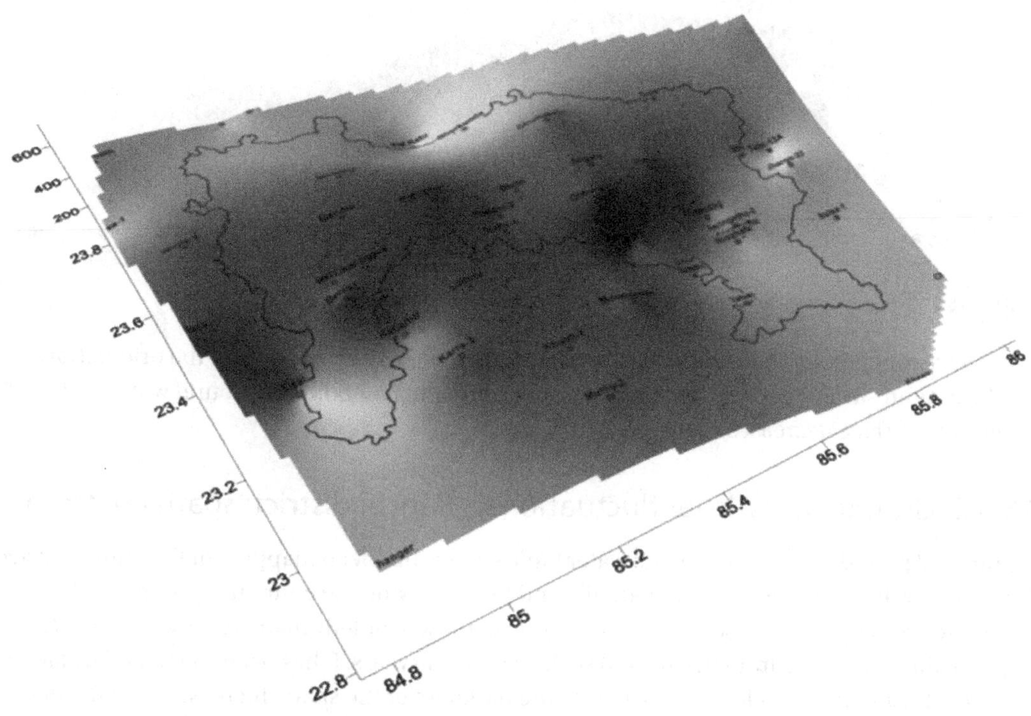

FIGURE 18.1 Surface topography variation in and around Ranchi district.

as regards future availability of this vital resource. These have been analyzed both spatially and temporally to get the areas of vulnerability and compute the related statistics. Then the volume in between these surfaces is calculated using the Trapezium Rule Function available in SURFER-11. Hence a quantitative picture of the actual groundwater resource is feasibly obtained, which can be used for planning a sustainable utilization strategy once both estimated demand and supply of groundwater for this region is available.

18.5 Results and findings

18.5.1 Groundwater level position in Ranchi district with respect to topography

a. Topography of Ranchi district and surroundings: The 3D surface generated (Fig. 18.1) represents the surface topography of Ranchi district and surroundings. The altitude of place increases with the darkness in the shade. The elevation is lowest in the eastern side of the district which lies at the boundary of Ranchi Plateau. The elevation is maximum in the central portion and it decreases towards the fringes.

b. Position of average groundwater level (1996–2013) with respect to the ground surface topography, Ranchi district and surroundings: This diagram (Fig. 18.2) represents the average position of water table of Ranchi district from 1996 to 2013 with respect to the actual ground

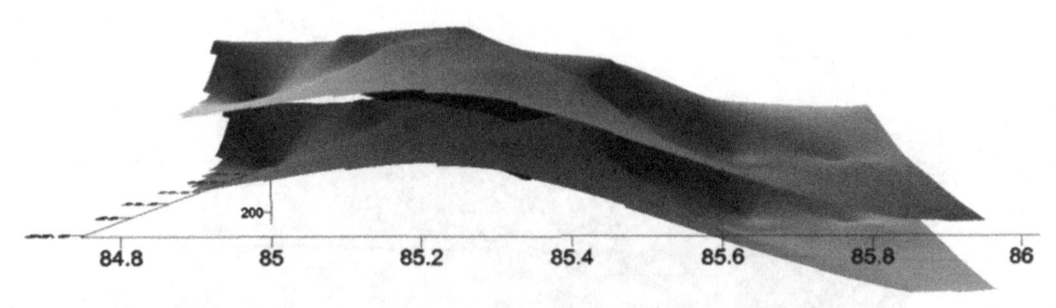

FIGURE 18.2 Groundwater level surface variation and relation with surface topography.

surface topography. It can be seen that the water table closely resembles the orientation of the ground surface above. The actual elevation from msl for both the ground water table and ground surface decreases eastwards.

18.5.2 Groundwater level fluctuation in Ranchi district: spatial pattern

The annual periodic movement of the water table surface has been mapped for the entire district for each year from 1993 to 2014 individually and then showing the entire range (Figs. 18.3–18.7), presenting a continuous assessment for 20 years. These isopleth maps represent the average depth of the water table from the ground surface for these years. It has been calculated by taking average of readings taken four times a year. The darkness of the shade increases with the depth of water table from the ground surface.

18.5.2.1 Depth to groundwater table from the surface for each year from 1994 to 2014

1996–2001: The water-level remained closest to the surface at Ranchi and adjoining areas. The depth increases as one moves towards the fringes, especially along the north-western and north-eastern boundaries. It reached up to a maximum depth of around 8–9 m at places like Burmoo and Berro.

2002–2007: The water-level remained closest to the surface in the central part of the district at places like Ratuchati and Kantitnar, Buti and at times along the north-central boundary. The depth increased towards the fringes along the entire boundary alternatively in different years. It reaches around 12 m at Hatia, reaching this mark here for the first time. The depth becomes about 10 m at places like Burmoo, Berrol and Berro. The overall depth from the surface to the groundwater table has thus increased compared to previous years, indicating the excess exploitation over recharge. Areas having depths greater than 8 m now covered more than 70% of the district. The level falls to its maximum depth of around 26 m at Bundu while being around 14 m north of Rratuchati.

2008–2013: The water-level remains closest to the surface in the central part, at places like Buti, Tatisilwai, Seringathu. It reaches down to depths of around 14 m in the north-central part of the district and up to 10 m depth at Burmu and Tamar. The maximum depth reached is around 22-24 m along the north-central boundary.

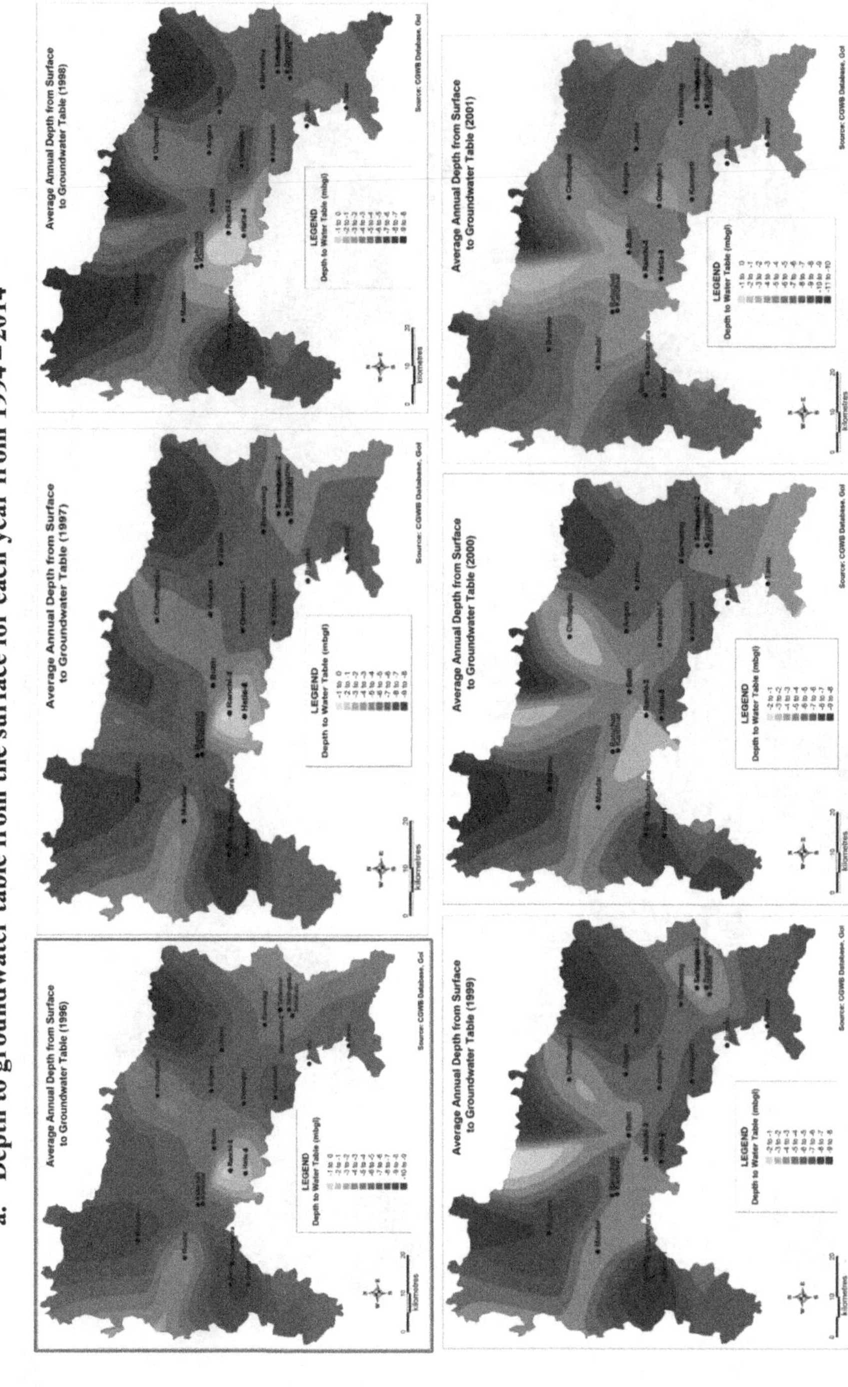

FIGURE 18.3 Fluctuations in the depth to the groundwater table in and around Ranchi district (1996–2001).

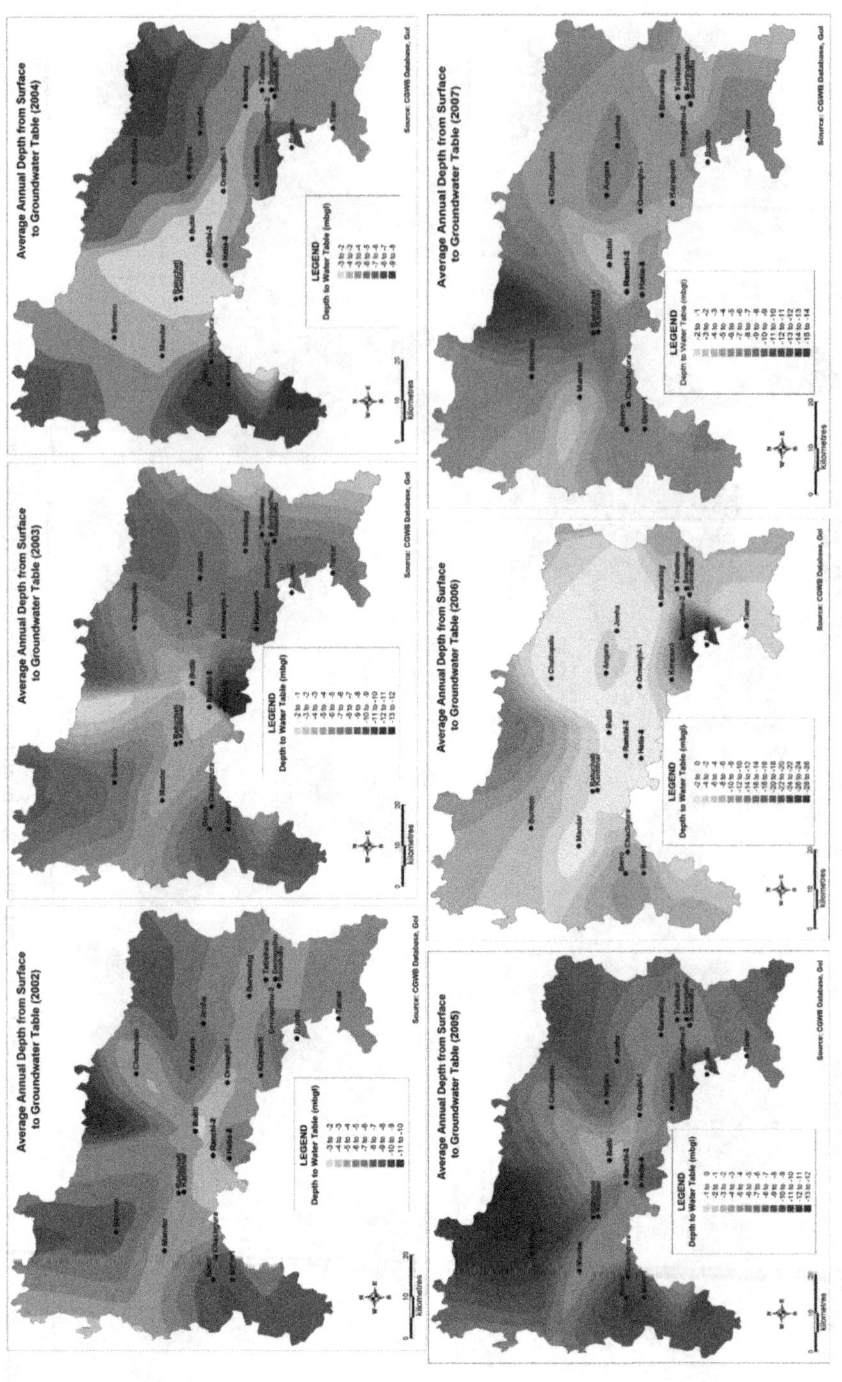

FIGURE 18.4 Fluctuations in the depth to the groundwater table in and around Ranchi district (2002–2007).

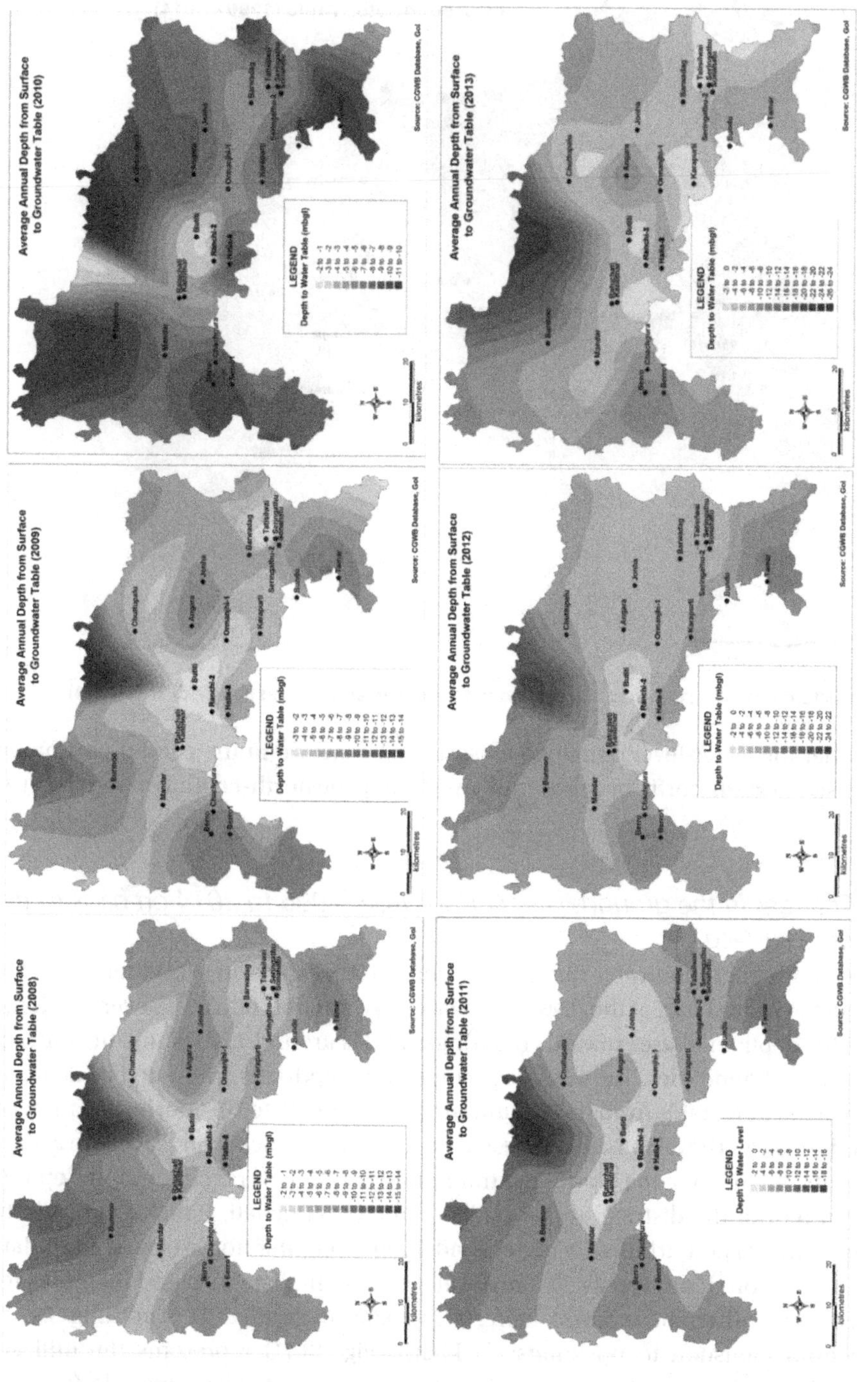

FIGURE 18.5 Fluctuations in the depth to the groundwater table in and around Ranchi district (2008–2013).

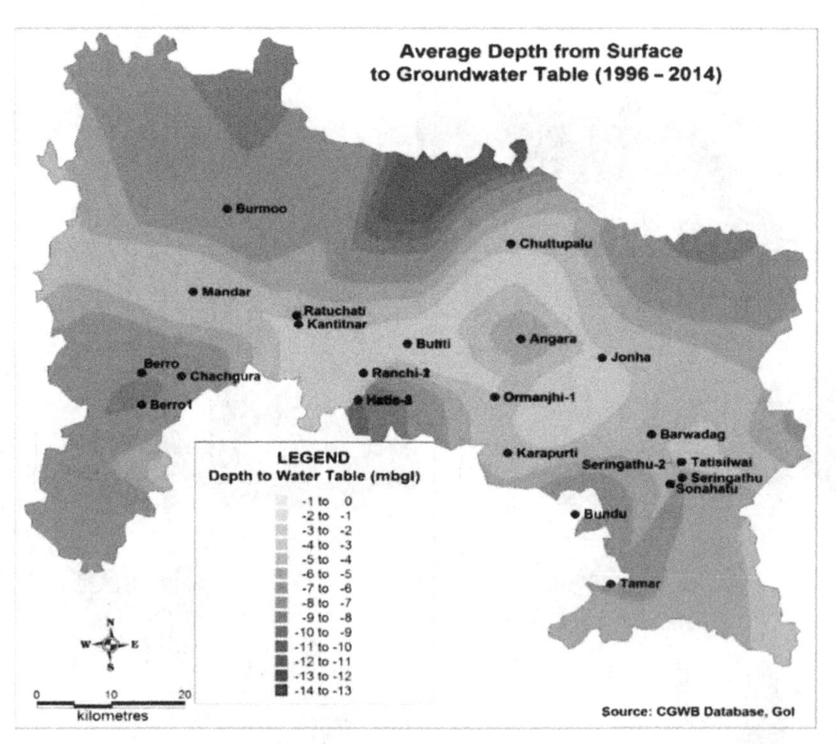

FIGURE 18.6 Fluctuations in the depth to the groundwater table in and around Ranchi district (2014).

In 2014, that the water-level remained closest to the surface in the central portion and the depth increased towards north, becoming the most along the north-central boundary, at around 20 m.

18.5.2.2 Changes in the groundwater surface from 1996 to 2014 on basis of depth from surface

The average depth to groundwater fluctuations showed that the water-level remains closest to the surface in the central part of the district, at places like Ranchi, Tatisilwai, over the duration of 1996–2014. This depth increases towards the fringes and is around 13 m along the north-central and south-central boundaries. The average premonsoon depth was computed by taking the average depth of the water table from the ground surface from 1996 to 2014 during the premonsoon (May) period. The water-level remains closest to the surface at Ranchi and reaches to a maximum depth of around 16 m along the North-Central Boundary. The average monsoon depth is least in the central part of the district, at places like Mandar, Ratuchati, Ranchi, Buti, Chuttupalu and increases towards the fringes, being around 10 m along the north-central boundary. The average postmonsoon depth is least at Tatisilwai and Ranchi-2 stations and increases towards the northern and southern boundaries, being at a maximum depth of 14 m at Hatia-3.

Change from monsoon to postmonsoon kharif: Fig. 18.7D represents the utilization of groundwater during the agricultural period (i.e., from monsoon to kharif period). It denotes the

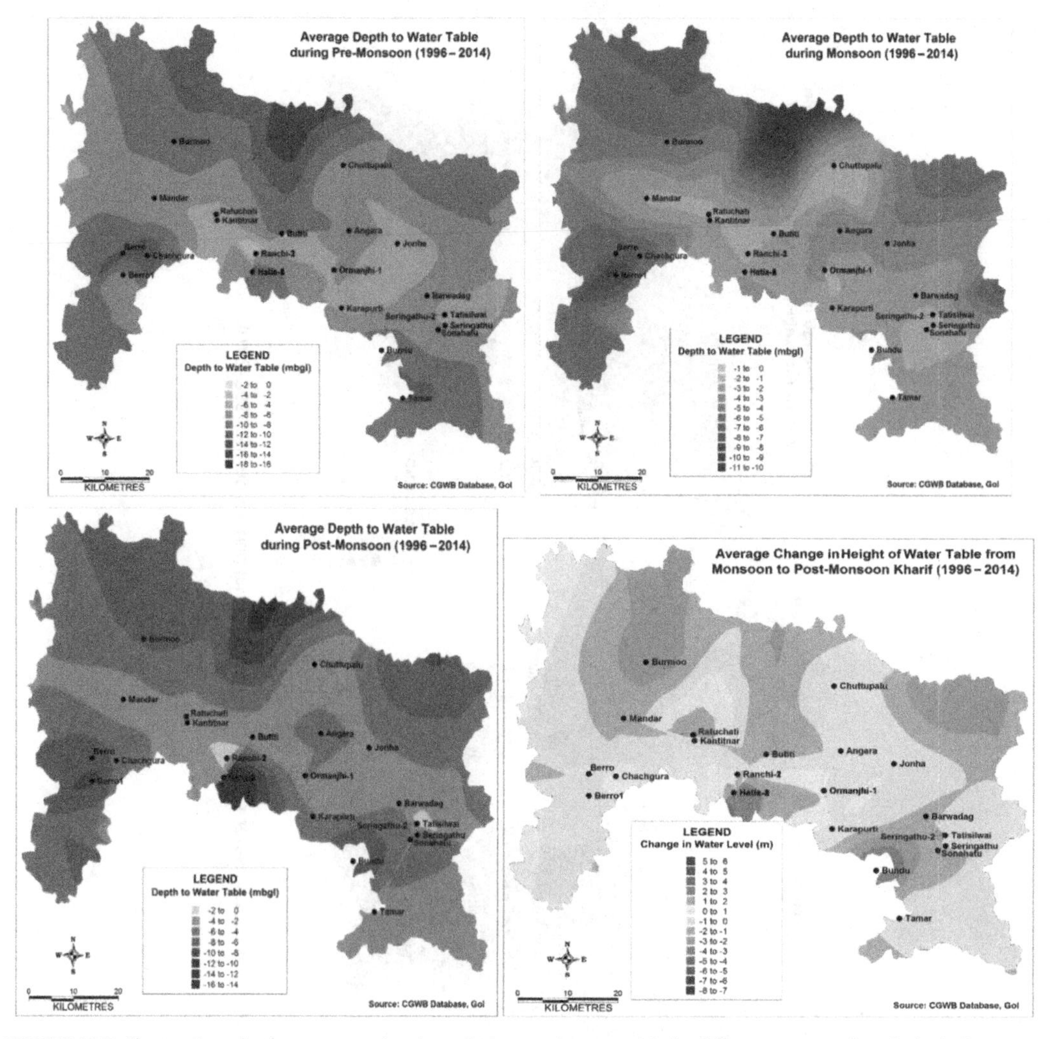

FIGURE 18.7 Fluctuations in the average depth to the groundwater table in different seasonal periods during 1996–2014.

average change in the height of the water table from the monsoon to postmonsoon kharif period. Shades of red indicate net withdrawal of water, while shades of blue indicate a net recharge during this period. It can be observed that the change in water level has been highest near the district headquarter Ranchi and its adjoining area, Hatia-2 and Ratuchati. Here, the ground water table has receded by around 5 m. The change in water level decreases as one moves towards the fringes. Negative values indicating net recharge were found in areas like Bundu, Burmoo, where there has been about 6 m rise in the water level.

Mean annual cumulative loss and gain: Fig. 18.8 shows the utilization of groundwater during a yearly cycle and was calculated as the average change in height of water table from monsoon of one year to premonsoon period of next year. The change in water level has been highest near the

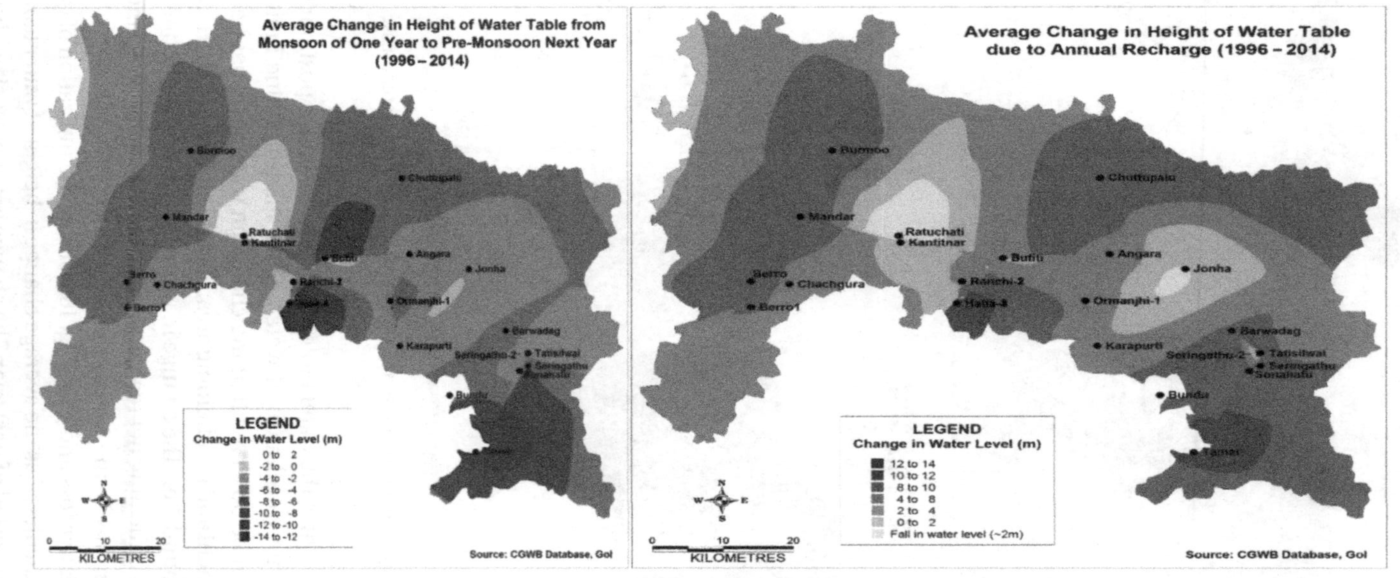

FIGURE 18.8 Fluctuations in the average depth to the groundwater table in different seasonal periods during 1996–2014.

district headquarter Ranchi and adjoining area, Hatia-2 and Tamar, wherein the water table has receded by around 12 m. Positive values indicating net recharge were found in around Ratuchati, where there has been about 0–2 m rise in the water level. Conversely, the average rise in the water table due to recharge of groundwater during the monsoon period was calculated as change in water table from premonsoon period of a year to the monsoon period of the same year. Water level changes are most near Ranchi and Hatia-2 where the water table has risen by around 12 m. This change lessens towards the fringes and negative values indicating net withdrawal lie around Ratuchati and Jonha, where there has been about 2 m decline in the water level.

18.5.3 Groundwater level changes across the years: fluctuation and volume estimates

The fluctuations noted in the depths to the groundwater table across the different years and seasonal extents are examined next.

The graphs (Fig. 18.9) represents the depth to groundwater in Ranchi district from 1996 to 2014 on a year wise basis and their variations. It has been calculated by taking average of depth at four times a year—February, May, August, and November for each year from 1996 to 2014. It can be observed that the average depth of water table has more or less fallen farther from the surface over the years. The depth to the water table has roughly varied from 5 to 7 mbgl. The right graph gives an idea of the utilization of groundwater in Ranchi district, during the agricultural season of Kharif, which lies in between August and November. This is calculated as average change in water table height from monsoon to postmonsoon Kharif period in each year throughout the district. Negative value indicates decline in water table, while positive value denotes increment of water table during the said period. Upward trend of the moving average curve indicates that the utilization has decreased over the years. Its value varies from decline of water table from 2 m to increment of 1 m.

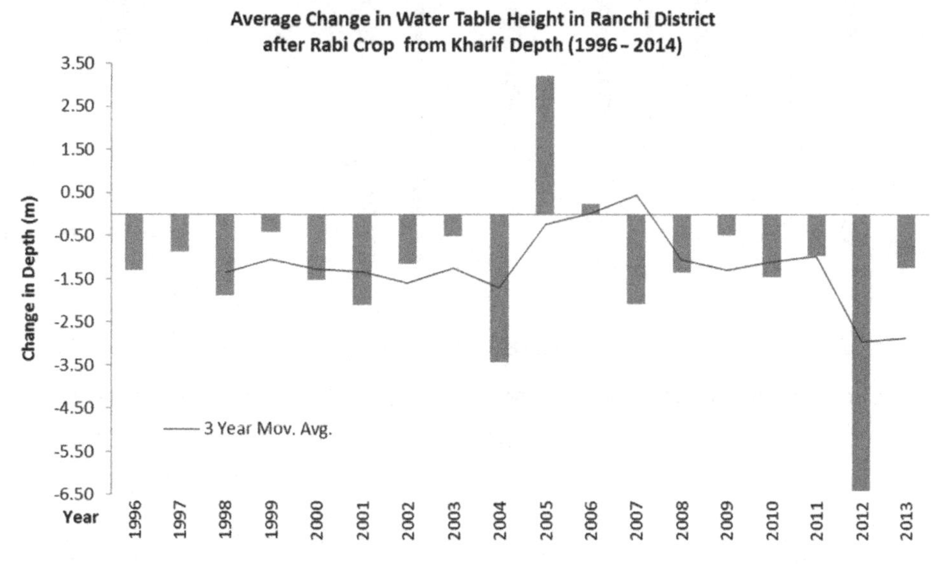

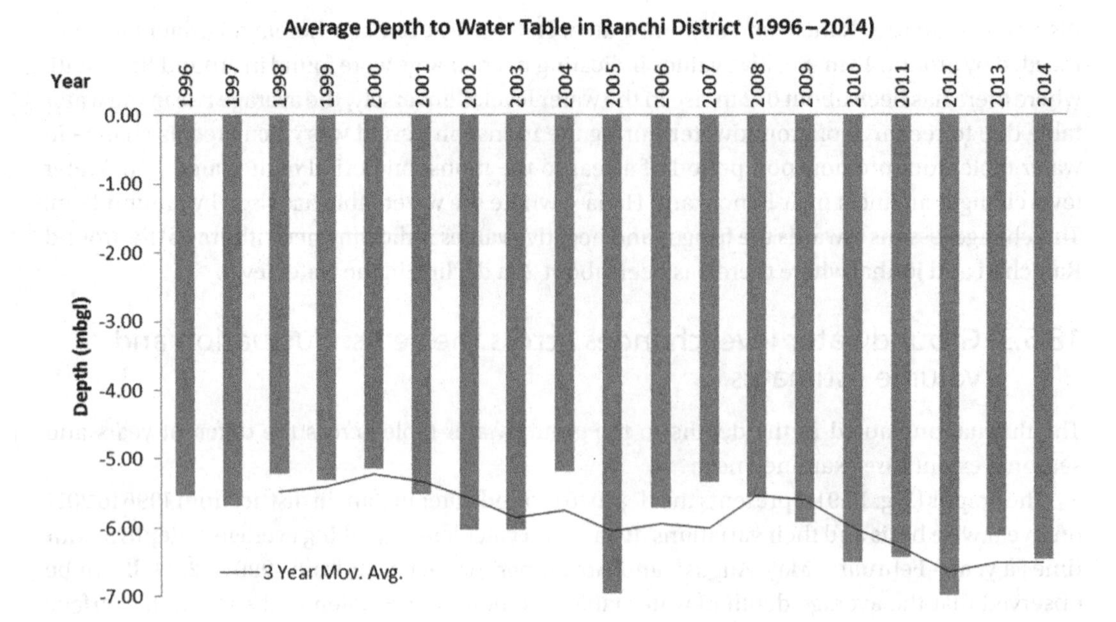

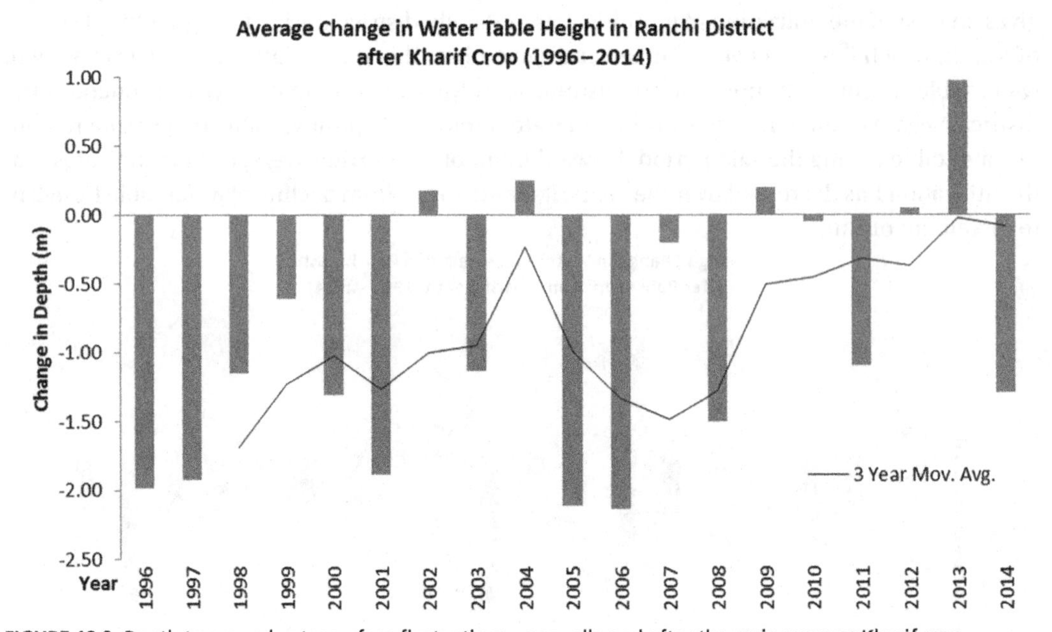

FIGURE 18.9 Depth to groundwater surface fluctuations—annually and after the main summer Kharif crop.

The left graph represents the utilization of groundwater in Ranchi district, during the agricultural season of Rabi, which lies in between November and March. This is calculated as average change in water table height from postmonsoon Kharif to postmonsoon Rabi period in each year throughout the district. Negative value indicates decline in water table, while positive value denotes increment of water table in the said period. Its value has roughly varied from decline of water table from 6.5 m to increment of 3.5 m.

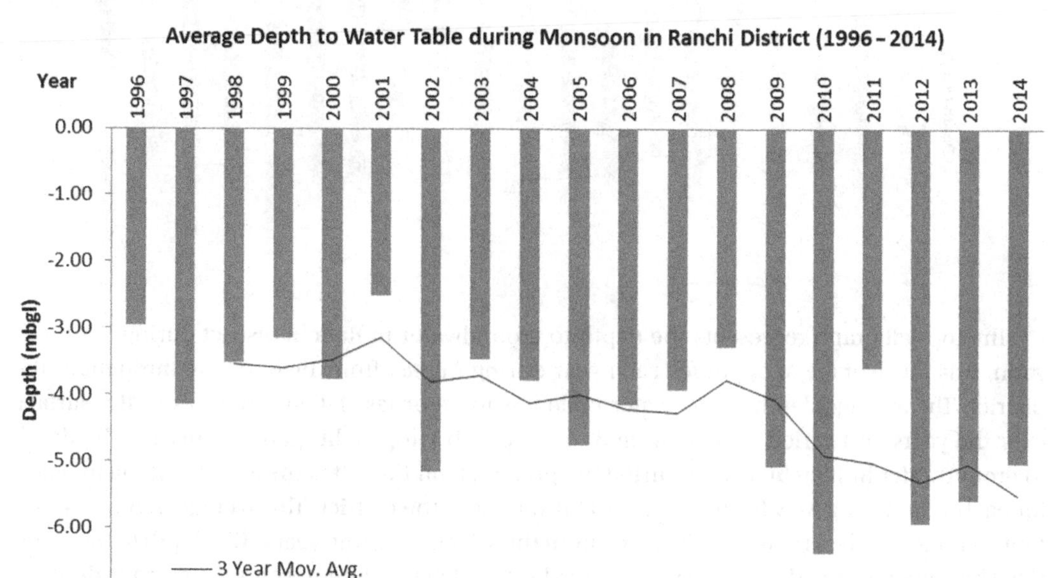

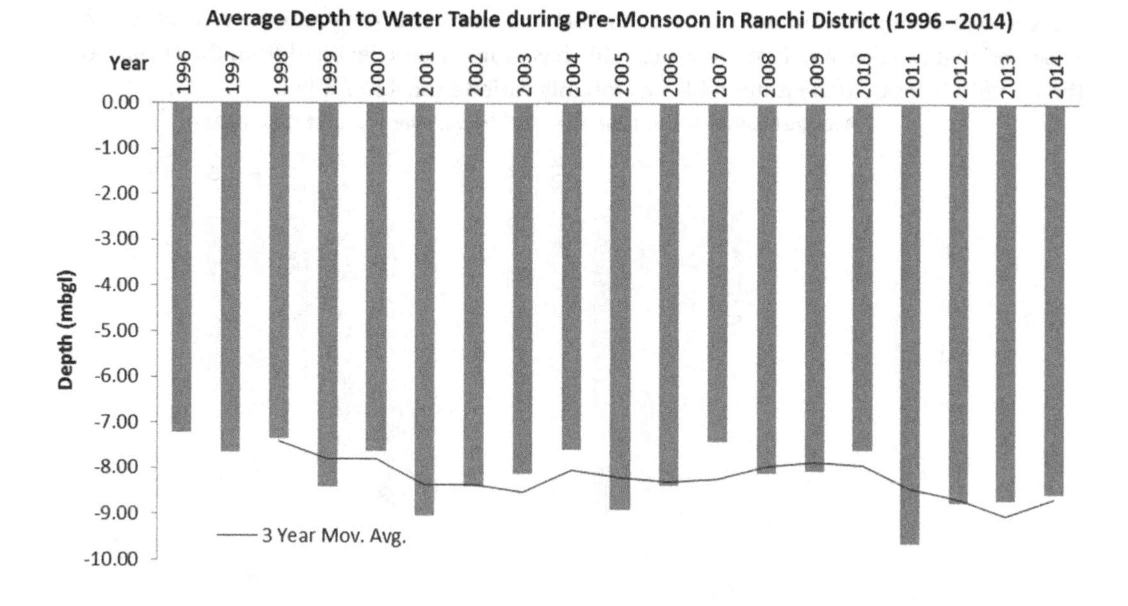

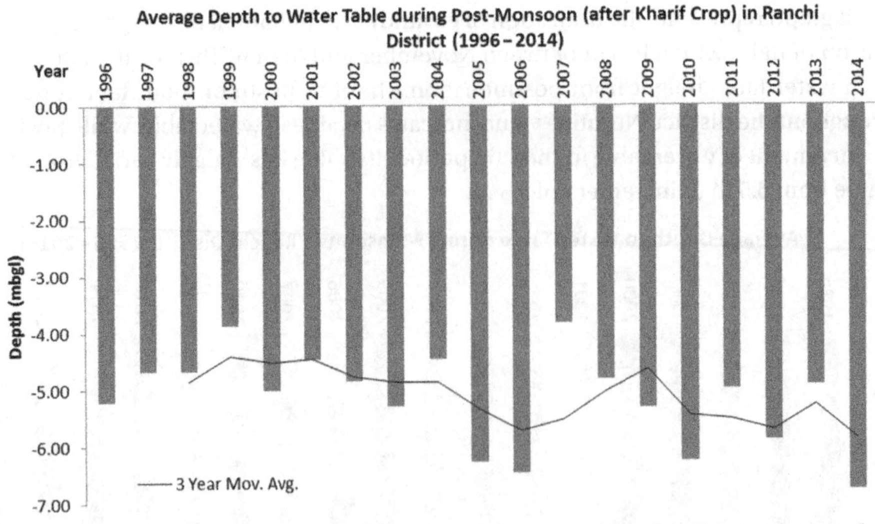

The top-left graph represents the depth to groundwater in Ranchi district during the monsoon. It is the average of depth for each year during August from 1996 to 2014 throughout the district. The average depth to the water table has more or less fallen farther from the surface over the years and varied roughly from 3 to 6 mbgl. The top-right graph represents the depth to groundwater in Ranchi district during the premonsoon time. It is the average of mean depth for each year during May from 1996 to 2014 throughout the district. The average depth of water table has more or less remained the same from the surface over the years. The depth to the water table has roughly varied from 7 to 9 mbgl. The bottom-left graph represents the average depth of groundwater in Ranchi district during the postmonsoon Kharif time. It is the average of depth for each station during November from 1996 to 2014 in each year throughout the district. It can be observed that the average depth of water table has more or less declined from the surface over the years. The depth to the water table has roughly varied from 4 to 7 mbgl.

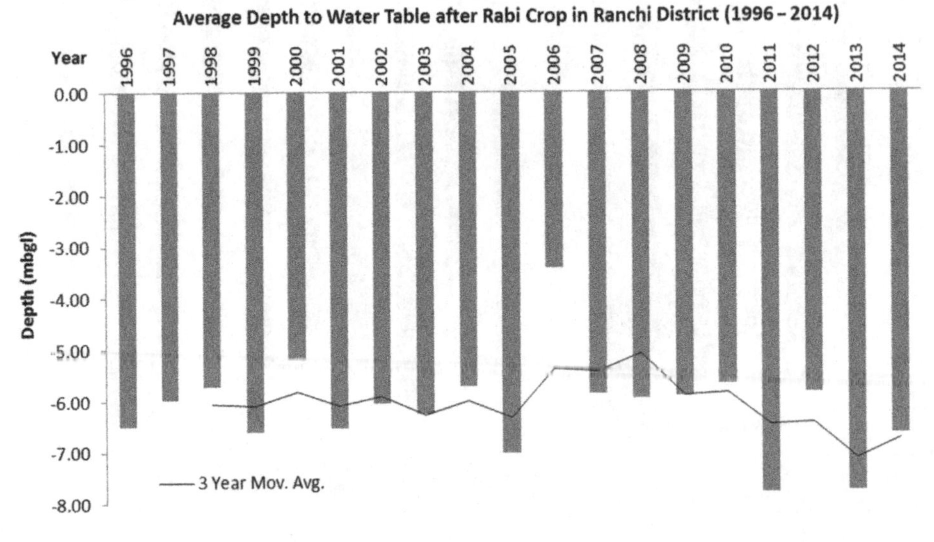

The other graph represents the average depth to groundwater surface in Ranchi district during the postmonsoon Rabi. It is the average of depth for each station during February from 1996 to 2014 in each year throughout the district. This average depth to the water table has more or less declined (fallen away) from the surface over the years. The depth to the water table has roughly varied from 3 to 8 mbgl.

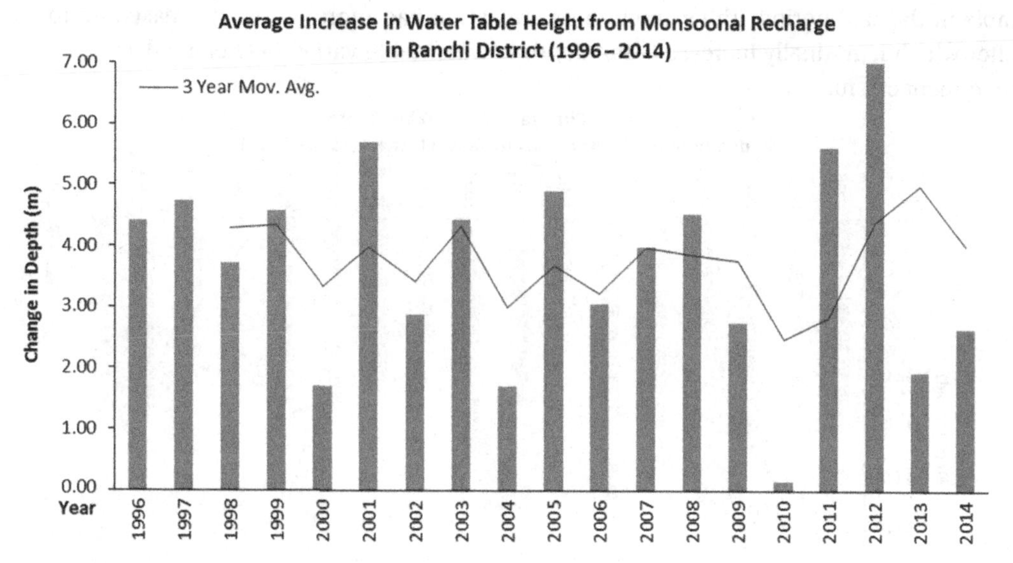

The left-side graph represents the recharge of groundwater in Ranchi district, between May and August. This is the average change in water table height from premonsoon to monsoon period during 1996–2014 throughout the district. Its value has roughly varied from 0.2 mbgl to 7 mbgl.

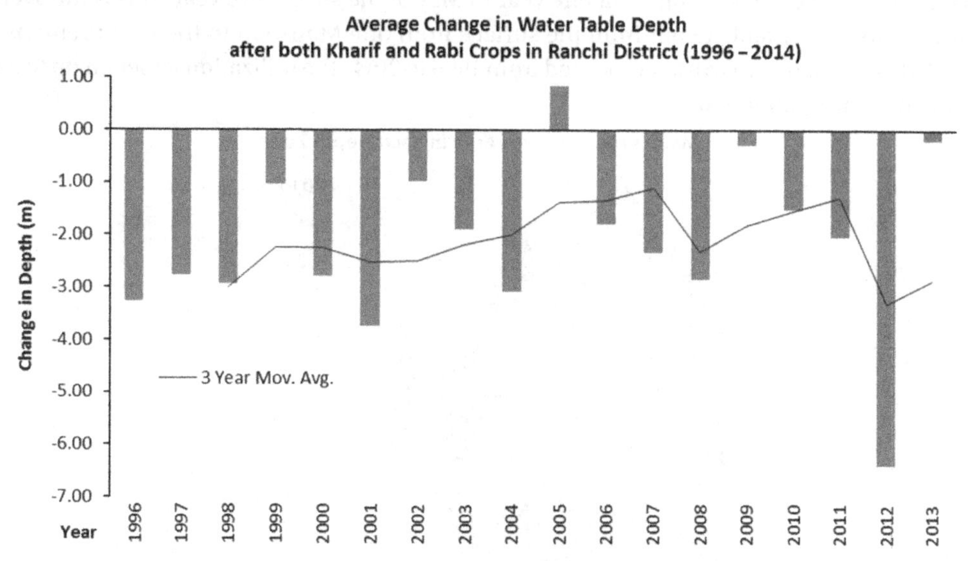

The graph represents the combined utilization of groundwater in Ranchi district, during the agricultural season of Rabi and Kharif, which lies in between August of one year to February of the successive year. This is the average change in water table height from the monsoon to postmonsoon Rabi period of the successive year, as measured for each year from 1996 to 2014. Negative values indicate decline in water table, while positive value denotes increment of water table in the said period. The extraction or utilization had more or less decreased up to 2005, after which it gradually increased. The depth to water table varied between a decline of 6m to increment of 1 m.

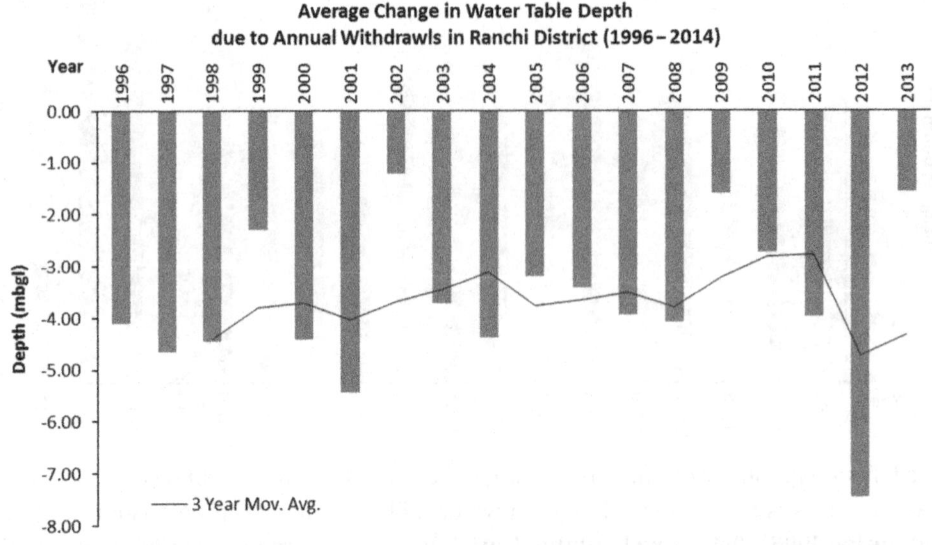

The graph represents the overall withdrawal or utilization of groundwater in Ranchi District, during an entire year, from August of one year to May of the successive year. This is the average change in the water table depth from the surface from one Monsoon to the next premonsoon period of the successive year, as computed from 1996 to 2014. The utilization lessened up to 2009, increasing sharply after that.

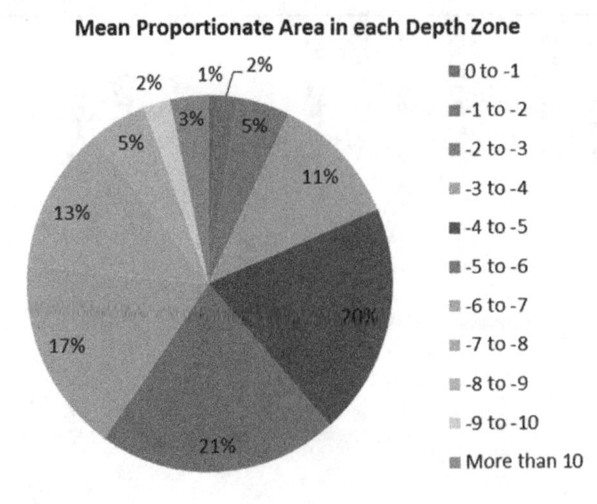

The percentage of total district area win each particular depth zone was enumerated. In the depth zone of 5–6 mbgl, the maximum proportion of area is placed. In all, with 50% of the district, the water table lies at 3–6 mbgl. The least proportion of area lies in the depth zone of more than 10 mbgl.

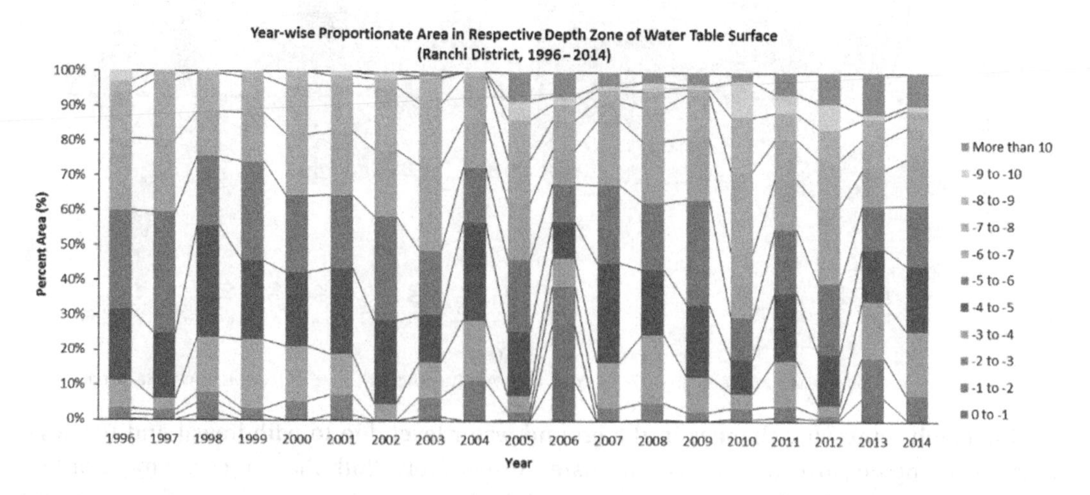

The above diagram represents the proportion of the district having a particular groundwater depth zone in a year-wise manner from 2013 to 2014. It can be seen that over time, more areas have come under the higher depth zone. The depth zone of 10 m and above appears for the first time in 2005 and after that the area under this depth zone has more or less increased. In other words, those areas where the depth to the water table is less than 10 m has consistently decreased. Hence, overall the water table has fallen in most part of the district.

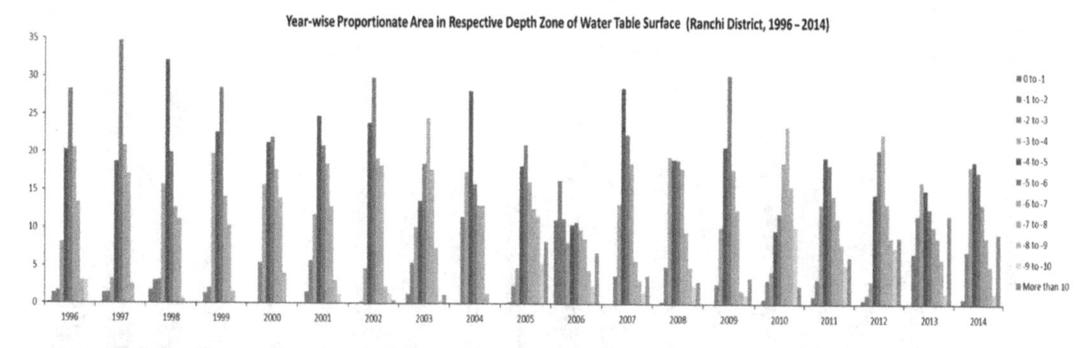

The above diagram represents the proportion of the area of a particular groundwater depth zone in a year-wise manner. It can be seen that the bar representing the 5–6 mbgl value remains the tallest in most of the years, indicating this depth zone covers most portion of the district. The 10 m and greater depth zone appears for the first time in 2005 and after that the area under this depth zone has more or less increased.

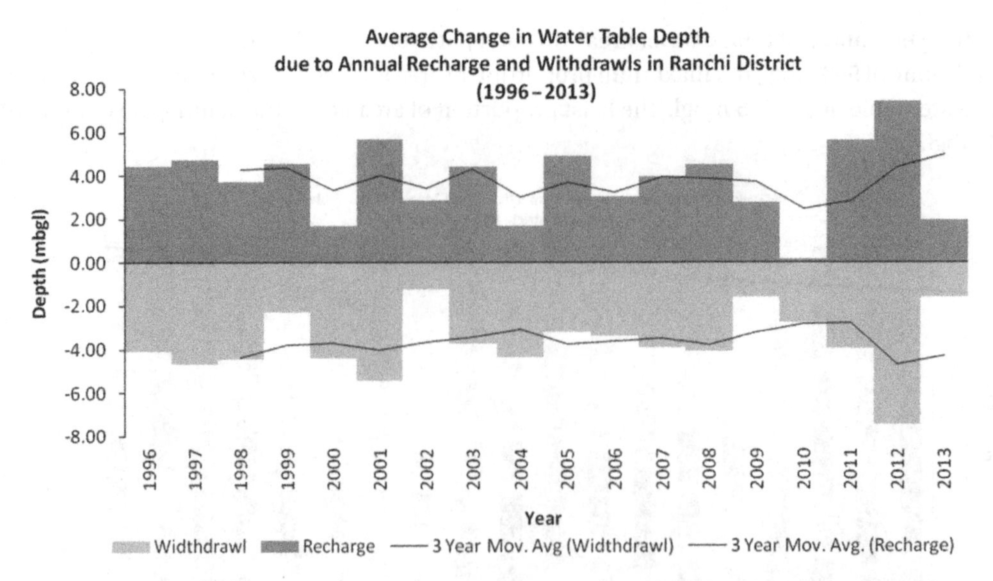

The graph represents changes in the ground water level due to withdrawal and recharge together, in the negative and positive ordinates, respectively. Both the curves are more or less mirror images of each other among the x axis. This shows that excess water recharge ends up being excessively withdrawn. Hence resulting in no addition in the net recharge. Moreover, the total area under orange is greater than in green. Hence withdrawal over these 18 years has surpassed recharge.

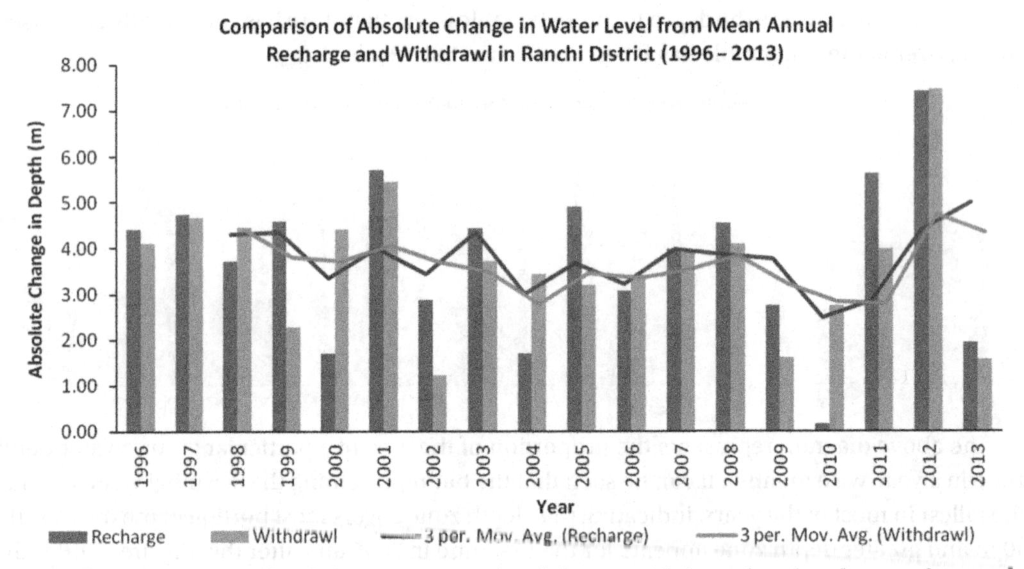

The absolute magnitude of change in water level due to withdrawal and recharge of groundwater was ascertained. The moving average curves closely follows one another representing very less net recharge over withdrawal.

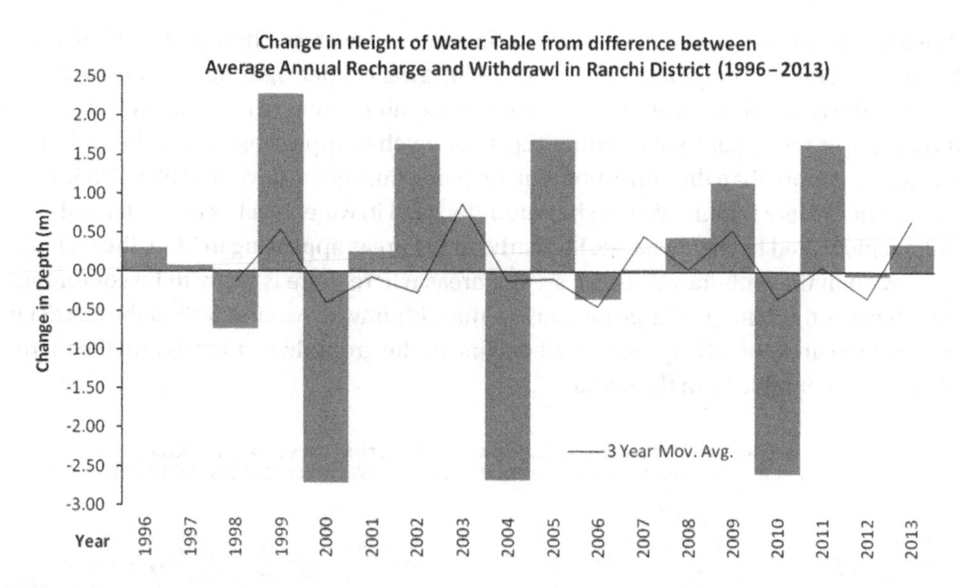

The bar diagram shows the net change in water level due to recharge and withdrawal. The bars show positive values for most of the years indicating actual increase in groundwater level. However, this increase is not much. Overall values range between +2.5 m and -3 m approximately. Hence, there is no significant change in average groundwater level across the years.

18.5.4 Wireframe surfaces for annual monsoonal recharge and annual withdrawal: overall and year-wise

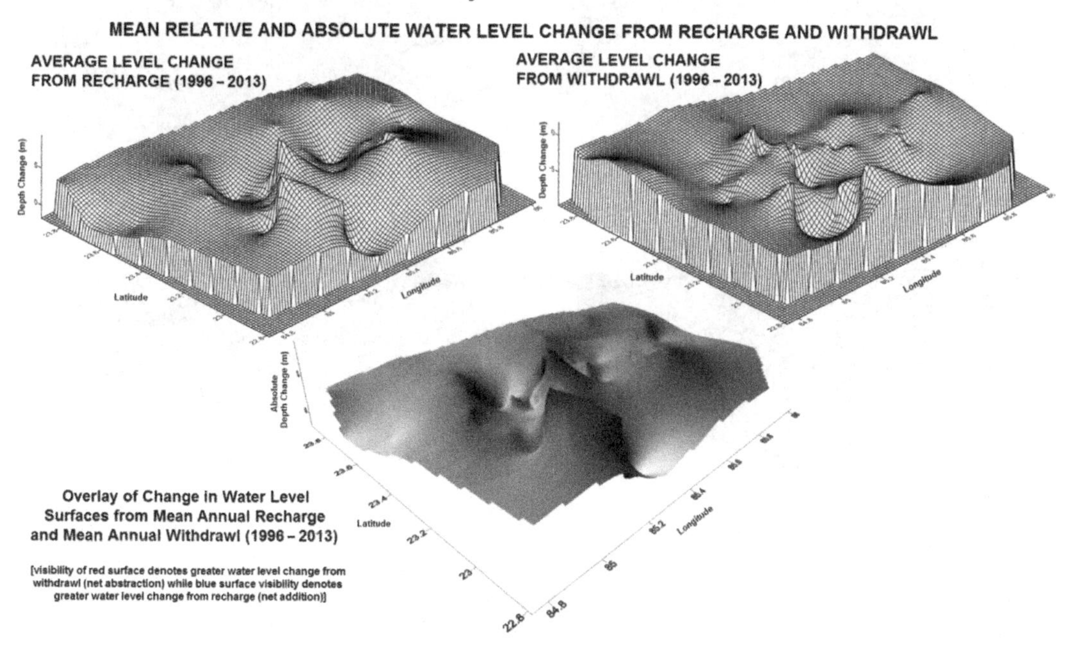

The wireframe surface generated above represents the average change of depth in the water table due to recharge (upper left figure) and withdrawal (upper right figure) during 1996–2013. The 3-D surface below is created by overlaying the above two wireframe surfaces, with the withdrawal over recharge. Hence, within the areal patches appearing in red, the withdrawal of water has been more than the corresponding recharge during the period of 1996–2013. These are the areas where there is more likely to be continuous fall in water level as the water withdrawn is not fully replenished by the recharge. Similarly, in the areas appearing in blue, the recharge has been more than the withdrawal. These are the areas where there is likely to be continuous rise in water level as the water recharge overpasses the withdrawal. Around 60% of the district lies in the red colored area, signifying the overall decline in the groundwater levels and the increasing depth to the water table from the surface.

COMPARISON OF WATER TABLE SURFACE FLUCTUATIONS DUE TO ANNUAL MONSOON-PERIOD RECHARGE AND SUBSEQUENT WITHDRAWL IN RANCHI DISTRICT (1996 - 1998)

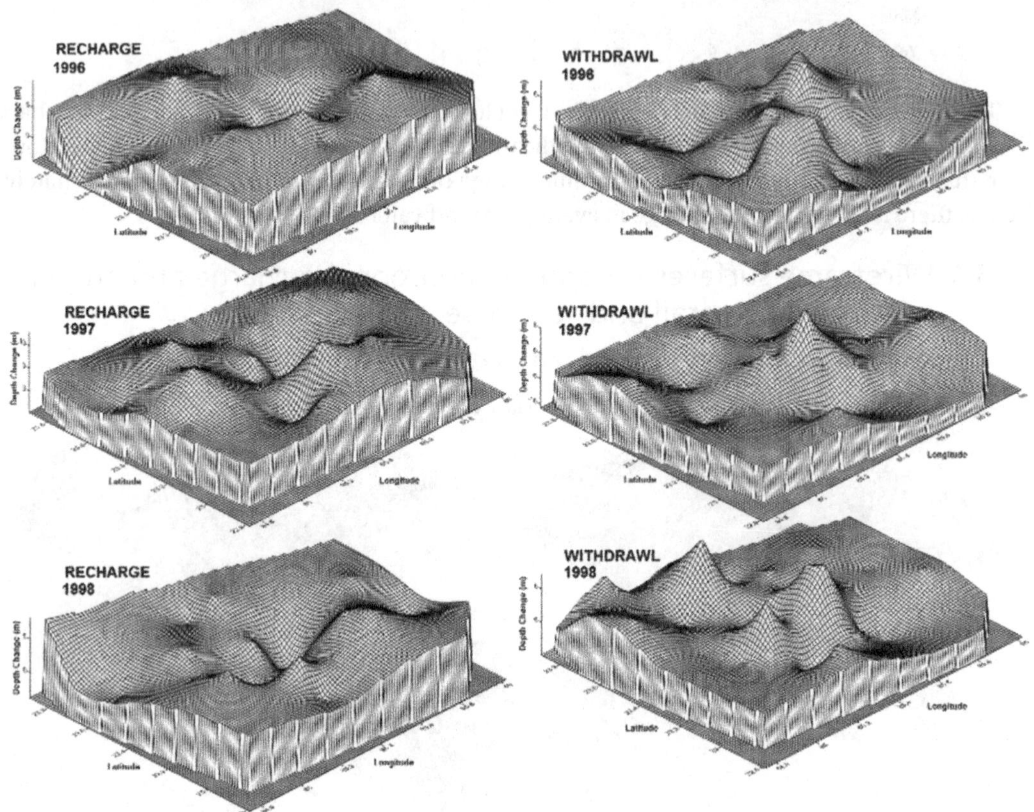

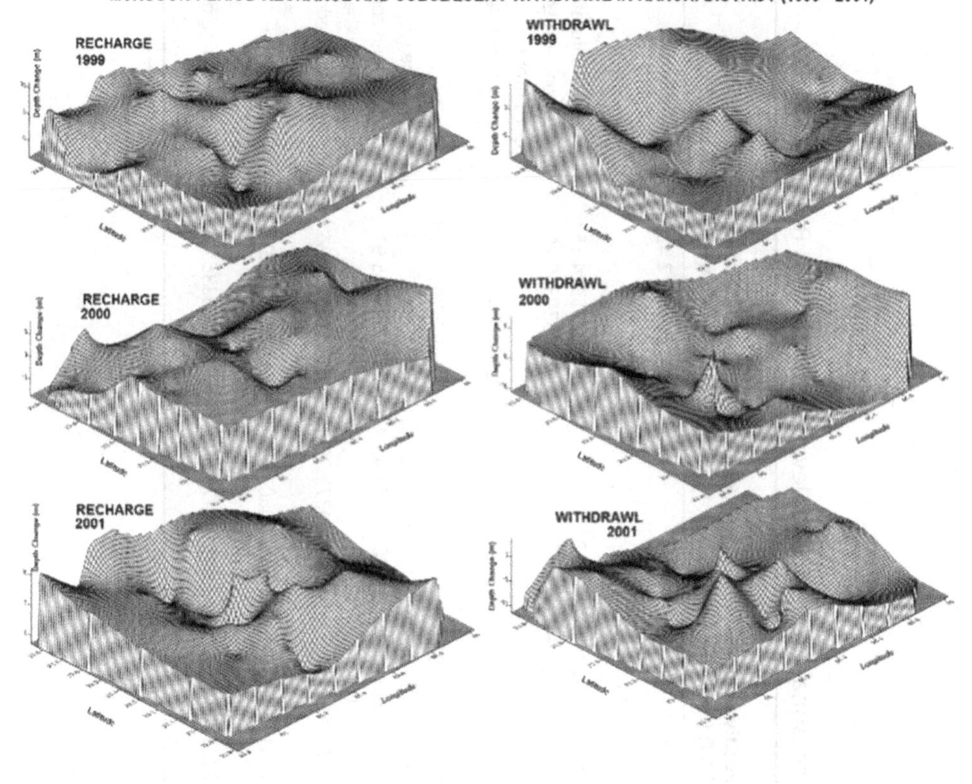

OVERLAYS OF ANNUAL CHANGES IN GROUNDWATER LEVEL FROM RECHARGE AND WITHDRAWL FOR RESPECTIVE YEARS (1996 - 2001) FOR RANCHI DISTRICT

1996

1997

1998

1999

2000

2001

(visibility of red surface denotes greater water level change from withdrawl (net abstraction) while blue surface visibility denotes greater water level change from recharge (net addition))

Table 18.1 Estimated volume computations from seasonal groundwater level fluctuations in Ranchi district.

Sl. no.	Year	Upper surface	Lower surface	Positive volume (Cut) (m^3)	Negative volume (Fill) (m^3)	Net volume (trapezoidal rule) (Cut – Fill) (m^3)	Remarks
1	1996–2014	Surface elevation	Average groundwater level	80710078472.43	0.00	80710078472.43	
2	1997–2014	Surface elevation	Monsoon level	56140275224.55	0.00	56140275224.55	Growth of unsaturated vadose zone across seasons (from monsoon of one year to premonsoon of next year)
3	1998–2014	Surface elevation	Postmonsoon (Kharif) level	69349330360.46	0.00	69349330360.46	
4	1999–2014	Surface elevation	Postmonsoon (Rabi) level	82192054818.41	0.00	82192054818.41	
5	2000–2014	Surface elevation	Premonsoon level	105597620053.89	0.00	105597620053.89	
6	2001–2014	Monsoon level	Premonsoon level	49530557979.90	73213150.56	49457344829.34	Mean recharge volume
7	2002–2014	Monsoon level	Postmonsoon (Kharif) level	13799522810.34	590467674.43	13209055135.91	Drop in saturated zone volume during Kharif and Rabi crop periods
8	2003–2014	Monsoon level	Postmonsoon (Rabi) level	26791693967.22	739914373.36	26051779593.86	
9	2004–2014	Postmonsoon (Kharif) level	PostMonsoon (Rabi) level	14513132484.06	1670408026.11	12842724457.95	Winter cultivation
10	2005–2014	Postmonsoon (Rabi) level	Premonsoon level	23850667411.48	445102176.00	23405565235.47	Further loss during dry season

In the upper black and white diagrams, the wireframe surface generated represents the average change of depth in the water table due to recharge and withdrawal, separately from 1996 to 2001, for each year (also see Table 18.1). Each blue and red surface below is created by overlaying the two wireframe surfaces from the particular year, with the withdrawal placed over the recharge. Hence, red zones denote more withdrawal of water than recharge during that particular year and may experience fall in water level as the water table. Similarly, in blue zones the recharge has been more than withdrawal and these places may see rise in water level as the water level (i.e., lessening depth to water table from surface). Such individual and overlay wireframe diagrams have also been made for the other years.

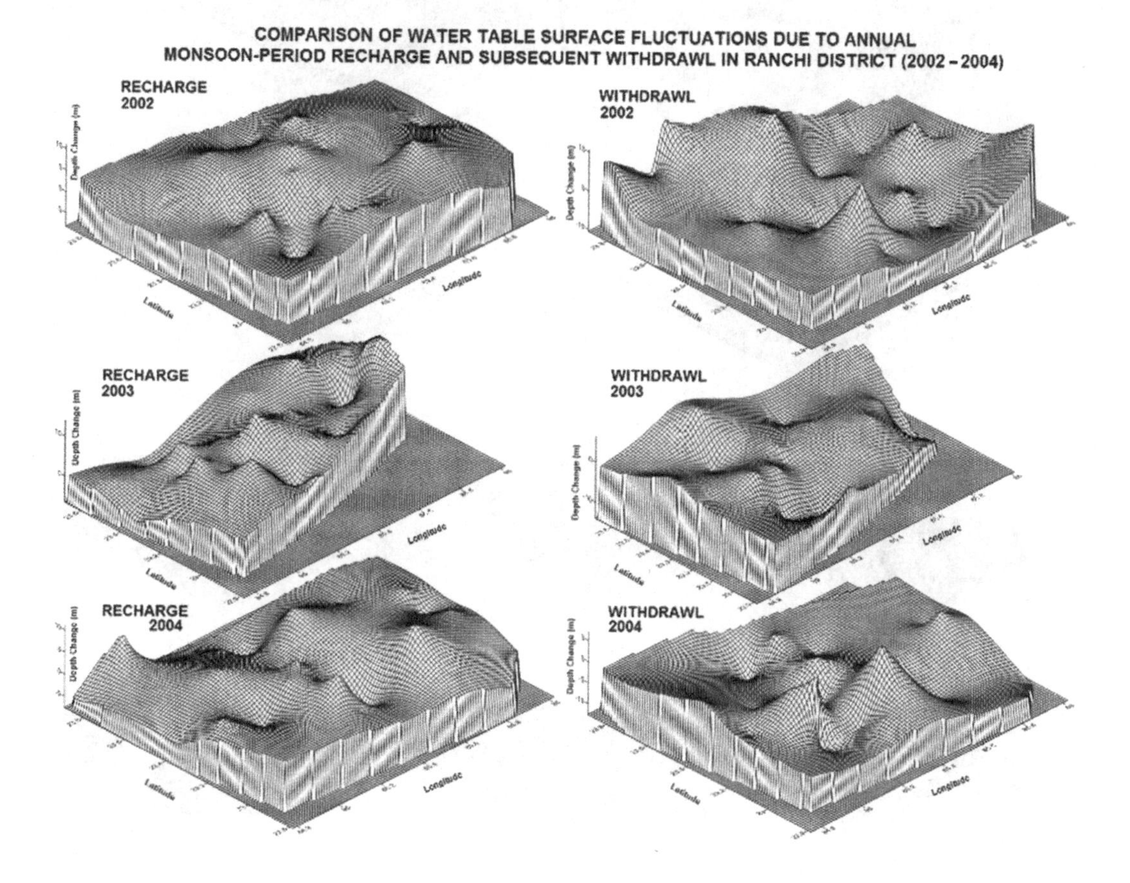

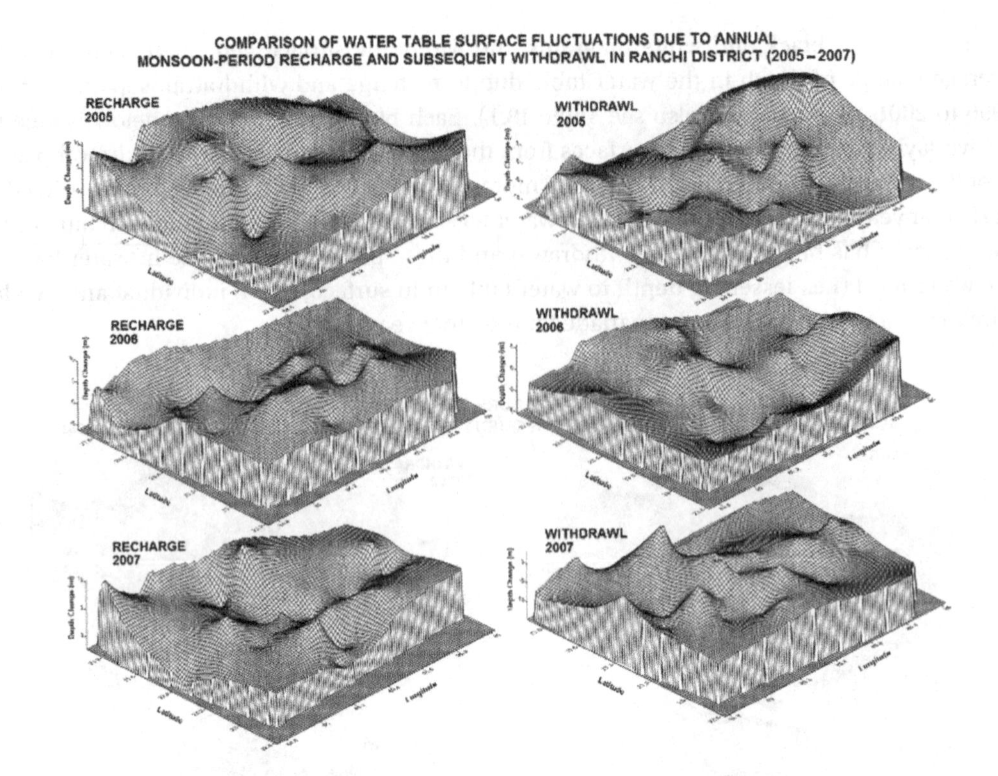
COMPARISON OF WATER TABLE SURFACE FLUCTUATIONS DUE TO ANNUAL
MONSOON-PERIOD RECHARGE AND SUBSEQUENT WITHDRAWL IN RANCHI DISTRICT (2005 – 2007)
RECHARGE
2005
WITHDRAWL
2005
RECHARGE
2006
WITHDRAWL
2006
RECHARGE
2007
WITHDRAWL
2007
Latitude
Longitude

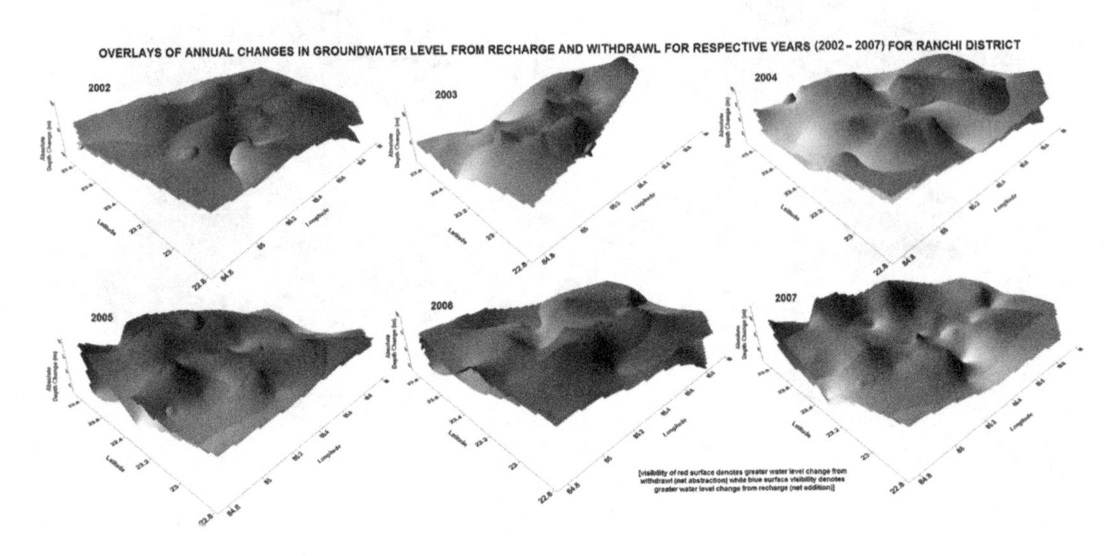
OVERLAYS OF ANNUAL CHANGES IN GROUNDWATER LEVEL FROM RECHARGE AND WITHDRAWL FOR RESPECTIVE YEARS (2002 – 2007) FOR RANCHI DISTRICT
2002
2003
2004
2005
2006
2007
[visibility of red surface denotes greater water level change from withdrawl (net abstraction) while blue surface visibility denotes greater water level change from recharge (net addition)]

COMPARISON OF WATER TABLE SURFACE FLUCTUATIONS DUE TO ANNUAL MONSOON-PERIOD RECHARGE AND SUBSEQUENT WITHDRAWL IN RANCHI DISTRICT (2008 – 2010)

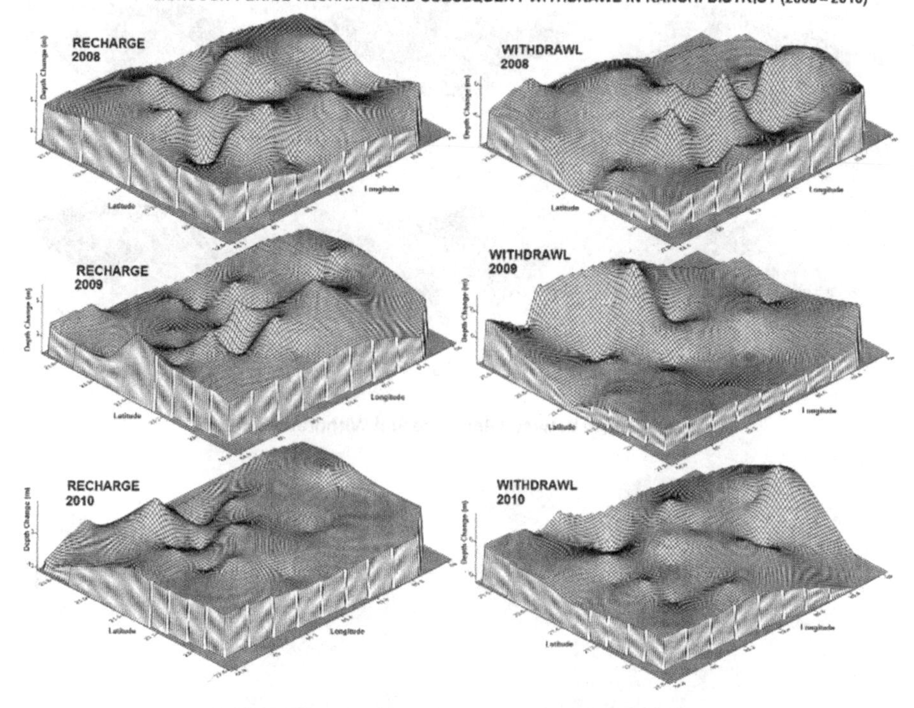

COMPARISON OF WATER TABLE SURFACE FLUCTUATIONS DUE TO ANNUAL MONSOON-PERIOD RECHARGE AND SUBSEQUENT WITHDRAWL IN RANCHI DISTRICT (2011 – 2013)

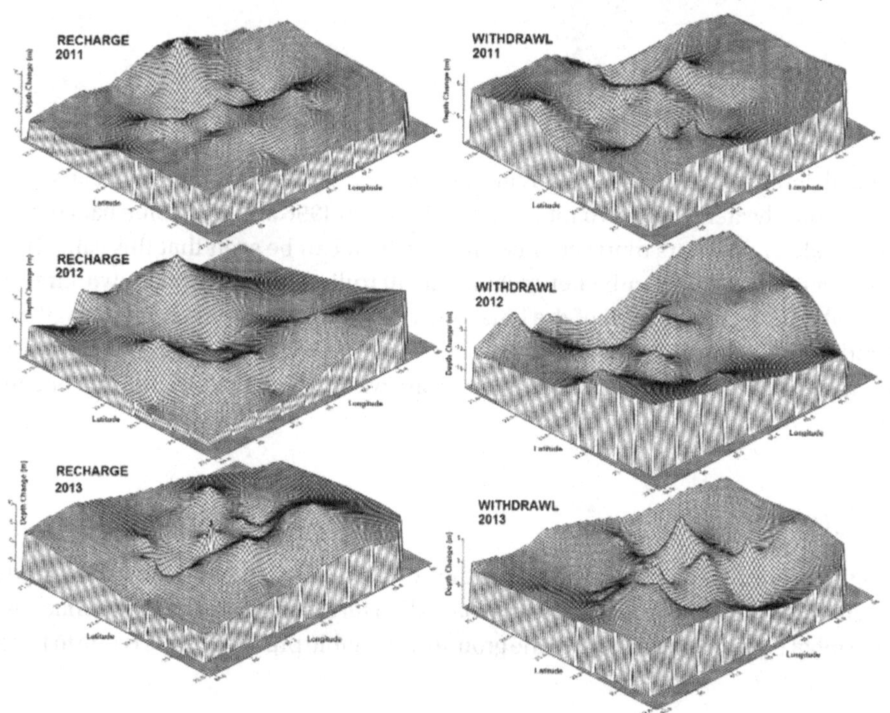

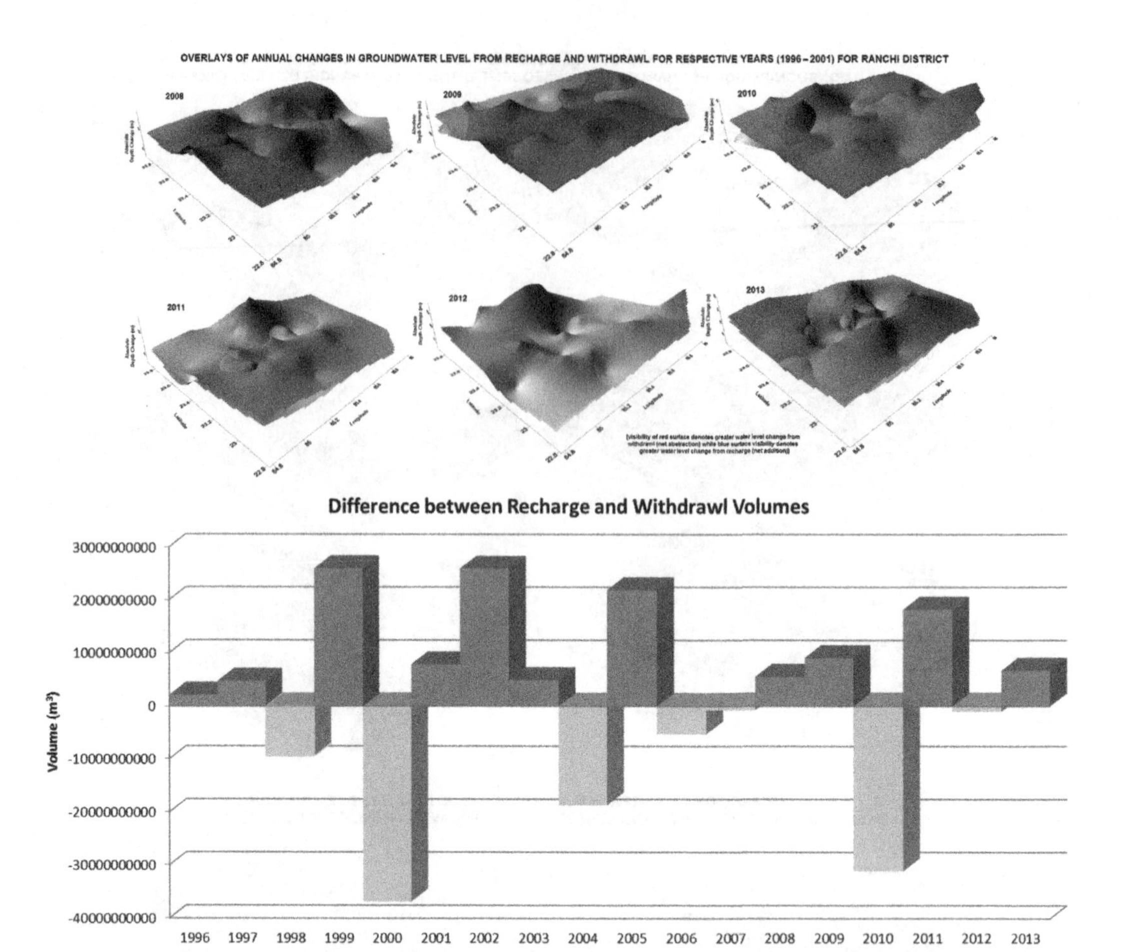

The bar diagram represents the difference between the volume of groundwater recharged and volume of groundwater withdrawn for each year between 1996 and 2013. Blue bars represent net recharge, while orange bars represent net withdrawal. It can be seen that this value has ranged in between 30 trillion meter cube net-recharge to 40 trillion meter cube net-withdrawal. It can be observed that the total length of the blue bars combined is more than total length of orange bars. It indicates that there has been possible net-recharge of ground water in between 1996 and 2013 for most of the years, even though in the years when the deficit occurs, this becomes very marked.

18.6 Conclusion

Through the various maps generated during the course of study, it has been established that the average groundwater table in Ranchi district has fallen further from the ground surface across the years. 5–6 mbgl is the average depth to the groundwater table presently. This is due to the demand for groundwater increasing exponentially year by year. Thus, rain water harvesting should be adopted in those areas where the postmonsoon depth to the water level is more than 7 mbgl and where the long-term water level trend is declining at the rate of more than 0.1 mbgl.

References

Babiker, I.S., Mohammed, M.A.A., Hiyama, T., Kato, K., 2005. A GIS-based DRATIC model for assessing aquifer vulnerability in Kakamigahara Heights, Gifu Prefecture, central Japan. Sci. Total Environ. 345, 127–140.

Bhunia, G.S., Samanta, S., Pal, D.K., Pal, B., 2012. Assessment of groundwater potential in Paschim Medinipur District, West Bengal: a meso-scale study using GIS and remote sensing. J Environ. Earth Sci. 2 (5), 41–59.

Boyer, R., McQueen, J., 1964. Comparison of mapped rock fractures and airphoto linear features. Photogramm. Eng. Remote Sens. 30 (4), 630–635.

Brown, N., 1994. Integrating structural geology with remote sensing in hydrogeological resource evaluation and exploration. In: Proceedings of Tenth Thematic Conference in Geologic Remote Sensing: Exploration, Environment and Remote Sensing, 9–12 May, San Antonio, Texas, pp. 144–154.

Das, S., Mukherjee, J., Bhattacharyya, S., Patel, P.P., Banerjee, A., 2022. Detection of groundwater potential zones using analytical hierarchical process (AHP) for a tropical river basin in the Western Ghats of India. Environmental Earth Sciences doi:10.1007/s12665-022-10543-1.

Dhakate, R., Singh, V.S., Negi, B.C., Chandra, S., Rao, V.A., 2008. Geomorphological and geophysical approach for locating favorable groundwater zones in granitic terrain, Andhra Pradesh, India. J. Environ. Manag. 88 (4), 1373–1383.

Fetter, C.W., 2001. Applied Hydrogeology. Prentice Hall, Englewood Cliffs, NJ, p. 598.

Foster, S.S.D., 1987. Fundamental concepts in aquifer vulnerability, pollution risk and protection strategy. In: Van Duijevenboden, W., Van Waegeningh, H.G. (Eds.), Vulnerability of Soil and Groundwater to Pollutants. Committee on Hydrological Research, The Hague, pp. 69–86.

Garza, L.D.L., Slade, R.M., 1986. Relations between areas of high transmissivity and lineaments: the Edwards aquifer, Barton springs segment, Travis and Hays counties. In: Abbott, P.L., Woodruff, Jr., C.M. (Eds.), The Balcones Escarpment: Geology, Hydrology, Ecology and Social Development in Central Texas. Geological Society of America, San Antonio, TX, pp. 131–144.

Ghazali, F.A., 1992. Poisoned waters, mindless industrialization polluting rivers. Nation and the World 28–29 15th Feb.

Gupta, D., Patel, P.P., 2021. Mapping groundwater level fluctuation and utilisation in Puruliya District, West Bengal. In: Adhikary, P.P., Shit, P.K., Santra, P., Bhunia, G.S., Tiwary, A.K., Chaudhary, B.S. (Eds.), Geostatistics and Geospatial Technologies for Groundwater Resources in India. Springer, Cham, pp. 413–442.

Jha, M.K., Sebastian, J., 2005. Vulnerability Study of Pollution Upon Shallow Groundwater Using Drastic/GIS. Map India, New Delhi.

Krishnamurty, J., Srinivas, 1996. Demarcation of geological and geomorphological features of parts of Dharwar Craton, Karnataka using IRS LISS-II data. Int. J. Remote Sens. 17, 3271–3288.

Naik, P.K., Awasthi, A.K., 2003. Groundwater resources assessment of the Koyna river basin, India. Hydrogeology 11, 582–594.

Parizek, R.R., Kardos, L.T., Sopper, W.E., Myers, E.A., Davis, D.E., Farrell, M.A., Nesbitt, J.B., 1967. Penn State Studies 23: waste water renovation and conservation. Administrative Committee on Research, The Pennsylvania State University. Am. J. Publ. Health 65, 71.

Rahman, A., 2003. Assessing water quality from Jal Nigam hand pumps in Aligarh city, India. In: Nature Environment amd Pollution Technology. KARAD, India, pp. 241–244.

Ravindran, K.V., 1997. Drainage morphometry analysis and its correlation with geology, geomorphology and ground water prospects in Zuvari basin, South Goa, using remote sensing and GIS. In: Proceedings of

National Symposium on Remote sensing for natural resource with special emphasis on water management, Pune, pp. 270–296.

Ravindran, K.V., Jeyaram, A., 1997. Groundwater prospectus of Shahba Teshil, Baran District, Eastern Rajasthan: a remote sensing approach. J. Indian Soc. Remote Sens. 25 (4), 239–246.

Reddy, K.R., Adams, J.A., 1996. 'In situ air sparging: a new approach for ground water remediation. Geotech News 14 (4), 27–32.

Rupert, M.G., 2001. Calibration of the DRASTIC groundwater vulnerability mapping method. Ground Water 39, 630–635.

Sree Devi, P.D., Srinivasulu, S., Raju, K.K., 2001. Hydrogeomorphological and groundwater prospects of the Pageru River Basin by using remote sensing data. Environ. Geol. 40, 1088–1094.

Tiwari, V.M., Wahr, J., Swenson, S., 2009. Dwindling groundwater resources in northern India, from satellite gravity observations. Geophys. Res. Lett. 36, L18401.

Todd, D.K., 1980. Groundwater Hydrology. John Wiley and Sons, New York, p. 535.

Villeneuve, J.P., Banton, O., Lafrance, P., 1990. A probabilistic approach for the groundwater vulnerability to contamination by pesticides: the VULPEST model. Ecol. Model 51, 47–58.

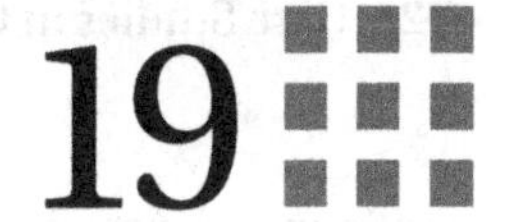

Groundwater conservation and management: Recent trends and future prospects

Gouri Sankar Bhunia[a], Pravat Kumar Shit[b] and Soumen Brahma[a]

[a]DEPARTMENT OF GEOGRAPHY, NALINI PRABHA DEV ROY COLLEGE, BILASPUR, CHHATTISGARH, INDIA [b]DEPARTMENT OF GEOGRAPHY, RAJA N. L. KHAN WOMEN'S COLLEGE (AUTONOMOUS), MIDNAPORE, WEST BENGAL, INDIA

19.1 Introduction

Water resources are becoming increasingly vulnerable across the world as a result of rising demand due to population expansion, the need for more food production, expanding industry due to higher living standards, pollution from different human activities, and climate change impacts (Manisalidis et al., 2020). The number of areas with a stable groundwater balance is dwindling all over the world. Groundwater is essential for the survival of rivers, lakes, wetlands, and ecological systems because it compensates 99% of the world's liquid fresh water (Dodds et al., 2019). However, because groundwater is tucked away beneath the land surface, few people see it. Groundwater use is dominated by three issues: depletion of groundwater because of over exploitation; waterlogging and salinization due to inadequate drainage and concurrent use; and contamination due to agricultural, industrial, as well as other anthropogenic activities (Shah et al., 2000). It is estimated that by 2050, one in every four people would live in a nation with a freshwater deficit due to the shortage and poor water quality. As a result, the United Nations Sustainable Development Goals, Transforming Our World: The 2030 Agenda for Sustainable Development, have included guaranteeing water access to safe and affordable management as one of the goals.

Weathering and geomorphologic procedures influence the earth's surface. Rivers, lakes, and wetlands are surface manifestations of groundwater that exchange flow with the groundwater reservoir that continues to feed them when they need water and tends to take some of their flow when there is an enormous amount of surface water (Leibowitz et al., 2018). Groundwater is a massive, slow-moving reservoir of freshwater resources (in its shallower arenas) that forms complexes with earth materials along its channel flow and interactions with river and atmosphere water transportation networks. This massive groundwater reservoir needs to serve as a: (1) fresh water hydrologic cycle controller by facilitating the flow of continental rivers and streams, (2) a chemical factory and conveyor belt for accessing Earth's material and transporting it from

Case Studies in Geospatial Applications to Groundwater Resources. DOI: https://doi.org/10.1016/B978-0-323-99963-2.00001-8

continents to oceans, (3) a waste repository/processing plant, and (4) a worldwide life support system.

Water scarcity can be caused by a variety of factors. Water scarcity can be caused by a physical deficiency (physical water lack of availability), or by poverty and underdevelopment or an absence of sufficient infrastructure to ensure a consistent supply, even in places where water is plentiful (economic water scarcity). Water use has increased at a rate more than twice that of global population growth over the last century, and it continues to rise in all sectors (Islam and Karim, 2019). Droughts will become more common as a result of climate change. Nowadays, over 2 billion people live in countries with severe water scarcity, and the number will continue to rise (Boretti and Rosa, 2019). By 2100, the earth's population will have risen to nearly 8 billion people. Humans must learn to produce enough food without depleting the soil, water, or climate. The biggest hurdle humanity has ever directly confronted has been labelled as this. The remedy is primarily focused on sustainable groundwater management (Jakeman et al., 2016). Groundwater needs to be better understood and managed scientifically, because it can help solve the problem if we seek liable use and sustainment through good governance practices.

19.2 Water, land, energy, and agriculture

Rapid urbanization, the growth of peri-urban areas, and wastewater treatment problems are all adding to the situation's complexity. To limit harmful effects on human health and the environment, it is clear that alternate water sources, suitable water supply methods, and wastewater treatment are needed (Sharma et al., 2017). As the world's population grows, so does the demand for important resources like water, electricity, and land. Innovative spatial-temporal solutions for managing the water-land-energy nexus should be pushed if future human demands are to be satisfied. Remote sensing techniques are progressively becoming more prevalent in efficient and effective of agriculture production related topics like transmission losses, irrigation increased efficiency, farming techniques, pest/disease attack, soil erosion, and moisture patterns, largely owing to the provision of high-resolution spatial data. Infrastructure that is resilient is essential for sustainable water resource management and development (Hutton et al., 2017). Evidence-based measures are needed to get a vital understanding of best practices in water-related architecture.

19.3 Groundwater prospects and geographical settings

Excessive consumption of groundwater is undeniably occurring in isolated areas, and it can have disastrous consequences for communities. The several parts of coastal areas have silky sediment deposits varying ages from Pleistocene to recent, which have resulted in multi-aquifer systems with high potential (Alfarrah and Walraevens, 2018). Groundwater progression from these kind of aquifer systems has a lot of room for growth. Moreover, developing groundwater from these aquifers must be treated with caution, and care must be taken to confirm that resource overexploitation does not result in saltwater water intrusion. As the multi-aquifer systems in coastal areas are likely to have all possible dispositions of fresh and saline waters, it is necessary to take-up detailed studies to establish the saline–fresh water interface and establish the replenishable discharge of groundwater to sea (Hussain et al., 2019). Moreover, the effective implementation of surface water irrigation systems without due regard for ecological sustainability, numerous canal command areas are experiencing water logging and soil salinity issues as groundwater levels gradually rise (Kumar and Sharma, 2020). These areas have a lot of potential for groundwater progression because the shallow water table can be dropped to six meters or more without causing any negative impacts on the environment. The challenges caused by poor water quality in these areas can be remedied by melding them with the accessible canal waters.

Flood plains of rivers are normally good repositories of groundwater and offers excellent scope for development of groundwater. Groundwater levels in these tracts are mostly shallow, leaving little room for accommodating the monsoon recharge, a major portion of which flows down to the river as surface (flood) and sub-surface runoff. Surplus monsoon runoff, which would be wasted, can be captured with careful water resource management in these areas. During the nonmonsoon season, groundwater is controlled removed from flood plains to generate extra space in the unconfined aquifer for later recharge/infiltration during the rainy season (Greene et al., 2016). The second instance occurs more frequently in rivers with irregular flows; the loose sediments in the flood plains are more or less concentrated, culminating in a shallower groundwater level (Ahuja and Tatsutani, 2009). During the nonmonsoon period, the massive pull-out of such flood plain aquifers generates adequate space in the groundwater reservoir, which is refreshed by the river during the raining season.

19.4 Advancement in groundwater data

The gravity recovery and climate experiment (GRACE) satellite, begun in 2002 by NASA, is the only one that expressly evaluates variation in water column by assessing monthly mass changes. GRACE does not differentiate between snow, surface water, soil moisture, or groundwater, instead depending on other data to figure out which portion is related to groundwater variations (Neves et al., 2020). GRACE has made it possible to comprehend evolving water budgets on a global level (Frappart and Ramillien, 2018). Since 1982, groundwater changes have been measured at 30 to 120 m spatial resolutions every 16 days using satellite data. The composite image contains the value of these data: vegetation index, land surface temperature, and a normalized difference water index.

Research team from Hamburg University compiled the global river chemistry dataset (GLORICH), which contains different water quality parameters for river locations throughout the world (Thorslund and van Vliet, 2020; Hartmann et al., 2019). This information is open to the public and can be retrieved as a zip file from PANGEA. 1.27 million samples of major compounds, nutrients, carbon species, and physical and mechanical properties are included in the dataset. For all stations that met our eligibility criteria, we retrieved Specific Conductivity data (another term for EC) from the "hydro-chemistry" csv file and teamed it with station relevant data ("Sampling locations" file).

The global groundwater information system (GGIS) is a web-based immersive portal for groundwater-related knowledge and information (https://groundwaterportal.net/global-groundwater-information-system). The system's main goal is to aid in the collection and analysis of data on groundwater resources, as well as the communicating of that data among water experts, decision makers, and the general public. Transboundary groundwater, global country data, project related information, managed aquifer recharge, small islands, and groundwater monitoring are among the modules/viewers currently available in the GGIS. The groundwater spatiotemporal data analysis tool (GWSDAT) is a user-friendly, open-source software application for visualizing and interpreting groundwater monitoring data (https://www.api.org/oil-and-natural-gas/environment/clean-water/ground-water/gwsdat). It also allows for the use of other forms of monitoring data gathered over time and space. The FREEWAT platform combines the geographical data evaluation power of GIS geo-processing and postprocessing tools with process-based simulation models. The FREEWAT environment supports the storage of large spatial datasets, data management and visualization, as well as the execution of distributed modelling codes (mainly belonging to the MODFLOW family) (Rossetto et al., 2018).

19.5 Opportunities of sensor

Satellites can now provide a repository of imagery that spans centuries. These historical records enable us to spot trends and patterns in various aspects of watershed management, including water supply, water conservation, and crop production. Information gathering costs are being drastically reduced thanks to recent advancements in water technology. Groundwater tracking needed comparatively expensive equipment and practitioners to acquire and interpret data until recently, leaving private citizens unaware of groundwater conditions and unable to appropriate measure with groundwater challenges. Cost-effective sensors and meters are being developed as a result of technological advancements, allowing private citizens access. Wellntel offers meters that record groundwater levels (supply) and cranking for a reasonable price (between $800 and $1200). One of the most important aspects of Wellntel's approach is that the data is not only obtained, but also addressed for the user: the data is sent directly to the cloud, where individual citizens can access it.

Remote sensing is extensively used to offer this implicit hydrogeological information, acquiring data on factors including geology, geomorphology, drainage patterns, vegetation, and land use owing to the absence of high permeability capacity (Agarwal et al., 2013). Thematic maps were generated using remotely sensed data, GIS, geophysical techniques, and ground-truth data to demarcate groundwater potential zones (GWPZ) and monitor groundwater vulnerability. Multi-spectral (enhanced thematic mapper (ETM)) and spatial (shuttle radar topography mission (SRTM)) data, radar technology, and thermal surveys are used in remote sensing for groundwater monitoring. Apart from radar and geophysical methods, most Earth Observation sensors do not permeate the earth's surface. Senay et al. (2012) used one of their models that required land surface temperature. This was calculated using data from the moderate resolution imaging spectroradiometer's thermal band (MODIS). Due to the capillarity in the soil zone, through which groundwater can float to the surface and modify the soil moisture content, reflectance, and scattering attributes can represent variability in groundwater levels (Huo et al., 2011). The tropical rainfall measurement mission (TRMM) microwave imager (TMI), the advanced microwave scanning radiometer (AMSR) satellite systems, the soil moisture ocean salinity (SMOS) mission and soil moisture active passive (SMAP) mission, as well as multiple other synthetic aperture radar (SAR) series, have all been deployed over the years (Jackson, 2002).

19.6 Advancement of space technology

Satellite remote sensing is becoming more widely used as a supplement to in-situ monitoring networks, and in some cases, it is the only feasible option. And almost all elements of the hydrological cycle can now be measured directly and indirectly using satellite-based sensors (McCabe et al., 2017). From the few early committed operations that concentrated on snow extent and land cover to the existing continuous monitoring of nearly all elements of the water balance and vegetation health, satellite data of dependent variable for water management and hydrological hazard controlling has improved markedly. Hydrometeorological and agricultural monitoring networks are often meagre and have a long latency in several regions, attempting to make real-time decision-making extremely difficult (Sheffield et al., 2018). As a result, remote sensing sensors are capable of delivering key information for water management and monitoring the progression of hazards and their effects (van Dijk and Renzullo, 2011). Satellite data also allows for risk evaluation in the context of regional water security, food production, retrieval, and trade, kudos to its extensive coverage (Dalin et al., 2017). The EUMETSAT satellite application facility on land surface analysis provides a number of products for Europe (MW, IR, and MW-IR merged precipitation; MW-based soil moisture; VIS/IR-based snow cover extent; and MW-based SWE) that are directed specifically at operations and maintenance hydrology and WRM (EUMETSAT, 2018).

19.7 Water treatment technology

Water treatment technological improvements have the opportunity to boost a variety of industries by increasing water efficiency and addressing concerns like ageing infrastructure, sustainable development, and smart water. Numerous water users, primarily utilities, are risk averse and hesitant to be early adopters of new technologies, which poses a significant dilemma for water treatment technology (Torresen, 2018). Moreover, the negative consequences (deteriorated water quality, lower groundwater levels, saltwater interferences, and so on) of untenable groundwater pumping, rising demand, regulatory issues, and more extreme weather events as an outcome of a warming climate are pressuring infrastructure to innovate out of necessity.

To achieve such a lofty goal, creative business models have been developed that incorporate various treatment technologies to make water fit-for-purpose in regards to quality, thereby enabling more freshwater resources while lowering treatment costs for nonpotable uses. Transitioning toward a fit-for-purpose treatment model necessitates legislative changes and the development of legal standards for what makes up "fit" for various applications, such as potable water quality thresholds and benchmarks. Establishing new technology is frequently accompanied by the adoption of new business models. Public-private partnerships (P3s) can help utilities resolve enhanced regulations, enact conservation and efficiency programs, and adjust to diminished public funding by facilitating the implementation of innovative business models and accumulating innovations.

There is a massive business opportunity, and facilities are being bombarded with providers of emerging technologies, forcing them to either overlook them or spend in determining how well they fulfil their needs. Most utilities cannot avail to take the risk of adopting new, untested technologies that may not produce the savings claimed or fail to meet regulatory requirements. It would be beneficial if a third party, including a private or nonprofit organization, streamlined and vetted the plethora of technologies. As part of the integrated conversation between utility services and industrial applications, this third-party entity could also recognize gaps and needs for technological advances, primarily ensure that the innovations are achieving a precise need. In addition, training staff to use emerging innovations costs the utility time and money. The upfront costs of implementing new technologies could be limited if the utility sector pooled its knowledge.

19.8 Integrating monitoring network

Groundwater resource problems are usually associated with long-term effects. Monitoring may offer essential data for characterizing, analyzing, and recording changes in aquifers, which can help us better understand the complexity and uncertainties connected with groundwater resources (Mogheir et al., 2006). The majority of data on changes in groundwater quantity and quality is dispersed, and important characteristics (such as heavy metals and organic matter) are even absent. Scientists and stakeholders have insufficient evidence to ensure that resources are used efficiently and the environment is adequately safeguarded. Under these circumstances, an interconnected network for analyzing the physical characteristics (e.g., temperature and water levels) as well as the chemical composition of groundwater and surface water bodies is urgently needed (lakes, rivers, and drains). The groundwater management method should focus on impurities in drinking water such as nitrate, fluoride, arsenic, and heavy metals. It's just as vital for rural communities to have clean drinking water as it is for city dwellers. Additionally, the variance in chemical components in restricted aquifers should be given more attention, and groundwater pumping should be strictly supervised, particularly near the two water cones.

When designing a monitoring network for groundwater quality and quantity, there are two key characteristics that differentiate groundwater from surface waters that must be taken into account (Little et al., 2016). The padding of observation wells in the groundwater quality network will be determined by the strategic approach for distinguishing between diffuse and point pollution stations, national and regional stations, and disparities between principal, specific, and temporary networks. According to Tuinhof et al. (2002), a monitoring network consists of a set of observation wells and a predetermined quantity of extraction wells. In addition, a ground-water monitoring program comprises both a quantity and a quality network (groundwater level and recharge rates). These networks must be optimized based on current and future requirements, as well as accessible financial resources.

The early warning system (EWS) must be attributed to the change in the groundwater system as groundwater is a dynamic system. The EWS may be used to evaluate sustainable abstraction quantities, development viability, and groundwater management strategies. The availability of data on fluctuations in water level and water quality in the surface water-soil-groundwater system, on the other hand, is one of the most important drivers of the EWS's effectiveness. The global groundwater monitoring network (GGMN) is a web-based participatory approach of networks created to enhance the performance and availability of groundwater monitoring data (https://ggis.un-igrac.org/view/ggmn) and, as a result, our understanding of the state of groundwater resources. GGMN is a UNESCO initiative that is being executed by IGRAC with the help of many global and regional relationships.

19.9 Strengthening scientific support

Groundwater management's ultimate goal is to provide a sufficient amount and quality of water for human health, livelihood, and productivity. Water resources, ecosystems, and water security are the three main components of modern management paradigms. Groundwater management is hampered by the need to bridge the theoretical and practical divide. As a result, the interactions between different bodies of water are poorly understood. Additionally, new technologies in the areas of water conservation, groundwater pumping, information collecting, and contaminant transport modeling are difficult to apply due to a lack of groundwater expertise. Scientific research should be devoted toward better explaining, communicating, and educating water managers and decision makers in these circumstances (Li, 2016).

By enabling a clear understanding of hydrological processes, evaluating the consequences of anthropogenic activities on the quality and variety of groundwater, enlightening the function of groundwater in the eco-environment, and delivering a holistic perspective, science can contribute to the resolution of conflicts, including those between groundwater development and preservation points of view. Groundwater management that is sustainable requires not just outside expertise and university assistance, but also the help of internal professionals. To determine the geographical and temporal distribution and movement of pollution in the aquifer system, modern technologies and modeling techniques should be introduced and utilized, particularly in drinking water protected zones, landfills, industrial parks, and mine sites. Concerned departments should provide professionals greater opportunity to improve their technical expertise through frequent education and training.

Global map of irrigated areas (GMIA) plots the percent of total area equipped for fertigation in 2005 on a raster with a spatial resolution of 5 arc-minutes (https://wbwaterdata.org/dataset-/global-map-of-irrigated-areas-gmia). Supplementary map layers show the percentage of irrigated agriculture that was definitely used for irrigation, as well as the percentages of irrigated land that was cultivated with groundwater, surface water, or nonconventional water sources. The global groundwater information system (GGIS) is a web-based interactive portal for groundwater-related knowledge and information. The GGIS is made up of several modules that are organized around different themes (https://www.un-igrac.org/). Global groundwater data by aquifer and country decentralization, global groundwater stress (based on GRACE data), and global groundwater quality data are among the data sets. India-WRIS aims to raise public and stakeholder awareness of the current state of water resources and the need for efficient management by enticing them to participate in achieving the overall goal of water security (https://indiawris.gov.in/wris/#/). New technologies such as GIS and remote sensing, combined with domain expertise in water resources, have proven to be effective in asset mapping, evaluation, and management strategizing. This involves data collection from a variety of sources, standardization, and storage of the complete gamete of data on a national scale.

19.10 Encouraged public engagement

The concerned with establishing efficient and productive interactions in order to achieve a shared understanding of goals or a mutual commitment to change is referred to as participation (Curtis et al., 2016). To encourage community participation, a variety of processes and models are employed. The public is actively involved in the development and execution of groundwater management programs. Contributing to collaborative research programs as a social researcher has had numerous advantages. Our awareness of ecology and hydrogeology, as well as the guessed links between property management and environmental consequences, has improved as a result of regular and structured interactions with scientists. Because public access to information on groundwater conservation and preservation is constrained, there is no effective way to promote public awareness of the need for groundwater protection (Jie et al., 2018). Efforts should be made to increase public involvement, which is beneficial for monitoring industrial and agricultural operations, along with improving water conservation in urban and rural regions using water-saving methods like spray and drop irrigation. Information transparency should be preserved for simple access to information and data related to the usage of and potable water in aquifers, typically via a web service or a newsletter. In addition, improved public knowledge is needed to implement water-saving measures and pollution-prevention initiatives in both urban and rural regions.

19.11 Future planned and conservation strategy

NOAA and NASA are collaborating on a new polar-orbiting operational environmental satellite system called JPSS. In the VIR spectra, the sensor measures in 22 bands. Based on the band wavelength, the spatial resolution varies between 375 and 750 m. LST and ET retrievals, as well as vegetation characteristics like fPAR, leaf water content, and LAI, and snow cover products, are all relevant. For example, based on Hain et al. (2017), a global 400-m VIIRS-based ET product is being developed, starting with regional data sets for the Middle East/North Africa, the United States, and Brazil. Through continuous spectral sampling of the VNIR and SWIR regions, numerous hyperspectral imaging missions are scheduled or suggested, which can improve lookup of snow and vegetation attributes, as well as bring new water quality and soil property products. At nadir and revisit times of 19 and 5 days, respectively, NASA's proposed HyspIRI has an imaging spectrometer evaluating from the VSWIR range in 10-nm consecutive bands and a multispectral imager assessing from 3 to 12 m in the TIR range with 60-m resolution. The German Space Agency's EnMAP mission (Guanter et al., 2015) is a hyperspectral imager that operates in the spectral range of 420–2450 nm, with bands of 10–40 nm width, at 30-m spatial resolution, and a revisit time of at least 4 days, with prospective research implementations in crop and forest monitoring, inland and coastal waterways, and soil science, among others.

Research and collaboration: By participating in the research that intersects sectors and disciplines, and by adjusting study goal that can be assigned to different problems, scientists in academia and the wider research community have the opportunity to put some of the future research around groundwater resources. The majority of data is presently used to illustrate where groundwater is diminished or has a negative impact on land surface. Academics can use data to measure progress and prospects, such as suitable groundwater recharge sites. Universities could let concerns drive research agendas and engage in more collaborative, implemented client research. By offering alternative approaches, researchers could play the role of interventionists. It is encouraging to have a focal point for dealing with problems. Arizona State University collaborated with the Earth Genome Project to develop a tool that equates the cost effectiveness of using fallow farmland for flood recharge. These kinds of tools have the potential to be ramped across the country and/or developed globally. Research could reveal how to better arrangement assets to define water conservation and economic career progression. Defining the most valuable use of land, such as converting an unviable piece of cropland into a solar farm that produces energy for the farm while conserving water that would otherwise result in a lower return. Academic institutions have the ability to form inventive collaborations with both the public and private sectors. These alliances provide an ability to build relationships and encourage future collaborations.

Organic combined efforts can be quick and effective, but they must be supported by a regulatory framework to ensure their long-term viability. Many successful ensembles are the result of a few key people's leadership; moreover, enshrining relationships can help ensembles survive personnel changes. Dedicated funding streams for cooperative relationships should be part of the institutionalization process. When it comes to establishing stringent regulations, the challenge is to find the sweet spot that approach was undertaken while enabling for process flexibility. The disadvantage of formalizing collaborations is that these organizations are slow to adapt to technological advancements and new revolutionary practices that allow for goldilocks extensibility.

Awareness and messaging: Groundwater messaging should create intelligent sense to people who are not familiar with the subject. To confirm that the relevance of groundwater was conveyed clearly, was pertinent, and appeared from a reliable source to various sectors, the messaging campaign around sustainable groundwater management had to meld the key messages with relevant messiahs and visualizations. Industrial partners can effectively communicate the magnitude of the problem. Slogans like "One Water" are useful for communicating complex ideas to the general public. The public will not comprehend where their water comes from or how different water sources are intertwined if the message is not clear. Social media could be used to educate young people about groundwater as well as the interconnected nature of water. Evolving regulations that impede integrated water management will be required to move toward a multifaceted "One Water" strategy.

Data acquisition from satellite and sensor: NOAA and NASA are collaborating on a new polar-orbiting operational environmental satellite system called JPSS. One of the components on the JPSS is the VIIRS. VIIRS is a daily time series of multispectral data based on AVHRR and MODIS that has implementations in energy and water balance, vegetation interactions, land cover land use change, and the cryosphere. LST and ET sequences, as well as vegetation criteria like fPAR, leaf water content, and LAI, and snow cover products, are all relevant. For instance, based on Hain et al. (2017), a global 400-m VIIRS-based ET product is being developed, starting with regional data sets for the Middle East/North Africa, the United States, and Brazil. NASA's proposed HyspIRI has an imaging spectrometer measuring from the VSWIR range in 10-nm contiguous bands and a multispectral imager measuring from 3 to 12 m in the TIR range with 60-m resolution. The sensor is used to derive the information on snow retrieval, vegetation monitoring, and ET (Lee et al., 2015). The EnMAP mission (Guanter et al., 2015) is a hyperspectral imager with a spectral range of 420–2450 nm, 30-m spatial resolution, and a revisit time of at least 4 days, with potential research applications in crop and forest monitoring, inland and coastal waterways, and soil science. With intended missions like the joint US-Indian NASA-ISRO Synthetic Aperture Radar (NISAR) polar-orbiting mission, SAR sensors will be used to estimate groundwater through small differences in surface topography. This will use an L-band and S-band SAR to deliver meter-scale land surface height retrievals that can document in groundwater storage (Sheffield et al., 2018).

19.12 Conclusion

Remote sensing has the potential to improve hydrological forecasts by updating key parameters. Remote sensing can provide data needed to initialize predictive model (e.g., soil moisture, snow, and river levels/discharge), and several research has demonstrated the perks of revamping hydrological forecasts (Lü et al., 2016), especially in snow-dominated and dry areas of the planet (Shukla et al., 2013). The most enticing way to use satellite data for groundwater management is to combine it with in situ data and hydrological models. Socialization of soil moisture data into hydrological models can enhance soil water, ET, and streamflow predictions (Ridler et al., 2014). The use of actual ET estimates based on remote sensing to confine hydrological models can also enhance computer models dramatically (Roy et al., 2017). Satellite based gridded products can benefit from the interconnection of in situ data from meteorological networks (Chaney et al., 2014). With just a few gauges, bias modification of satellite-based precipitation can drastically enhance hydrological conjectures (Serrat-Capdevila et al., 2014). Over the last few decades, climate forecasting has improved in terms of technical ability and resolution to the point where it can now provide some consistency in areas with sturdy teleconnections to sea surface temperatures, and at spatial resolutions (order of 1–10 km) compelled for decision-making. For instance, dynamical climate model estimates of the El Nino–Southern Oscillation are now outpacing statistical projections (Barnston, 2012), and there is conceivable for proficient global climatic forecasting in the tropics and subtropics (Kumar et al., 2013).

One of the most difficult tasks is to provide services that allow users to access and use the massive amounts of data engendered by current and planned missions. The Sentinel-1 satellite delivers about 1 TB of raw data per day, equating to 10 PB (1016 bytes) over the mission's long lifespan (Wagner et al., 2009). When deduced data sets for variety of applications (hydrological, water resources, cultivation, health, and so forth) are incorporated, the number grows even larger, posing a logistical and computational dilemma that is likely beyond the scope of any single data facility. To address these challenges, high-performance computing, which includes supercomputing, cluster computing, and distributed/grid computing, is obligated. However, despite the massive strides in the stipulation of data management and visualization tools by data centers, classical infrastructures at scientific government entities are out of scope for the number of customers. Cloud computing which provides "pervasive, convenient, on-demand network access to a shared pool of configurable computing resources," are probably to be the alternative (Mell and Grance, 2011).

Amazon's Elastic Compute Cloud (EC2), which is usable via Amazon Web Services, Google's Cloud Platform, and Microsoft's Azure are some examples of business pay-as-you-go cloud computing technology. These facilities provide scalable computing infrastructure (e.g., virtual machines like Google's Compute Engine) that requires the user to insert, configure, and run their own processing software, or computing platforms that coherent the software and file system firm the user (e.g., Google Earth Engine). Despite the availability of dependable and scalable access to dispersed computing power, the cost of training users should not be overlooked, particularly for implementations that are not simple parallel computation of subsets of satellite imagery.

Near-real-time retrievals of nearly all elements of the terrestrial water cycle are now possible thanks to satellite remote sensing, despite numerous challenges such as accuracy, consistency, coherency, and efficiency. While much work remains to empower and ability to improve for fetching groundwater, water quality, surface water levels, and river flows, the majority of these lookups are global in scope and at spatial, temporal, and spectral resolutions sufficient to settle hydrological parameters and their interactions with anthropogenic activities. Satellites can offer data in areas where on-the-ground relevant information is scarce, unusable, or inaccessible as a source of real-time information. Moreover, fully realizing the prospective presupposes an insight into the various, independent, supplementary, and contending data products, as well as their functionality for a variety of management applications, such as flood/drought risk evaluation and water availability monitoring. Capacity must be created interested in working with satellite data and translate it into knowledge that can be used to inform decision-making. Moreover, there is still a significant disconnect between the accessibility of these products and their use in decision-making. As a result, there is still a privilege to work more assertively with national stakeholders to fortify capacity to use these remote sensing products, particularly in data-scarce zones, in order to develop remedies and possibilities for natural-hazard monitoring and early warning systems in provision of efficient hazard mitigation policies at the national level.

References

Agarwal, E., Agarwal, R., Garg, R.D., Garg, P.K., 2013. Delineation of groundwater potential zone: an AHP/ANP approach. J. Earth Syst. Sci. 122 (3), 887–898. doi:10.1007/s12040-013-0309-8.

Ahuja, D., Tatsutani, M. Sustainable energy for developing countries. S.A.P.I.E.N.S, 2.1, 2009, http://journals.openedition.org/sapiens/823.

Alfarrah, N., Walraevens, K., 2018. Groundwater overexploitation and seawater intrusion in coastal areas of arid and semi-arid regions. Water 10, 143. doi:10.3390/w10020143.

Barnston, A.G., 2012. Skill of real-time seasonal ENSO model predictions during 2002–2011: Is our capability increasing? Bull. Am. Meteorol. Soc. 93, 631–651.

Boretti, A., Rosa, L., 2019. Reassessing the projections of the World Water Development Report. NPJ Clean Water 2, 15. doi:10.1038/s41545-019-0039-9.

Chaney, N.W., Sheffield, J., Villarini, G., Wood, E.F., 2014. Development of a High Resolution Gridded Daily Meteorological Data Set over Sub-Saharan Africa: Spatial Analysis of Trends in Climate Extremes. J. Clim. 27, 5815–5835.

Curtis, A., Mitchell, M., Mendham, E., 2016. Social science contributions to groundwater governance. In: Jakeman, A.J., Barreteau, O., Hunt, R.J., Rinaudo, JD., Ross, A. (Eds.), Integrated Groundwater Management. Springer, Cham doi:10.1007/978-3-319-23576-9_19.

Dalin, C., Wada, Y., Kastner, T., Puma, M.J., 2017. Groundwater depletion embedded in international food trade. Nature 543, 700–704.

Dodds, W.K., Bruckerhoff, L., Batzer, D., Schechner, A., Pennock, C., Renner, E., Tromboni, F., Bigham, K., Grieger, S., 2019. The freshwater biome gradient framework: predicting macroscale properties based on latitude, altitude, and precipitation. Ecosphere 10 (7), e02786. doi:10.1002/ecs2.2786.

EUMETSAT, 2018. Satellite application facility on support to operational hydrology and water management (H-SAF), http://hsaf.meteoam.it (last accessed May 2018).

Frappart, F., Ramillien, G., 2018. Monitoring groundwater storage changes using the gravity recovery and climate experiment (GRACE) satellite mission: a review. Remote Sens. 10, 829. doi:10.3390/rs10060829.

Greene, R., Timms, W., Rengasamy, P., Arshad, M., Cresswell, R., 2016. Soil and aquifer salinization: toward an integrated approach for salinity management of groundwater. In: Jakeman, A.J., Barreteau, O., Hunt, R.J., Rinaudo, JD., Ross, A. (Eds.), Integrated Groundwater Management. Springer, Cham doi:10.1007/978-3-319-23576-9_15.

Guanter, L., Kaufmann, H., Segl, K., Foerster, S., Rogass, C., Chabrillat, S., et al., 2015. The EnMAP spaceborne imaging spectroscopy mission for earth observation. Remote Sens. 7 (7), 8830–8857. doi:10.3390/rs70708830.

Hartmann, J., Lauerwald, R., Moosdorf, N., 2019. GLORICH - global river chemistry database. PANGAEA doi:10.1594/PANGAEA.902360.

Hain, C., Anderson, M.C., Schull, M.A., Neale, C.M.U., 2017. A framework for mapping global evapotranspiration using 375-m VIIRS LST. Abstract H52G-02, presented at 2017 Fall Meeting, AGU 11–15 December 2017.

Hussain, M.S., Abd-Elhamid, H.F., Javadi, A.A., Sherif, MM., 2019. Management of seawater intrusion in coastal aquifers: a review. Water 11 (12), 2467. doi:10.3390/w11122467.

Huo, A., Xunhong, C., Huike, L., Ming, L., Xiaojing, L., 2011. Development and testing of a remote sensing-based model for estimating groundwater levels in Aeolian Desert areas of China. Can. J. Soil Sci. 91, 29–37. doi:10.4141/CJSS10044.

Hutton, G., Chase, C., 2017. Water supply, sanitation, and hygiene. In: Mock, CN, Nugent, R, Kobusingye, O, et al. (Eds.), Injury Prevention and Environmental Health, third ed. The International Bank for Reconstruction and Development/The World Bank, Washington (DC) https://www.ncbi.nlm.nih.gov/books/NBK525207.

Islam, S.M.F., Karim, Z., 2019. World's Demand for Food and Water: The Consequences of Climate Change, Desalination—Challenges and Opportunities, Ed. M.H.D. AbadiFarahani, V. Vatanpour and A.H. Taheri, IntechOpen, doi:10.5772/intechopen.85919. https://www.intechopen.com/chapters/66882.

Jackson, T., 2002. Remote sensing of soil moisture: implications for groundwater recharge. Hydrol. J. 10 (1), 40–51. doi:10.1007/s10040-001-0168-2.

Jakeman, A.J., et al., 2016. Integrated groundwater management: an overview of concepts and challenges. In: Jakeman, A.J., Barreteau, O., Hunt, R.J., Rinaudo, JD., Ross, A. (Eds.), Integrated Groundwater Management. Springer, Cham doi:10.1007/978-3-319-23576-9_1.

Jie, C., Hao, Wu, Hui, Q., Xinyan, Li, 2018. Challenges and prospects of sustainable groundwater management in an agricultural plain along the Silk Road Economic Belt, north-west China. Int. J. Water Resour. Dev. 34 (3), 354–368. doi:10.1080/07900627.2016.1238348s.

Kumar, P., Sharma, P.K., 2020. Soil salinity and food security in India. Front. Sustain. Food Syst. 4, 533781. doi:10.3389/fsufs.2020.533781.

Kumar, D.S., Tony, D.E., Kumar, A.P., Kumar, K.A., Rao, D.B.S., Nadendla, R., 2013. A review on: abelmoschus esculentus (OKRA). Int. Res. J. Pharm. App. Sci. 3 (4), 129–132.

Lee, C.M., Cable, M.L., Hook, S.J., Green, R.O., Ustin, S.L., Mandl, D.J., Middleton, E.M., 2015. An introduction to the NASA Hyperspectral InfraRed Imager (HyspIRI) mission and preparatory activities. Remote Sens. Environ. 167, 6–19. doi:10.1016/j.rse.2015.06.012.

Leibowitz, S.G., Wigington Jr, P.J., Schofield, K.A., Alexander, L.C., Vanderhoof, M.K., Golden, H.E, 2018. Connectivity of streams and wetlands to downstream waters: an integrated systems framework. J. Am. Water Resour. Assoc. 54 (2), 298–322. doi:10.1111/1752-1688.12631.

Li, P., 2016. Groundwater quality in Western China: challenges and paths forward for groundwater quality research in Western China. Expo Health 8 (3), 305–310. doi:10.1007/s12403-016-0210-1.

Little, K.E., Hayashi, M., Liang, S., 2016. Community-based groundwater monitoring network using a citizen-science approach. Ground Water 54 (3), 317–324. doi:10.1111/gwat.12336, May.

Manisalidis, I., Stavropoulou, E., Stavropoulos, A., Bezirtzoglou, E., 2020. Environmental and health impacts of air pollution: a review. Front. Public Health 8, 14. doi:10.3389/fpubh.2020.00014.

McCabe, M.F., Rodell, M., Alsdorf, D.E., Miralles, D.G., Uijlenhoet, R., Wagner, W., Lucieer, A., Houborg, R., Verhoest, N., Franz, T., Shi, J., Gao, H., Wood, E.F., 2017. The future of earth observation in hydrology. Hydrol. Earth Syst. Sci. Discuss. 21, 3879–3914.

Mell, P., Grance, T., 2011. The NIST Definition of Cloud Computing. National Institute of Standards and Technology Special Publication 53, 1–7.

Mogheir, Y., Singh, V., de Lima, J., 2006. Spatial assessment and redesign of a groundwater quality monitoring network using entropy theory, Gaza Strip. Hydrogeol. J. 14, 700–712. doi:10.1007/s10040-005-0464-3.

Neves, M.C., Nunes, L.M., Monteiro, JP., 2020. Evaluation of GRACE data for water resource management in Iberia: a case study of groundwater storage monitoring in the Algarve region. J. Hydrol. Reg. Stud. 32, 100734.

Ridler, M.-E., Madsen, H., Stisen, S., Bircher, S., Fensholt, R., 2014. Assimilation of SMOS-derived soil moisture in a fully integrated hydrological and soil-vegetation-atmosphere transfer model in Western Denmark. Water Resour. Res. 50, 8962–8981.

Rossetto, R., De Filippis, G., Borsi, I., Foglia, L., Cannata, M., Criollo, R., Vázquez-Suñé, E., 2018. Integrating free and open source tools and distributed modelling codes in GIS environment for data-based groundwater management. Environ. Modell. Softw. 107, 210–230. doi:10.1016/j.envsoft.2018.06.007.

Roy, S.S., Malik, A., Gulati, R., Obaidat, M.S., Krishna, P.V., 2017. A Deep Learning Based Artificial Neural Network Approach for Intrusion Detection, Communications in Computer and Information Science. Int. J. Appl. Math. Comput. doi:10.1007/978-981-10-4642-1_5.

Senay, G.B., Stefanie, B., Verdin, JP., 2012. Remote sensing of evapotranspiration for operational drought monitoring using principles of water and energy balance. In: Remote Sensing of Drought: Innovative Monitoring Approaches. Taylor & Francis, US, pp. 123–144.

Serrat-Capdevila, A., Valdes, J.B., Stakhiv, E.Z., 2014. Water Management Applications for Satellite Precipitation Products: Synthesis and Recommendations. J. Am. Water Resour. Assoc. (JAWRA) 50 (2), 509–525. doi:10.1111/jawr.12140.

Shah, T., Molden, D., Sakthivadivel, R., Seckler, D., 2000. The Global Groundwater Situation: Overview of Opportunities and Challenges. International Water Management Institute, Colombo.

Sharma, D.A., Rishi, M.S., Keesari, T., 2017. Evaluation of groundwater quality and suitability for irrigation and drinking purposes in southwest Punjab, India using hydrochemical approach. Appl. Water Sci. 7, 3137–3150. https://doi.org/10.1007/s13201-016-0456-6.

Sheffield, J., Wood, E.F., Pan, M., Beck, H., Coccia, G., Serrat-Capdevila, A., Verbist, K., 2018. Satellite remote sensing for water resources management: potential for supporting sustainable development in data-poor regions. Water Resour. Res. 54, 9724–9758. doi:10.1029/2017WR022437.

Shukla, S., Sheffield, J., Wood, E. F., Lettenmaier, D. P., 2013. On the sources of global land surface hydrologic predictability. Hydrol. Earth Syst. Sci. 17, 2781–2796. https://doi.org/10.5194/hess-17-2781-2013.

Thorslund, J., van Vliet, M.T.H., 2020. A global dataset of surface water and groundwater salinity measurements from 1980–2019. Sci. Data 7, 231. doi:10.1038/s41597-020-0562-z.

Torresen, J., 2018. A review of future and ethical perspectives of robotics and AI. Front. Robot. AI 4, 75. doi:10.3389/frobt.2017.00075.

Tuinhof, A., Dumars, C., Foster, S., Kemper, K., Garduño, H., Nanni, M., 2002. Groundwater Resource Management: An introduction to its Scope and Practice. Paris, France.

van Dijk, A.I.J.M., Renzullo, L.J., 2011. Water resource monitoring systems and the role of satellite observations. Hydrol. Earth Syst. Sci. 15, 39–55. doi:10.5194/hess-15-39-2011.

Wagner, K., Chessler, M., York, P., Raynor, J., 2009 Sep. Development and implementation of an evaluation strategy for measuring conservation outcomes. Zoo Biol. 28 (5), 473–487. doi:10.1002/zoo.20270. PMID: 19725124.

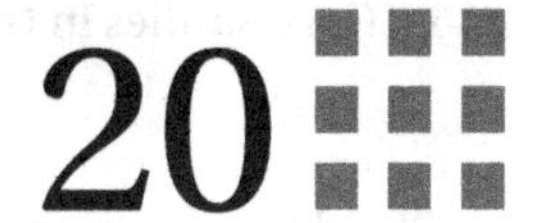

Seasonal fluctuation of groundwater table and its impact on rural livelihood: A village level study at coastal belt of Purba Medinipur District, India

Subrata Jana[a] and Sriparna Jana[b]

[a]DEPARTMENT OF GEOGRAPHY, BELDA COLLEGE, BELDA, PASCHIM MEDINIPUR, INDIA
[b]DEPARTMENT OF GEOGRAPHY, BAJKUL MILANI MAHAVIDYALAYA, BAJKUL, PURBA MEDINIPUR, INDIA

20.1 Introduction

In the 21st century, human beings are intensively depending on groundwater rather than for uses of surface water resources. Like the urban areas, the rural people also adapted themselves to exploit groundwater in every demand of water (Kumar et al., 2005; Shah, 2009). The installation of shallow and deep tube wells as per easy requirements for groundwater extraction is now a fashion that becomes intensified with the rural electrification (Srivastava et al., 2009; MacDonald et al., 2015). The supply of irrigation water, urban-industrial water, and water in the fisheries, in conjunction with bathing, washing, and drinking purposes, freshwater is being used from the groundwater sources (Meinzen-Dick and Appasamy, 2002). In connection with that the groundwater table is gradually depleted year after year. This depletion is intensifying due to the rural-urban infrastructural development by the concretization and minimizing the surface water percolation limit even up to the subsurface zone (Punjabi and Johnson, 2019). Moreover, the natural storage capacity of surface water is reducing due to the shrinking of ponds, tanks, canals, and rivers by human encroachment (Bassi et al., 2014; Jana, 2021a). This effect also emphasized over-dependence on the groundwater even in rural areas.

Case Studies in Geospatial Applications to Groundwater Resources. DOI: https://doi.org/10.1016/B978-0-323-99963-2.00005-5

The demands of freshwater are related to the rate of extraction of groundwater is gradually increasing from time to time. The natural fluctuation of the water table varies seasonally depending on the monsoonal rainfall and quantity of groundwater extraction throughout the year (Rangarajan and Athavale, 2000; Zencich et al., 2002). In general, most of the rural people used ponds, rivers, canal water for household activities. Therefore, some percentage of groundwater extraction has been less and that will create the same save of groundwater. But, in the recent decade, in rural areas, people are also using groundwater for all kinds of household activities and more and more construction of submersible pumps gradually extract more groundwater from the deeper parts of the water table (Dixit and Upadhya, 2005; Narain, 2014). So, the other tube wells in the surrounding areas of the submersible pumps are not surviving up to the end of the dry season and creating water scarcity conditions (Vij and Narain, 2016; Das et al., 2021). In the same way, recently, a huge amount of water scarcity has been happening all over the world. In the 21st century, throughout the global perspective, all people are not able to get the required quantities of water for their survival and also for sustainability (Gleick, 1996; Loucks, 2000; Rockstrom et al., 2009).

The people of coastal areas suffer most regarding the quality and quantity of drinking water (Sarkar et al., 2021; Jabed et al., 2020) as they do not have proper drinking water facilities. This miserable condition gradually undergoes to a panic situation during the summer months almost every year. During summer months the groundwater table falls far downward and most people are not able to get water, as the depth of the water table falls more than the depth of tube wells (Chinnasamy and Agoramoorthy, 2015; Mukherjee et al., 2018). So, people choose other options to get drinking water like constructing submersible pump and extracting groundwater from the far depths (Mondal, 2021).

Basically, during the dry seasons or summer months, the groundwater table falls far downward and thus blackish water or sediment mixed water is extracted from the tube wells (Prusty et al., 2018; Kumar et al., 2020). The situation of groundwater depletion is quite different in the coastal belt compared with the other areas. The increasing water demand and extraction are exaggerating groundwater depletion. The piezometric pressure of saline water is increasing and promoting saline water encroachment in the fresh groundwater table (Das and Mukherjee, 2019; Sarkar et al., 2021). The surface water is also contaminated with saline seawater during heavy storm surge (Rani et al., 2021; Sarkar et al., 2021). Therefore, people are compelled to entirely depend on the groundwater sources which emphasized the dramatic increase of groundwater vulnerability in the coastal belt.

In the present study area of Jahanabad village, Khejuri-I CD Block, Purba Medinipur coastal tract the dependency on groundwater water is ever-increasing in contrasting association with the increasing needs of socio-economic development and increasing coastal flooding vulnerability. In earlier, many studies have been done in the coastal areas of Purba Medinipur district regarding the groundwater uses and its potentiality (Sar et al., 2015; Das, 2017; Acharya et al., 2019; Chakraborty et al., 2020; Halder et al., 2021). However, those studies did not emphasize the groundwater fluctuation and its vulnerability in the Medinipur coastal tract. Therefore, the present study emphasizes the estimation of groundwater source-wise population pressure, water availability, demand, and scarcity along with the level of vulnerability.

20.2 Data base and methodology

20.2.1 Study area

The present study has been conducted over the Jahanabad village of Khejuri-I CD Block situated in the low-lying coastal belt of Purba Medinipur district, West Bengal. This village is situated about 7 km from the Rasulpur river in the south, 14 km from the Haldi river mouth and only 10 km from the Hugli estuary mouth. This area is located in the plain land of the Haldi-Rasulpur interfluves zone with an average elevation of 5.50 m. The latitudinal extension of this village is between 21°56′32.42″N to 21°58′03.08″N and longitudinal extension 87°53′30.97″E to 87°55′00.22″E (Fig. 20.1). During the middle Holocene period, this area was formed under Panskura formation with a combination of fine sand, silt, and clay (Geological Quadrangle Map, 73O). At the initial stage, this area was formed as Khejri Island likewise the other islands of Hijili at the head of the Bay of Bengal (O'Malley, 1911). Then, those isolated islands were merged as a single island and connected with the Midnapore littoral tract (O'Malley, 1911) with due effects from gradual sedimentation from Hugli, Haldi, and Rasulpur rivers. Very deep, poorly drained, fine cracking soils, occurring on nearly level to very gently sloping coastal plain with the clayey surface, moderate flooding, and moderate salinity (moderate extent) associated with deep, poorly drained, fine soils are found within this study area (National Bureau of Soil Survey and Land Use planning, https://www.nbsslup.in/). From the initiation of civilization, the agrarian society made land reclamation under different revenue-earning agents of Britishers. Since then, frequent natural calamities like floods, cyclones, and storm surges strikes frequently and smashed this area till now. It is observed that almost every year about 5–6 cyclone strikes in the Midnapore littoral tract (Singh et al., 2001). This area is dominated by tropical monsoon climatic regions and about 75% of the rainfall receives from June to September. Being located in the coastal zone and nearer to the head of the Bay of Bengal, some amount of rainfall receives in October and November due to cyclonic disturbance. The annual average rainfall is near about 1629 mm. During the months of winter (November to January), the minimum temperature variations in between 4°C and 12°C and maximum temperature in summer months (April to June) between 35°C and 38°C (IMD, 2008).

20.2.2 Data base and data processing

20.2.2.1 Maps and satellite data processing

In this study, the cadastral maps of the Jahanabad village (Sheet No. 2 and 3) were collected from the Khejuri-I CD block annual report of 2011. The map sheets were merged and rectified based on the universal transverse mercator (UTM) projection after 45 N zone and world geodetic survey 1984 (WGS84) datum. The vector layer of the village boundary was extracted from the rectified map. The Landsat 8 (OLI sensor) satellite image of 03.12.2016 was collected from the United States Geological Survey (USGS) and used for land use and land cover (LULC) classification. The radiometric correction technique was applied by converting the Digital Numbers (DNs) into radiance values based on the method proposed by Mishra et al. (2014). The image was re-projected in the UTM projection after 45 N zone and WGS84 datum and resampled with

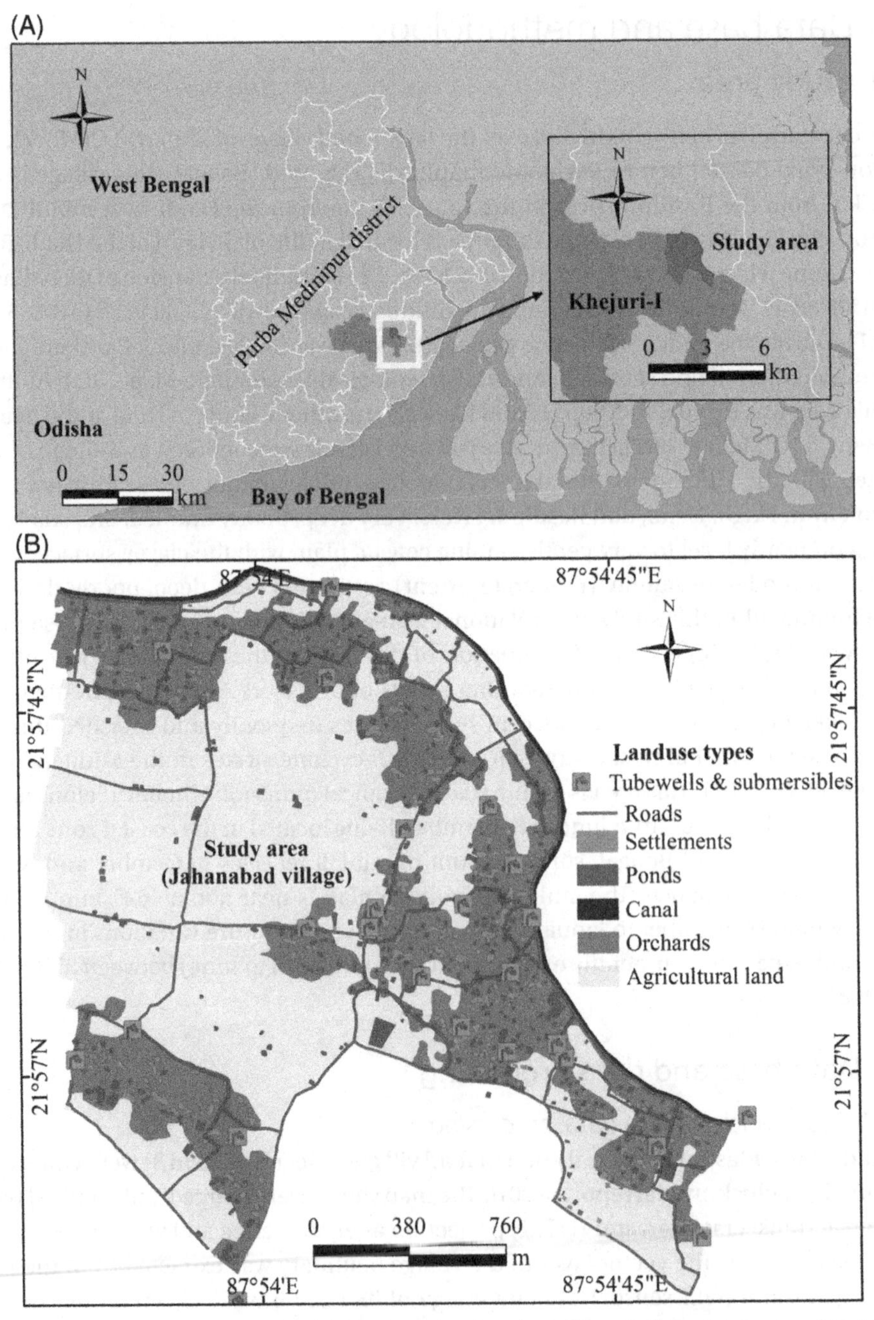

FIGURE 20.1 The study area indicating the (A) location of Janhanabad village in the coastal areas of Purba Medinipur district, West Bengal and (B) the existing land use types in the Jahanabad village.

<0.5 pixel accuracy of root mean square error (RMSE). The processed image was clipped as per the vector layer of the study area. The supervised classification technique and support vector machine (SVM) algorithm were applied over the clipped image for the LULC classification of the study area (Bouaziz et al., 2017). For the LULC classification, 10-point specific land use features of each five classes were demarcated based on the field observation. Also, the classified image was validated with ground truth verification and considered after >90% accuracy based on the Kappa coefficient. The village roads were extracted from the Google Earth image, while, the location of drinking water sources (tube wells and submersible pump) and location of individual households were acquired using the global positioning system (GPS) survey which was imported into the geographic information system (GIS) platform using ArcGIS 10.4 software for the detailed land use map preparation (Fig. 20.1B).

20.2.2.2 Primary survey and water demand estimation

The information of population and dwelling units were collected from the questionnaire survey in every family. The location-specific seasonal water level fluctuation data were collected from the prior knowledge of local people and tube well constructing persons, which have been verified with the available data accessed from the local panchayat office. Based on the surveyed outcome, the water demand was estimated by the following equation (Eq. 20.1).

$$N = (D + C + O) \tag{20.1}$$

$D = (F_n \times 4\ lit.)$ and $C = (F_n \times 3\ lit.)$.

where N is the daily need for water (lit.), D represent water used for drinking purpose, C stands for cooking purpose; O represents other purposes, and F_n is the number of family members. The water volume of 4 lit. and 3 lit. is selected as standard required water per person respectively for drinking and cooking purposes.

20.2.2.3 Delineation of sub-zones

Within the study area total 28 numbers of drinking water sources are available, that is, 19 tube wells (T-1 to T-19) and 9 submersible pumps (S-1 to S-9). For the detailed study, the entire study area was divided into a similar number of sub-zones according to the seasonally functioning number of water sources and those zones having households or populations. Other areas that have not any population or those people are living in the remote areas from the source of drinking water or collecting drinking water from the other village are not considered under these sub-zones which area mentioned as not applicable (NA) in the figures. Depending on this situation, 22 sub-zones have been delineated (T to T-19 and S-1 to S-3) in case of the rainy season, whereas, 19 sub-zones have been delineated (T-1 to T-9 and Sub-1 to Sub-10) for the summer season.

20.2.2.4 Estimation of aggregate distance index (ADI)

The ADI has been estimated for each of the sub-zones corresponding to the position of tube wells and submersible pumps. The aggregate travel distance of each household to access drinking water from the nearby water source is measured from the GPS-based collected locations of households and water sources. In this case of the study, the ADI of each of the water sources has been estimated following the existing formula used for the nearest neighbour index (NNI). The number of households under each of the functioning water sources has been varied seasonally. Therefore, the NNI-based ADI of each water source has been estimated for rainy and summer seasons following Eq. (20.2).

$$ADI = \frac{\bar{D}_O}{\bar{D}_E} \tag{20.2}$$

where $\bar{D}_O$ is the observed mean distance between each household and its nearest water source from where people collect their required drinking water, and $\bar{D}_E$ is the expected mean distance of the households to the respective water source. The $\bar{D}_O$ and $\bar{D}_E$ are calculated followed by Eqs. 20.3 and 20.4, respectively.

$$\bar{D}_O = \frac{\sum_{i=1}^{n} d_i}{n} \tag{20.3}$$

$$\bar{D}_E = \frac{0.5}{\sqrt{n/A}} \tag{20.4}$$

where d_i is the distance between household i and its nearer water source, n corresponds to the total number of households under respective water sources, and A is the area under corresponding water sources.

The estimated ADI values have been assigned against each of the respective water sources of both seasons and the raster layer has been prepared by inverse distance weighting (IDW) interpolation method in ArcGIS 10.4 software.

20.2.2.5 Vulnerability analysis

The seasonal vulnerability of accessing groundwater has been assessed after considering the 9 risk variables, that is, household distribution (x_1), household density (x_2), population distribution (x_3), population density (x_4), ADI (x_5), depth of water table (x_6), water demand (x_7), water availability (x_8), and water deficit (x_9). As mentioned earlier, all the data was collected from the household and other questionnaire surveys of each family. The collected data have been imported into the GIS platform against point features at each of the sub-zones. Subsequently, the IDW interpolation method (with 0.5 m cell size) has been adopted to extract the spatial distribution of different aspects. All the maps have been prepared considering the five equal classes. The zone-wise actual status of the 9 different risk variables has been assigned after applying the zonal statistics in the GIS platform. Based on these zone-wise values of each risk variable, the five categorical risk-rating values have been assigned as very low (1), low (2), moderate (3), high (4), and very high (5). Among these 9 risk variables, only for the water availability the higher values having lower vulnerability and lower values having higher vulnerability. Whereas, in the case of the other 8 risk variables the reverse situation has been observed. Finally, the sub-zone wise vulnerability index (VI) has been estimated based on the product of square root of risk variables ($x_1 - x_9$)divided by the total number of variables (9) as per Eq. (20.5), which have been further calculated as per Eq. (20.6) to get the standardized VI (VIs) (Jana, 2020, 2021b). The VIs have been

Table 20.1 Quantitative aspect of different land uses within the study area.

Land use types	Number	Area (km^2)	Length (km)
Settlement patches	321	0.07	
Canal			3.96
Pond	460	0.11	
Orchards		1.53	
Agricultural land		2.39	
Road	23		23.45
Tube wells	19		
Submersible pumps	4		

assigned in the zone-wise corresponding location of tube wells and submersible pumps, and the IDW interpolation method (with 0.5 m cell size) has been adopted to prepare the raster layer of vulnerability level. Finally, the five equal classes of vulnerability maps of both seasons have been prepared based on the respective raster layers.

$$VI = \sqrt{\frac{x_1 + x_2 + ...x_9}{9}} \tag{20.5}$$

$$VIs = \frac{x - x_{min}}{x_{max}} \tag{20.6}$$

20.3 Results and discussion

20.3.1 Land use types

In the entire study area (3.99 km^2) six types of land uses have been observed in association with the location of tube wells and submersible pumps (Fig. 20.1B). A detailed account of different land uses has been mentioned in Table 20.1. Within this area, there have 321 settlement patches and 460 ponds covering an area of 0.07 km^2 and 0.11 km^2 respectively (Table 20.1). Total 23.45 km of major interconnecting village roads is playing a key role in the connectivity. On the northern side of the village, a 3.96 km long canal exists which is playing an important role in irrigational water supply and rainwater drainage system. The irrigational water is also supplied from the existing ponds during the dry summer. The settlements are sparsely distributed in the northern, southern, and eastern sides of the village, which have been surrounded by orchards covers (1.53 km^2). The rest of the areas (2.39 km^2) are utilized for agricultural purposes. Total 28 water sources (19 tube wells and 9 submersible pumps) are publicly or privately supplying the required drinking water and irrigational water on demand.

20.3.2 Demographic structures

20.3.2.1 Household distribution

The seasonal diversities in household distribution have been observed as per the diversities in the number of functioning water sources and regarding assortment in sub-zones. Basically,

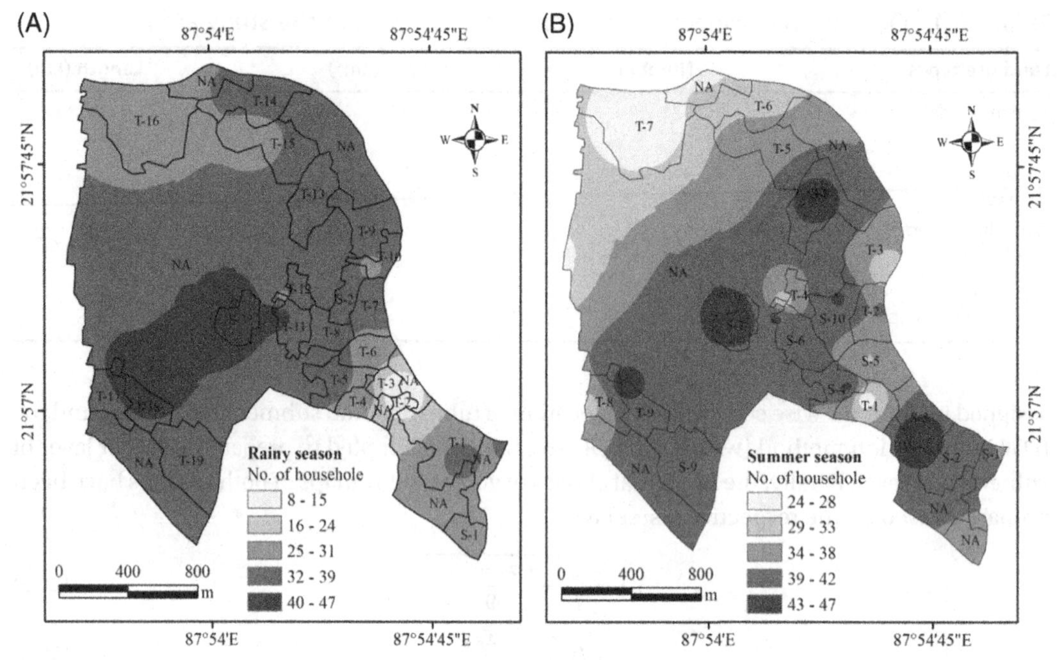

FIGURE 20.2 Categorical diversities in household distribution of the (A) rainy season and (B) summer season.

people are collecting daily required water from the nearby water sources. As per the seasonal status of functioning and malfunctioning of the water sources, the number of households and allied number of the population have been varied in different zones. In the rainy season, the number of household distribution has been varied from 8 to 47 which are increases from 24 to 47 in the summer season (Fig. 20.2). In summer, the number of functioning tube wells reduce due to the increasing depth of accessible groundwater water levels. Therefore, the number of sub-zones is also minimized in the summer season in comparison to the rainy season, which has a direct influence on the result of the spatial distributional pattern of household number. In the case of the rainy season (Fig. 20.2A), the household distribution shows very low (8–15) in T-1 and T-2 zones due to the plenty of available functioning water sources. In such a way, the very high (40–47) household distribution is observed in the middle portion of the area. However, most of the area is resulting in high (32–39) household distribution. Similarly, in the summer season (Fig. 20.2B), very low (24–28) household distribution has been found in T-1, T-7 and in some parts of T-3, T-4, and T-8 zones. The very high (43–47) household distribution has been recorded in S-3, S-7, S-8 and in parts of T-9 zones due to the reduction of the number of functioning water sources. The eastern and middle portion of the village has resulted in high (39–42) household distribution.

20.3.2.2 Household density

The zone-wise household density is resulted according to the size of each sub-zones. Therefore, the spatial distributional pattern of household density is quite different in comparison to the number of household distributional patterns for both seasons. The household density varies

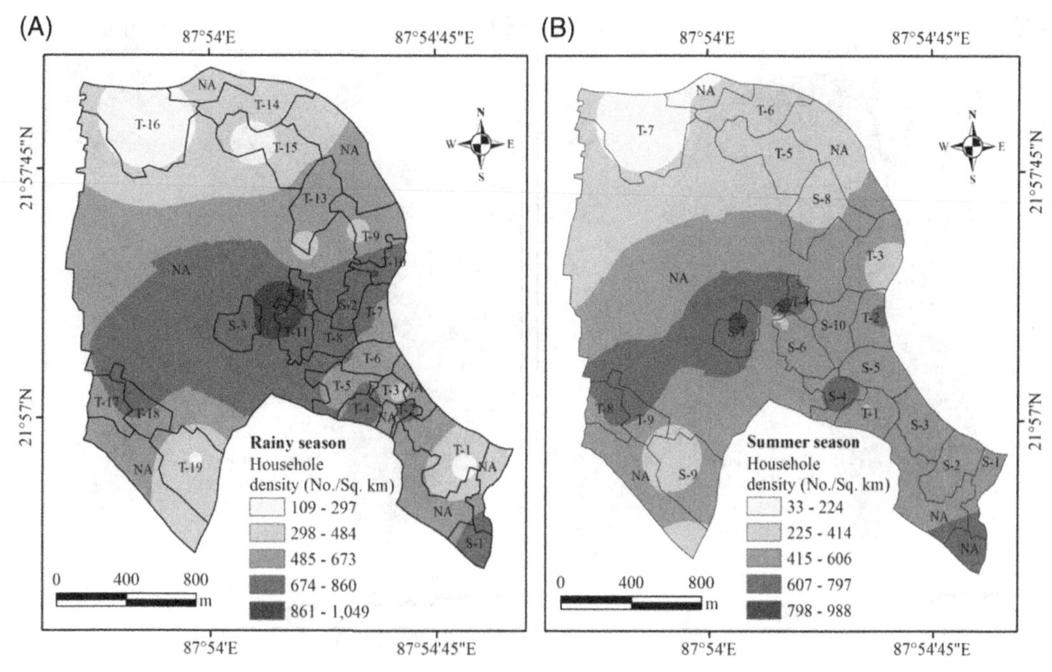

FIGURE 20.3 Seasonal variations of household density in (A) rainy season and (B) summer season.

from 109 to 1049 household/km^2 and 33 to 988 household/km^2, respectively in the rainy season and summer season (Fig. 20.3). In the rainy season, the very low (109–297 household/km^2) and low (298–484 household/km^2) densities are observed in the areas of T-1, 14, 15, 16, and 19 zones (Fig. 20.3A), which is comprised to nearly similar areas (zones of T-5, 6, 7, 8, and 9) of the summer season (Fig. 20.3B). The aerial coverage of higher (674–1049 household/km^2) household density has resulted in a higher percentage of total area (zones of T-2, 4, 8, 10, 11, 12, 17, 18, and S-2, 3) in the rainy season in comparison with the areas of the summer season (T-4, 8, 9 and S-4, 7). The moderate density (41–606 household/km^2) is observed in most of the areas in the summer season (Fig. 20.3B).

20.3.2.3 Population distribution

Likewise, the household distribution, the population distribution is showing a similar type of result as the number of population is almost alike to the number of the household of each sub-zone. The result of population distribution is also corresponding to the size of the sub-zones. The population distribution varies from 55 to 232 and 127 to 236 for the rainy and summer season, respectively (Fig. 20.4). In the rainy season, the lower population distribution (55–125 population) is observed in most of the areas of T-2 and T-3 zones due to the highest number of available functioning water sources (Fig. 20.4A). Whereas, the higher population distribution is observed in T-5, 9, 11, 12, 13, 14, 17, 18, 19, and S-3 zones as these zones have higher relative water dependency in respect to the number of functioning water sources (Fig. 20.4A). Most of the

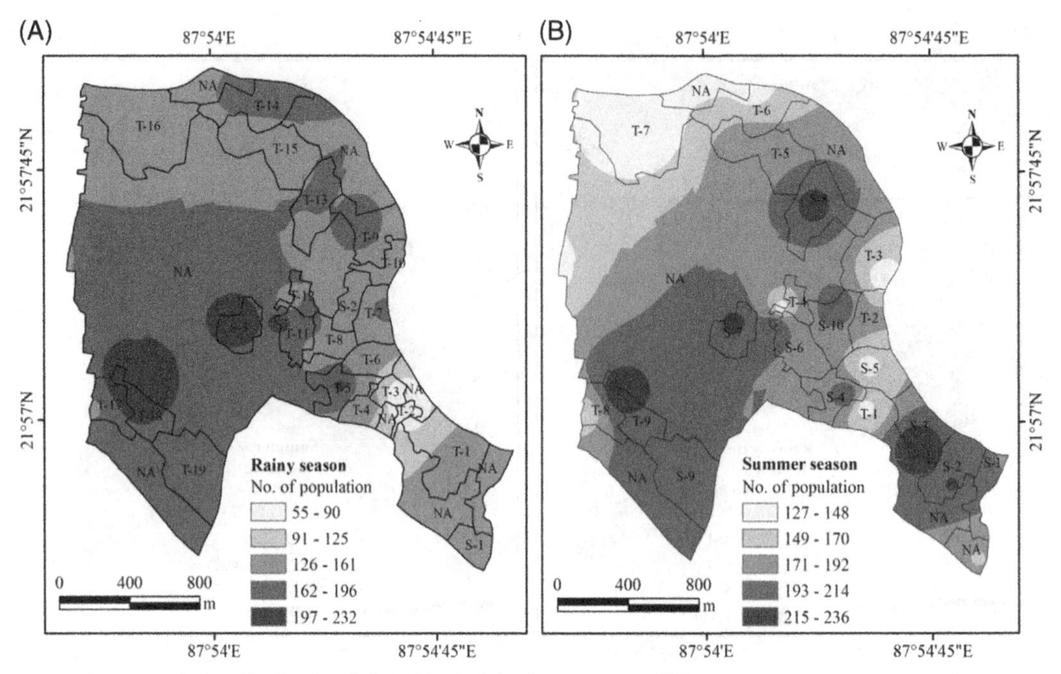

FIGURE 20.4 Population distributional diversities in (A) rainy season and (B) summer season.

zones (T-1, 4, 6, 8, 10, 15, 16, and S-1, 2) are resulted with moderate population distribution (126–161 population) for rainy season (Fig. 20.4A), while, T-2, 4, 5 and S-4, 6 zones having moderate population distribution (171–192 population) in summer season (Fig. 20.4B). In the summer season, the lower (127–170 population) population distribution is found in T-1, 3, 4, 6, 7, and S-5 zones, and higher (193–236 population) population distribution is observed in T-9 and S-1, 2, 3, 7, 8, 9 zones (Fig. 20.4B).

20.3.2.4 Population density

The resultant population density is also corresponding to the product of population number and size of respective sub-zones. The seasonal variation of population density is varied from 577 to 4805/km^2 and 168 to 4429/km^2 respectively for the rainy and summer season (Fig. 20.5). In the rainy season, the lower population density (577–2268/km^2) is observed in T-1, 14, 15, 16, 19 zones (Fig. 20.5A), which is almost similar in the T-5, 6, 7, 8 zones in the northern side of the village for the summer season (Fig. 20.5B). The higher density (3114–4805/km^2) has resulted in T-2, 4, 5, 8, 10, 11, 12, 17, 18 and S-1, 2, 3 zones of the rainy season, while, T-4, 8, 9 and S-1, 2, 3, 4, 7 zones have a higher density (2725–4429/km^2) in the summer season. Most of the zones (T-1, 2, 3 and S-5, 6, 8, 9, 10) have remained under moderate population density (1873–2724/km^2) in the summer season, whereas, T-7, 9 and 13 zones exist under moderate population density (2269–3113/km^2) in the rainy season.

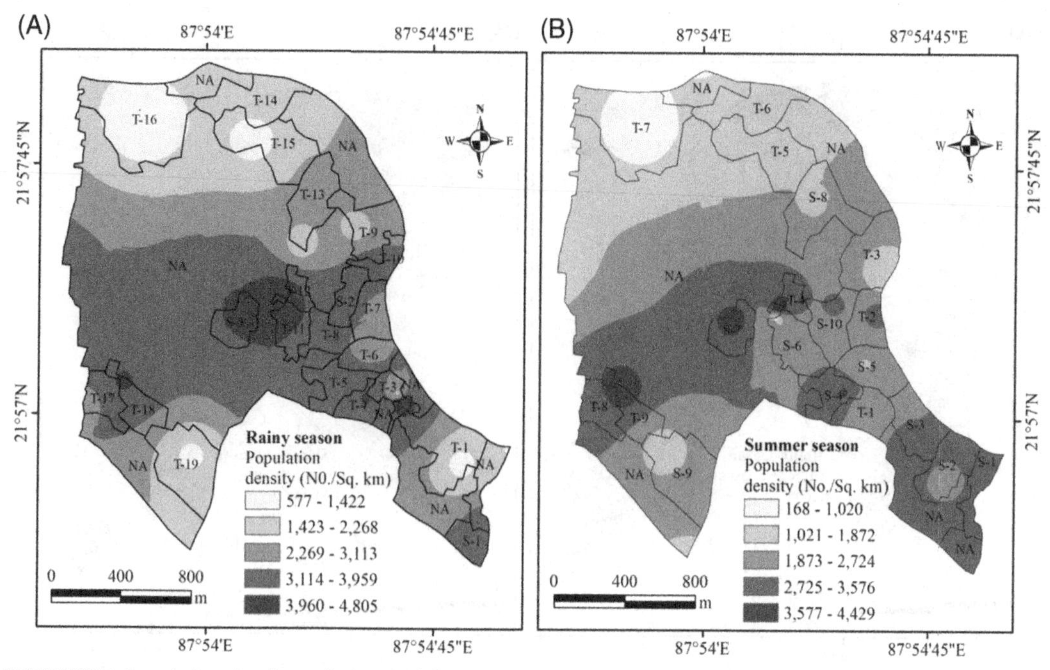

FIGURE 20.5 Population density variations in (A) rainy season and (B) summer season.

20.3.2.5 Aggregate distance travel to get drinking water

This village has plenty of water sources. Therefore, the people do not have to go much distance to collect their daily required drinking water. Still, the ADI has been estimated to understand the seasonal diversity of the travel distances. The ADI varies from 0.77 to 11.59 and 0.68 to 10.33 for rainy and summer seasons, respectively (Fig. 20.6). In the rainy season, the very low (0.77–2.93) ADI has been observed in T-2, 3, 5, 6, 11 and S-3 zones where the location of water sources is closely distributed; very high (9.43–11.59) ADI has been observed in S-1 zone; and most of the areas resulted with moderate (5.10–7.25) ADI (Fig. 20.6A). Whereas, in the summer season, very low (0.68–2.61) ADI has been observed in the zones of T-4 and S-5, 6, 7, 8; high (6.47–8.39) ADI in T-6, 7 and S-1, 9 where the water source points are sparsely distributed; and moderate (4.55–6.46) ADI has resulted in T-1, 2, 3, 5, 8, 9 and S-2, 3, 4, 10 (Fig. 20.6A).

20.3.3 Seasonal groundwater status

20.3.3.1 Depth of water level

The groundwater level is the adjusted status of the rate of groundwater extraction and the rate of water recharge into the groundwater table. The fluctuating status of the groundwater table is clearly indicating a miserable condition within the study area. In the rainy season, the depth of groundwater varies from 143.42 mbgl (below ground level) to 167.00 mbgl, which is reasonably indicating a better situation in the summer season (171–185 mbgl) (Fig. 20.7). About 4 m of water level depletion has resulted between the highest depth (167 mbgl) in the rainy season and the

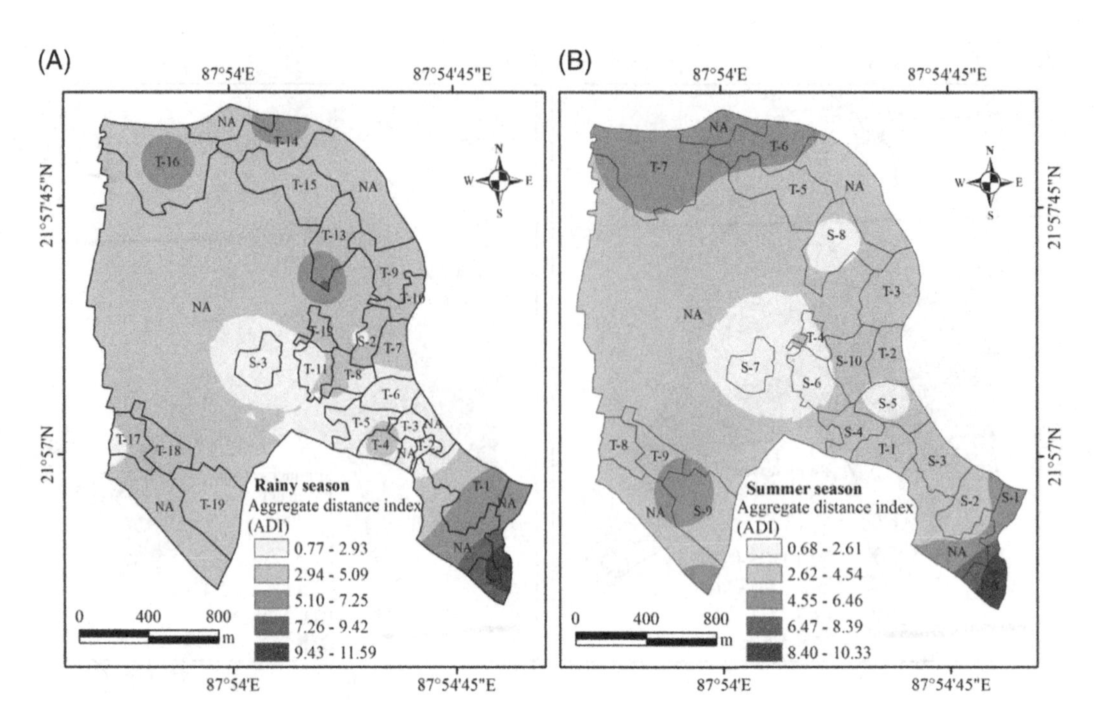

FIGURE 20.6 Categorical diversities in pattern of ADI in (A) rainy season and (B) summer season.

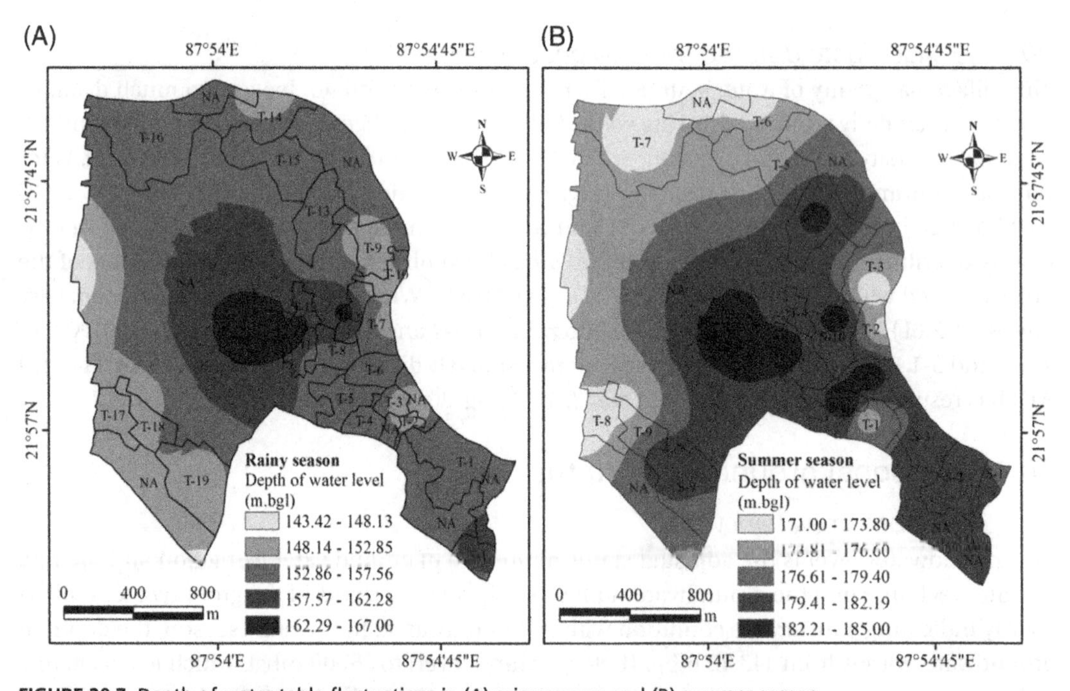

FIGURE 20.7 Depth of water table fluctuations in (A) rainy season and (B) summer season.

lowest depth (171 mbgl) in the summer season. The mean depth of groundwater level has resulted with 155.65 mbgl and 178.23 mbgl for the rainy season and summer seasons respectively, which indicates about 22 m of seasonal groundwater depletion within the study area. This seasonal groundwater depletion status in this village is very high in respect to the average depletion (15.7 m) in the Purba Medinipur district (Chakraborty et al., 2020). The areal coverage of the depth of water level is increased about twice in the summer season in response to the rainy season. The positions of tube wells have resulted in a relatively minimum depth of water level than the positions of submersible pumps for both seasons. The corresponding result shows that the highest depth of water level, as well as the highest level of depletion is associated with the positions of submersible pumps.

Such type of groundwater depletion is associated with the transformation of economic activities and associated changes in the lifestyle of the people. The traditional paddy cultivation is now exhausted or transformed into high yielding varieties which require much more water. In earlier, the traditional rice paddy cultivation was only done during the rainy season depending on seasonal rainfall and meagerly required irrigational water can be supplied from the nearby ponds or water bodies. Recently, the existing ponds and water bodies become converted into the larger earning fisheries sector which cannot able to supply irrigational water to the agricultural fields, whereas, it extracted required groundwater for its own sustainability. Moreover, the single crop agricultural fields become transformed into two or multi-cropped agricultural fields which also extracts more and more groundwater for irrigational water supply. In these circumstances, the depth of available groundwater tables is increasing year by year. The existing tube wells become malfunctioning and rural people are facing challenges to access the required drinking water. Moreover, most of the people are now modernized themselves and they do not want to use pond's water even for bathing purposes. Therefore, the rural livelihoods are completely dependents on groundwater, which increases the rate of groundwater extraction and allied depletion of water level. Moreover, the existing tube wells were not able to supply water during the summer season and many submersible pumps have been installed to overcome the problem of drinking water supply. However, the rich families and some businesspeople personally install submersible pump for their own purpose and to supply water into the fisheries and agricultural fields. Hence, the rate of groundwater depletion is to be increased in near the future which will be a more threatening condition with due effects from saline water encroachment into the freshwater aquifer (Sarkar et al., 2021).

20.3.3.2 Water demand

Water sources wise water demand of respective seasons is virtually identical to the number of population under each sub-zone. The water demand is estimated based on the basic daily needs of water for drinking, cooking, and other purposes like washing and bathing. Moreover, the rate of groundwater depletion is accelerated due to the over-extraction of groundwater for irrigation and fisheries purposes. Therefore, the freshwater demand is ever-escalating day by day as a result of the enhanced socioeconomic status of the people. The water demand under each of the water sources varies from 291 lit. to 1130 lit. and 569 lit. to 1278 lit., respectively for rainy and summer seasons (Fig. 20.8). The results of water demand for the rainy season shows that the higher level of

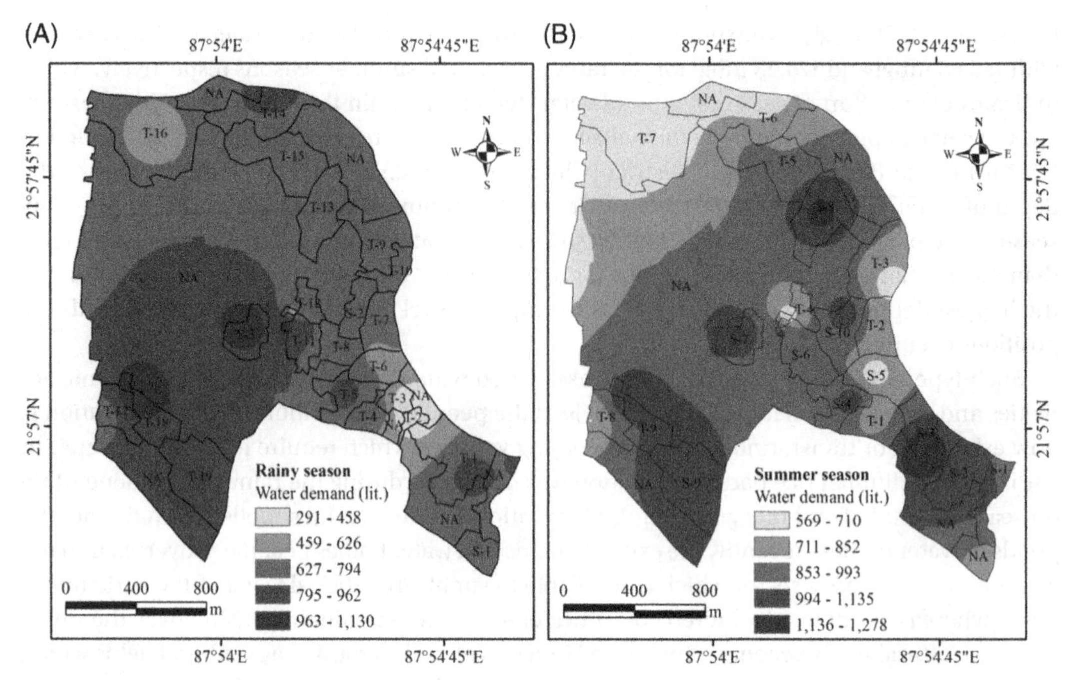

FIGURE 20.8 Changing spatial pattern of water demand in (A) rainy season, and (B) summer season.

water demand (795–1130 lit.) is observed mainly in the southern part of the village, whereas, the lower level of water demand (291–626 lit.) is observed in a narrow stretch of the eastern part of the study area where the number of water sources is maximum (Fig. 20.8A). The majority portion of the study area resulted in a moderate level of water demand in the rainy season. In the case of the summer season, a higher level of water demand (994–1278 lit.) is observed mainly in the zones where submersible pumps are the main water sources as most of the tube wells remained malfunctioning (Fig. 20.8B). The lower level of water demand (569–852 lit.) is resulted in the northern and eastern sides of the study area, whereas, the middle portion of the area is evidenced by a moderate level of water demand (853–993 lit.) in the summer season (Fig. 20.8B).

20.3.3.3 Water availability

The available volume of underground freshwater is significantly reducing time by time as a product of the balance between recharge of surface water into the groundwater table and extraction of groundwater. The human behavioral transformations, socioeconomic needs, and mental set-up against groundwater resources are the key factor for the huge depletion of groundwater table and concerning the availability of fresh groundwater. The sub-surface water table is seasonally fluctuated depending on the seasonal recharge and discharge of water. However, the groundwater table is always prone to be depleted as recharging capacity is very negligible with respect to the water extraction rate. Likewise, the water demand (Fig. 20.8), the resulted water availability (Fig. 20.9) is also indicating a nearly similar categorical pattern of spatial distributional as it is

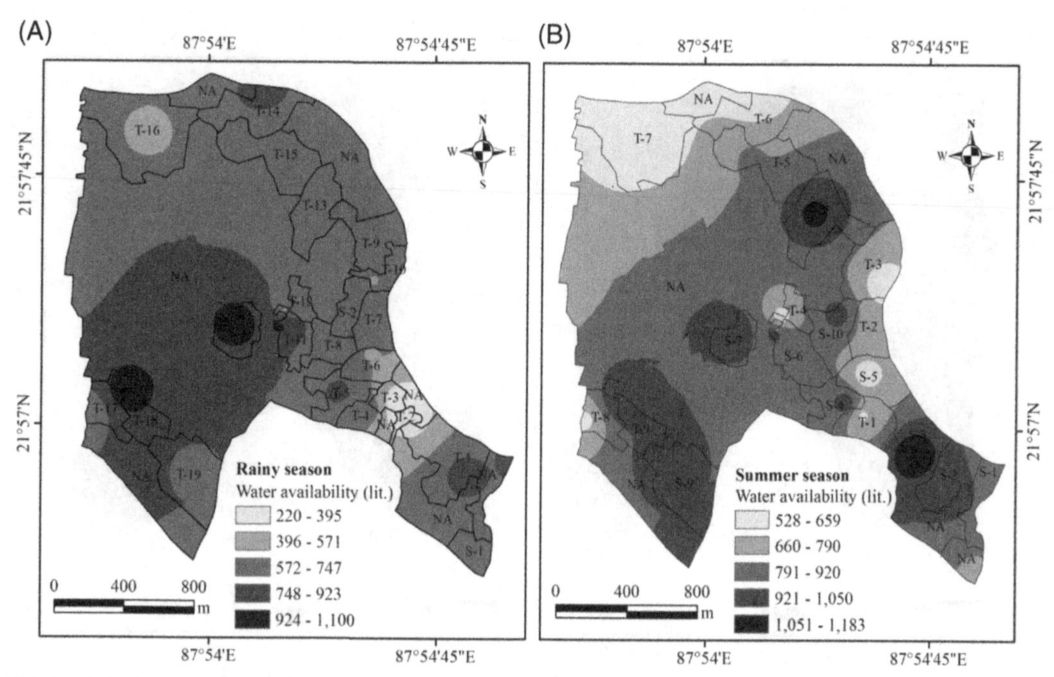

FIGURE 20.9 Seasonal status of water availability in (A) rainy season, and (B) summer season.

alike to the number of population under respective sub-zones. However, the range of the resulted water availability (Fig. 20.9) is different from the result of water demand (Fig. 20.8). The resulted water availability varies from 220 lit. to 1100 lit. and 528 lit. to 1183 lit. for rainy and summer seasons, respectively (Fig. 20.9). The higher level of water availability is observed nearer to the positions of maximum depth of water sources for both seasons. Hence, people are compelled to install new submersible pumps to a comparatively more depth than the depth of surrounding existing water sources.

20.3.3.4 Water deficit

The status of the water deficit is the product of the balance between water demand and water availability. The resulted water deficit is varied from 19 to 88 lit. and 26 to 115 lit., respectively for the rainy and summer seasons (Fig. 20.10). In the rainy season, the higher level of water deficit (61–88 lit.) has observed in T-3, 4, 5, 9, 10, 13, 17, 18 and 19 zones, whereas, the lower level of water deficit observed in T-6, 12, 16 and S-1, 2, 3 zones (Fig. 20.10A). The rest of the zones have resulted in a moderate level of water deficit (47–60 lit.) in the rainy season (Fig. 20.10A). Similarly, in the summer season (Fig. 20.10B), the higher level of water deficit (80–115 lit.) is observed in T-1 and S-1, 2, 3, 4, 8, 9, 10 zones, whereas, the zone T-7 and some parts of T-4, 8, 9 and S-5 zones have a lower level of water deficit (26–61 lit.). The moderate level of water deficit (62–79 lit.) is observed in T-2, 3, 5, 6 and some parts of T-8, 9 and S-5, 6, 10 zones in the summer season (Fig. 20.10B). From the result, it is clear that this village is already facing an ample water deficit despite lots of

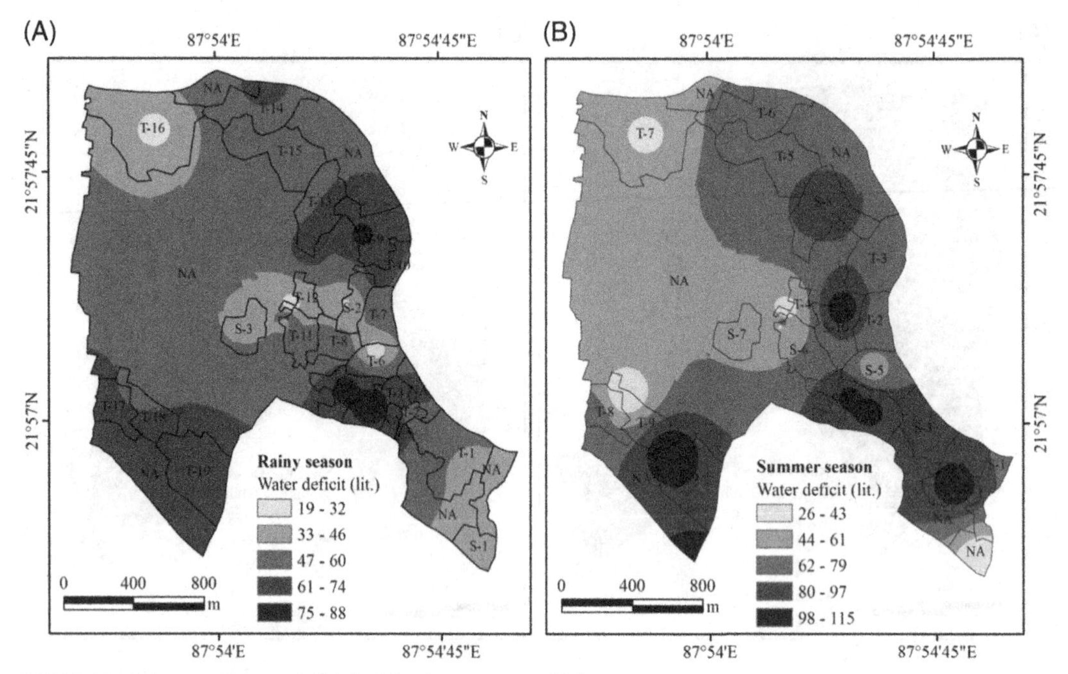

FIGURE 20.10 Status of water deficit in (A) rainy season and (B) summer season.

submersible pumps and tube wells. Very recently submersible pumps are installed even up to 250 m depth to get the fresh groundwater and also to minimize the water deficit. Moreover, such greater depth for getting fresh drinking water is increasing in a competitive way in the village and its surrounding areas which also enhances the vulnerability for getting the freshwater in future.

20.3.4 Groundwater vulnerability

The nine risk variables and their categorical result-based risk-rating scores have been assigned for the rainy season (Table 20.2) and summer season (Table 20.3). Moreover, the zone-wise actual vulnerability nature has been assigned by computing the mean score of each zone of every risk variable for both seasons (Tables 20.4 and 20.5). The risk-rating scores (among 1–5) have been assigned depending on the individual score of vulnerability nature (Tables 20.4 and 20.5) coupled with respective categorical risk-rating scores of Table 20.2 and Table 20.3 respectively for rainy and summer seasons. The VIs based vulnerability analysis indicates the five categorical vulnerability levels as very low, low, moderate, high, and very high for both seasons (Fig. 20.11).

The categorical vulnerability classes of the rainy season show that among the 22 sub-zones the six sub-zones (T-4, 8, 11, 12, 18 and S-1) resulted in very high (0.64–0.89) level of vulnerability and seven sub-zones (T-5, 7, 10, 13, 17 and S-2, 3) remained under high (0.53 - 0.63) level of vulnerability (Fig. 20.11A and Table 20.4). While, six sub-zones (T-1, 3, 6, 14, 15, 19) are resulted with low (0.25–0.39) level of vulnerability, whereas, two sub-zones (T-2, 16) remained under very low (0–0.24) level of vulnerability and only T-9 zone exists under moderate (0.40–0.52) class

Table 20.2 Risk variables and assigned risk-rating for vulnerability estimation of rainy season.

Risk variables	Very low (1)	Low (2)	Moderate (3)	High (4)	Very high (5)
Household distribution (No.)	≤15	16–24	25–31	32–39	>39
Household density (No./km^2)	≤297	298–484	485–673	674–860	>860
Population distribution (No.)	≤90	91–125	126–161	162–196	>196
Population density (No./km^2)	≤1422	1423–2268	2269–3113	3114–3959	>3959
Depth of water level (mbgl)	≤148.13	148.14–152.85	152.86–157.56	157.57–162.28	>162.28
Water demand (Lit.)	≤458	459–626	627–794	795–962	>962
Water availability (Lit.)	>923	748–923	572–747	396–571	≤395
Water deficit (Lit.)	≤32	33–46	47–60	61–74	>75
ADI	≤2.93	2.94–5.09	5.10–7.25	7.26–9.42	>9.42

Table 20.3 Risk variables and assigned risk-rating for vulnerability estimation of summer season.

Risk variables	Very low (1)	Low (2)	Moderate (3)	High (4)	Very high (5)
Household distribution (No.)	≤28	29–33	34–38	39–42	>42
Household density (No./km^2)	≤224	225–414	415–606	607–797	>797
Population distribution (No.)	≤148	149–170	171–192	193–214	>214
Population density (No./km^2)	≤1020	1021–1872	1873–2724	2725–3576	>3576
Depth of water level (mbgl)	≤173.80	173.81–176.60	176.61–179.40	179.41–182.19	>182.19
Water demand (Lit.)	≤710	711–852	853–993	994–1135	>1135
Water availability (Lit.)	>1050	921–1050	791–920	660–790	≤659
Water deficit (Lit.)	≤43	44–61	62–79	80–97	>97
ADI	≤2.61	2.62–4.54	4.55–6.46	6.47–8.39	>8.39

of vulnerability level (Fig. 20.11A and Table 20.4). Only the northern and eastern parts of the village remained under a lower level of vulnerability (Fig. 20.11A). In the case of the summer season, out of 19 zones, five sub-zones (S-1, 2, 3, 4, 9) remained under very high (0.73–0.99) level of vulnerability, and two sub-zones (S-7, 10) resulted in high (0.54–0.72) level of vulnerability (Fig. 20.11B and Table 20.5). A higher (0–0.36) level of vulnerability is observed in the extreme eastern and southern sides of the study area (Fig. 20.11B). Seven sub-zones (T-3, 4, 5, 6, 7, 8 and S-5) are resulted with very low (0–0.21), three sub-zones (T-1, 2 and S-6) remained under low (0.22–0.36) level of vulnerability, and two sub-zones (T-9 and S-8) remained under moderate (0.37–0.53) level of vulnerability for summer season (Fig. 20.11B and Table 20.5). Therefore, about 60% and 37% area of the entire sub-zones indicates a very high and high level of vulnerability, respectively in the rainy and summer seasons.

The nine risk variables based risk-rating scores are varied within 1-3 for the T-16 sub-zone of the rainy season (Table 20.4) and T-5, 6 and 7 of the summer season (Table 20.5) which indicates the very low level of vulnerability in the corresponding zones (Figs. 20.11A and 20.11B). In the case of the rainy season, the very high risk-rating score (5)

Table 20.4 Nature of vulnerability, assigned risk-rating and vulnerability index in different zones of rainy season.

Sub-zones	Risk variables	Household distribution (No.)	Household density (No./km^2)	Population distribution (No.)	Population density (No./km^2)	Depth of water level (mbgl)	Water demand (Lit.)	Water availability (Lit.)	Water deficit (Lit.)	ADI	VI	VIs	Vulnerability level
T-1	VN	27	442	130	2309	154.68	684	634	50	4.72	31	0.30	Low
	RR	3	2	3	3	3	3	3	3	2			
T-2	VN	13	702	74	4075	152.84	367	308	59	1.99	8	0.00	Very low
	RR	1	4	1	5	2	1	5	3	1			
T-3	VN	17	548	92	3145	152.65	467	397	69	1.96	29	0.28	Low
	RR	2	3	2	4	2	2	5	4	2			
T-4	VN	26	696	138	3711	154.19	695	617	78	3.06	66	0.76	Very high
	RR	3	4	3	4	3	3	3	5	2			
T-5	VN	33	663	168	3400	155.11	770	699	71	2.81	48	0.53	High
	RR	4	3	4	4	3	3	3	4	1			
T-6	VN	30	646	137	3047	154.34	628	582	47	2.76	27	0.25	Low
	RR	3	3	3	3	3	3	3	3	1			
T-7	VN	34	689	151	3095	153.81	685	636	49	3.19	51	0.56	High
	RR	4	4	3	3	3	3	3	3	2			
T-8	VN	34	740	154	3369	156.42	704	655	50	2.97	59	0.67	Very high
	RR	4	4	3	4	3	3	3	3	2			
T-9	VN	36	583	159	2614	152.45	729	662	67	3.80	42	0.44	Moderate
	RR	4	3	3	3	2	3	3	4	2			
T-10	VN	32	758	145	3384	152.56	672	611	62	3.96	55	0.62	High
	RR	4	4	3	4	2	3	3	4	2			
T-11	VN	37	838	177	3960	161.49	824	775	49	2.86	58	0.66	Very high
	RR	4	4	4	5	4	4	2	3	1			

(*continued on next page*)

T-12	VN	35	863	160	3974	161.33	734	694	40	3.27	69	0.81	Very high
	RR	4	5	3	5	4	3	3	2	2			
T-13	VN	35	531	161	2419	155.64	736	676	61	4.92	51	0.56	High
	RR	4	3	3	3	3	3	3	4	2			
T-14	VN	33	344	169	1756	153.45	769	711	58	5.02	34	0.34	Low
	RR	4	2	4	2	3	3	3	3	2			
T-15	VN	31	353	153	1693	156.27	696	640	56	4.57	29	0.28	Low
	RR	4	2	3	2	3	3	3	3	2			
T-16	VN	27	228	139	1157	155.69	628	590	38	5.01	10	0.03	Very low
	RR	3	1	3	1	3	3	3	2	2			
T-17	VN	37	662	188	3353	151.89	860	795	65	3.30	52	0.58	High
	RR	4	3	4	4	2	4	2	4	2			
T-18	VN	41	673	205	3434	152.23	950	889	61	3.81	65	0.75	Very high
	RR	5	3	5	4	2	4	2	4	2			
T-19	VN	39	390	175	1816	152.41	816	747	69	4.24	37	0.38	Low
	RR	4	2	4	2	2	4	3	4	2			
S-1	VN	30	713	141	3404	158.25	760	723	38	9.51	76	0.89	Very high
	RR	3	4	3	4	4	3	3	2	5			
S-2	VN	33	757	146	3323	158.27	668	622	46	3.30	55	0.62	High
	RR	4	4	3	4	4	3	3	2	2			
S-3	VN	43	843	204	3979	163.80	989	943	46	2.31	53	0.59	High
	RR	5	4	5	5	5	5	1	2	1			

Table 20.5 Nature of vulnerability, assigned risk-rating and vulnerability index in different zones of summer season.

Sub-zones	Risk variables	Household distribution (No.)	Household density (No./km²)	Population distribution (No.)	Population density (No./km²)	Depth of water level (mbgl)	Water demand (Lit.)	Water availability (Lit.)	Water deficit (Lit.)	ADI	VI	VIs	Vulnerability level
T-1	VN	33	502	165	2532	177.92	862	767	95	3.08	24	0.22	Low
	RR	2	3	2	3	3	3	2	4	2			
T-2	VN	37	555	171	2553	176.87	805	731	74	3.12	25	0.23	Low
	RR	3	3	3	3	3	2	2	3	2			
T-3	VN	35	417	164	1936	175.60	793	717	76	3.50	17	0.15	Very low
	RR	3	3	2	3	2	2	2	3	2			
T-4	VN	36	656	171	3029	181.39	805	752	53	2.57	23	0.21	Very low
	RR	3	4	3	4	4	2	2	2	1			
T-5	VN	36	322	186	1630	176.58	881	811	73	4.03	17	0.15	Very low
	RR	3	2	3	2	2	3	2	3	2			
T-6	VN	31	285	158	1417	174.16	725	662	70	4.95	8	0.07	Very low
	RR	2	2	2	2	2	2	1	3	3			
T-7	VN	27	185	139	930	173.74	634	594	48	4.86	1	0.00	Very low
	RR	1	1	1	1	1	1	1	2	3			
T-8	VN	37	630	189	3198	173.35	877	819	59	3.52	20	0.18	Very low
	RR	3	4	3	4	1	3	2	2	2			
T-9	VN	41	585	211	3000	176.01	1023	961	63	4.26	55	0.52	Moderate
	RR	4	3	4	4	2	4	3	3	2			

(continued on next page)

S-1	VN	39	597	194	3028	179.93	972	871	74	6.37	96	0.90	Very high
	RR	4	3	4	4	4	3	3	3	4			
S-2	VN	41	543	209	2754	180.86	1066	962	95	4.07	91	0.85	Very high
	RR	4	3	4	4	4	4	3	4	2			
S-3	VN	42	550	208	2741	179.79	1093	999	91	3.19	105	0.99	Very high
	RR	4	3	4	4	4	4	4	4	2			
S-4	VN	39	622	188	3021	182.24	962	871	91	2.91	88	0.83	Very high
	RR	4	4	3	4	5	3	3	4	2			
S-5	VN	35	498	160	2292	181.61	773	704	69	2.55	12	0.10	Very low
	RR	3	3	2	3	4	2	1	3	1			
S-6	VN	39	519	188	2462	182.22	906	842	64	2.43	33	0.30	Low
	RR	4	3	3	3	5	3	2	3	1			
S-7	VN	44	720	208	3386	183.74	1018	960	58	2.06	65	0.61	High
	RR	5	4	4	4	5	4	3	2	1			
S-8	VN	42	411	204	1987	181.20	1049	967	83	2.59	45	0.42	Moderate
	RR	4	2	4	3	4	4	3	4	1			
S-9	VN	41	399	203	1965	179.68	1041	944	98	4.60	88	0.83	Very high
	RR	4	2	4	3	4	4	3	5	3			
S-10	VN	39	551	190	2634	181.61	940	854	86	2.87	59	0.55	High
	RR	4	3	3	3	4	3	3	4	2			

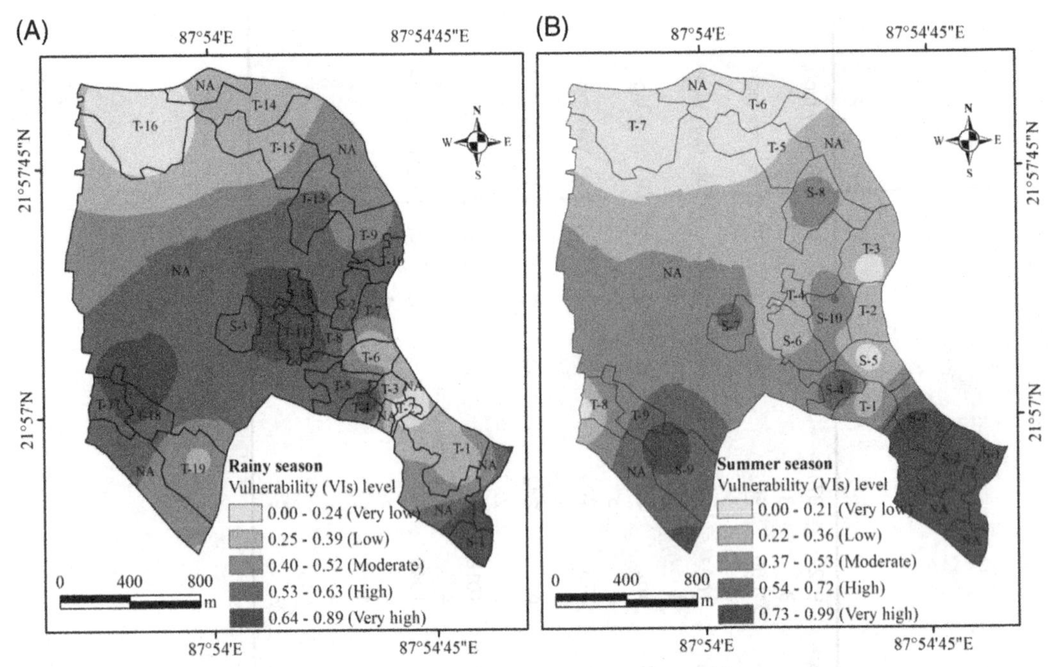

FIGURE 20.11 Seasonal pattern of vulnerability levels in (A) rainy season and (B) summer season.

have been assigned for very low water availability (308 lit.) and very high population density (4075/km^2), high risk-rating score (4) has been assigned for high household density, whereas, other risk-rating scores (1, 2, and 3) have been assigned for other remaining risk variables for T-2 sub-zone (Table 20.4) which is also resulted under very low level of vulnerability (Fig. 20.11A). Consequently, the T-1 sub-zone consists on 2–3 risk-rating scores (Table 20.4) and exists under a low level of vulnerability (Fig. 20.11A). T-11, 12, 18 and S-1 sub-zones have a very high level of vulnerability (Fig. 20.11A) with 2–5 risk-rating scores for the different risk variables, except the very low risk-rating score (1) for ADI of T-1 sub-zone (Table 20.4). In the summer season, a very high level of vulnerability has been observed in the sub-zones of S-1, 2, 3, 4 and 9 (Fig. 20.11B) with risk-rating scores of 2, 3, 4, and 5 for the different risk variables (Table 20.5).

20.4 Conclusion

The study indicates that the present scenario of seasonal fluctuating groundwater table and associated status of functioning and malfunctioning water sources have a huge impact on the water demand, water availability, and water deficit within the village. Moreover, the shape and sizes of the sub-zones have been differentiated with due effects from the number of seasonally functioning and malfunctioning water sources. Therefore, the number of households and population to the corresponding sub-zones has also been varied which stressed upon the spatial distributional patterns of categorical classes of different results of household distribution and density, population distribution and density, ADI, depth of water level, water demand, water availability, and water deficit for both the rainy and summer seasons. Moreover, the categorical vulnerability levels have also been seasonally varied depending on the results of nine risk variables. Local people are compelled to go to the different water sources in different seasons for accessing required water as some of the water sources of the rainy season become malfunctioning during the summer season. In recent years lots of new submersible pumps have been installed publicly and privately even up to 250 m depth to overcome the water accessibility issue. However, through such kind of efforts, the future water accessibility can only be minimized for two or three years, as it was observed that every year any of the existing water sources was not working properly due to water level issues. Therefore, from this study, it can be said that the depth of extractable groundwater table become increased by year after year and associated vulnerability level will also be increased proportionately if people will not be responsible for the present vulnerability scenario.

Acknowledgement

The co-author of this chapter conveys her sincere gratitude and thanks to the Kalagachia panchayat office, Khejuri-I CD block, Purba Medinipur district for providing necessary data regarding demographic status and tube wells of the studied village. She is also like to thank Mr. Jayanta Das, the tube well boring and repairing mechanic, for providing valuable information about the groundwater table, depth of functioning and malfunctioning tube wells and submersible pumps of the study area.

References

Acharya, T., Kumbhakar, S., Prasad, R., Mondal, S., Biswas, A., 2019. Delineation of potential groundwater recharge zones in the coastal area of north-eastern India using geoinformatics. Sustain. Water Resour. Manage. 5 (2), 533–540.

Bassi, N., Kumar, M.D., Sharma, A., Pardha-Saradhi, P., 2014. Status of wetlands in India: A review of extent, ecosystem benefits, threats and management strategies. J. Hydrol. Reg. Stud. 2, 1–19.

Bouaziz, M., Eisold, S., Guermazi, E., 2017. Semiautomatic approach for land cover classification: a remote sensing study for arid climate in southeastern Tunisia. Euro-Mediterr. J. Environ. Integr. 2 (1), 24. doi:10.1007/s41207-017-0036-7.

Chakraborty, S., Maity, P.K., Das, S., 2020. Investigation, simulation, identification and prediction of groundwater levels in coastal areas of Purba Midnapur, India, using MODFLOW. Environ. Dev. Sustain. 22 (4), 3805–3837.

Chinnasamy, P., Agoramoorthy, G., 2015. Groundwater storage and depletion trends in Tamil Nadu State, India. Water Resour. Manage. 29 (7), 2139–2152.

Das, G.K., 2017. A geo-spatial analysis and assessment of groundwater potential zones by using remote sensing and GIS techniques: a micro level study of Bhagwanpur-I CD Block in Purba Medinipur District, West Bengal, India. Int. J. Exp. Res. Rev. 14, 9–19.

Das, K., Mukherjee, A., 2019. Depth-dependent groundwater response to coastal hydrodynamics in the tropical, Ganges river mega-delta front (the Sundarbans): Impact of hydraulic connectivity on drinking water vulnerability. J. Hydrol. 575, 499–512.

Das, P., Mukherjee, A., Lapworth, D.J., Das, K., Bhaumik, S., Layek, M.K., ... Sen, J., 2021. Quantifying the dynamics of sub-daily to seasonal hydrological interactions of Ganges river with groundwater in a densely populated city: implications to vulnerability of drinking water sources. J. Environ. Manage. 288, 112384. doi:10.1016/j.jenvman.2021.112384.

Dixit, A., Upadhya, M., 2005. Augmenting groundwater in Kathmandu Valley: challenges and possibilities. Nepal Water Conservation Foundation 5–40. https://assets.publishing.service.gov.uk/media/57a08c7ced915d3cfd001404/R8169-NepalpaperJan05final.pdf. Accessed on 26 October 2021.

Gleick, P.H., 1996. Basic water requirements for human activities: meeting basic needs. Water Int. 21 (2), 83–92.

Halder, S., Dhal, L., Jha, M.K., 2021. Investigating groundwater condition and seawater intrusion status in coastal aquifer systems of eastern India. Water 13 (14), 1952. doi:10.3390/w13141952.

, 2008. Climate of West Bengal. National Climate Centre, India Meteorological Department, Pune, p. 160.

Jabed, M.A., Paul, A., Nath, T.K., 2020. Peoples' perception of the water salinity impacts on human health: a case study in south-eastern coastal region of Bangladesh. Exposure Health 12 (1), 41–50.

Jana, S., 2020. Micro-level coastal vulnerability assessment in relation to post-Aila landscape alteration at the fragile coastal stretch of the Sagar Island, India. Reg. Stud. Mar. Sci. 33, 100908. doi:10.1016/j.rsma.2019.100908.

Jana, S., 2021a. Groundwater and society in India: challenging issues and adaptive strategies. In: Shit, P.K., et al. (Eds.), Groundwater and Society: Applications of Geospatial Technology. Springer Nature Switzerland, pp. 11–28. doi:10.1007/978-3-030-64136-8_2.

Jana, S., 2021b. An automated approach in estimation and prediction of riverbank shifting for flood-prone middle-lower course of the Subarnarekha river, India. Int. J. River Basin Manag. 19 (3), 359–377.

Kumar, P., Tiwari, P., Biswas, A., Acharya, T., 2020. Geophysical and hydrogeological investigation for the saline water invasion in the coastal aquifers of West Bengal, India: a critical insight in the coastal saline clay-sand sediment system. Environ. Monit. Assess. 192 (9), 1–22.

Kumar, R., Singh, R.D., Sharma, K.D., 2005. Water resources of India. Curr. Sci. 89 (5), 794–811.

Loucks, D.P., 2000. Sustainable water resources management. Water Int. 25 (1), 3–10.

MacDonald, A.M., Bonsor, H.C., Taylor, R., Shamsudduha, M., Burgess, W.G., Ahmed, K.M., ... Moench, M., 2015. Groundwater Resources in the Indo-Gangetic Basin. BGS, Keyworth, UK, p. 58 British Geological Survey Open Report, OR/15/047.

Meinzen-Dick, R., Appasamy, P.P., 2002. Urbanization and intersectoral competition for water. In: Dabelko, G.D. (Ed.), Finding the Source: The Linkages Between Population and Water. The Woodrow Wilson Institute, Washington, DC, pp. 27–51.

Mishra, N., Haque, M.O., Leigh, L., Aaron, D., Helder, D., Markham, B., 2014. Radiometric cross calibration of Landsat 8 operational land imager (OLI) and Landsat 7 enhanced thematic mapper plus (ETM+). Remote Sens. 6 (12), 12619–12638.

Mondal, D., 2021. Coastal Urbanization and Population Pressure with Related Vulnerabilities and Environmental Conflicts A Case Study at Medinipur Littoral Tract West Bengal. Unpublished Ph.D. thesis submitted to Vidyasagar University, Midnapore, West Bengal, India.

Mukherjee, A., Bhanja, S.N., Wada, Y., 2018. Groundwater depletion causing reduction of baseflow triggering Ganges river summer drying. Sci. Rep. 8 (1), 1–9.

Narain, V., 2014. Whose land? Whose water? Water rights, equity and justice in a peri-urban context. Local Environ. 19 (9), 974–989.

O'malley, L.S.S., 1911. Bengal District Gazetteers: Midnapore. The Bengal Secretariat Book Depot, Calcutta.

Prusty, P., Farooq, S.H., Zimik, H.V., Barik, S.S., 2018. Assessment of the factors controlling groundwater quality in a coastal aquifer adjacent to the Bay of Bengal, India. Environ. Earth Sci. 77 (22), 1–15.

Punjabi, B., Johnson, C.A., 2019. The politics of rural–urban water conflict in India: untapping the power of institutional reform. World Dev. 120, 182–192.

Rangarajan, R., Athavale, R.N., 2000. Annual replenishable ground water potential of India: an estimate based on injected tritium studies. J. Hydrol. 234 (1-2), 38–53.

Rani, N.S., Satyanarayana, A.N.V., Bhaskaran, P.K., Rice, L., Kantamaneni, K., 2021. Assessment of groundwater vulnerability using integrated remote sensing and GIS techniques for the West Bengal coast, India. J. Contam. Hydrol. 238, 103760. doi:10.1016/j.jconhyd.2020.103760.

Rockstrom, J., Falkenmark, M., Karlberg, L., Hoff, H., Rost, S., Gerten, D., 2009. Future water availability for global food production: the potential of green water for increasing resilience to global change. Water Resour. Res. 45 (7). doi:10.1029/2007WR006767.

Sar, N., Khan, A., Chatterjee, S., Das, A., 2015. Hydrologic delineation of ground water potential zones using geospatial technique for Keleghai river basin, India. Model. Earth Syst. Environ. 1 (3), 1–15.

Sarkar, B., Islam, A., Majumder, A., 2021. Seawater intrusion into groundwater and its impact on irrigation and agriculture: evidence from the coastal region of West Bengal, India. Reg. Stud. Mar. Sci. 44, 101751. doi:10.1016/j.rsma.2021.101751.

Shah, T., 2009. India's ground water irrigation economy: The challenge of balancing livelihoods and environment. In: Chopra, K., Dayal, V. (Eds.), Handbook on Environmental Economics in India. Oxford University Press, New Delhi, pp. 21–37.

Singh, O.P., Khan, T.M.A., Rahman, M.S., 2001. Has the frequency of intense tropical cyclones increased in the north Indian Ocean? Curr. Sci. 80 (4), 575–580.

Srivastava, S.K., Kumar, R., Singh, R.P., 2009. Extent of groundwater extraction and irrigation efficiency on farms under different water-market regimes in Central Uttar Pradesh. Agric. Econ. Res. Rev. 22, 87–98.

Vij, S., Narain, V., 2016. Land, water & power: The demise of common property resources in periurban Gurgaon, India. Land Use Pol. 50, 59–66.

Zencich, S.J., Froend, R.H., Turner, J.V., Gailitis, V., 2002. Influence of groundwater depth on the seasonal sources of water accessed by Banksia tree species on a shallow, sandy coastal aquifer. Oecologia 131 (1), 8–19.

Index

Page numbers followed by "*f*" and "*t*" indicate, figures and tables respectively.

Printed in the United States
by Baker & Taylor Publisher Services